Reflectarray Antennas: Theory, Designs, and Applications

Reflectarray Antennas: Theory, Designs, and Applications

Payam Nayeri
Colorado School of Mines
USA

Fan Yang
Tsinghua University
China

Atef Z. Elsherbeni
Colorado School of Mines
USA

WILEY

IEEE PRESS

This edition first published 2018

Registered Offices
John Wiley & Sons, Inc., 111 River Street, Hoboken, NJ 07030, USA
John Wiley & Sons Ltd, The Atrium, Southern Gate, Chichester, West Sussex, PO19 8SQ, UK

Editorial Office
The Atrium, Southern Gate, Chichester, West Sussex, PO19 8SQ, UK

For details of our global editorial offices, customer services, and more information about Wiley products visit us at www.wiley.com.

Wiley also publishes its books in a variety of electronic formats and by print-on-demand. Some content that appears in standard print versions of this book may not be available in other formats.

Library of Congress Cataloging-in-Publication data applied for

ISBN: 978-1-118-84676-6

Cover design by Wiley
Cover images: (Background) © Andrey Prokhorov/Gettyimages;
(Foreground) Courtesy of Payam Nayeri, Fan Yang, and Atef Z. Elsherbeni

Set in 10/12pt Warnock by SPi Global, Pondicherry, India

10 9 8 7 6 5 4 3 2 1

To my parents who I am eternally grateful for their love, support, and encouragement throughout my career

Payam Nayeri

To my colleagues and students, and to my family

Fan Yang

To my wife, Magda, daughters, Dalia and Donia, son, Tamer, and the memory of my parents

Atef Z. Elsherbeni

Contents

Foreword *xiii*
Preface *xv*
Acknowledgments *xvii*

1 Introduction to Reflectarray Antennas *1*
1.1 Reflectarray Concept *1*
1.2 Reflectarray Developments *2*
1.3 Overview of this Book *5*
References *7*

2 Analysis and Design of Reflectarray Elements *9*
2.1 Phase-Shift Distribution on the Reflectarray Aperture *9*
2.2 Phase Tuning Approaches for Reflectarray Elements *13*
2.2.1 Elements with Phase/Time-Delay Lines *14*
2.2.2 Elements with Variable Sizes *15*
2.2.3 Elements with Variable Rotation Angles *16*
2.3 Element Analysis Methods *18*
2.3.1 Periodic Boundary Conditions and Floquet Port Excitation *19*
2.3.2 Metallic Waveguide Simulators *19*
2.3.3 Analytical Circuit Models *21*
2.3.4 Comparison of Element Analysis Techniques *22*
2.3.4.1 Comparison between PBC and Metallic Waveguides *23*
2.3.4.2 Comparison between PBC and the Circuit Model *24*
2.4 Examples of Classic Reflectarray Elements *26*
2.4.1 Rectangular Patch with Phase-Delay Lines *26*
2.4.2 Variable Size Square Patch *30*
2.4.3 Single Slot Ring Elements *33*
2.5 Reflectarray Element Characteristics and Design Considerations *37*
2.5.1 Frequency Behavior of Element Reflection Coefficients *37*
2.5.2 Effects of Oblique Incidence Angles on Element Reflection Coefficients *37*
2.5.3 Sources of Phase Error in Reflectarray Element Design *41*
2.6 Reflectarray Element Measurements *43*
References *46*

3 System Design and Aperture Efficiency Analysis *49*
3.1 A General Feed Model *49*
3.1.1 Models of Linearly Polarized and Circularly Polarized Feeds *50*
3.1.2 Balanced Feed Models *51*
3.2 Aperture Efficiency *53*
3.2.1 Spillover Efficiency *53*
3.2.2 Illumination Efficiency *54*
3.2.3 Effects of Aperture Shape on Efficiency *55*
3.2.4 Effects of Feed Location on Efficiency *59*
3.3 Aperture Blockage and Edge Diffraction *60*
3.3.1 Aperture Blockage and Offset Systems *60*
3.3.2 Edge Taper and Edge Diffraction *63*
3.4 The Analogy between a Reflectarray and a Parabolic Reflector *70*
3.4.1 The Offset System Configurations *71*
3.4.2 Analogous Offset Reflector *72*
3.4.2.1 Transformation from Reflector to Reflectarray System *72*
3.4.2.2 Transformation from Reflectarray to Reflector System *75*
3.4.3 Example of Analogous Offset Systems *76*
References *77*

4 Radiation Analysis Techniques *79*
4.1 Array Theory Approach: The Robust Analysis Technique *80*
4.1.1 Idealized Feed and Element Patterns *80*
4.1.2 Element Excitations and Reflectarray Radiation Pattern *81*
4.2 Aperture Field Approach: The Classical Analysis Technique *82*
4.2.1 Complex Feed Patterns *82*
4.2.2 Field Transformations from Feed to Aperture and Equivalent Surface Current *83*
4.2.3 Near-Field to Far-Field Transforms and Reflectarray Radiation Pattern *85*
4.3 Important Topics in Reflectarray Radiation Analysis *87*
4.3.1 Principal Radiation Planes *87*
4.3.2 Co- and Cross-Polarized Patterns *89*
4.3.3 Antenna Directivity *90*
4.3.4 Antenna Efficiency and Gain *91*
4.3.5 Spectral Transforms and Computational Speedup *94*
4.4 Full-Wave Simulation Approaches *96*
4.4.1 Constructed Aperture Currents Under Local-Periodicity Approximation *96*
4.4.2 Complete Reflectarray Models *96*
4.5 Numerical Examples *98*
4.5.1 Comparison of the Array Theory and Aperture Field Analysis Techniques *98*
4.5.1.1 Example 1: Reflectarray Antenna with a Broadside Beam *99*
4.5.1.2 Example 2: Reflectarray Antenna with an Off-Broadside Beam *100*

4.5.1.3 Comparison of Calculated Directivity versus Frequency *103*
4.5.2 Consideration in the Array Theory Technique: Element Pattern Effect *105*
4.5.3 Consideration in the Aperture Field Technique: Variations of Equivalence Principle *106*
4.5.4 Comparisons with Full-Wave Technique *107*
References *110*

5 Bandwidth of Reflectarray Antennas *113*
5.1 Bandwidth Constraints in Reflectarray Antennas *113*
5.1.1 Frequency Behavior of Element Phase Error *113*
5.1.2 Frequency Behavior of Spatial Phase Delay *115*
5.1.3 Aperture Phase Error and Reflectarray Bandwidth Limitations *118*
5.2 Reflectarray Element Bandwidth *121*
5.2.1 Physics of Element Bandwidth Constraints *121*
5.2.2 Parametric Studies on Element Bandwidth *122*
5.3 Reflectarray System Bandwidth *135*
5.3.1 Effect of Aperture Size on Reflectarray Bandwidth *135*
5.3.2 Effects of Element on Reflectarray Bandwidth *140*
References *144*

6 Reflectarray Design Examples *147*
6.1 A Ku-band Reflectarray Antenna: A Step-by-Step Design Example *147*
6.1.1 Feed Antenna Characteristics *147*
6.1.2 Reflectarray System Design *150*
6.1.3 Reflectarray Element Design *153*
6.1.4 Radiation Analysis *156*
6.1.5 Fabrication and Measurements *159*
6.2 A Circularly Polarized Reflectarray Antenna using an Element Rotation Technique *165*
6.3 Bandwidth Comparison of Reflectarray Designs using Different Elements *169*
References *176*

7 Broadband and Multiband Reflectarray Antennas *179*
7.1 Broadband Reflectarray Design Topologies *179*
7.1.1 Multilayer Multi-Resonance Elements *179*
7.1.2 Single-Layer Multi-Resonance Elements *181*
7.1.3 Sub-Wavelength Elements *184*
7.1.4 Reflectarrays Employing Single-Layer and Double-Layer Sub-Wavelength Elements *188*
7.1.5 Broadband Design Methods for Large Reflectarrays *197*
7.2 Phase Synthesis for Broadband Operation *197*
7.2.1 A Phase Synthesized Broadband Reflectarray *200*
7.2.2 A Dual-Frequency Broadband Reflectarray *203*

7.3 Multiband Reflectarray Designs *206*
7.3.1 A Single-Layer Dual-Band Circularly Polarized Reflectarray *210*
7.3.2 A Single-Layer Tri-Band Reflectarray *213*
References *221*

8 Terahertz, Infrared, and Optical Reflectarray Antennas *227*
8.1 Above Microwave Frequencies *227*
8.2 Material Characteristics at Terahertz and Infrared Frequencies *228*
8.2.1 Optical Measurements and Electromagnetic Parameters *228*
8.2.2 Measured Properties of Conductors and Dielectric Materials *229*
8.2.3 Calculating Drude Model Parameters for Conductors *229*
8.3 Element Losses at Infrared Frequencies *234*
8.3.1 Conductor Losses *234*
8.3.1.1 Effect of Conductor Thickness *234*
8.3.1.2 Effect of Complex Conductivity *237*
8.3.2 Dielectric Losses *240*
8.3.3 Effect of Losses on Reflection Properties of Elements *241*
8.3.4 Circuit-Model Analysis *242*
8.3.4.1 Circuit Theory and Loss Study *242*
8.3.4.2 Zero-Pole Analysis of Element Performance *243*
8.4 Reflectarray Design Methodologies and Enabling Technologies *245*
8.4.1 Reflectarrays with Patch Elements *245*
8.4.2 Dielectric Resonator Reflectarrays *248*
8.4.3 Dielectric Reflectarrays *251*
8.4.3.1 Dielectric Property and 3D Printing Technique *251*
8.4.3.2 Dielectric Reflectarray Design *253*
8.4.3.3 Dielectric Reflectarray Prototypes and Measurements *259*
8.5 Future Trends *261*
References *264*

9 Multi-Beam and Shaped-Beam Reflectarray Antennas *267*
9.1 Direct Design Approaches for Multi-Beam Reflectarrays *268*
9.1.1 Geometrical Aperture Division *268*
9.1.2 Superposition of Aperture Fields *271*
9.1.3 Comparison of Direct Design Approaches *272*
9.2 Synthesis Design Approaches for Shaped- and Multi-Beam Reflectarrays *275*
9.2.1 Basics of Synthesis Techniques *275*
9.2.2 Local-Search Techniques *276*
9.2.3 Global-Search Techniques *279*
9.2.4 Full-Wave Optimization Design Approaches *280*
9.3 Practical Reflectarray Designs *281*
9.3.1 Single-Feed Reflectarray with Multiple Symmetric Beams *281*
9.3.2 Feed Reflectarrays with Multiple Asymmetric Beams *286*
9.3.3 Shaped-Beam Reflectarrays *294*
9.3.4 Multi-Feed Multi-Beam Reflectarrays *297*
References *300*

10 Beam-Scanning Reflectarray Antennas *303*
10.1 Beam-Scanning Approaches for Reflectarray Antennas *304*
10.1.1 Design Methodologies *304*
10.1.2 Classifications Based on Reflector Type *306*
10.2 Feed-Tuning Techniques *307*
10.2.1 Fully Illuminated Single-Reflector Configurations *307*
10.2.1.1 Parabolic-Phase Apertures *307*
10.2.1.2 Non-Parabolic-Phase Apertures *313*
10.2.2 Partially Illuminated Single-Reflector Configurations *324*
10.2.2.1 Parabolic Cylindrical-Phase Reflectarray Antennas (PCPRA) *324*
10.2.2.2 Parabolic Torus-Phase Reflectarray Antennas (PTPRA) *329*
10.2.2.3 Spherical-Phase Reflectarray Antennas (SPRA) *331*
10.2.3 Dual-Reflector Configurations *334*
10.2.3.1 Parabolic Reflector/Reflectarray Antennas *334*
10.2.3.2 Non-Parabolic Reflector/Reflectarray Antennas *336*
10.2.4 Summary of Feed-Tuning Techniques *337*
10.3 Aperture Phase-Tuning Techniques *339*
10.3.1 Basics of Aperture Phase Tuning *339*
10.3.2 Enabling Technologies *341*
10.3.2.1 Mechanical Actuators/Motors *341*
10.3.2.2 Electronic Devices *343*
10.3.2.3 Functional Materials *352*
10.4 Frontiers in Beam-Scanning Reflectarray Research *355*
10.4.1 Active Reflectarrays *355*
10.4.2 Comparison Between Analog and Digital Phase Control *355*
10.4.3 Sub-Array Techniques *358*
10.4.4 Hybrid Configurations *359*
References *359*

11 Reflectarray Engineering and Emerging Applications *367*
11.1 Advanced Reflectarray Geometries *367*
11.1.1 Conformal Reflectarrays *367*
11.1.1.1 Analysis of Conformal Reflectarrays *367*
11.1.1.2 Radiation Characteristics of Conformal Reflectarrays on Cylindrical Surfaces *369*
11.1.2 Dual-Reflectarrays *375*
11.2 Reflectarrays for Satellite Applications *379*
11.2.1 An L-Band Reflectarray for the Beidou Satellite System *381*
11.2.2 Reflectarrays Integrated with Solar Cells *384*
11.3 Power Combining and Amplifying Reflectarrays *388*
11.4 A Perspective on Reflectarray Antennas *393*
11.4.1 Large-Aperture Planar Reflectarray Antennas *393*
11.4.2 Reflectarray Antennas with Broad Bandwidth, Beam-Scanning Capability, and Low Cost *396*
11.4.3 From Reflectarray Antennas to Transmitarray Antennas *396*
References *397*

Index *401*

Foreword

Although the concept of the reflectarray antenna was first introduced in 1963, the vast interest in it did not come about until in the late 1980s with the development of low-profile microstrip antennas. From the word *reflectarray*, it can be deduced that this is an antenna that combines the unique features of a parabolic *reflector* and a phased *array*. Thus, a low-profile reflectarray consists of an array of microstrip elements that are provided with a set of pre-adjusted phases to form a focused beam when illuminated by a feed, in a similar way to a parabolic reflector. The array elements can be printed onto either a flat surface or a slightly curved surface and have been demonstrated to have the ability to produce a high-gain pencil beam, a contour-shaped beam, multiple beams, or an electronically scanned beam. Because the array elements in a reflectarray are not physically interconnected, it can produce a high-gain beam with relatively high efficiency similar to that produced by a parabolic reflector. There were several pioneers that initiated the study of printed reflectarrays during the late 1980s. I thought about the idea of a reflectarray due to my earlier work experiences with microstrip antennas and frequency selective surfaces (FSS). At certain resonant frequencies, the FSS can only reflect as a nearly perfect conductor since all elements are identical. It cannot cause the reflected waves to form a phase-coherent beam. However, if each FSS element is designed differently with appropriate phase delay, a coherent beam can then be formed and a printed reflectarray is consequently formed.

This book gives a comprehensive presentation of reflectarray antennas. Chapter 1 is a general overview of the operating principles as well as the developmental history of reflectarray antennas. Chapters 2 through 5 provide very complete and detailed design and analysis techniques, including the important element characterization and selection, radiation efficiency analysis and system design, various radiation analysis approaches and tradeoffs, and the most critical bandwidth issues and analysis. Chapter 6 gives a few specific design examples; in particular, a Ku-band step-by-step design example and a circularly polarized reflectarray design. It is well known that the bandwidth limitation generally presents critical issues in reflectarray design. Chapter 7 is devoted to broadband solutions by presenting several bandwidth widening techniques and multiband approaches. The Terahertz, infrared, and optical frequencies have been found to be the frontier of research and application for antennas. Reflectarray antennas have also found applications in these extremely high frequency areas and are presented in Chapter 8, where the critical issues of material characterization and element loss are discussed. Low-loss dielectric resonators, used as elements, are also presented in this chapter. A single reflectarray antenna can not only be designed to produce a high-gain pencil beam, but, due to its many array elements, also has the ability to generate a specifically contour-shaped beam as well as multiple beams. Chapter 9 gives a thorough presentation of the design approaches, which include direct design approaches and synthesis design approaches for a single reflectarray to radiate a contour-shaped beam or multiple beams. Chapter 10 engages in discussion about a reflectarray's beam scanning capability and design approaches. One of the key advantages of the reflectarray is its ability to achieve fast electronic beam scanning by implanting a low-loss phase shifter into each of its elements without the need for expensive transmit/receive modules and high-loss power division network. Thus, the reflectarray, owing to the hybrid nature of reflector and array, can behave like an efficient high-gain parabolic reflector and a relatively low-cost phased array. Finally, Chapter 11 discusses several emerging and future applications of reflectarray antennas, such as a reflectarray conformally mounted on curved surfaces, satellite applications, integration with solar cells, amplifying reflectarrays, dual-reflectarrays, very large aperture applications, and so on.

By comparing this book with the very first reflectarray book published by the Wiley-IEEE Press (Huang and Encinar) in 2008, this book not only gives more updated information, but also gives more detailed analysis and design presentations. The authors of that 2008 book also presented their own pioneering contribution in the areas such as broadband design using sub-wavelength patch elements, a special phase synthesis approach, and single as well as multilayer approaches. In particular, a single layer design with tri-band circular polarization performance was achieved. It was a cooperative effort that fulfilled my contractual request from the Jet Propulsion Laboratory while the authors were teaching at the University of Mississippi. A unique split-square ring element was also used to achieve excellent circular polarization for this single-layer multiband reflectarray. In that book, the authors presented their own contributions in the area of Terahertz and infrared reflectarray applications. In addition, the synthesis technique for a single reflectarray to achieve multiple beams and specifically shaped beams was presented as well. The electronic beam scanning capability of the reflectarray was also fully discussed with several well-presented new design approaches.

This book is well organized and has significant amount of information in design and analysis with many practical application results augmented with adequate number of references to help the readers to comprehend. Undoubtedly, I believe this book is not only well suited as a university text book but also is an excellent source of design and analysis information for antenna engineers for many years to come.

Dr. John Huang
Principle Engineer, retiree of
The Spacecraft antenna research group
Jet Propulsion Laboratory
California Institute of Technology
Pasadena, California, USA

Preface

High-gain antennas are an essential part of long-distance wireless communications, radar, and remote sensing systems, which vary with frequency, coverage, resolution, and flexibility of operation. The conventional choices for antennas in these systems were typically reflectors, lenses, or arrays. In recent years, however, a new generation of high-gain antennas has emerged that combines the favorable features of both printed arrays and reflector antennas and creates a high-gain antenna with low-profile, low-mass, and low-cost features. This antenna is known as the reflectarray.

The reflectarray is an antenna with a flat reflecting surface consisting of hundreds of elements and an illuminating feed antenna. The hybrid nature of the reflectarray antenna offers more flexibility in aperture phase control and can provide advantages over both reflectors and array antennas for many applications. The elements of the reflectarray are individually designed to reflect the electromagnetic wave with a certain phase to compensate for the phase delay caused by the spatial feed. The phase shift of the elements is realized using various methods such as variable-size elements. Single and multilayer reflectarrays have been designed to achieve broadband and multiband performance from microwave frequencies up to the THz range. Meanwhile, the direct control of the phase of every element in the array allows multi-beam or shaped beam performance with single or multiple feeds. Another advantage of reflectarrays is the ability of the antenna to scan the main beam to large angles off broadside. The advantages of reflectarrays, such as being low-profile, lightweight, and having conformal geometry, make it desirable for various communication systems, especially for mobile platforms. Its applications in space exploration, satellite communications, remote sensing, and radar systems are rising, and will continue to increase in the future. In addition, the current printed circuit board (PCB) fabrication technology and available low-cost commercial laminates, allows for low-cost rapid prototype fabrication. This is also leading to commercial implementation and large-scale fabrication of reflectarray antennas. The potential of reflectarray capabilities has not yet been fully exploited. Researchers in this field are constantly presenting new ideas and designs ranging from advanced materials to multifunctional system designs. As such, it is expected that this field will remain an active area of research, and there is no doubt that reflectarrays will become an important member of the antenna family.

The aim of the book is to provide scientists and engineers in the fields of antenna, microwave, and electromagnetics, with up-to-date knowledge of reflectarray antenna theories, designs, and applications. This book will provide the reader with an overview of the reflectarray antenna research history and state-of-the-art, good knowledge of the basic theories for design and analysis of reflectarray antennas, and detailed design procedures for a wide range of diversified and advanced applications.

The prerequisite for this book is that the readers should be familiar with the basics of antenna engineering. The first part of this book includes the fundamental theories of reflectarrays, and is intended for engineers that know the basics of antenna theory and are becoming familiar with this new generation of high-gain antennas. Chapter 1 introduces the reflectarray concept and historical backgrounds, and provides an overview of this book. Chapter 2 provides a comprehensive coverage of aperture phase requirements in reflectarray systems, phasing element design methodologies, and element analysis techniques. Reflectarray system design and efficiency analysis are introduced in Chapter 3. A detailed coverage of the various methods to compute the radiation characteristics of reflectarray antennas is presented in Chapter 4. The bandwidth characteristics of reflectarray antennas are studied in detail in Chapter 5. A variety of reflectarray designs are presented in Chapter 6 that can serve as a useful reference for interested readers.

The second part of the book is intended for researchers and specialists that have a good knowledge of the basic theories in reflectarrays, and aim to design reflectarray antennas for specific applications/operations. It starts with a comprehensive overview of broadband and multiband reflectarray antennas in Chapter 7. Reflectarrays operating above microwave frequencies such as in the terahertz, infrared, and optical spectrums are introduced in Chapter 8. A detailed coverage of multi-beam and shaped-beam reflectarrays is presented in Chapter 9. Chapter 10 presents beam-scanning reflectarray antennas, where the extensive research on these types of reflectarrays is summarized and analyzed in a comprehensive fashion. The final chapter of this book, Chapter 11, is devoted to advanced reflectarray antenna configurations.

Acknowledgments

The work presented in this book was supported in part by the following institutions:

National Aeronautics and Space Administration (NASA)
National Science Foundation (NSF)
Tsinghua National Laboratory for Information Science and Technology
Chinese High-Technology Research and Development Program (863-Program)

The authors would like to thank the technical reviewers for their insightful feedback which enhanced the clarity and efficacy of this work. We would also like to express our sincere gratitude to our colleagues and students over these years: Dr. John Huang, Prof. Yahya Rahmat-Samii, Prof. Jianhua Lu, Prof. Shenheng Xu, Prof. Glenn Boreman, Prof. Hao Xin, Prof. Maokun Li, Dr. Ang Yu, Dr. Chye Hwa Loo, Dr. Wenxing An, Dr. Yilin Mao, Dr. Ahmed Hassan Abdelrahman, Dr. Huanhuan Yang, Dr. Ruyuan Deng, Mr. Yanghyo Kim, Ms. Bhavani Devireddy, Mr. Tamer Elsherbeni, Ms. Fang Guo, Mr. Xiaolin Zhu, Mr. Xiao Liu, Mr. Jun Luo, Mr. Xiangfei Xu, Mr. Lin Gao, Ms. Xue Yang, Mr. Martye Hickman, Mr. Junling Zhao, and Mr. Lin Xiong. We also greatly appreciate the generous contributions of ANSYS Inc. in providing us with HFSS and Designer simulation software, and Rogers Corp. for providing us with high quality laminates that have been used for many of the designs that are presented in this book.

1

Introduction to Reflectarray Antennas

1.1 Reflectarray Concept

Communicating over long distances had long been a dream for mankind until 1901 when Marconi demonstrated the first cross Atlantic wireless signal transmission. Since then, long distance communications have evolved to a degree where mankind can communicate wirelessly across the Solar System and beyond. Long distance communication requires large antennas in order to establish the wireless link between the transmitter and receiver. One of the most practical types of electrically large antennas are reflectors. While reflectors were originally built as optical devices [1], the discovery of electromagnetic waves by Maxwell, began a new era for communication with these antennas. The first experimental demonstration of wireless communication by Hertz in 1887, used a dipole-fed cylindrical parabolic antenna, which is believed to be the first reflector antenna operating at non-optical frequencies. Since then, reflectors have become the most widely used high-gain antenna in communications, radio astronomy, remote sensing, and radar [2].

An alternative approach to realization of a large antenna is by using several smaller antennas in the form of an array [3]. The first antenna array was built over 100 years ago [4]. In order to increase the directivity of a single monopole, Brown used two vertical antennas separated by half a wavelength and fed them out of phase [5]. He and several other notable scientists such as Marconi, Braun, and Adcock explored the unique characteristics of antenna array over the years [6]–[8]. Antenna array engineering evolved rapidly thereafter, particularly during the Second World War; however, it was the development of semiconductor technology in the 1960s and the printed circuit board technology in the 1970s that had the largest impact on their development. In particular the microstrip patch antenna proposed by Deschamps in 1953 [9] and later made practical by Munson in 1972 [10], revolutionized array engineering. Microstrip antenna arrays have since then played an important role in modern phased array systems.

While reflectors and arrays still compete for large aperture jobs in many types of systems, in the recent years, a new generation of high-gain antennas has emerged, which have attracted increasing interest from the antenna/electromagnetic community because of their low-profile, low-mass, and in many cases, low-cost features. This antenna is known as the reflectarray antenna [11]–[13]. The reflectarray antenna is a hybrid design, which combines many favorable features of reflectors and printed arrays, and as

Reflectarray Antennas: Theory, Designs, and Applications, First Edition.
Payam Nayeri, Fan Yang, and Atef Z. Elsherbeni.

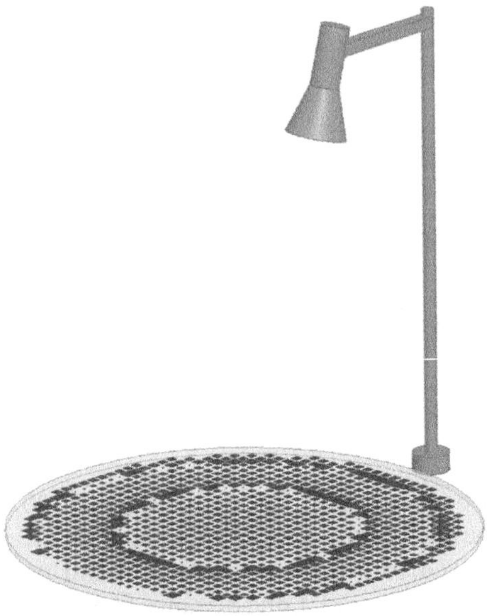

Figure 1.1 The geometry of an offset-fed reflectarray antenna.

such can provide advantages over these two conventional antennas. The parabolic reflector is difficult to manufacture in many cases due to its curved surface that requires expensive custom molds and also become more difficult to manufacture at higher microwave frequencies. On the other hand, while antenna arrays offer the advantages of flexible design freedoms and versatile radiation performance, its feeding network suffers from the energy loss and design complexity, and the cost of the T/R modules [14] in active phased arrays becomes prohibitively high for many applications. As such, the reflectarray has fast been gaining attention as an alternative to these more mature technologies as it is able to mitigate the disadvantages associated with both of these high-gain antennas.

The reflectarray is an antenna with a flat reflecting surface consisting of hundreds of elements on its aperture and an illuminating feed antenna, as shown in Figure 1.1.

The feed antenna spatially illuminates the aperture where the elements are designed to reflect the incident field with certain phase shifts in order to collimate the beam of the antenna in the desired direction and with the preferred shape. Its operation principle is similar in concept to reflector antennas with respect to the spatial illumination, and again similar in concept to antenna arrays with respect to phase synthesis and beam collimation.

1.2 Reflectarray Developments

The concept of reflectarray antennas was initially introduced in the early 1960s using short-ended waveguide elements with variable lengths [11]. The feed antenna illuminated the waveguides where the lengths of the shorted waveguides were designed such that the phase of the reradiated signals would form a collimated beam in the desired far-field direction. While the concept was very interesting, the bulky and heavy

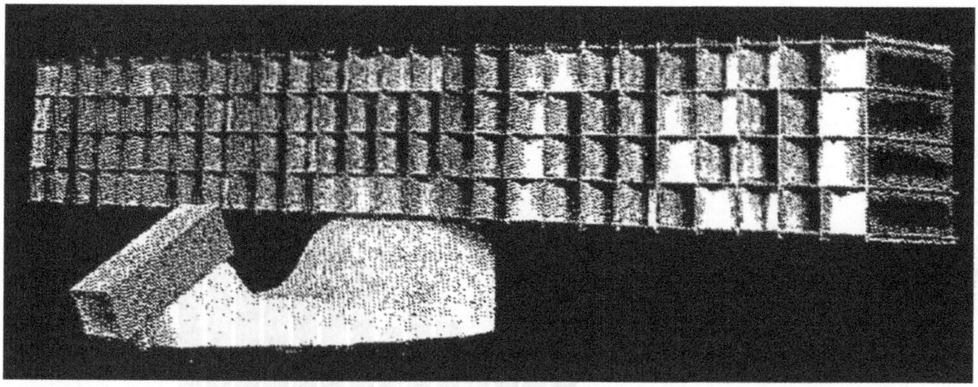

Figure 1.2 The first reflectarray antenna using waveguide technology. *Source:* Berry 1963 [11]. Reproduced with permission from IEEE.

waveguide structure of this first reflectarray antenna was a major drawback. The experimental model of the waveguide reflectarray is shown in Figure 1.2.

Although some work on spiralphase reflectarrays was reported by Phelan in the mid-1970s [15], the reflectarray antenna did not receive much attention after that until the revolutionary breakthrough of printed microstrip antenna technology in the 1980s. Since then, research on reflectarray antennas has been on the rise, and several diversified applications such as multi-beam antennas for point-to-point communication, beam-scanning antennas for radar applications, and spatial power combining reflectarray systems have been demonstrated. In particular, over the past 10 years, an increased interest in reflectarray antenna research has been observed in both academic and industrial sectors of the antenna community, which is also propelled by advances in fabrication technologies as well as computational resources.

Since 2006, the *IEEE Antennas and Propagation International Symposium* (APS) has included sessions dedicated to reflectarray antennas in the general conference proceedings, and several sessions and special sessions have been held since then. Most notably a full-day special session on reflectarray antennas was held at the 2011 APS. Several hundred papers have been presented in these sessions, and many researchers are now interested in joining this active research area. In 2012, the *International Journal of Antennas and Propagation* published a special issue on Reflectarray Antennas: Analysis and Synthesis Techniques, which further stimulated the research interest in this area. A literature search on IEEE Xplore using the keyword "reflectarray" showed more than 1200 articles have been published in IEEE in this area, as shown in Figure 1.3. The majority of the articles, however, have been published in the recent years, and in particular, there has a notable increase in the number of papers over the last 10 years.

The reflectarray antenna offers a multitude of capabilities that has encouraged continuous development and exciting applications in recent years. The elements of the reflectarray are designed to reflect the electromagnetic wave with a certain phase to compensate for the phase delay caused by the spatial feed. The phase shift of the elements is realized using various methods such as variable size elements, phase-delay lines, and element rotation techniques. The infinite array approach is used to calibrate the element phase versus element change [12]. Due to the very large number of elements

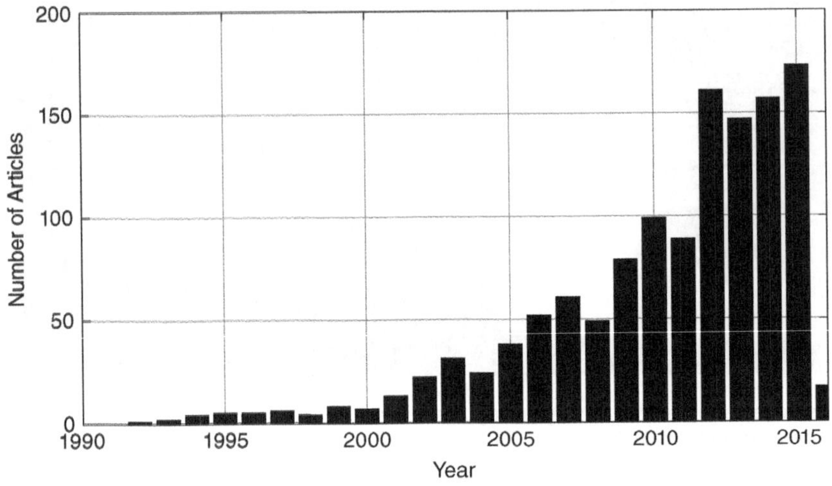

Figure 1.3 The number of articles on reflectarray antennas published in IEEE. Data obtained from IEEE Xplore on April 1, 2016.

involved in a reflectarray, full-wave simulation of the entire reflectarray antenna is still challenging. On the other hand, different theoretical models have been developed for the analysis of reflectarrays, such as the array theory formulation and the aperture field analysis technique, which show a good agreement with measured results. Moreover, implementing the spectral transform in these calculations allows for fast calculation of the radiation characteristics of the antenna, which is a considerable advantage for synthesis design problems using iterative procedures.

Single and multilayer reflectarrays have been designed to achieve broadband and multiband performance from microwave frequencies up to the THz range [16], [17]. Considerable improvements have been made to these designs over the years, and many practical designs have been demonstrated. One of the main challenges in reflectarray designs is improving the bandwidth of the antenna, which is the major drawback of printed resonator-type structures [18]. Different bandwidth improvement techniques such as using multilayer designs [19], [20], true time-delay lines [21], and sub-wavelength elements [22] have been studied and bandwidths of more than 20% have been reported.

Meanwhile, the direct control of the phase of every element in the array allows multi-beam performance with single or multiple feeds. The design of contoured beam reflectarrays is also a challenging field [23]. A phase-only synthesis process is used to obtain the required element phase shift from any given mask. Multi-feed multi-beam contoured beam designs have been demonstrated [24]; however, the performances of these designs are slightly inferior to the shaped-beam parabolic reflectors. Another advantage of reflectarrays is the ability of the antenna to scan the main beam to large angles off broadside. Beam-scanning reflectarrays are designed by using low-loss phase shifters integrated in every element of the array [25]. These beam-scanning reflectarrays require a switch board to control the main beam direction and are well suited for radar applications, and some models have been demonstrated; however, considerable challenges lie in improving the performance of these beam-scanning antennas.

In addition to the numerous capabilities and potentials that reflectarray antennas have demonstrated, a great deal of interest is now in the practical implementation of reflectarray antennas for space applications. Since the common considerations for space antennas are size, weight, and power (SWaP), because of limitations imposed by the satellite launch capabilities [26], the reflectarray antenna shows significant advantages over conventional high-gain space antennas, which are typically reflectors/lenses and arrays. These momentous promises make the reflectarray antenna a suitable low-cost choice for the new generation of space antennas.

The advantages of reflectarrays, such as low-profile, lightweight, and conformal geometry, make it desirable for various communication systems, especially for those mobile platforms. Its applications in space exploration, satellite communications, remote sensing, and radar systems are rising up within the last decade, and will continue to increase in the future. In addition, the current printed circuit board (PCB) fabrication technology and available low-cost commercial laminates, allows for low-cost rapid prototype fabrication. This is also leading to commercial implementation and large-scale fabrication of reflectarray antennas for commercial applications.

Terahertz and optical applications are also a very promising future of reflectarrays. With advances in fabrication technologies such as 3D printing devices and nanotechnology, the practical implementation of THz and optical reflectarray designs at a competitive cost is not far away. The full potential of reflectarray capabilities has not yet been fully exploited. Researchers in this field are constantly presenting new ideas and designs ranging from advanced materials to multifunctional system designs. As such it is expected that this field will remain an active area of research in the next decade, and there is no doubt that reflectarrays will become an important member in the antenna family.

1.3 Overview of this Book

The aim of the book is to provide scientists and engineers in the fields of antenna, microwave, and electromagnetic, with an up-to-date knowledge of reflectarray antenna theories as well as the design and analysis techniques. This book will provide the reader with:

- An overview of the reflectarray antenna research history, including various implementations and state-of-the-art.
- A good knowledge of the basic theories for design and analysis of reflectarray antennas, which will help to build up the fundamental capabilities for reflectarray research. In addition, a wealth of design examples along with numerical and experimental results are presented, which serves as a reference for researchers to verify their own developed programs.
- Detailed design procedures for a wide range of diversified applications, such as broadband designs, multiband operation, multi-beam performance, contour-beams, beam-scanning systems, and conformal reflectarray antennas, along with illustrative examples for each design.

The prerequisite for this book is that the readers should be familiar with the basics of antenna engineering. An introductory course to antenna engineering is typically offered as a senior level course for a bachelor student in the field of electrical engineering.

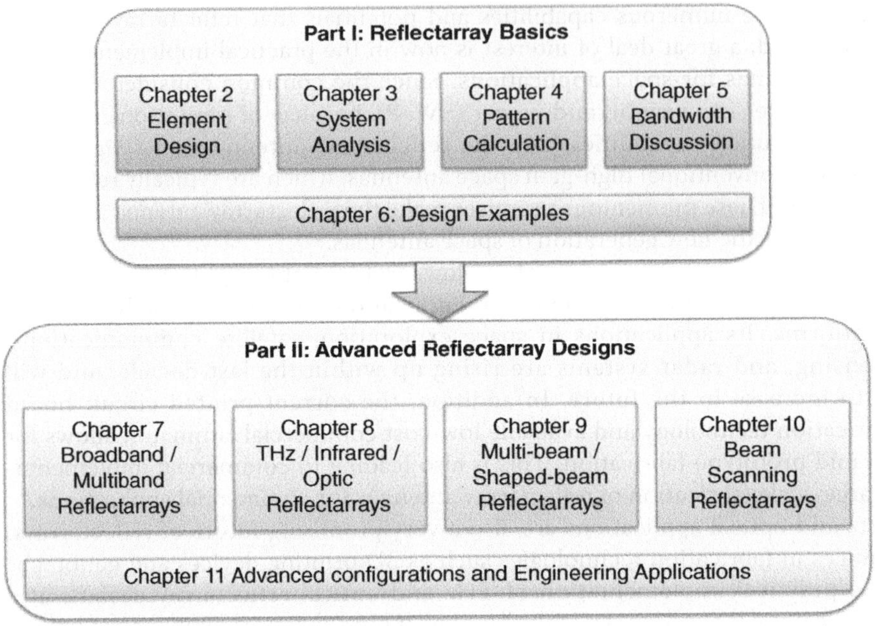

Figure 1.4 Organization of this reflectarray book.

As such any student in this field will be able to benefit from this book. However, this book is intended for both beginners and specialists in the field of electrical engineering. This is achieved by organizing and preparing this book in two parts and in 11 chapters, as illustrated in Figure 1.4.

The first part, which includes the fundamental theories of reflectarrays, is intended for engineers that know the basics of antenna theory and are starting to become familiar with this new generation of high-gain antennas. The second part of the book is intended for researchers and specialists that have a good knowledge of the basic theories in reflectarray, and aim at designing reflectarray antennas for specific applications/operations.

The first part includes the basic theories for analysis and design of reflectarray antennas. This section of the book builds the fundamental knowledge one needs to have in order to understand the governing dynamics of a reflectarray antenna system, and efficiently design and analyze reflectarray antennas. Chapter 2 is devoted to analysis and design of reflectarray phasing elements, and provides a comprehensive coverage of aperture phase requirements in reflectarray systems, phasing element design methodologies, element analysis techniques, as well as design examples. The reflectarray system design is introduced in Chapter 3, where the readers will learn the basics of the reflectarray systems and efficiency analysis for practical designs. A detailed coverage of the various methods to compute the radiation characteristics of reflectarray antennas is presented in Chapter 4. The bandwidth characteristics of reflectarray antennas is studied in detail in Chapter 5. The last chapter of the first part of this book is devoted to design examples, where a variety of reflectarray designs are presented that can serve as a useful reference for interested readers.

The second part of the book is intended for researchers that have a good knowledge of the basic theories in reflectarray, and aim at designing reflectarray antennas for specific applications/operations. This part starts with a comprehensive chapter on broadband and multiband reflectarray antennas in Chapter 7. Reflectarrays operating above microwave frequencies such as in the terahertz, infrared, and optical spectrums are introduced in Chapter 8. After discussion of the frequency behaviors of reflectarrays, advanced designs on the radiation patterns are followed. A detailed coverage of multi-beam and shaped-beam reflectarrays is presented in Chapter 9. Chapter 10 is devoted to beam-scanning reflectarray antennas, where the extensive research on these types of reflectarrays are summarized and analyzed in a comprehensive fashion. The final chapter of this book is devoted to advanced configurations of reflectarray antennas, such as conformal geometries and dual-reflector configurations, and applications such as satellite communications and spatial power combining.

References

1 A. Dodd, "An acre of glass – The history of the telescope," [Online]. Available: http:/ ezinearticles.com/?An-Acre-of-Glass–The-History-of-the-Telescope&id=2601009 (accessed July 1, 2017).
2 Y. Rahmat-Samii and R. L. Haupt, "Reflector antenna developments: A perspective on the past, present and future," *IEEE AP-S Mag*, Vol. 57, No. 2, Apr 2015.
3 R. L. Haupt and Y. Rahmat-Samii, "Antenna array developments: A perspective on the past, present and future," *IEEE AP-S Mag*, Vol. 57, No. 1, Feb 2015.
4 J. A. Fleming, *The Principles of Electric Wave Telegraphy and Telephony*, 3rd Edn, New York: Longmans, Green, and Co., 1916.
5 S. G. Brown, Brit. Patent No. 14,449, 1899.
6 G. Marconi, "On methods whereby the radiation of electric waves may be mainly confined to certain directions, and whereby the receptivity of a receiver may be restricted to electric waves emanating from certain directions," *Proc. Roy. Soc. Lond., Ser. A.*, Vol. 77, p. 413, 1906.
7 F. Braun, "Electrical oscillations and wireless telegraphy," Nobel Lecture, Dec. 11, 1909.
8 F. Adcock,Improvement in means for determining the direction of a distant source of electro-magnetic radiation, UK Patent 130,490, Aug. 7, 1919.
9 G. A. Deschamps, "Microstrip microwave antennas," presented at the *3rd USAF Symp. on Antennas*, 1953.
10 R. E. Munson, "Microstrip phased array antennas," *Proc. of 22nd Symp. on USAF Antenna Research and Development Program*, Oct 1972.
11 D. G. Berry, R. G. Malech, and W. A. Kennedy, "The reflectarray antenna," *IEEE Trans. Antennas Propagat.*, Vol. AP - 11, Nov. 1963, pp. 645–651.
12 D. M. Pozar, S. D. Targonski, and H. D. Syrigos, "Design of millimeter wave microstrip reflectarrays," *Proc. IEEE Trans. Antennas Propagat.*, Vol. 45, pp. 287–295, Feb. 1997.
13 J. Huang and J. A. Encinar, *Reflectarray Antennas*. New York, NY, USA: Wiley-IEEE, 2008.
14 R. C. Hansen, *Phased Array Antennas*, 2nd Edn, Chichester, UK: John Wiley & Sons, Ltd, 2009.
15 H. R. Phelan, "Spiralphase reflectarray for multitarget radar," *Microwave Journal*, Vol. 20, July 1977, pp. 67–73.

16 A. Yu, "Microstrip reflectarray antennas: Modeling, design and measurement," Ph.D. dissertation, Dept. Elect. Eng., University of Mississippi, Oxford, MS, 2010.

17 P. Nayeri "Advanced design methodologies and novel applications of reflectarray antennas," Ph.D. dissertation, Department of Electrical Engineering, University of Mississippi, MS, 2012.

18 D. M. Pozar, "Bandwidth of reflectarrays", *Electronics Letters*, Vol. 39, No. 21, Oct. 2003.

19 J. A. Encinar, "Design of two-layer printed reflectarrays using patches of variable size", *IEEE Trans. Antennas Propag.*, Vol. 49, No. 10, pp. 1403–1410, Oct. 2001.

20 J. A. Encinar and J. A. Zornoza, "Three-layer printed reflectarrays for contoured beam space applications," *IEEE Trans. Antennas Propag.*, Vol. 52, No. 5, pp. 1138–1148, May 2004.

21 E. Carrasco, J. A. Encinar, and M. Barba, "Bandwidth improvement in large reflectarrays by using true-time delay", *IEEE Trans. Antennas Propag.*, Vol. 56, No. 8, pp. 2496–2503, Aug. 2008.

22 P. Nayeri, F. Yang, and A. Z. Elsherbeni, "A broadband reflectarray using sub-wavelength patch elements," *IEEE Antennas and Propagation Society International Symposium*, South Carolina, U.S., 2009.

23 D. M. Pozar, S. D. Targonski, and R. Pokuls, "A shaped-beam microstrip patch reflectarray," *IEEE Trans. Antennas Propag.*, Vol. 47, pp. 1167–1173, July 1999.

24 M. Arrebola, J. A. Encinar, and M. Barba, "Multifed printed reflectarray with three simultaneous shaped beams for LMDS central station antenna", *IEEE Trans. Antennas Propag.*, Vol. 56, No. 6, pp. 1518–1527, June 2008.

25 S. V. Hum, M. Okoniewski, and R. J. Davies, "Modeling and design of electronically tunable reflectarrays," *IEEE Trans. Antennas Propagat.*, Vol. 55, No. 8, pp. 2200–2210, Aug. 2007.

26 R. B. Dybdal, "Satellite antennas," in *Antenna Engineering Handbook*, J. Volakis (ed.), McGraw-Hill, 2007.

2

Analysis and Design of Reflectarray Elements

A reflectarray antenna consists of a planar or conformal array of elements that are excited with a feed antenna [1]–[3]. A typical model of a reflectarray antenna is given in Figure 2.1. Each element is designed such that when it is illuminated by the feed antenna, it incorporates a certain reflected phase. The phase distribution over the reflectarray aperture is then synthesized so the reflectarray can realize a collimated or shaped beam in the desired direction. As such, analysis and design of the reflectarray elements, typically referred to as phasing elements, is of paramount importance.

There are two steps in the design of a reflectarray, namely, the element design and the system design. The element design will be discussed in this chapter, and the system design will be discussed in the following chapters. In this chapter, we will first study the basics of designing the phase distribution on the reflectarray aperture. Next, we will outline the phase tuning approaches for reflectarray elements. In other words, how the individual elements are designed to scatter electromagnetic waves with the desired phases. Moreover, numerical and analytical approaches for analysis of reflectarray phasing elements will be outlined, and several examples of reflectarray phasing elements will be presented. Some discussions on frequency behavior, effects of oblique excitation, and sources of phase error for reflectarray elements will also be presented.

2.1 Phase-Shift Distribution on the Reflectarray Aperture

In classic planar antenna arrays, a uniform phase distribution on the aperture will yield a collimated beam in broadside direction, that is, normal to the plane of the array. To focus the beam in a certain direction, a progressive phase distribution is assigned to the elements [4], [5]. For reflectarrays, the basic operating principle is similar, however, one also needs to account for the feed antenna position [6]. The feed antenna is located at a certain position with respect to the reflectarray coordinate system, as shown in Figure 2.1. Typically, the elements of the reflectarray are assumed to be in the far field of the feed antenna; therefore, the incident electromagnetic field on each reflectarray element can be approximated by a plane wave that excites the element with a certain incident angle. The electromagnetic fields emanating from the feed, propagate as a spherical wave which originate from the phase center of the feed antenna. The incident electromagnetic fields on the reflectarray aperture have a phase proportional

Reflectarray Antennas: Theory, Designs, and Applications, First Edition.
Payam Nayeri, Fan Yang, and Atef Z. Elsherbeni.

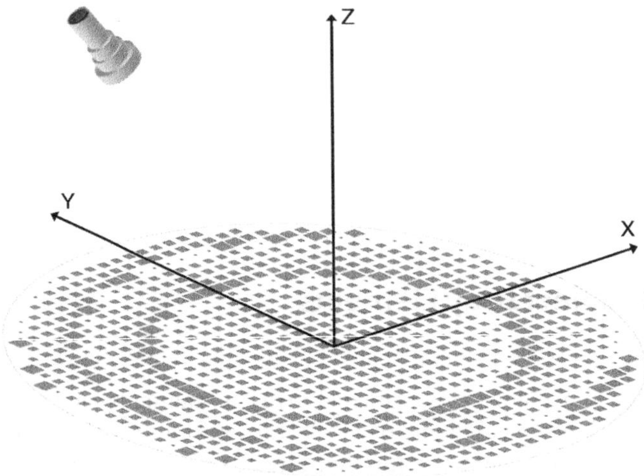

Figure 2.1 Typical geometry of a planar reflectarray antenna.

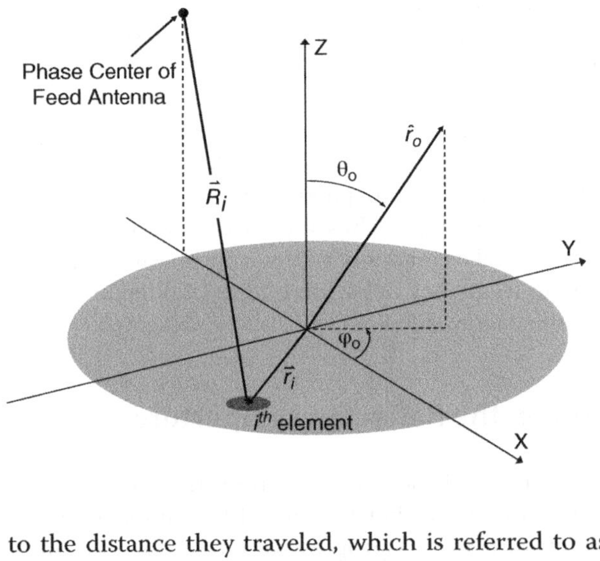

Figure 2.2 Typical geometrical parameters of a planar reflectarray antenna.

to the distance they traveled, which is referred to as spatial phase delay. As such, in order to achieve a collimated beam, the phasing elements of the reflectarray have to compensate for this phase. A geometrical model of the reflectarray system, showing the position of the feed phase center, and the reflectarray coordinate system is given in Figure 2.2.

The reflection phase of a reflectarray element should compensate for the spatial phase delay (spd) from the feed phase center to that element. Mathematically this is given by

$$\phi_{spd} = -k_0 R_i, \tag{2.1}$$

where R_i is the distance from the feed phase center to the ith element, and k_0 is the wavenumber at the center frequency. Such a phase distribution converts the spherical wave radiated by the feed antenna, to a collimated beam in the broadside direction, that is, in the Z-direction with respect to Figure 2.2. To scan this collimated beam in any direction, a progressive phase (pp) can be added to the aperture, which in vector form is given by

$$\phi_{pp} = -k_0 \vec{r}_i . \hat{r}_o, \tag{2.2}$$

where $\vec{r}_i$ is the position vector of the ith element, and $\hat{r}_o$ is the direction of the main beam as shown in Figure 2.2. In the Cartesian coordinate system of Figure 2.2, the position of each element can be expressed as (x_i, y_i), thus for a beam directed in a certain spherical direction (θ_o, φ_o), this equation simplifies to

$$\phi_{pp} = -k_0 \left(x_i \sin\theta_o \cos\varphi_o + y_i \sin\theta_o \sin\varphi_o \right). \tag{2.3}$$

The required phase shift on the reflectarray aperture therefore needs to compensate for the spatial delay (-spd) and add the progressive phase to the aperture, which is given by

$$\phi_{RA} = k_0 \left(R_i - \sin\theta_o \left(x_i \cos\varphi_o + y_i \sin\varphi_o \right) \right) + \phi_0, \tag{2.4}$$

where ϕ_0 is a phase constant, indicating that a relative phase is needed for the reflectarray elements. It should be noted that the required phase distribution given in (2.4), will produce a pencil beam in the desired direction. For shaped beams, however, the progressive phase in (2.3) should be replaced by an appropriate phase distribution, which is obtained using a phase-only synthesis and will be discussed in later chapters. In all designs, however, the phasing elements of the reflectarray antenna must compensate for the spatial phase delay associated with the feed.

The reflectarray antenna is basically analogous to a parabolic reflector antenna, however, unlike the parabola, the reflectarray consists of a certain number of elements, typically arranged in a Cartesian grid. As such, the phase distribution on the reflectarray aperture is essentially pixelated. As an example, the grid layout of a Ka-band reflectarray antenna with a circular aperture is given in Figure 2.3, where each element is assumed to have a constant phase over its own aperture.

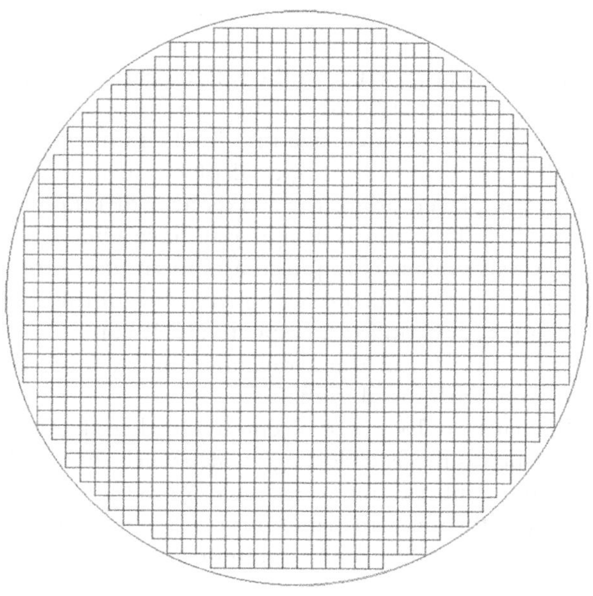

Figure 2.3 Grid layout of the unit-cells of a reflectarray antenna with circular aperture.

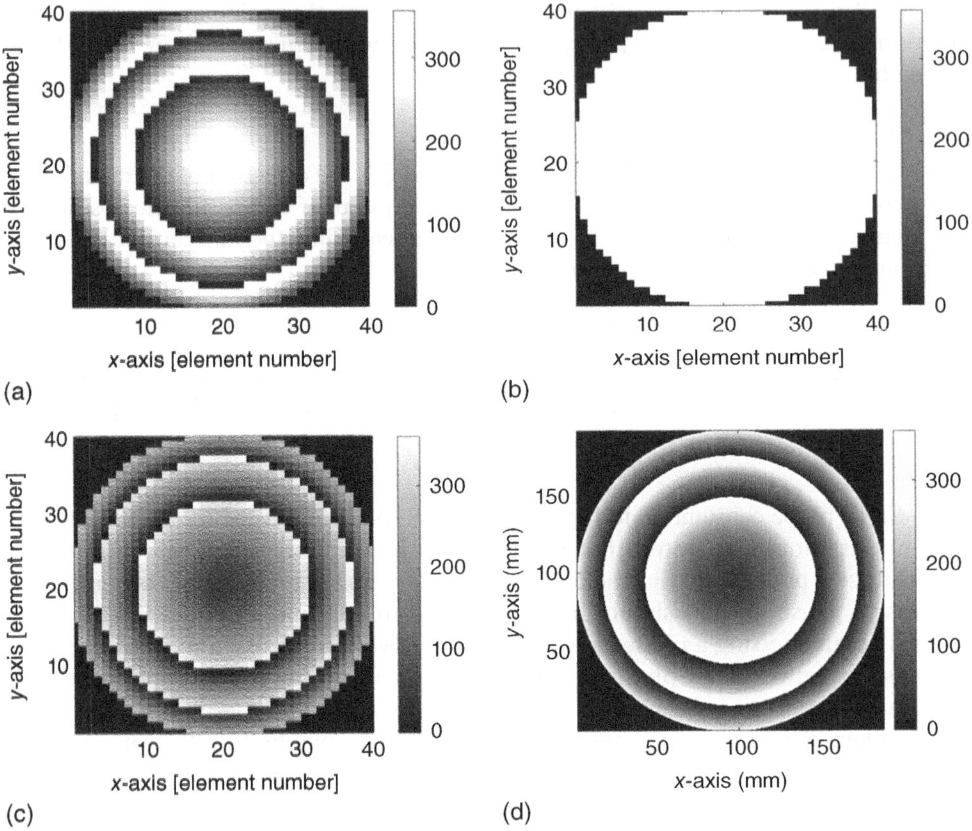

Figure 2.4 Phases on the aperture with a centered feed and broadside beam: (a) spatial delay, (b) progressive phase, (c) phase distribution on the reflectarray antenna, and (d) phase distribution on the continuous aperture.

The circular aperture of the reflectarray has a diameter of 190 mm, which is about 20 wavelengths at 32 GHz. The element spacing is 4.7 mm, that is, each element of the array has a lattice size of $4.7 \times 4.7\,\text{mm}^2$. In total, this array has 1184 elements. Figure 2.4(a) shows the spatial phase delay on the aperture when the feed phase center is placed along the z-axis at a distance of 170 mm from the aperture, see Figure 2.2. The array is designed to have a beam in the broadside direction, hence no progressive phase is required as shown in Figure 2.4(b). The phase distribution on the reflectarray aperture along with the continuous phase distribution are also given in Figure 2.4(c) and (d), respectively. It can be seen that the phase moves outwards from the center of the array, and almost three complete phase cycles (0 to 2π) is observed with this design.

Similar to reflectors, the centered feed configuration exhibits a number of disadvantages such as the feed blockage, which will be outlined in a later chapter. As such, typically an offset configuration is desirable. With the same aperture parameters as the previous example, the phase distributions of an offset configuration with an off-broadside beam given in Figure 2.5. The feed antenna phase center is placed in the xz-plane, and has an

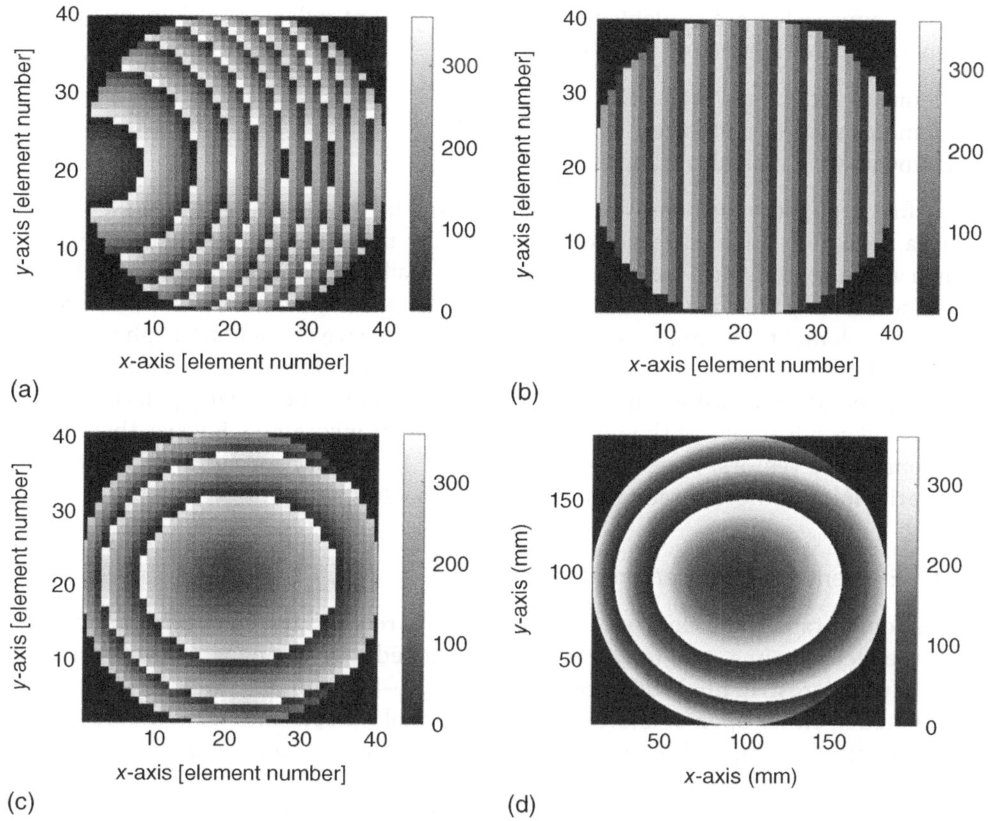

Figure 2.5 Phases on the aperture with an offset feed and an off-broadside beam: (a) spatial delay, (b) progressive phase, (c) phase distribution on the reflectarray antenna, and (d) phase distribution on the continuous aperture.

offset angle of 30°, that is, $x = -85$ mm and $z = 147.22$ mm with respect to the coordinate system of Figure 2.2. The feed points to the geometrical center of the aperture, and the array is designed to focus the beam at ($\theta_o = 30°$, $\varphi_o = 0°$).

The phase distributions presented here clearly show that the elements of a reflectarray must be able to provide a complete phase tuning range. In other words, a mechanism is needed to change the reflection phase of the elements from 0 to 2π. In Section 2.2, we will outline three different phase tuning techniques that have been developed to address this fundamental requirement.

2.2 Phase Tuning Approaches for Reflectarray Elements

A critical step in designing a reflectarray antenna, is selecting a phase tuning methodology that enables the reflectarray elements to achieve the desired phase tuning range. Once the phase tuning approach is selected, the element characteristics can be determined, which are ultimately used to realize the radiation characteristics of the antenna.

Different approaches are available for tuning the phase of reflectarray elements, which in general can be categorized into three groups:

1) Elements with phase/time-delay lines,
2) Elements with variable sizes,
3) Elements with variable rotation angles.

In the first group, the incident wave is first received by the element, then phase shifted using a delay-line microwave network, and finally re-radiated. This methodology is known as the guided-wave technique, and is essentially the concept that Berry [1] used when introducing the first reflectarray antenna in 1963 that utilized short-circuited waveguide elements. With the second group, variable size elements, the phase of the scattered field is controlled by changing the dimensions of the element, for example, the patch length. The last group, which is applicable only to circularly polarized (CP) designs, takes advantage of the unique property of an incoming CP wave, that is, the rotational wave vectors, to achieve phase tuning. In this section, we will outline the basic principles of these three phase tuning approaches.

2.2.1 Elements with Phase/Time-Delay Lines

In this configuration, the element (typically a patch) receives the electromagnetic wave from the feed antenna, and transfers it into a guided wave along a transmission line (typically a microstrip line) with a certain length [7]–[9]. The transmission line can be terminated in either open or short circuit. The signal reflects back from the transmission line termination, and is re-radiated by the element. A schematic model of this type of element is given in Figure 2.6.

In this approach, the phase shift is proportional to twice the line length, which is given by

$$\phi_{delay\,line} = 2kl, \tag{2.5}$$

where l is the length of the line, and k is the propagation constant of the signal along the transmission line. The differential phase shift between any two elements of the reflectarray, $\Delta\phi$, can then be achieved by setting the physical length difference, Δl, equal to $\Delta\phi/2k$. In the ideal case, the reflection phase of these types of elements will be perfectly linear, however, the specular reflection from the ground plane, as well as the stub

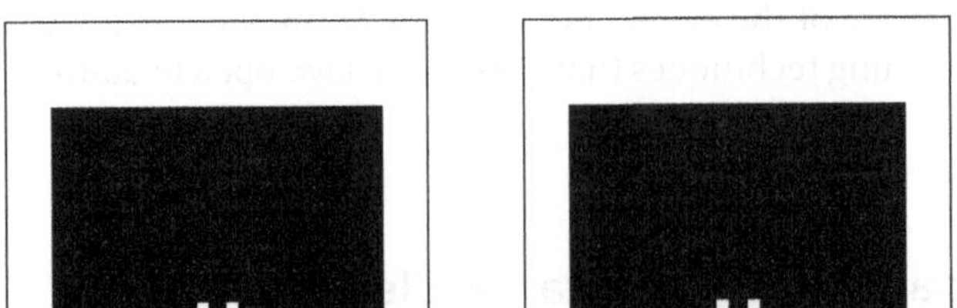

Figure 2.6 Schematic models of reflectarray patch elements with attached phase/time-delay lines of two different lengths.

resonances which occur for certain lengths, and the dissipative losses of the bends in the transmission line, will distort the phase response.

The design principle for this approach is similar to classic microstrip antennas, where the patch dimensions must be suitably chosen such that it resonates at the frequency of incident wave. The line (stub) should also be matched to the patch to allow for proper guided-wave propagation. This is particularly important, since if there is mismatch between the patch and the line, part of the signal will be reflected by the patch before going through the line. In this case, the reflected wave will be the sum of two components, and the reflection phase will no longer be proportional to twice the length of the line. It is also critical to note here that to account for mutual coupling between the elements of the reflectarray antenna, design of the patch and delay line should be carried out when the element is placed in the array environment, not as an isolated element. Detailed discussions on analysis techniques for reflectarray elements will be given in Section 2.3.

2.2.2 Elements with Variable Sizes

In this approach, the physical size of the element is changed to provide phase tuning. Theoretically, it is known that changing the length of a resonant element changes the resonance frequency of the antenna, which corresponds to a change in radiated phase at a certain frequency. The operating principle of the variable size technique is therefore based on the fact that the reflected phase from resonant elements with different sizes would be different. This unique method was first introduced in [10] for printed crossed dipoles, and in [11] for rectangular patches. The variable element size technique, although conventionally adopts square or circular patch geometry, provides the most diversified choices of element geometries [12].

A printed antenna is a highly resonant element with a large quality factor, therefore a small change in its size produces a wide range of phase change in the reflected wave. Ideally, a single resonance can provide a complete phase cycle of 360°, however, in practice, the total achievable phase range depends on a number of factors such as the separation between patches and the substrate thickness, and is less than a complete cycle. For small substrate thickness, typically less than one tenth of the wavelength, a phase tuning range above 300° can usually be achieved, which is sufficient for most reflectarray designs. A schematic model of this type of element is given in Figure 2.7.

It is important to note that the phase variation obtained using this approach is highly non-linear primarily because of the high-Q resonant nature of printed antennas with thin substrates. This results in a rapid phase variation near resonance, and slow

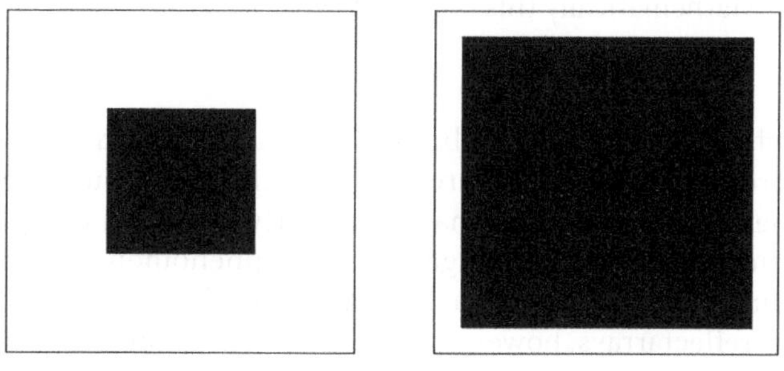

Figure 2.7 Schematic models of variable size reflectarray square patch elements of two different sizes.

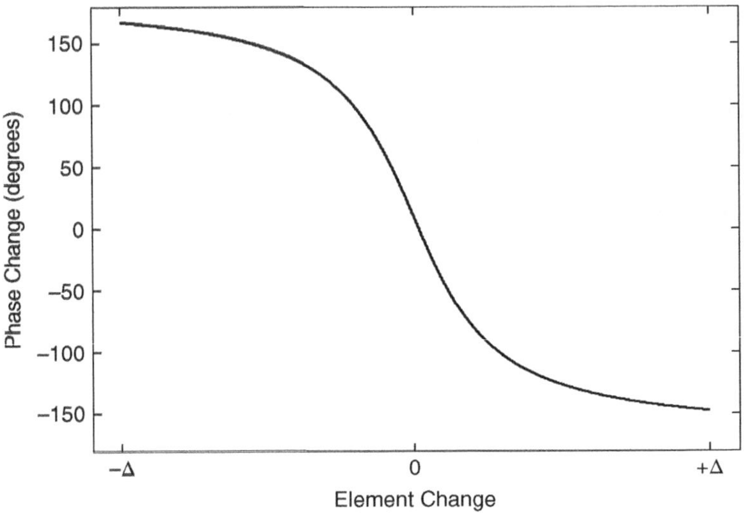

Figure 2.8 A typical S-curve for variable size reflectarray elements.

variations at the extreme dimensions. While the shape of the curve generally depends on the element design scheme, typically it is in the form of an S-shaped curve. A typical S-curve for reflectarray elements is given in Figure 2.8.

2.2.3 Elements with Variable Rotation Angles

A very clever phase tuning technique, which is restricted to CP designs, is the element rotation approach. This technique is based on the fact that rotating a CP antenna element about its origin by ψ°, will change the radiated phase by the same amount where the advance or delay of the phase depends on the direction of rotation [13]. This technique was first applied to reflectarrays using patch elements with attached phase-delay lines [14]. The patches were rotated about their origin (center of the unit-cell), resulting in a phase shift for the reflected wave, which is a linear function of the rotation angle. To utilize this technique for phase tuning of reflectarray elements, one needs to determine the direct relationship between rotation angle and phase of the reflected wave. The mathematical formulation is outlined here.

Without the loss of generality, we consider a right-hand circularly polarized (RHCP) wave traveling in the $-z$-direction impinging on the unit-cell, as schematically shown in Figure 2.9.

Mathematically this is given by

$$\vec{E}_i = E_0\left(\hat{x} + j\hat{y}\right)e^{jk_0 z}e^{j\omega t}, \tag{2.6}$$

where k is the wavenumber and is the angular frequency of the wave. The reflected wave propagates in the + z-direction, therefore in the usual manner, the reflected wave will be converted to a left-hand circularly polarized (LHCP) wave, due to the reversal of the direction of propagation. This phenomenon is observed in reflectarrays with dual-linear elements (such as square patches) and also in reflector antennas. In the case of reflectarrays, however, if the element can introduce a 180° phase difference between

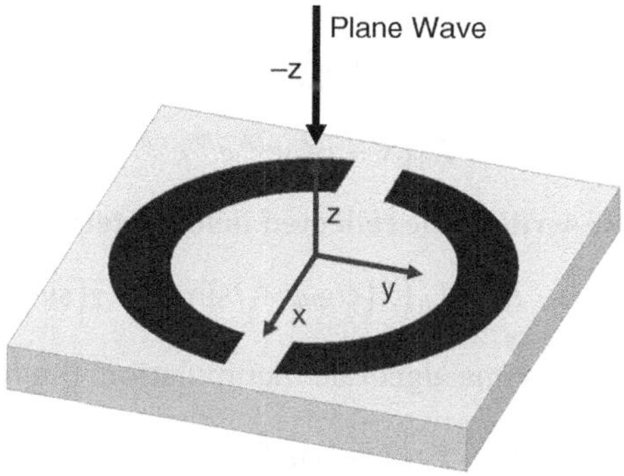

Figure 2.9 A 3D schematic model of a CP reflectarray element.

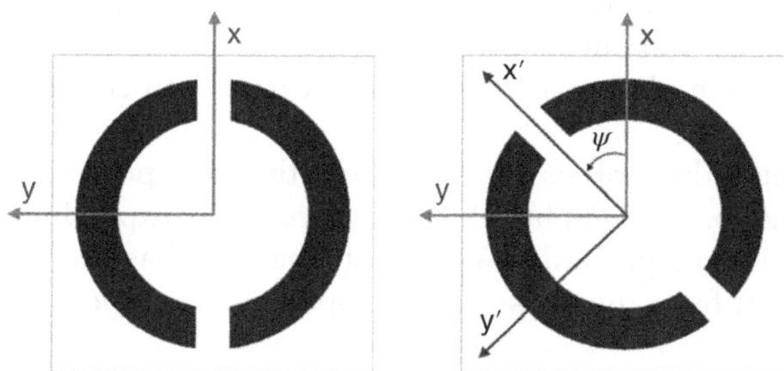

Figure 2.10 A schematic model of a CP reflectarray element showing the reference element with 0° phase shift (left) and the ψ rotated element with 2ψ phase shift (right).

the two orthogonal components of the reflected field, the sense of polarization can be maintained. In this setup, this requires the phase in the y-direction (ϕ_y) to lead by 180°, that is, $\phi_y = \phi_x + 180°$. This can be realized with phase-delay lines, or elements that can intrinsically create this phase difference, such as split rings. For the reference element in Figure 2.10(a), the reflected field can be written as

$$\vec{E}_r = E_0\left(\hat{x}e^{j\phi_x} + j\hat{y}e^{j\phi_y}\right)e^{-jk_0z}e^{j\omega t} = E_0\left(\hat{x} - j\hat{y}\right)e^{j\phi_x}e^{-jk_0z}e^{j\omega t}, \tag{2.7}$$

where it is assumed that there is no loss due to reflection from the element.

Now let us consider the case when the element is rotated by ψ, as shown in Figure 2.10(b). The incident and reflected fields can now be easily obtained in the new coordinate system (x',y'), which is aligned with the rotated element. The incident field is given by

$$\vec{E}_i = E_0\left(\left(\hat{x}'\cos\psi - \hat{y}'\sin\psi\right) + j\left(\hat{x}'\sin\psi + \hat{y}'\cos\psi\right)\right)e^{jk_0z}e^{j\omega t}, \tag{2.8}$$

which after some algebraic manipulation can be written as

$$\vec{E}_i = E_0\left(\hat{x}' + j\hat{y}'\right)e^{j\psi}e^{jk_0z}e^{j\omega t}. \tag{2.9}$$

The reflected field for the rotated element with $\varphi'_y = \varphi'_x + 180°$ is now given by

$$\vec{E}_r = E_0\left(\hat{x}'e^{j\phi_{x'}} + j\hat{y}'e^{j\phi_{y'}}\right)e^{j\psi}e^{-jk_0z}e^{j\omega t} = E_0\left(\hat{x}' - j\hat{y}'\right)e^{j\phi_{x'}}e^{j\psi}e^{-jk_0z}e^{j\omega t}. \tag{2.10}$$

Note that since the element is symmetric with rotation, such as the circular split ring in Figure 2.10, $\phi_x = \phi_{x'}, \phi_y = \phi_{y'}$, and the reflected field can then be written as

$$\vec{E}_r = E_0\left(\hat{x}' - j\hat{y}'\right)e^{j\phi_x}e^{j\psi}e^{-jk_0 z}e^{j\omega t}. \tag{2.11}$$

Re-writing the reflected fields in terms of the original (x, y) coordinates yields

$$\vec{E}_r = E_0\left(\hat{x}\left(\cos\psi + j\sin\psi\right) + \hat{y}\left(\sin\psi - j\cos\psi\right)\right)e^{j\phi_x}e^{j\psi}e^{-jk_0 z}e^{j\omega t}. \tag{2.12}$$

With some algebraic manipulation, the reflected field can be written as

$$\vec{E}_r = E_0\left(\hat{x} - j\hat{y}\right)e^{j\phi_x}e^{j2\psi}e^{-jk_0 z}e^{j\omega t}. \tag{2.13}$$

Comparing (2.13) to (2.7), the reflected field is delayed in phase by 2ψ when the element is rotated counter clockwise by ψ about its origin. Similarly, it can be shown that a LHCP wave would be advanced in phase when the element is rotated counter clockwise.

It is also important to point out that in the analysis presented here, the cross-polarized reflected fields were assumed to be zero. This can be taken into account by expanding (2.7) with proper reflection magnitudes and phases, however, note that the phase of the cross-polarized component is different from the phase of the co-polarized component discussed previously.

2.3 Element Analysis Methods

In general, for analysis and design of the elements of an antenna array, one needs to take into account the electromagnetic interaction between the elements, which is known as mutual coupling. The effect of mutual coupling is serious and needs to be accounted for, if the element spacing is small compared to the wavelength, which is usually the case for arrays. On the other hand, for large antenna arrays with several hundreds of elements, such as reflectarrays, it is very challenging to conduct the analysis in the full array model.

In an array with a very large number of regularly spaced elements, all the elements except those near the edge have approximately the same behavior. In an infinite array [2], every element has identical behavior because there is no edge or other variation in the arrangement of the elements. The basic properties of nearly all the elements in a very large regular array can therefore be represented by an element in the corresponding infinite array, which has firmly been established as the best basis for the element design in large finite arrays. It should be noted here that the infinite array approach assumes all surrounding elements are identical, in other words the structure that is analyzed is assumed to be periodic. In the case of reflectarrays, however, the elements are not identical, and the array is quasi-periodic in nature. Nonetheless, in most cases, this periodic approximation is adequate enough to characterize the properties of the reflectarray elements.

In this section, we will study different techniques that have been developed over the years for efficient analysis of reflectarray elements.

2.3.1 Periodic Boundary Conditions and Floquet Port Excitation

One technique that can mimic a plane wave impinging normally to the element aperture is the H-wall waveguide simulator, also known as the parallel waveguide simulator [15]–[17]. In most cases, however, it is desirable to study the reflection characteristics of the element, when it is excited with a plane wave that impinges in a general direction (θ_i, φ_i) and polarization (*TE, TM*). The wave then reflects back in the specular direction with a set of amplitude and phase distributions. Although such an H-wall waveguide simulator can accurately mimic a normal plane wave excitation, a more generalized approach will be outlined here that can simulate both normal and oblique excitations.

In this approach, the element is placed at the end of a rectangular cuboid where the width and length of the cross-section is the same as the unit-cell size [17]. A geometrical model of this element excitation setup is given in Figure 2.11. The four sidewalls of the unit-cell are linked together as periodic boundaries, which enables one to model planes of periodicity where the field on one surface matches the field on another with a certain phase difference. The unit-cell is then excited with a Floquet port, which is located on the top surface of the cuboid, typically at a distance of $\lambda/2$ at the minimum frequency. The Floquet port excites a set of Floquet modes, which are fundamentally plane waves with propagation direction set by the frequency and geometry of the periodic structure. This is the general setup for analysis of a periodic array element for any angle of incidence or sense of polarization. The reflected field quantities can then obtained directly at the port. It is important to note that several commercial electromagnetic simulation software are available for this analysis such as Ansys HFSS [18], CST Microwave Studio [19], and FEKO [20].

2.3.2 Metallic Waveguide Simulators

The metallic waveguide simulator can accurately characterize the reflection properties of the element under certain conditions. In this technique, the element is placed at the end of a real waveguide and is excited with the dominant TE_{10} mode. This waveguide setup essentially creates an infinite array scenario for the element, which can be explained using image theory [16]. The elements in the waveguide will be excited with an oblique angle given by

$$\psi = \cos^{-1}\left(\sqrt{1-\left(\frac{f_c}{f}\right)^2}\right) = \cos^{-1}\left(\sqrt{1-\left(\frac{\lambda}{2a}\right)^2}\right). \tag{2.14}$$

Here f_c is the cutoff frequency of the fundamental mode, which is directly related to a, the larger dimension of the rectangular waveguide. Each waveguide can only simulate one incident angle at one frequency, therefore, several different waveguides need to be constructed in order to analyze the element with several different excitation angles at multiple frequencies. Due to these limitations, this approach is seldom used for the element simulation. However, it is particularly useful for measuring the element performance.

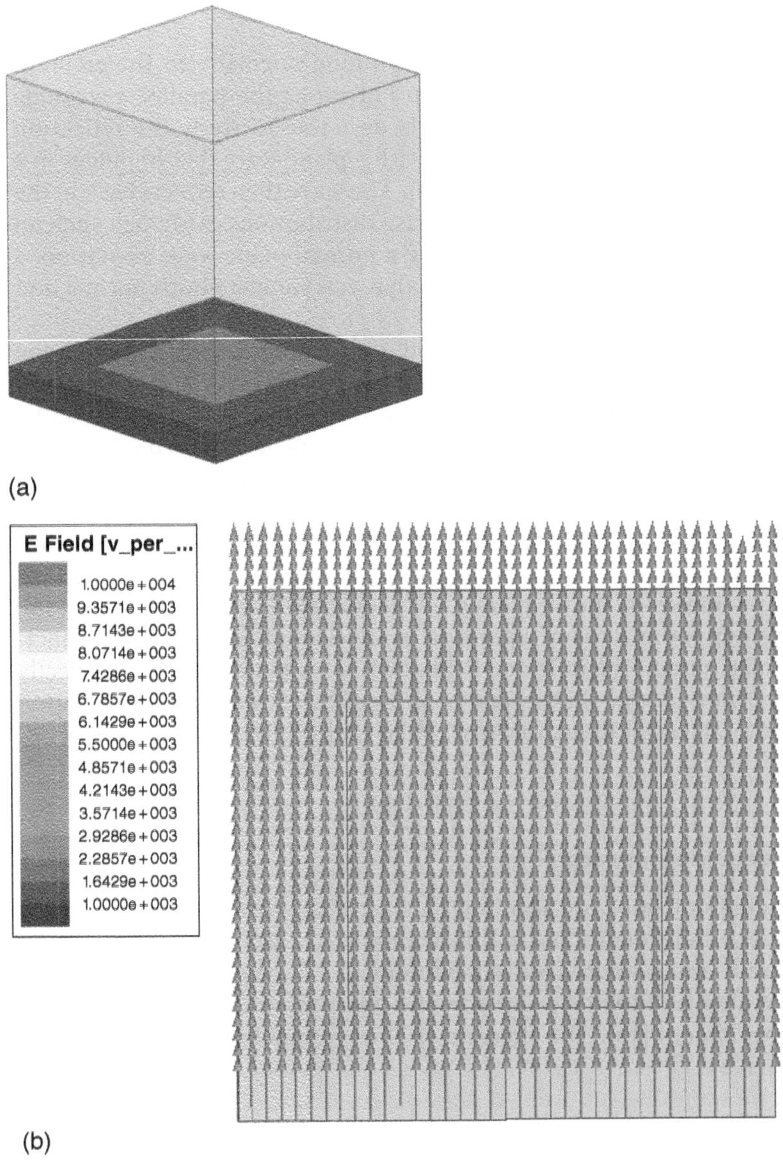

Figure 2.11 A unit-cell analysis setup: (a) 3D model in Ansys HFSS, (b) vector electric fields of the Floquet port excitation.

The standard rectangular waveguides typically have an aspect ratio about 2:1. By placing two elements in the waveguide, a unit-cell with an almost square lattice can be simulated. A simulation setup for a square patch element inside a standard rectangular waveguide is shown in Figure 2.12(a). The vector electric fields at the waveguide port are also shown in Figure 2.12(b). Note that the waveguide excitation is not a uniform plane wave, however, it will allow one to accurately characterize the reflection properties of the reflectarray element. More discussion on this issue, and comparison with other approaches will be given in Section 2.3.4.

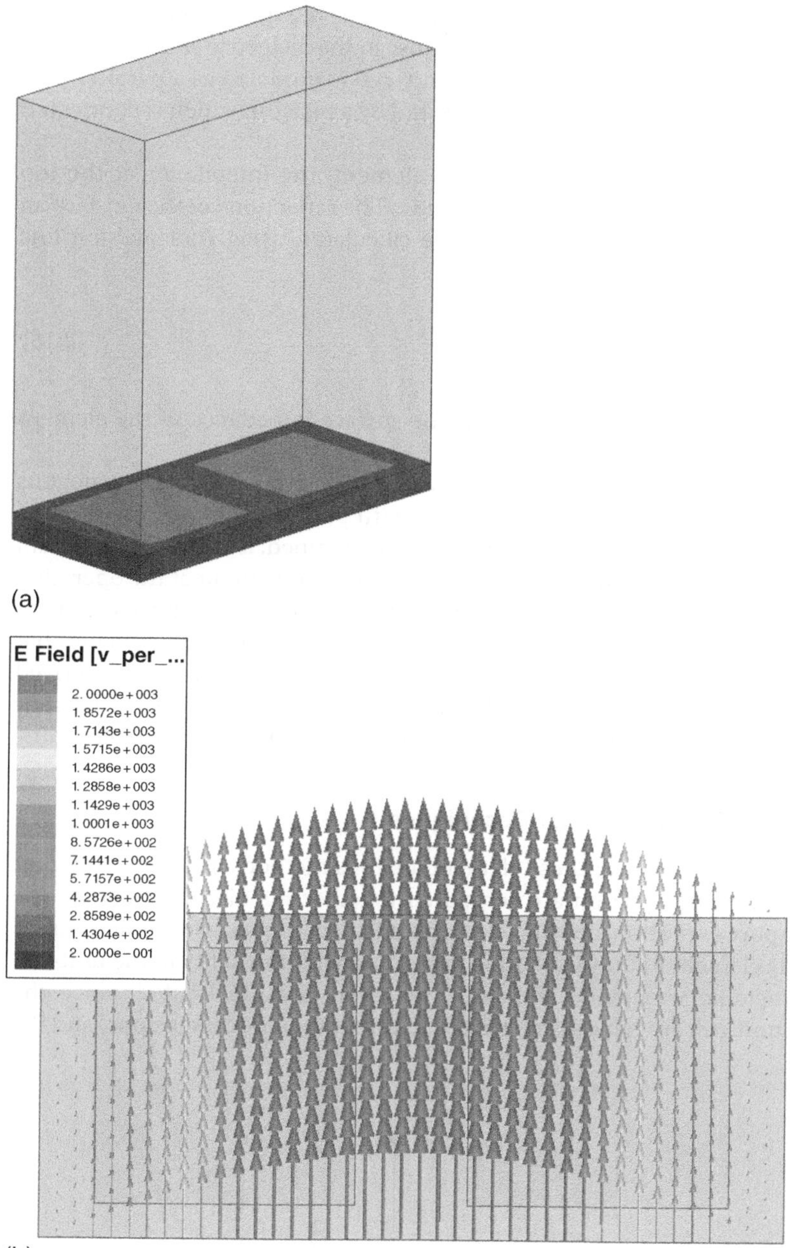

Figure 2.12 A metallic waveguide simulator: (a) 3D model in Ansys HFSS and (b) vector electric fields of the wave port excitation.

2.3.3 Analytical Circuit Models

The unit-cell can also be analyzed by viewing the structure as a transmission line circuit [21]. Depending on the geometrical features of the element, an analogous circuit model has to be derived for the analysis. For the square patch studied in this section, the metallic patches act as parallel capacitors with the adjacent cells while the short-circuited unit-cell ground can be modeled as a parallel inductor. The loss in the dielectric substrate can be modeled by a parallel resistor, that is, the total unit-cell is modeled as a parallel RLC circuit. The conductor losses can also be represented by a resistor, which is connected in series with the capacitor as shown in [22].

Using the circuit model representation of the element, the impedance at the top surface of the element (Zs) can be calculated directly. The reflection coefficient Γ of an incident plane wave on the unit-cell can then be calculated using transmission line theory, that is,

$$\Gamma = \frac{Z_S - Z_o}{Z_S + Z_o}. \tag{2.15}$$

where Z_o is the free space impedance and Z_S is the surface impedance of the element obtained by the circuit model as shown in Figure 2.13.

It is implicit that the main challenge in this analysis is deriving a correct analogous circuit model representation of the element, along with proper values for the lumped elements. Alternatively, once the circuit model has been obtained, it is possible to obtain accurate values for the lumped elements by designing a resonant circuit at the operating frequency and then tuning these values so that the circuit model reflection curves fit the curves obtained by the full-wave unit-cell simulations. It should also be pointed out that the circuit model analysis is general and the transmission line parameters can be derived for oblique incidence angles as well. The advantage of the circuit models is that they provide a simple and direct physical insight on the operation mechanism of the elements.

2.3.4 Comparison of Element Analysis Techniques

In order to observe the accuracy of these element analysis methods, the three different techniques described in this section are compared together. As discussed earlier, the PBC with Floquet port excitation, and the circuit model techniques allow one to simulate the reflectarray element for any excitation angle, and with flexibility in the unit-cell dimension. The metallic waveguide simulator, however, is more limited, since the dimensions of the unit-cell are typically restricted to the waveguide dimensions, and the

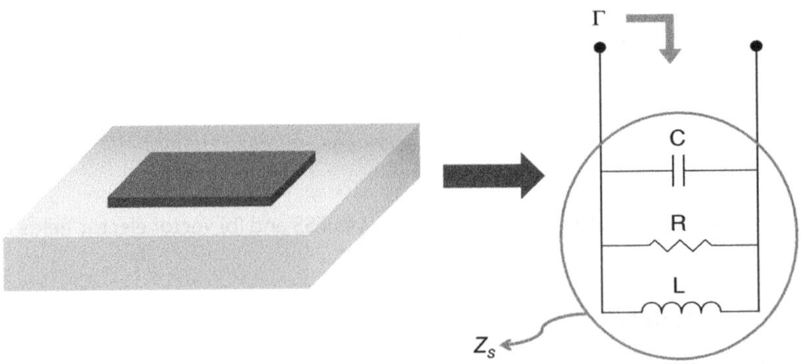

Figure 2.13 The circuit model for the unit-cell of a reflectarray antenna.

waveguide can only simulate one incident angle at one frequency. To mitigate the restrictions with the waveguide simulator, and without the loss of generality, we compare the reflection characteristics of a square patch element for two different cases.

2.3.4.1 Comparison between PBC and Metallic Waveguides

In the first study, we compare the reflection coefficients of a square patch element as a function of frequency in a standard X-band (WR-90) waveguide. The inner dimensions of the waveguide are 22.86 × 10.6 mm; therefore, we select a unit-cell size of 11.43 × 10.16 mm for the element that has an aspect ratio of 1.125. For the reflectarray element, a 62 mil Rogers 5880 laminate is selected for the substrate, and the square patch has a side length of 8 mm, which is the resonance dimension for the patch. Note that two elements are placed in the waveguide simulator as shown in Figure 2.12. The angle of incidence is frequency dependent and varies from 53° to 32° across 8.2–12.4 GHz, as shown in Figure 2.14. At 10 GHz the angle of incidence is 40.97°. Also note that the waveguide is excited in the fundamental mode, thus the incident wave is TE polarized.

The simulation setup for the PBC method with Floquet port excitations carried out as described earlier. However, note that for this comparison one has to define the excitation angle in the simulations at each frequency accordingly. The simulations are carried out using Ansys HFSS, and reflection coefficients obtained using these two methods are given in Figure 2.15. It can be seen that a very good agreement is observed between these methods. It is important to note here that in reflectarray designs it is necessary to compute the reflection phase at a certain reference plane. Typically, in printed microstrip reflectarrays, the reference plane is the top surface of the patch, and one has to de-embed the port in order to obtain the reflection phase at the desired reference plane.

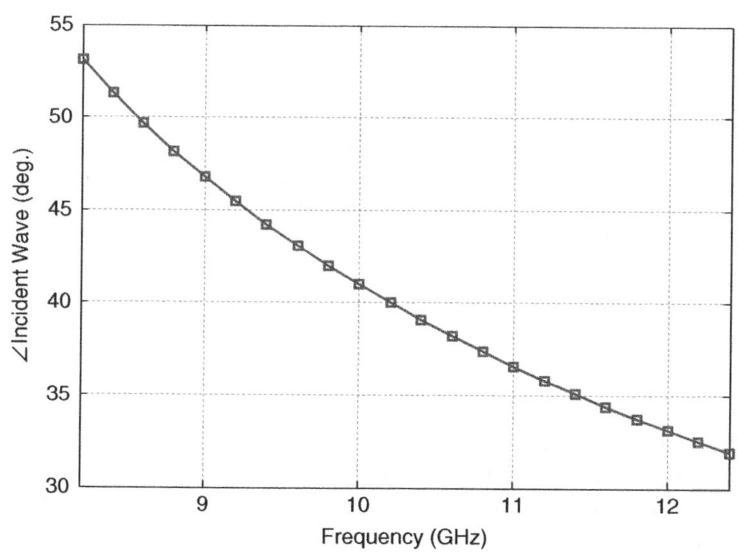

Figure 2.14 The angle of incidence as a function of frequency in the standard X-band waveguide.

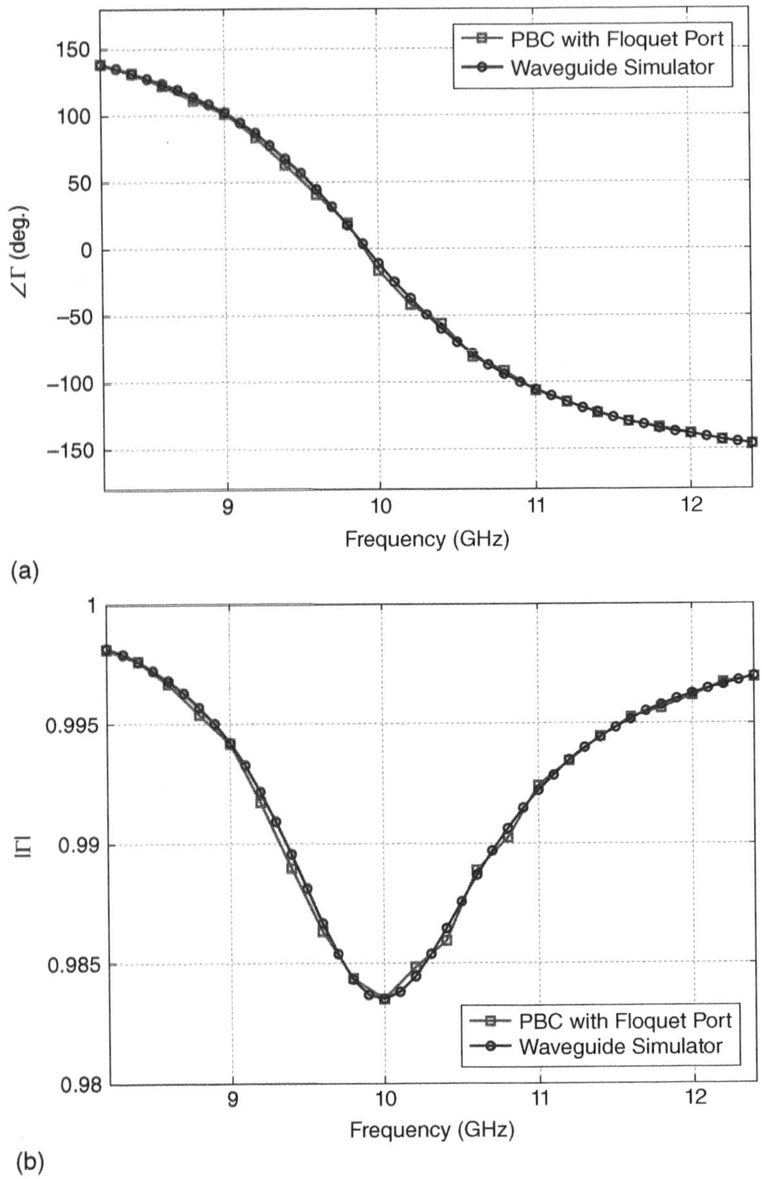

Figure 2.15 Comparison between the reflection coefficients of an X-band square patch reflectarray element using a waveguide simulator and a PBC model with Floquet port excitation: (a) phase, (b) magnitude.

2.3.4.2 Comparison between PBC and the Circuit Model

The comparative study presented here can also be conducted using the circuit model. However, since the excitation angle is a function of frequency, one has to derive the values of the lumped elements for each frequency independently. As such it will be more convenient to study a case where the angle of incidence is frequency independent. Here we study the reflection coefficients of the same element under normal incidence using the circuit model, and the PBC method with Floquet port excitation. As discussed earlier, for the patch circuit model, one needs to determine the values of the capacitor, inductor and resistors. Typically, the conductor losses are less dominant in this frequency range and the majority of material losses is attributed to the dielectric substrate. In the circuit model for this element, the admittance of the parallel resistor is 14.286 μS. For the reactive elements, the values of capacitor and inductor are determined to be 0.1685pF and 1.4448nH, respectively. Comparison between the reflection coefficients obtained using these two methods are given in Figure 2.16, where again it can be seen that a very good agreement is observed between these methods.

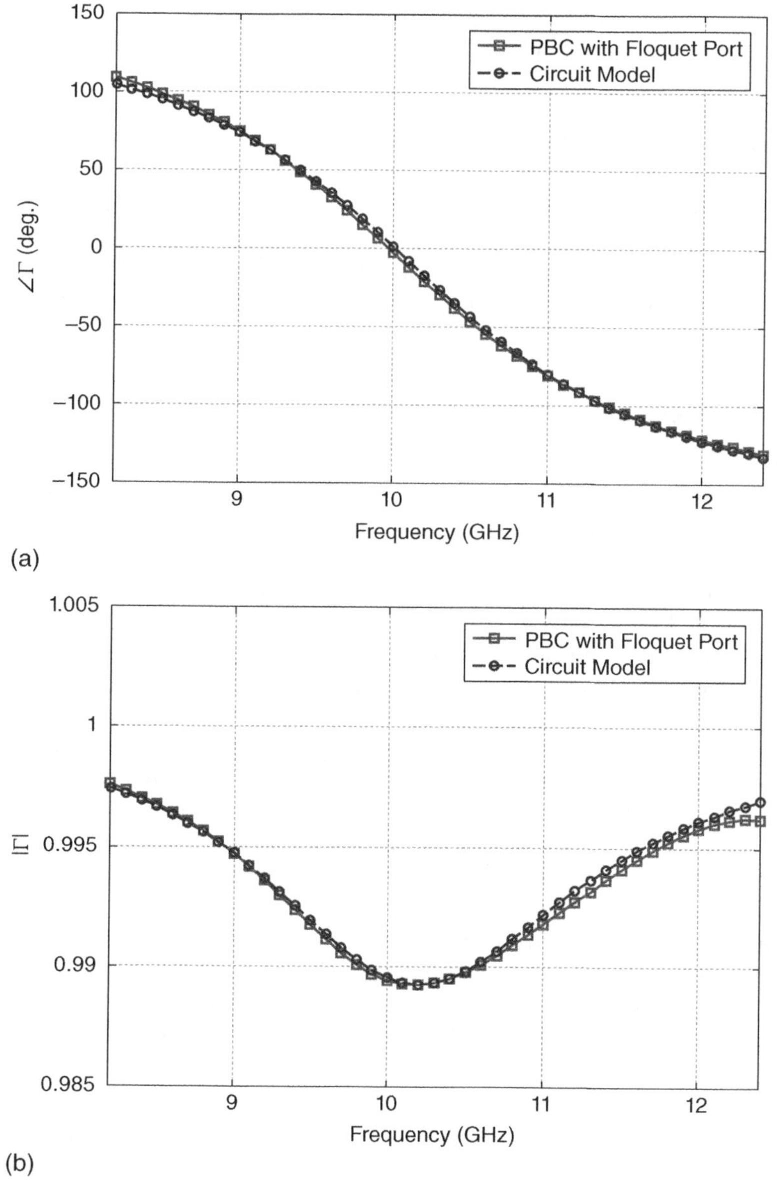

Figure 2.16 Comparison between the reflection coefficients of an X-band square patch reflectarray element using a circuit model and a PBC model with Flouqet port excitation: (a) phase, (b) magnitude.

In summary, a variety of techniques are available for analysis of reflectarray elements, which can accurately determine the reflection characteristics of the elements. While each of these methods have their own advantage, in general, the PBC method with Floquet port excitation is the most suitable approach for designing the element, since it can be conducted directly with a full-wave simulator and also allows one to study the element properties under different excitation angles. The other two techniques are more suitable to use after the element has been designed. In particular, the advantage of the waveguide simulator is that the element characteristics can be obtained experimentally for verification purpose. The circuit model, on the other hand, provides a good insight into the physical characteristics of the element. In the next section, we will study a few classic examples of reflectarray elements.

2.4 Examples of Classic Reflectarray Elements

In Section 2.2, three different methodologies were introduced which enable the reflectarray elements to achieve phase tuning. In this section, we provide one classic example for each of these phase tuning techniques, and outline the basic design principles. The unit-cell analysis is carried out using the periodic boundary conditions with Floquet port excitation as described in Section 2.3.

2.4.1 Rectangular Patch with Phase-Delay Lines

In this approach, the patch element receives the space wave and converts the power into a guided wave that is transmitted through an attached stub. The signal reflects back from the stub termination, travels backward on the transmission line, and is then re-radiated by the patch. Here we consider a Ka-band microstrip reflectarray element, designed for 32 GHz operation. A 20 mil Rogers 5880 laminate is used for the element which has a dielectric constant of 2.2 and a loss tangent of 0.0009. The unit-cell size is $4.7 \times 4.7\,mm^2$, which is about half-wavelength at the center frequency.

The design is carried out in two stages. First, a rectangular microstrip antenna element is designed which is excited by a microstrip feed line. The dimensions of the element are tuned to achieve an efficient radiation with a good impedance match at 32 GHz. According to reciprocal theory, it ensures that the signal received by the patch will be efficiently converted to a guided wave. The model of the element in Ansys HFSS along with the simulated reflection coefficient magnitude are given in Figure 2.17. The patch has length and width of 2.75 and 3.1 mm, respectively. The inset is 0.2 mm long, and 0.15 mm wide on each side of the feed line. The width of the feed line is 0.2 mm, which corresponds to an impedance of 140 Ω for the transmission line. It should be noted that unlike traditional array elements, that need to be designed for a certain input impedance, one typically has more freedom here, since in the reflectarray mode the line will be terminated with an open circuit. Also note that the reflection coefficient magnitude shown in Figure 2.17(b) is referenced to a 140 Ω port impedance.

In the second stage, the element is studied for reflectarray operation, that is, the element is excited by a plane wave using a Floquet port. As discussed earlier, the phase shift provided by the delay line is proportional to twice the length of the line. As such, one needs to bend the line in order to realize a sufficient long physical length in the lattice.

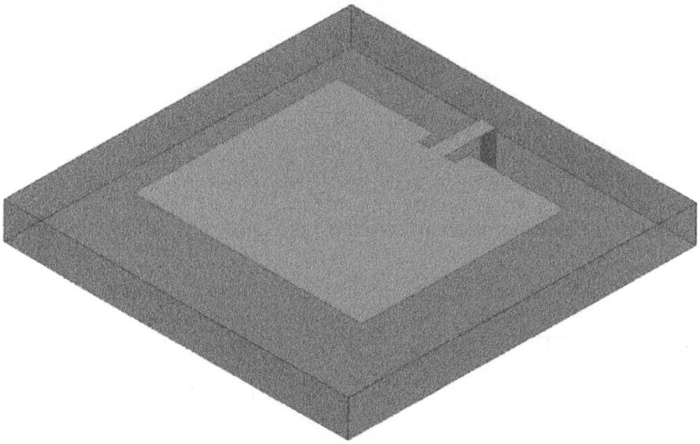

(a)

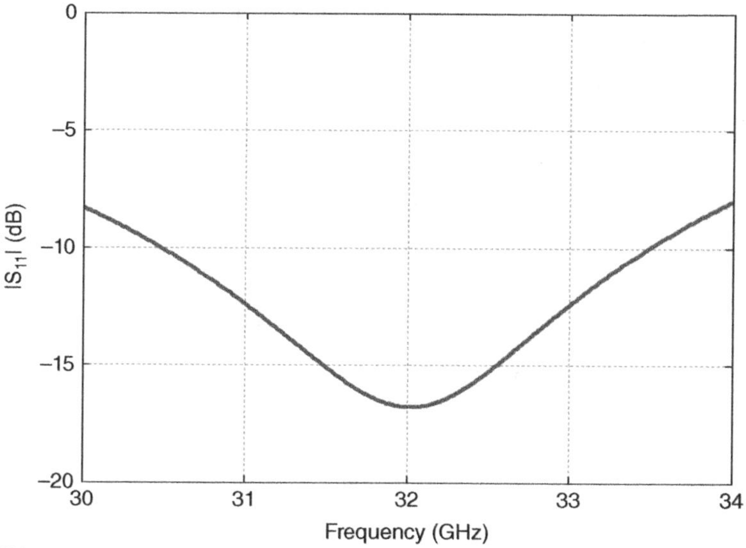

(b)

Figure 2.17 A Ka-band microstrip patch antenna array element: (a) 3D model of the element, (b) magnitude of reflection coefficient across the band.

The model of this reflectarray element in Ansys HFSS is shown in Figure 2.18, where the microstrip line with two mitered corner bends provides the necessary phase shift.

The reflection coefficient of this Ka-band reflectarray element is given in Figure 2.19, where it can be seen that this element yields an almost linear reflected phase variation as a function of the phase-delay line length. With a total line length of 5.15 mm, the phase tuning range of this element is more than 450°. The average reflection coefficient magnitude is close to 0.98. The effects of the bends around 1.1 and 3.2 mm can be seen on the phase plot as small regions with almost constant phase.

The design presented here can easily achieve a wider phase range by increasing the transmission line length; however, if several phase cycles are needed, it would be advantageous to use aperture coupled designs. More discussion on this will be given in

Figure 2.18 A Ka-band microstrip reflectarray element with attached phase-delay line stubs modeled in Ansys HFSS.

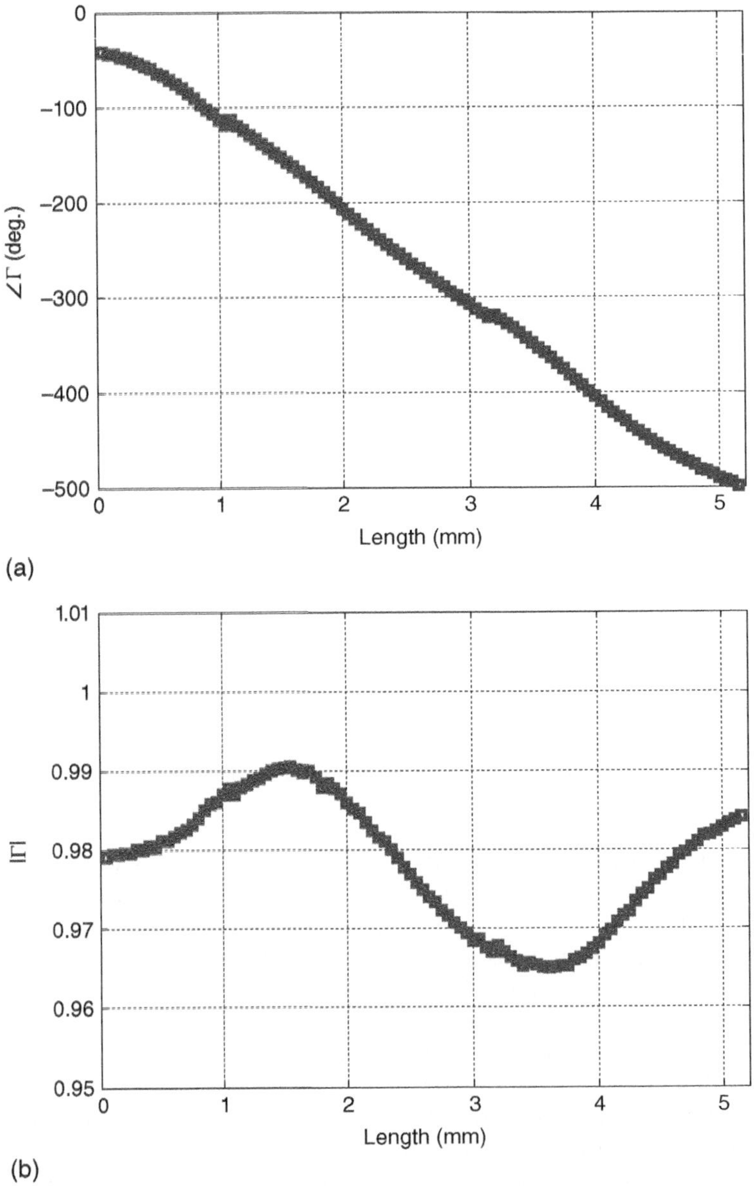

Figure 2.19 Reflection coefficient of the Ka-band microstrip reflectarray element as a function of the phase-delay line length: (a) phase, (b) magnitude.

a later chapter. It is also interesting to compare the radiation characteristics of this element in antenna and reflectarray operation. The surface currents at 32 GHz are given in Figure 2.20, where it can be seen that while there is some difference in the current distribution on the microstrip lines, the surface current distribution on the patches have an almost identical form. As such, the element radiation patterns for these two modes are similar. The far-field radiation pattern of the reflectarray element is given in Figure 2.21, where it can be seen that an almost omnidirectional radiation performance is obtained. More discussion on reflectarray element patterns and the effects of element beam-width on the radiation patterns of the reflectarray will be given in a later chapter.

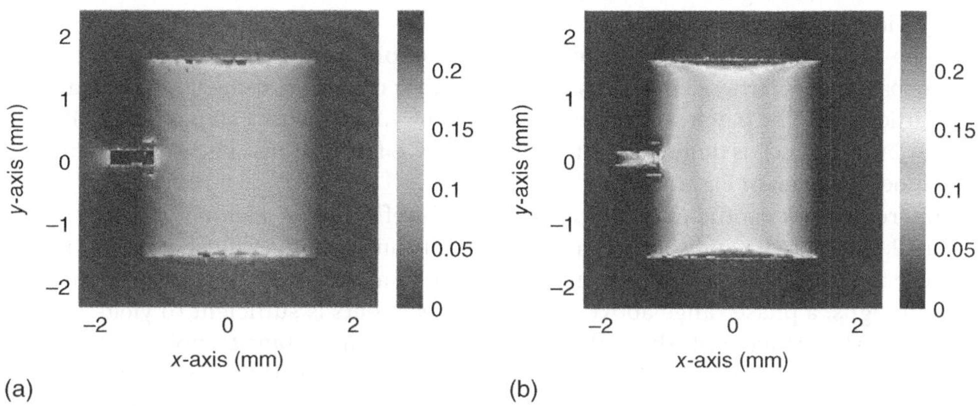

Figure 2.20 Magnitude of surface currents ($|J_x|$ in A/mm) on the unit-cells: (a) inset-fed microstrip patch element, (b) reflectarray element with open-circuit stub.

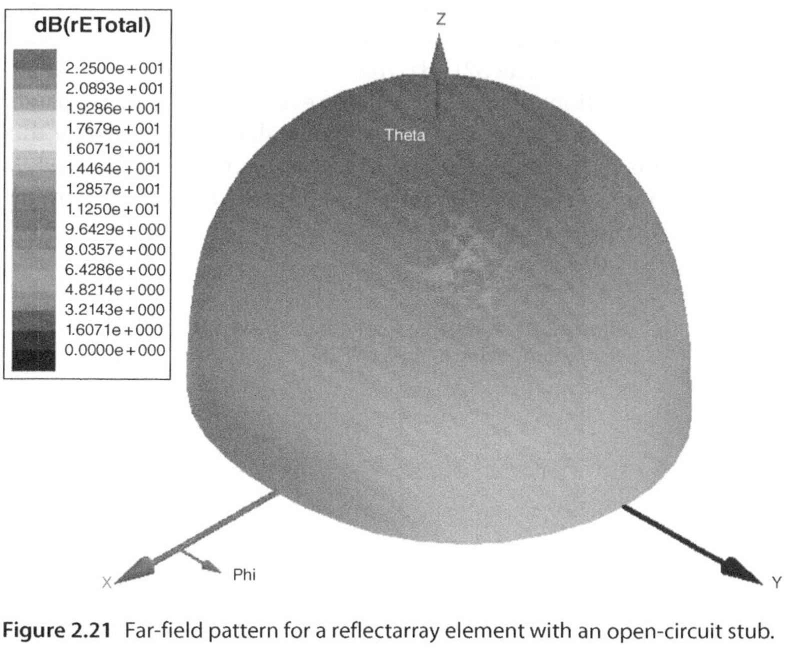

Figure 2.21 Far-field pattern for a reflectarray element with an open-circuit stub.

2.4.2 Variable Size Square Patch

The variable size square patch is perhaps the most popular element for reflectarrays. The design procedure is fairly simple, and the simplicity of the element shape is very advantageous for low-cost fabrication, particularly at very high frequencies. In this design, the only parameter is the width of the patch. For a very small patch width, the reflection phase is basically that of the grounded dielectric slab which depends on the thickness of the laminate. The maximum patch width where the patch almost covers the entire unit-cell essentially approaches that of an infinite conductor sheet, thus in the limit, the reflected phase is about −180°. The reflection phase varies with an S-shaped curve as shown in Figure 2.8, from a maximum value to minimum as a function of the patch width.

To be consistent with the previous design example, here we study a variable size square patch reflectarray element designed for 32 GHz operation, using the same material and lattice size. The patch width varies from 0.5 to 4.5 mm with a step size of 0.05 mm (about 2 mm), which is quite typical for low-cost PCB fabrications. The element simulation model is given in Figure 2.22.

The reflection coefficients of this Ka-band reflectarray element are given in Figure 2.23, where it can be seen that this element can achieve a reflection phase range of more than 310°. While this element cannot realize a complete phase range of 360°, in most designs, a phase range above 300° for the elements is sufficient to yield a satisfactory performance with the reflectarray. Also it is important to point out that in general, elements with dimensions close to resonance (patch widths ranging from about 2–3.5 mm in this design) provide the largest variation in phase range, but they also exhibit the highest loss. Nonetheless, the minimum reflection coefficient magnitude is around 0.975 (about 0.2 dB loss), which is quite sufficient for the element.

To get a better insight about the operating mechanism of the element, here we study the field properties of the resonant patch element which has a width of 2.73 mm. As discussed earlier, Floquet port excitation is used for the element analysis. The port is placed at a distance of 5 mm (about 0.53λ at 32 GHz) from the top surface of the element. For normal incidence, two polarizations can be excited; however, since the square patch is a dual-linear element, these excitations are essentially identical. Here we consider the mode where the electric field is in the x-direction (corresponding to the TM_{00} mode), and sample the fields at a distance of 0.1 mm (about 0.01λ at 32 GHz) from the top of the patch surface. The total electric fields are shown in Figure 2.24.

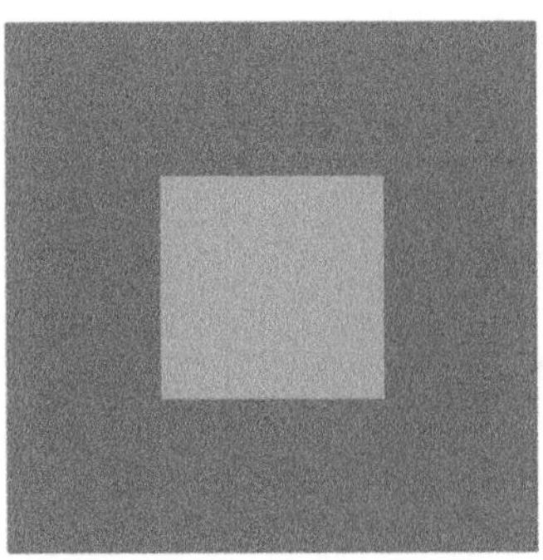

Figure 2.22 A variable size Ka-band microstrip reflectarray element modeled in Ansys HFSS.

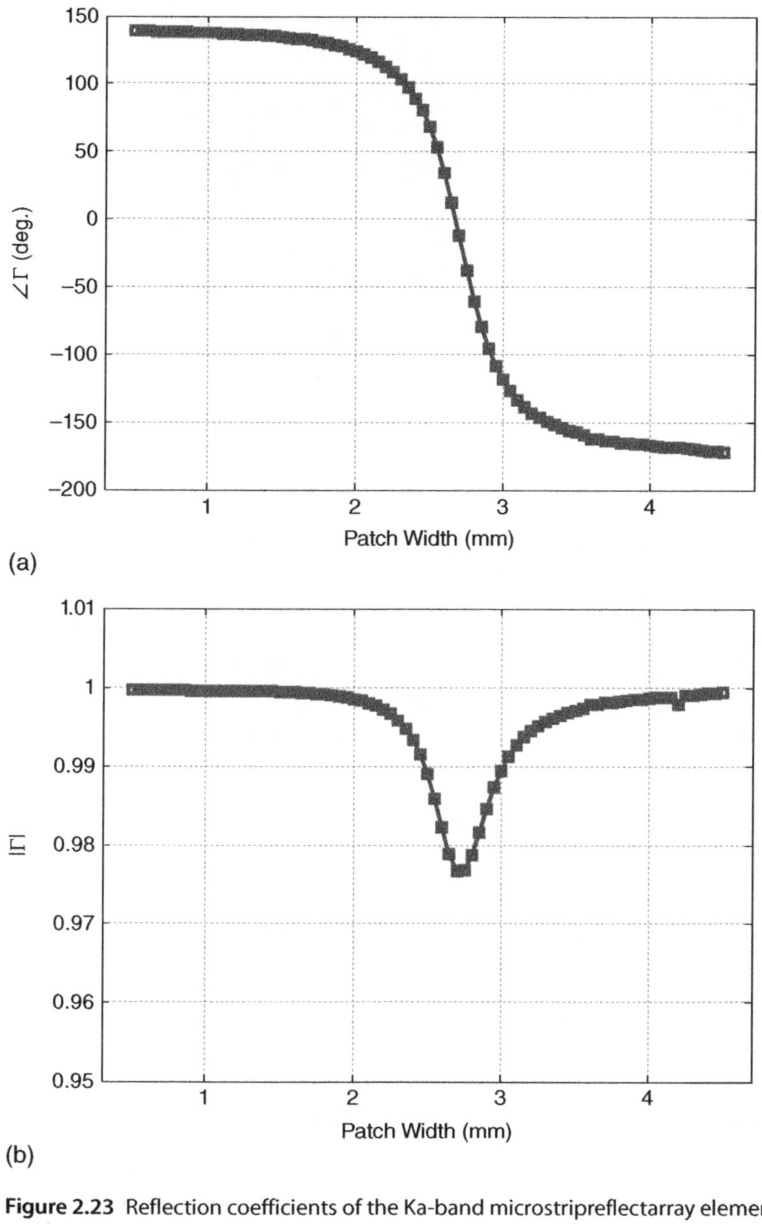

Figure 2.23 Reflection coefficients of the Ka-band microstripreflectarray element as a function of the patch width: (a) phase, (b) magnitude.

The magnitude of the tangential E_x fields clearly show an electric field distribution that is identical to a microstrip square patch antenna that is excited in the fundamental mode, where the radiating slots are located at the two edges of the patch in the *y*-direction. Similar to the patch antenna case, the cross-polarized components, E_y, will cancel out in the principal planes since they are out of phase. The normal components, E_z, will also be shorted out on the conductive patch.

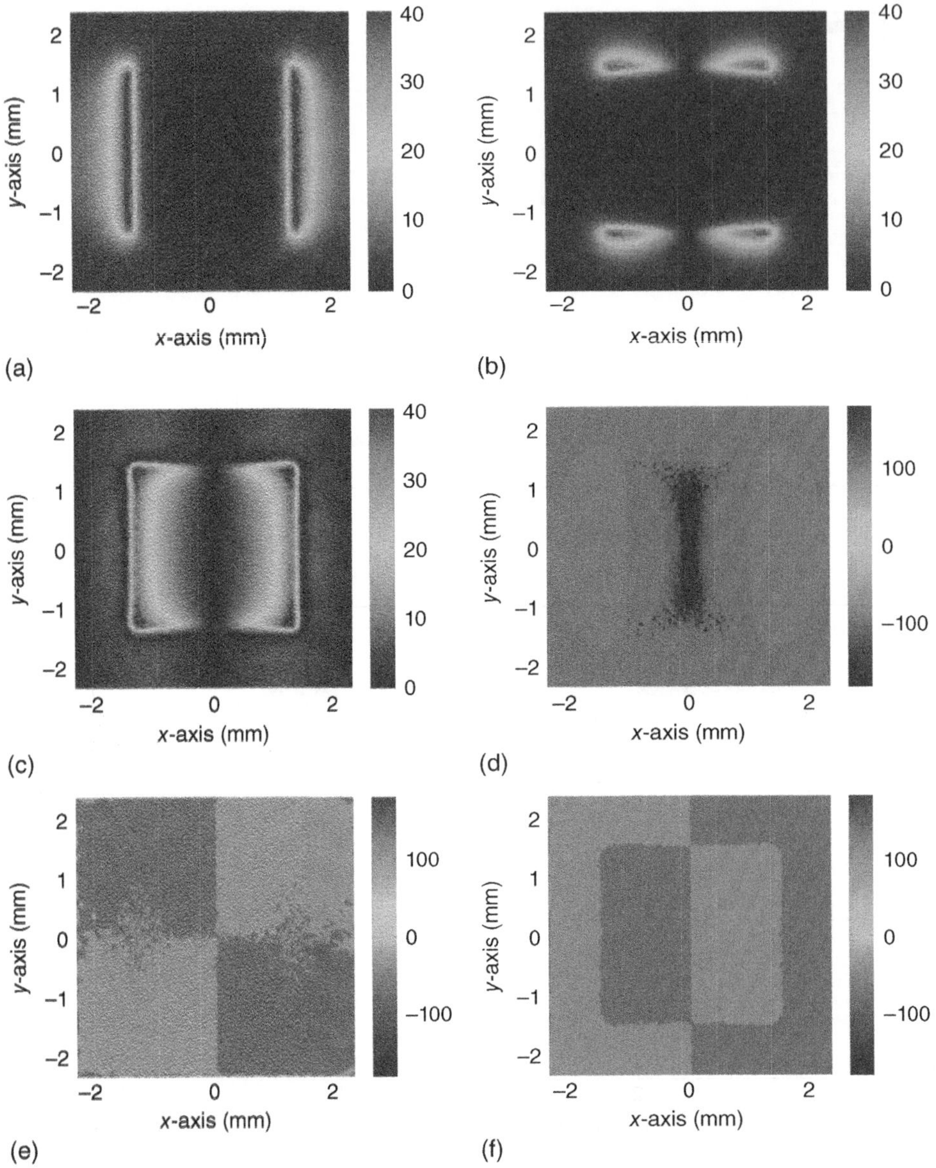

Figure 2.24 Total electric fields at a distance of 0.1 mm from the surface of the patch element: (a) $|E_x|$, (b) $|E_y|$, (c) $|E_z|$, (d) phase of E_x, (e) phase of E_y, and (f) phase of E_z. The unit for the magnitude plots is V/mm, and the unit for phase plots is degrees.

In the case of reflectarrays, the primary aim of the unit-cell analysis is to determine the reflection phase characteristic of the element. In this design, the co-polarized fields are in the x-direction. From Figure 2.24(d) it can be seen that the phase of the tangential E_x field is almost constant on its aperture. As discussed earlier, the element is excited with a plane wave, therefore, the magnitude and phase of the incident field is essentially

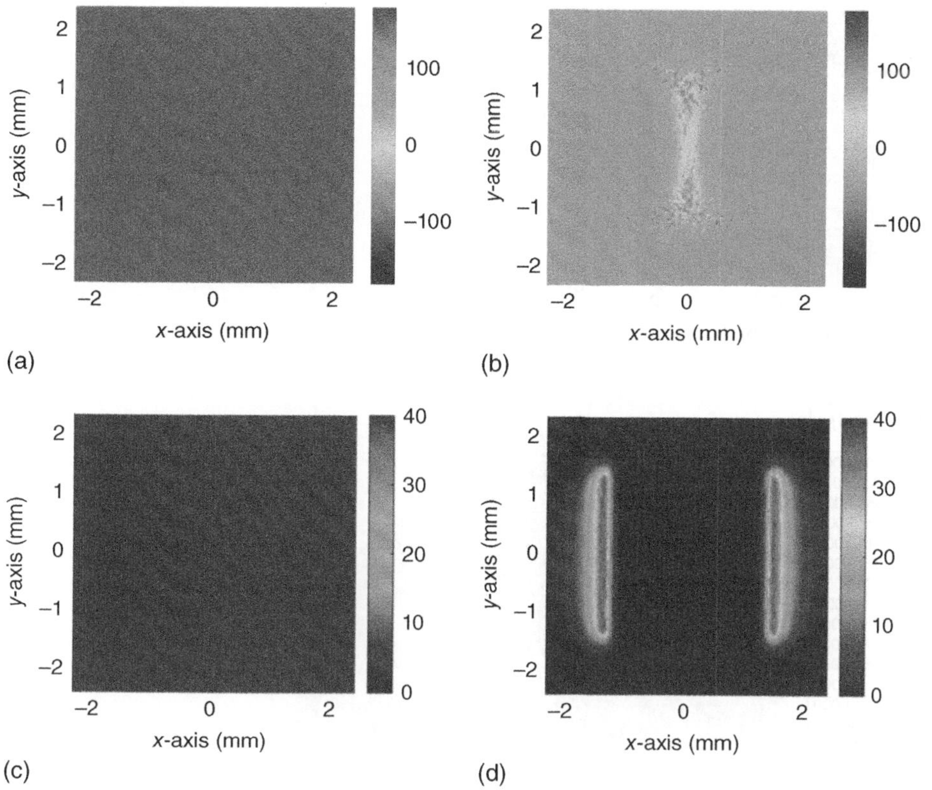

Figure 2.25 Phase of E_x: (a) incident fields, (b) scattered fields – amplitude of E_x, (c) incident fields, (d) scattered fields.

uniform on the element aperture. As such, the scattered field components would resemble the total fields. Nonetheless, the scattered fields can be computed directly using

$$\vec{E}_s = \vec{E}_t - \vec{E}_i. \tag{2.16}$$

The phases of the incident and scattered E_x fields are given in Figure 2.25. As expected, the scattered fields produced by the reflectarray element, have an almost uniform phase on the aperture. This is basically the reflection phase obtained by the elements that are about 0° in this case for the resonant patch. It should be noted that in the port analysis, the reflection phase is averaged over the port surface, resulting in a single number for the reflection phase.

The co-polarized surface currents on the resonant patch, and the resulting far-field electric fields are also shown in Figure 2.26, where it can be seen that the radiation pattern of the patch element is almost similar to the rectangular patch using the delay-line element studied earlier.

2.4.3 Single Slot Ring Elements

The element rotation technique is specially devised for circular polarization applications. Square patches and the dual-split ring geometry have been the most

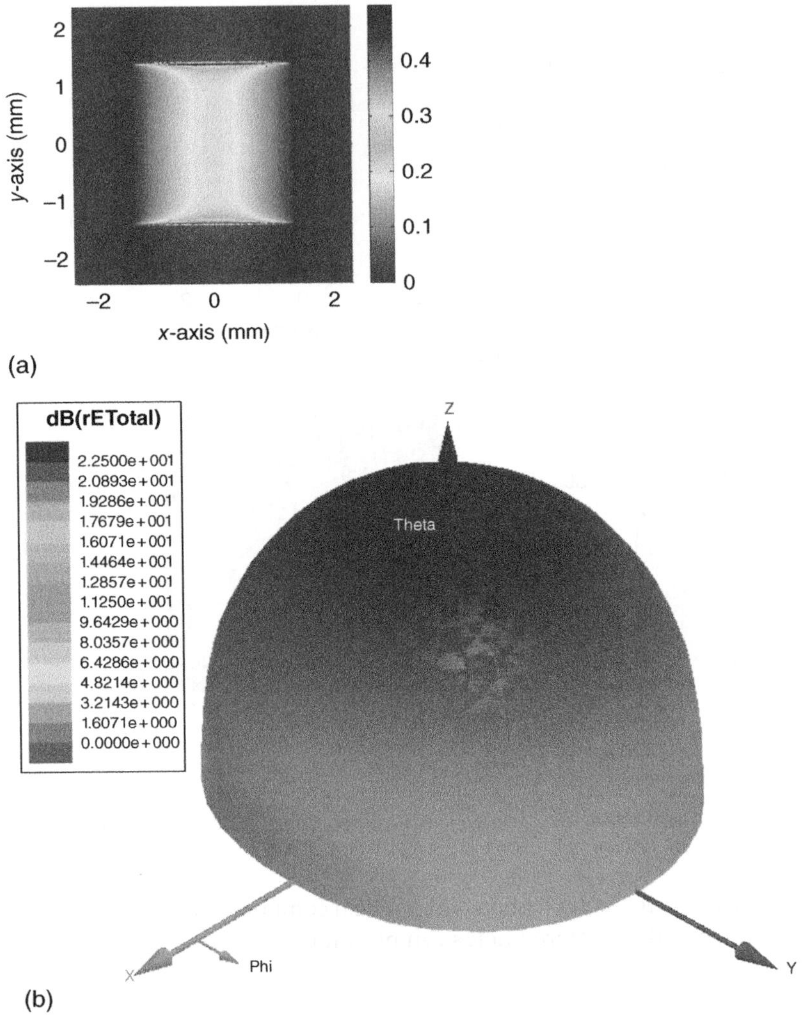

Figure 2.26 (a) Surface current magnitude ($|J_x|$) on the patch in A/mm, (b) Far-field pattern.

frequently used elements. This phase tuning technique provides several advantages over the other two approaches. In particular, fabrication tolerances of elements are generally not as stringent as the other techniques. This is due to the frequency independent nature of this phase compensation scheme. On the other hand, one important consideration for the element rotation technique is the cross-polarized component of the reflection coefficient, which in general is significantly affected by the incident angle.

In this section, we study the performance of a Ka-band reflectarray element using the element rotation technique. The element is a single slot ring (SSR). The geometrical model of the SSR element labeled with design parameters and the element simulation model in Ansys HFSS are shown in Figure 2.27.

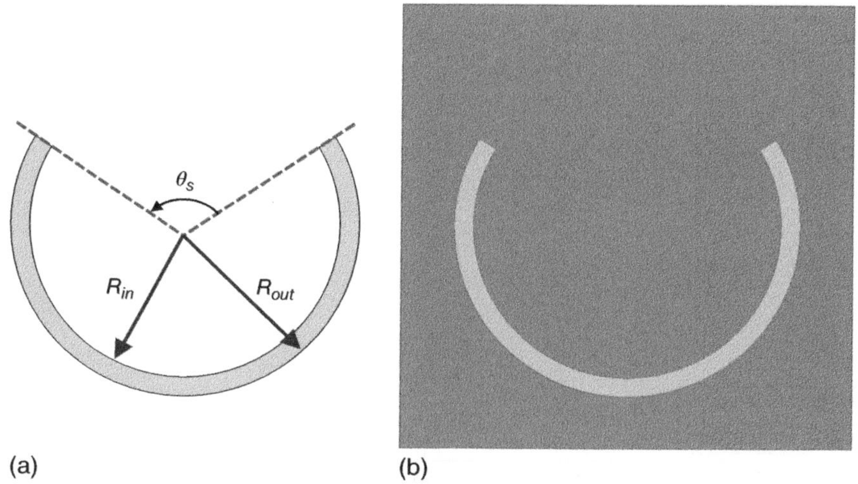

Figure 2.27 (a) Design parameters for an SSR element. (b) SSR element modeled in Ansys HFSS.

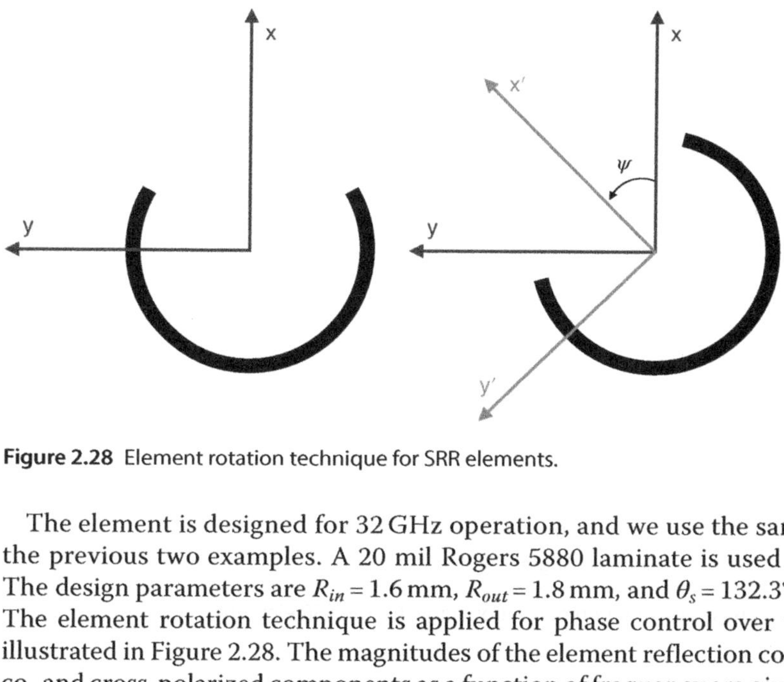

Figure 2.28 Element rotation technique for SRR elements.

The element is designed for 32 GHz operation, and we use the same lattice size as the previous two examples. A 20 mil Rogers 5880 laminate is used for the element. The design parameters are $R_{in} = 1.6$ mm, $R_{out} = 1.8$ mm, and $\theta_s = 132.3°$. Figure 2.29 (b). The element rotation technique is applied for phase control over the elements, as illustrated in Figure 2.28. The magnitudes of the element reflection coefficient for both co- and cross-polarized components as a function of frequency are given in Figure 2.29, where we notice that at the center design frequency of 32 GHz, a perfect CP performance is obtained with this element.

The element analysis is performed with a RHCP plane wave source. By rotating the element from −90° to +90°, one can obtain a full 360° phase tuning range. These results are given in Figure 2.30, where it can be seen that this element provides an almost ideal linear phase tuning as a function of rotation angle. As discussed earlier,

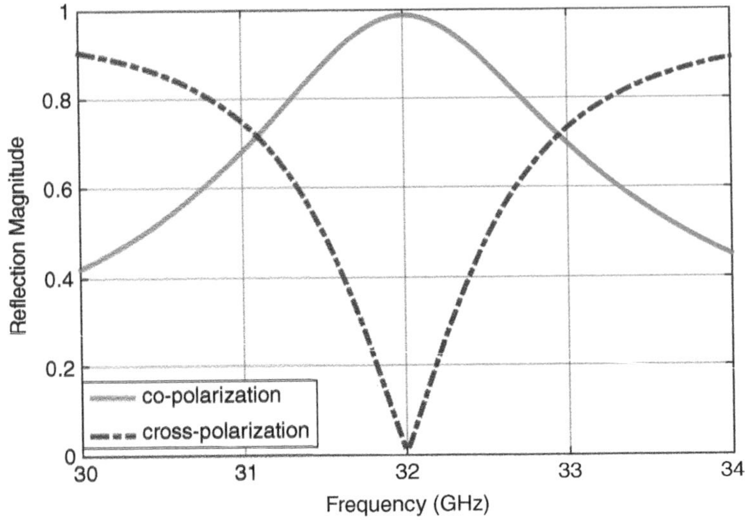

Figure 2.29 Magnitude of the reflection coefficient as a function of frequency for the SSR element.

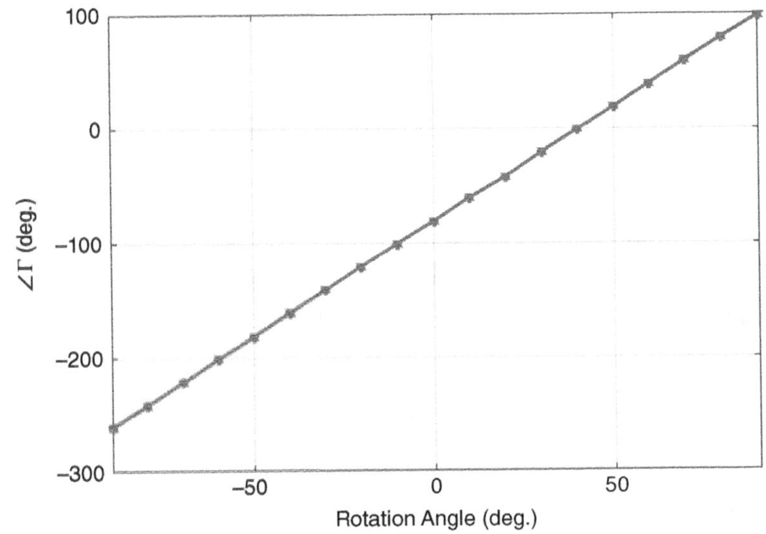

Figure 2.30 Reflection phase as a function of rotation angle for the SSR element.

theoretically there exists a linear dependence of the phase difference on the element rotation angle. Because the incident wave is RHCP, it is noted here that this curve is with a positive slope.

The element rotation technique is a very practical approach for phase tuning; however, it is restricted to CP designs. At the center design frequency and for normal incidence, the cross-polarized component of the reflection coefficient is almost zero. However, this value increases at off-center frequencies, which limits the element bandwidth. More discussion on this topic will be given later on in this chapter.

2.5 Reflectarray Element Characteristics and Design Considerations

In the previous sections, the basic design methodologies for reflectarray elements, along with the procedures for efficient design and analysis were outlined. Here we will look at some of the important characteristics of reflectarray elements.

2.5.1 Frequency Behavior of Element Reflection Coefficients

The reflection coefficients of reflectarray elements are frequency dependent [23], [24]. As such, the element bandwidth is generally considered to be an important design parameter. The bandwidth of elements is usually related to the substrate thickness, and in most cases, a thicker substrate leads to a better bandwidth. However, depending on the phase tuning technique, different phenomena limit the bandwidth of the element.

In the delay-line technique, the element bandwidth is generally considered to be the return loss bandwidth, as shown in Figure 2.17. Outside this frequency band, the element will not receive the RF power efficiently, hence cannot provide the phase tuning capability. In addition to this, for phase-delay lines, the phase shift provided by the line is also a function of frequency, which will result in phase errors at off-center frequencies.

For the element rotation technique, bandwidth is primarily limited by the tolerable level of the cross-polarized component. While the desirable level or threshold depends on the design requirements, a practical definition of the element bandwidth for the element rotation technique is the range where the loss of the cross-polarization level is at a tolerable level. This generally depends on the design. One option is to use the classic definition of 3 dB axial ratio bandwidth. Alternatively more stringent requirements for the array may limit the cross-polarization to be 20 dB below the co-polarization level.

For the variable element size technique, bandwidth definition is a bit more complicated. While some figures of merit have been proposed [25, 26], in general the linearity of the phase curve, and curves with small slopes, prove to be a practical measure to analyze the bandwidth performance of the element. Since the phase curve relies on the element size, whereas the size is linked to the resonant frequency, the phase curve implies the element phase stability with respect to frequency variation. Unlike the phase-delay line technique, there is no matching problem with the variable element size technique.

As discussed earlier, the variable size technique is the most popular choice for reflectarray elements. Therefore in this section we will take a closer look at the frequency response of this technique. The variable size square patch element designed in the previous section is considered for this study. The reflection coefficients at three different frequencies as a function of patch width are given in Figure 2.31. It can be seen that as expected, as the frequency increases, the size of the resonance patch decreases. This non-linear variation in phase as a function of frequency is the primary reason for the narrow bandwidth of reflectarrays with variable size patch elements. Several broadband techniques, however, exist that can mitigate this problem. These will be discussed in Chapters 6 and 7.

2.5.2 Effects of Oblique Incidence Angles on Element Reflection Coefficients

In many cases reflectarray elements are designed under normal incidence approximation. In practice however, elements on the aperture of the reflectarray will be excited

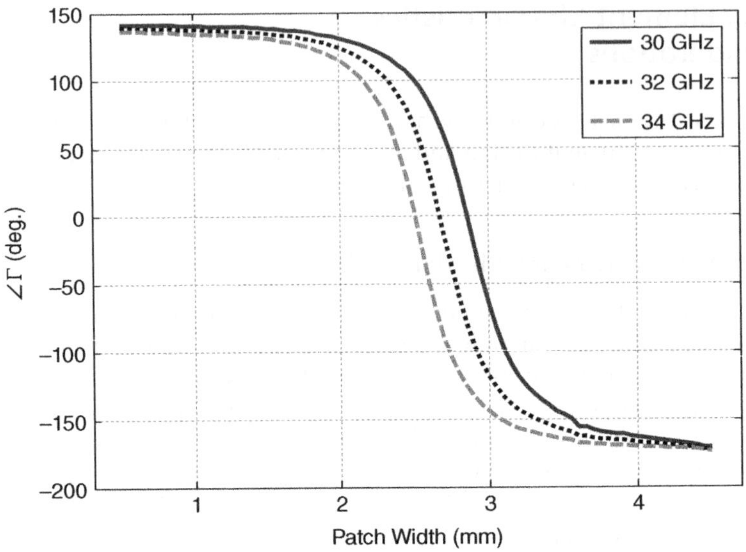

(a)

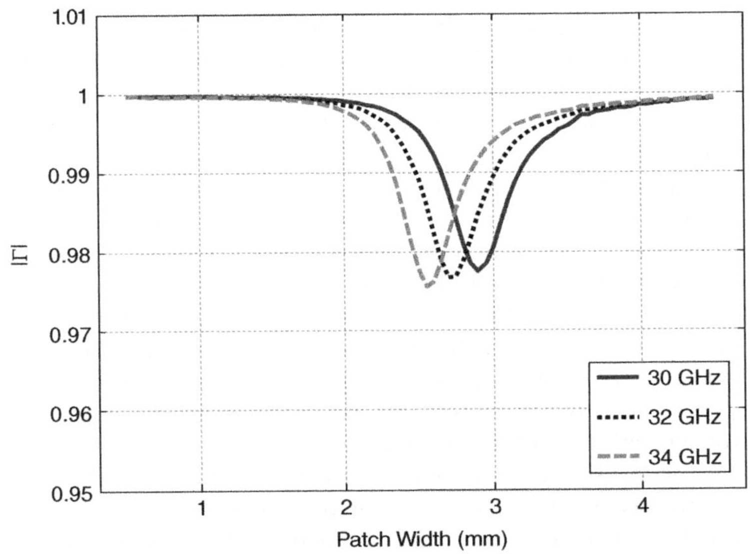

(b)

Figure 2.31 Reflection coefficients of the variable size square patch reflectarray element as a function of the patch width at three frequencies across the band: (a) phase, (b) magnitude.

with different oblique angles. While in many cases, especially for thin substrates, the effect of the oblique excitation angle on the reflection coefficients is small, depending on the design, one may need to consider this issue. The effects of oblique excitation angles generally depends on the choice of phase tuning method. For elements with delay lines, typically oblique excitation only manifests as a reduced received power for the element, the level of which depends on the element pattern. For the rotated element technique, the effect is primarily an increase in the level of cross-polarized component. Effects of oblique angles of incidence are more complicated for variable size elements, since both phase and magnitude of the element are dependent on the excitation angle.

As an example, here we will study the effects of oblique incident angles for the variable size square patch element designed in the previous section. The reflection coefficients are shown in Figure 2.32 and Figure 2.33 for several different excitation angles, for perpendicular (or H) and parallel (or V) polarizations, respectively.

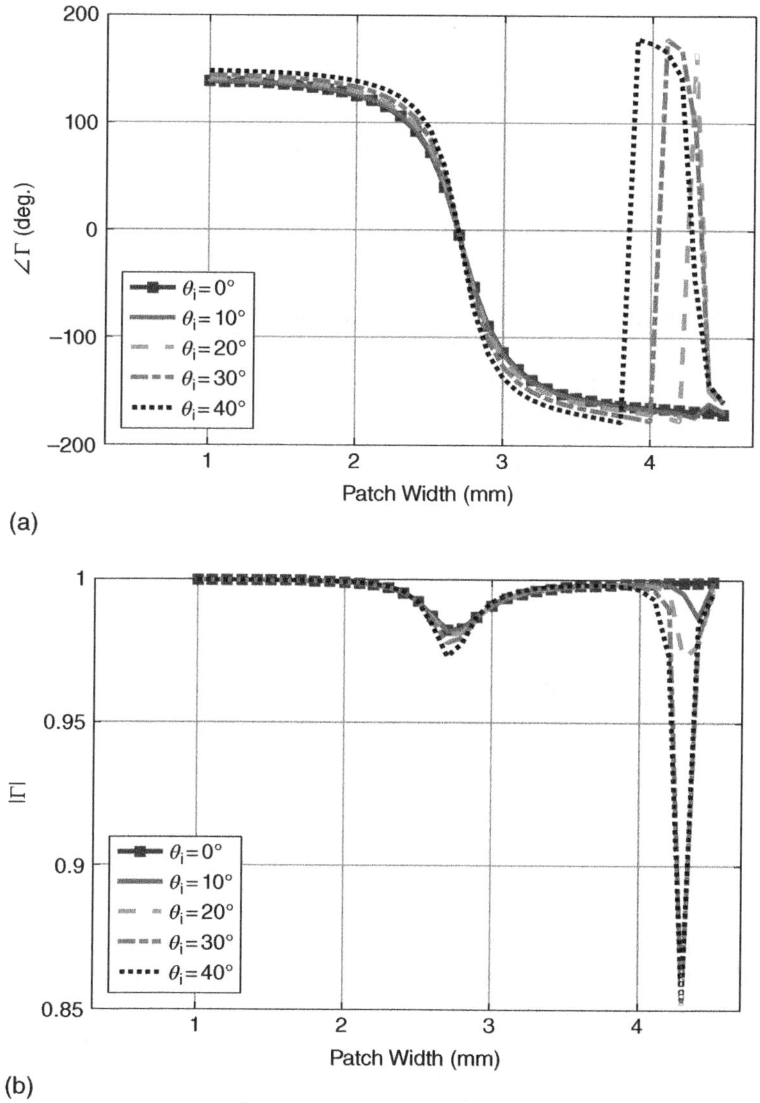

Figure 2.32 Reflection coefficients of the variable size square patch reflectarray element as a function of the patch width for different angles of incidence with perpendicular polarization: (a) reflection phase for $\varphi=0°$, (b) (a) reflection magnitude for $\varphi=0°$, (c) reflection phase for $\varphi=45°$, (c) and (d) reflection magnitude for $\varphi=45°$.

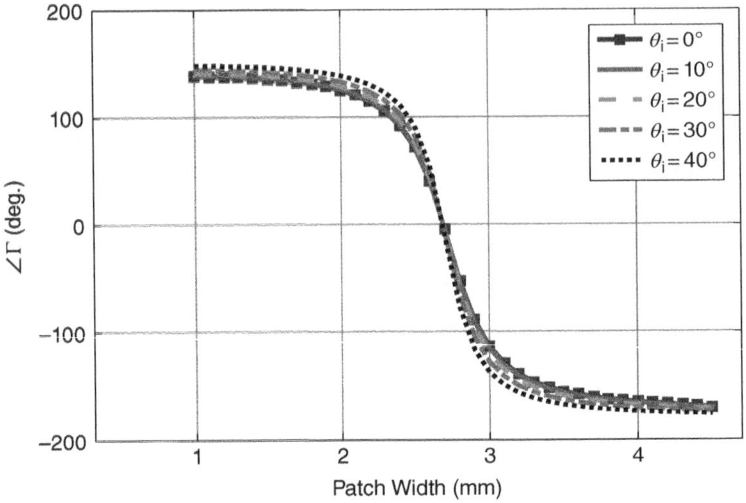

(c)

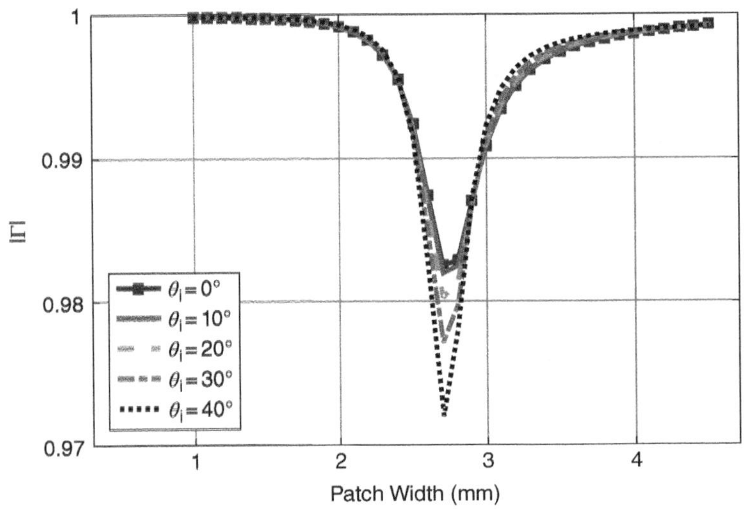

(d)

Figure 2.32 (Continued)

It can be seen that, as expected, while the effects of oblique excitation are not significant for most cases, they result in both phase and magnitude errors in the reflection coefficient. This may be accounted for in the design by generating data tables to select the appropriate element size based on the position of the element on the aperture and its true angle of incidence [2]. It is also important to note that for large patch sizes, that is, when the patch covers most of the lattice, higher order resonances are observed for oblique angles. As such depending on the design, one may need to avoid using these larger patches.

2.5.3 Sources of Phase Error in Reflectarray Element Design

The phasing elements are a key component in the reflectarray antenna design. Therefore, it is important to note that the accuracy of the unit-cell data used for the design is susceptible to errors arising from the design method, fabrication, or approximations in the analysis. A brief description of each of these element errors is given here.

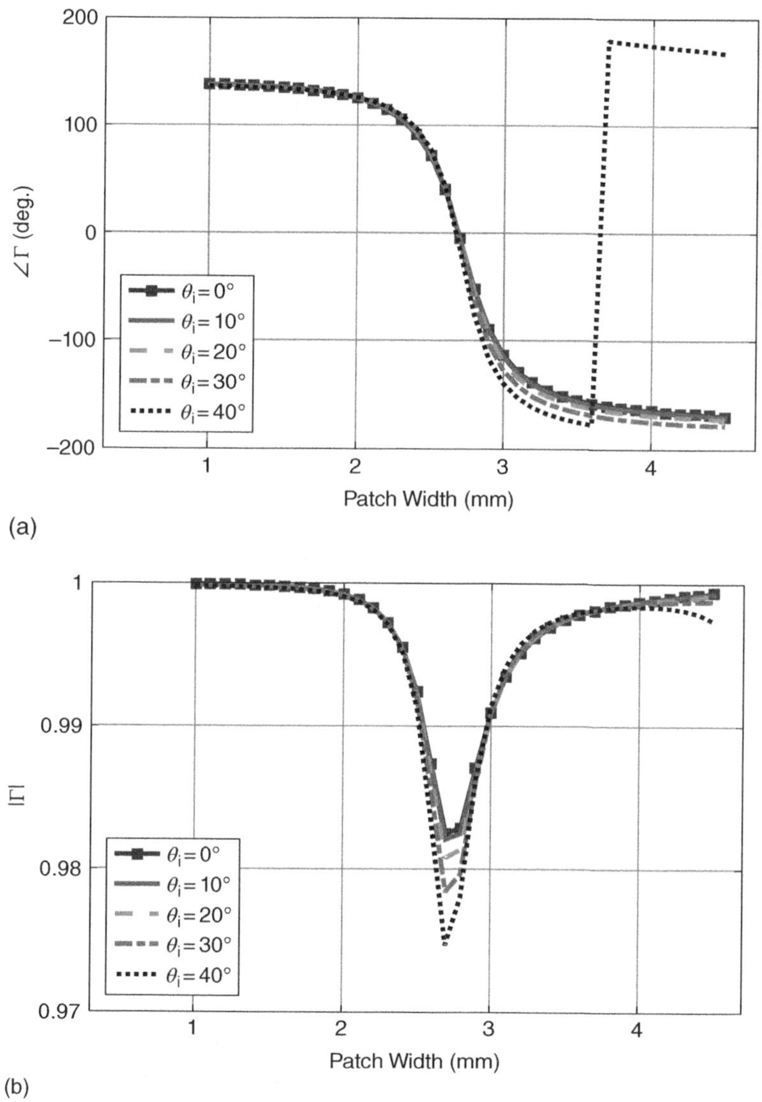

Figure 2.33 Reflection coefficients of the variable size square patch reflectarray element as a function of the patch width for different angles of incidence with parallel polarization: (a) reflection phase for φ = 0°, (b) (a) reflection magnitude for φ = 0°, (c) reflection phase for φ = 45°, (c) and (d) reflection magnitude for φ = 45°.

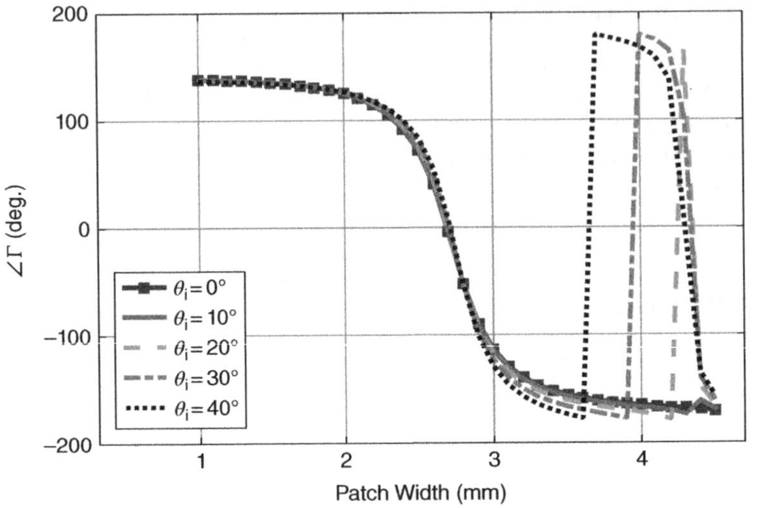

(c)

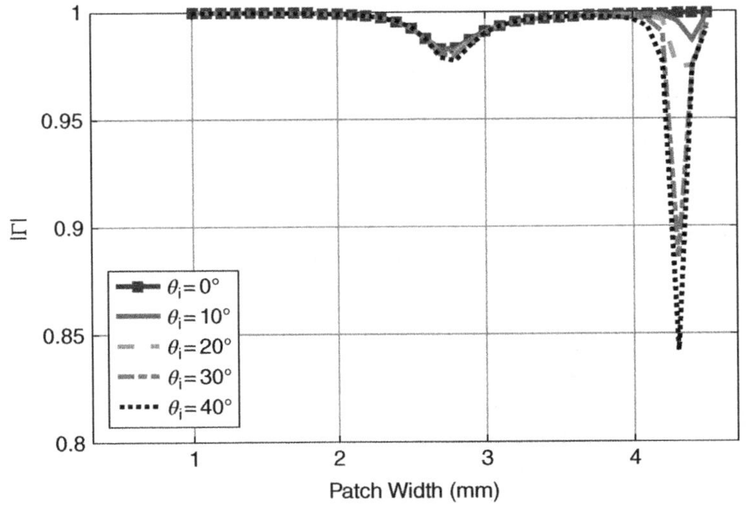

(d)

Figure 2.33 (Continued)

Quantization Errors: In practice, fabrication accuracy plays a significant role in the element's performance. The dimensions or rotations of the elements are changed by a certain amount depending on the fabrication precision, and as such, a continuous phase control is not possible. For each element on the aperture, a phasing element is selected that provides the closest quantized phase with respect to the ideal phase shift. The difference between the ideal element phase and the quantized phase of the selected element is categorized as a quantization error.

Phase Range Errors: Another important factor in element selection is the available phase range of the elements. Typically single-resonant phasing elements have a phase range below 360°. Although this problem can be avoided by using multi-resonance elements, if the phase range is smaller than the complete cycle, some elements will inevitably have unattainable phase shifts. While the element selection routine minimizes these errors by selecting the closest quantized values, these errors are in nature different from the quantization errors and are categorized as errors due to limited phase range.

Infinite Array Approximation: As discussed earlier, the infinite array approach is used to obtain the unit-cell data, which assumes that the dimensions (or rotations) of the adjacent elements should be identical or at least very close. In conventional single-beam reflectarrays with variable size elements, the required phase, and therefore dimensions of adjacent elements, grow gradually in each Fresnel zone. When the zone is complete, a phase wrap is observed and the element dimension jumps from a maximum size to a minimum size or vice versa, which violates the periodic approximation and results in some error. Similar conditions are observed for other phase control schemes.

2.6 Reflectarray Element Measurements

As discussed earlier in this chapter, one of the great advantages of the metallic waveguide simulation technique is the fact that it can be verified experimentally. In this section, we will present experimental and simulated results for a variable size square patch reflectarray element using a WR75 standard waveguide. An image of the components used for the experiment is given in Figure 2.34, where the reflectarray element is placed in a

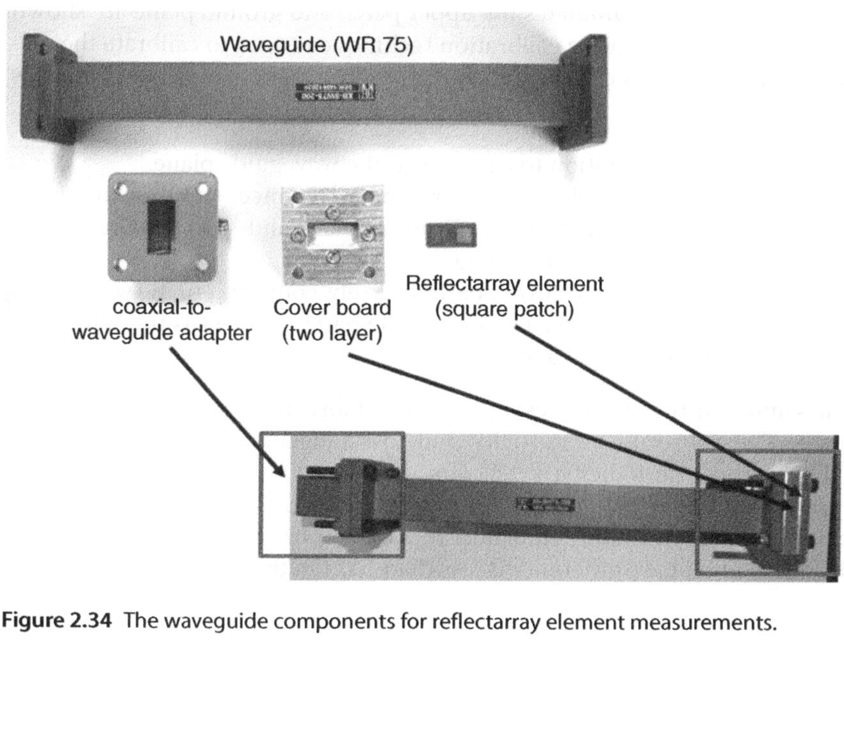

Figure 2.34 The waveguide components for reflectarray element measurements.

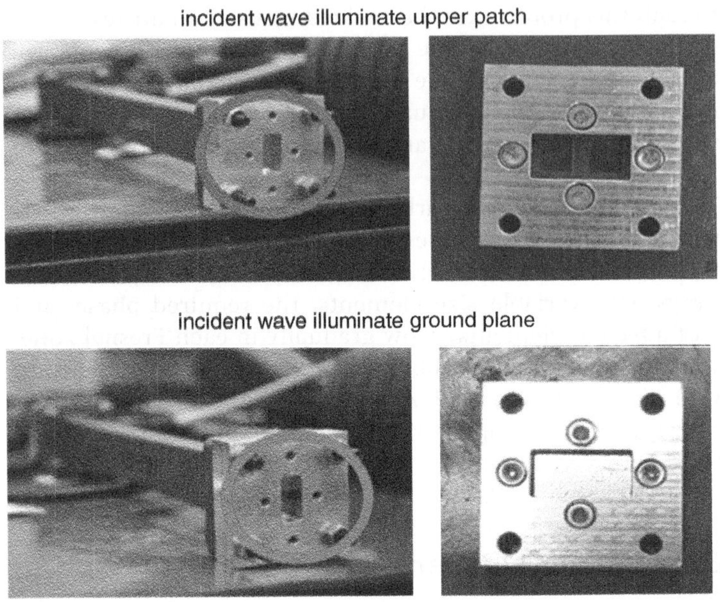

Figure 2.35 Measurement setups for the reflectarray element.

cover board for measurements. In this setup, when the element is placed such that the incident wave illuminates the patch, the element loss can be measured. Alternatively, when the element is placed such that the incident wave illuminates the ground plane, the ground plane loss can be measured. Measurement setups for the reflectarray element when the incident wave illuminates the upper patch and ground plane are shown in Figure 2.35. A two-port waveguide calibration technique is used to calibrate the systems at the waveguide plane. The measurement procedure used in this experiment is summarized here.

1) Use two-port waveguide calibration to calibrate at the waveguide plane.
2) Insert a reflectarray element sample into the cover board, connect it to the waveguide, and measure reflection coefficient of the element (phase and loss). The reflection phase of the element is obtained in this stage.
3) Insert the reflectarray element into cover board reversely, connect it to the waveguide, and measure the ground plane loss. The loss of the patch element is the difference between losses measured in step 2 and step 3.

The element sample in the study presented here is fabricated on a 62 mil Taconic TLX-8 laminate. The substrate permittivity and loss tangent are 2.55 and 0.0019, respectively. The element periodicity is 9.52 mm. Measured and simulated reflection coefficients for two different patch sizes are given in Figure 2.36. Note that a good agreement between measured and simulated reflection phase is observed in these results and the maximum phase difference is less than 10°. On the other hand measured and simulated losses show a larger discrepancy. This is primarily due to the fact that several factors such as stability of the coaxial line, the gap between cover board and waveguide, the gap between the element and cover board, and so on, influence the losses in the system. In general element loss is relatively hard to measure, particularly at higher frequencies. Nonetheless, the difference between measured and simulated losses is less than 0.3 dB at 12.5 GHz, and no more than 0.4 dB across the 11.5–13.5 GHz frequency range.

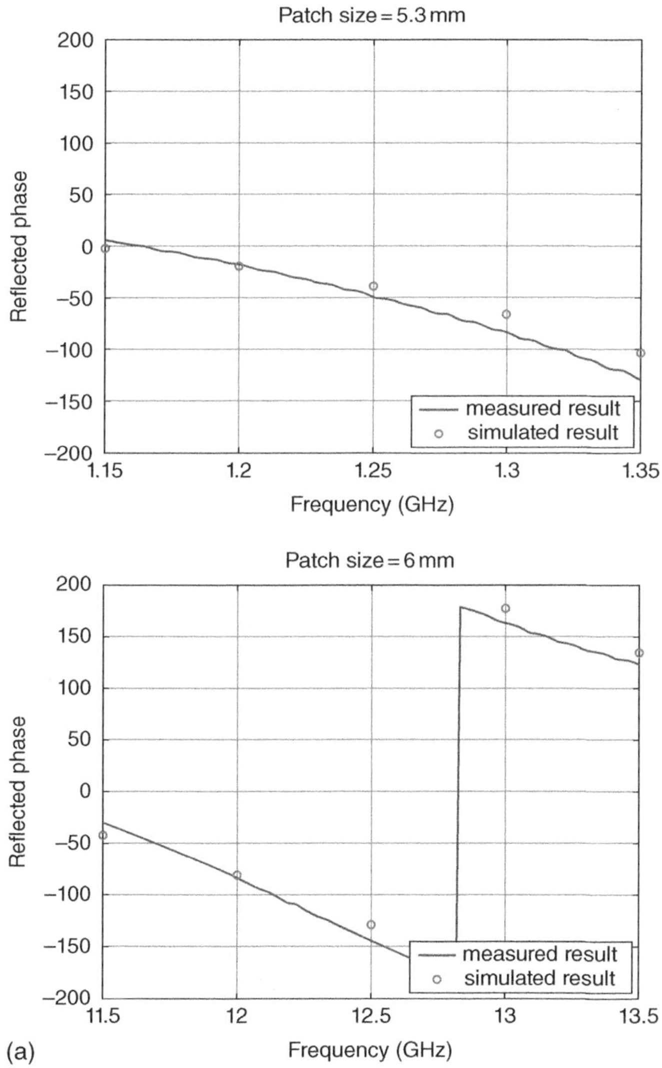

Figure 2.36 Measured and simulated reflection coefficients for variable size reflectarray elements: (a) element phase, (b) element loss.

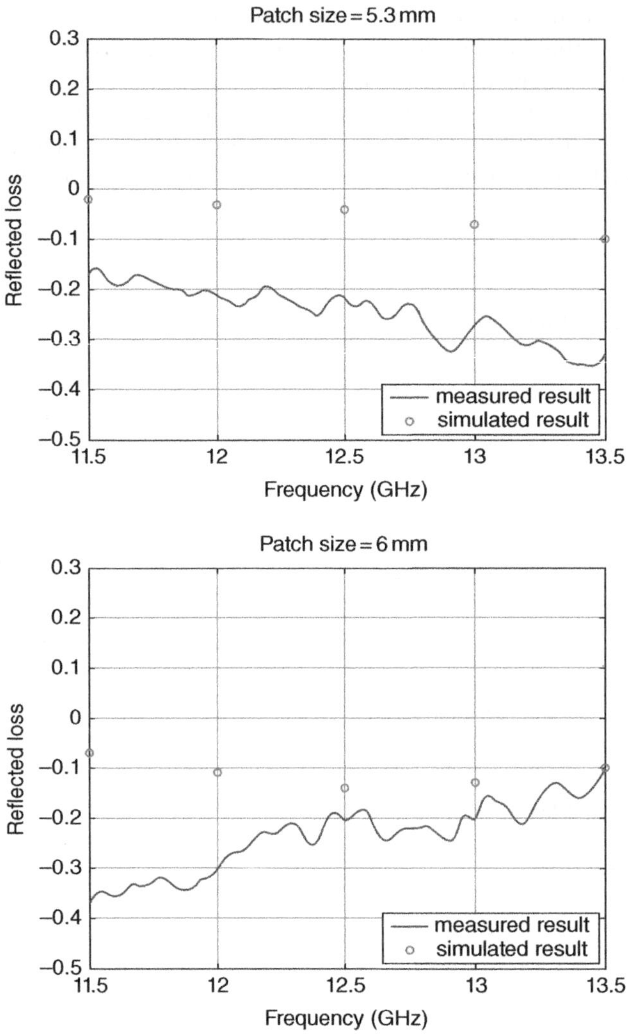

(b)

Figure 2.36 (Continued)

References

1 D. G. Berry, R. G. Malech, and W. A. Kennedy, "The reflectarray antenna," *IEEE Transactions on Antennas and Propagation*, Vol. 11, No. 6, pp. 645–651, Nov., 1963.

2 J. Huang and J. A. Encinar, *Reflectarray Antennas*, IEEE press by John Wiley & Sons Ltd. 2008.

3 R. Munson, H. A. Haddad, and J. W. Hanlen, "Microstrip reflectarray for satellite communication and radar cross-section enhancement or reduction," U.S. patent 4684952, Washington D. C., Aug. 1987.

4 R. J. Mailloux, *Phased Array Antenna Handbook*, 2nd Edn, Artech House, 2005.

5 M. I. Skolnik, *Radar Handbook*, 3rd Edn, McGraw-Hill, 2008.

6 M. E. Cooley, and D. Davis, "Reflector antennas," in *Radar Handbook*, 3rd Edn, McGraw-Hill, 2008.

7 D. C. Chang and M. C. Huang, "Multiple-polarization microstrip reflectarray antenna with high efficiency and low cross-polarization," *IEEE Transactions on Antennas and Propagation*, Vol. 43, No. 8, pp. 829–834, Aug. 1995.

8 R. D. Javor, X. Wu, and K. Chang "Design and performance of a microstrip reflectarray antenna," *IEEE Trans. Antennas and Propagation*, Vol. 43, No. 9, pp. 932–939, Sept., 1995.

9 K. Chen, C. C. Tzuang, and J. Huang, "A higher order microstrip reflectarray at Ka-band," in *IEEE Antennas and Propagation Society International Symposium Digest*, pp. 566–569, July, 2001.

10 D. G. Gonzalez, G. E. Pollon, and J. F. Walker, "Microwave phasing structures for electromagnetically emulating refl ective surfaces and focusing elements of selected geometry," Patent US 4905014, Feb. 1990.

11 D. Pozar and T. A. Metzler, "Analysis of a reflectarray antenna using microstrip patches of variable size," *IEE Electronics Letters*, Vol. 29, No. 8, pp. 657–658, Apr. 1993.

12 M. E. Bialkowski and K. H. Sayidmarie, "Investigations into phase characteristics of a single-layer reflectarray employing patch or ring elements of variable size," *IEEE Trans. Antennas and Propagation*, Vol. 56, No. 11, pp. 3366–3372, Nov, 2008.

13 H. R. Phelan, "Spiralphase reflectarray for multitarget radar," *Microwave Journal*, Vol. 20, pp. 67–73. July 1977.

14 J. Huang and R. J. Pogorzelski, "A Ka-band microstrip reflectarray with variable rotation angles," *IEEE Trans. Antennas and Propagation*, Vol. 46, No. 5, pp. 650–656, May, 1998.

15 H. Rajagopalan and Y. Rahmat-Samii, "On the reflection characteristics of a reflectarray element with low-loss and high-loss substrates," *IEEE Trans. Antennas and Propagation*, Vol. 52, No. 4, pp. 73–89, Apr. 2010.

16 P. W. Hannan and M. A. Balfour, "Simulation of a phased-array antenna in waveguide," *IEEE Trans. Antennas and Propagation*, Vol. 13, No. 3, pp. 342–353, May, 1965.

17 A. K. Bhattacharyya, *Phased Array Antennas: Floquet Analysis, Synthesis, BFNs and Active Array Systems*, John Wiley & Sons, Ltd, 2006.

18 ANSYS HFSS, ANSYS Inc. Website available at /www.ansys.com/ (accessed July 2017).

19 CST Microwave Studio, CST Studio Suite, Computer Simulation Technology AG. Website available at https://www.cst.com/ (accessed July 2017).

20 FEKO, Altair Engineering Inc. Website available at https://www.feko.info/ (accessed July 2017).

21 H. Mosallaei and K. Sarabandi, "Antenna miniaturization and bandwidth enhancement using a reactive impedance substrate," *IEEE Trans. Antennas Propag.*, Vol. 52, No. 9, pp. 2403–2414, Sep. 2004.

22 M. Bozzi, S. Germani, and L. Peregrini, "Modeling of losses in printed reflectarray elements," *European Microwave Conference*, Oct 2004.

23 J. Huang, "Bandwidth study of microstrip reflectarray and a novel phased reflectarray concept," *in IEEE Antennas and Propagation Society International Symposium Digest*, Vol. 1, pp. 582–585, July 1995.

24 D. Pozar, "Bandwidth of reflectarrays," *IEE Electronics Letters*, Vol. 39, No. 21, pp. 1490–1491, Oct. 2003.

25 M. Bozzi, S. Germani, and L. Perregrini, "Performance comparison of different element shapes used in printed reflectarrays," *IEEE Antennas and Wireless Propagation Letters*, Vol. 2, pp. 219–222, 2003.

26 H. Rajagopalan and Y. Rahmat-Samii, "Dielectric and conductor loss qualification for microstrip reflectarray: simulations and measurements," *IEEE Trans. Antennas and Propagation*, Vol. 56, No. 4, pp. 1192–1196, Apr. 2008.

3

System Design and Aperture Efficiency Analysis

Design of a reflectarray antenna typically consists of two steps, namely, the phasing element design and the system configuration design. In the previous chapter, the analysis of phasing elements is presented, and this chapter focuses on the analysis of the reflectarray system configuration.

Usually, the system design begins with a specified gain requirement. Similar to traditional reflector antennas [1], [2], the gain of a reflectarray antenna is proportional to the electrical size of the aperture, and is calculated as the product of the aperture efficiency and maximum aperture directivity [3], [4]. Through the aperture efficiency analysis, one could estimate the required size of the reflectarray antenna aperture in order to achieve the desired gain. More importantly, it is desirable to design the system such that the aperture efficiency is maximized. Through the aperture efficiency analysis, one could optimize the parameters of a system configuration, such as feed specifications, f/D ratio, aperture shape, and so on. Hence, aperture efficiency analysis is critical in the system design.

In addition to aperture efficiency considerations, several other factors influence the system design, including feed blockage and edge taper. Feed blockage is generally minimized by offsetting the feed antenna, which usually decreases the aperture efficiency and increases the cross-polarization. Typically, edge taper has to be kept below a certain level to ensure that strong diffractions don't occur at the reflectarray edge. Consequently, in order to realize an overall optimum performance some compromise has to be made in the system design.

In this chapter, we will study the basic considerations for an efficient reflectarray system design. Generalized formulations for computing the aperture efficiency and taper of planar reflectarray antennas are outlined, and several reflectarray system designs are analyzed. Based on these studies some general guidelines for reflectarray design are presented.

3.1 A General Feed Model

In reflectarray antenna systems, the feed antenna position and radiation characteristics are one of the most influential measures in the design. The aperture illumination is directly dependent on the radiation pattern of the feed antenna and its position, as such, the efficiency and other system characteristics of the reflectarray antenna cannot be

Reflectarray Antennas: Theory, Designs, and Applications, First Edition.
Payam Nayeri, Fan Yang, and Atef Z. Elsherbeni.

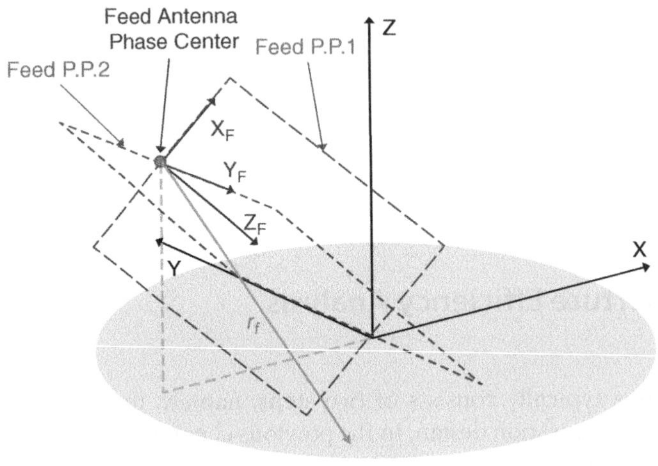

Figure 3.1 Coordinate system of the reflectarray and feed system.

determined without a proper description of the feed characteristics. In this section, we will present a generalized model for the radiation pattern of the feed antenna. Based on this feed model, we will carry out aperture efficiency analysis in the following sections.

Without the loss of generality, let's consider the geometrical setup of the reflectarray system shown in Figure 3.1. The reflectarray aperture is lying on the *xy*-plane and the aperture center is located at the center of the coordinate system. The feed antenna is pointing toward the geometrical center of the aperture. With the geometrical system setup specified, if one can now describe the radiation pattern of the feed antenna with a closed form mathematical expression, it would then be possible to analyze the system efficiency directly.

3.1.1 Models of Linearly Polarized and Circularly Polarized Feeds

Various pattern models have been developed to simulate the radiation properties of reflector feed antennas. Among these models, the $\cos^q$ pattern is perhaps the most common approach, because of its simplicity and versatility to describe different pattern types [3], [5]. In this model, the radiation pattern of the ideal feed antenna with a fixed phase center is typically described by the two principal plane patterns, and is given by

$$E^F(r_f,\theta_f,\varphi_f)=A_0\left[\hat{\theta}C_E(\theta_f)\cos\varphi_f-\hat{\varphi}C_H(\theta_f)\sin\varphi_f\right]\frac{e^{-jkr_f}}{r_f}, \text{ for } x \text{ polarized,} \quad (3.1)$$

$$E^F(r_f,\theta_f,\varphi_f)=A_0\left[\hat{\theta}C_E(\theta_f)\sin\varphi_f+\hat{\varphi}C_H(\theta_f)\cos\varphi_f\right]\frac{e^{-jkr_f}}{r_f}, \text{ for } y \text{ polarized.} \quad (3.2)$$

Here $C_E(\theta_f)$ and $C_H(\theta_f)$ are the radiation patterns of the feed antenna in the *E*- and *H*-planes, respectively, and r_f is the vector from the feed phase center (focal point) to any point on the reflectarray aperture. Note that *E*- and *H*-planes depend on the orientation and polarization of the feed antenna, and correspond to the feed principal planes, P.P.1 or P.P.2 in Figure 3.1, accordingly. Also note that θ_f and φ_f are the spherical angles of the feed coordinate system, that is (X_F, Y_F, Z_F) shown in Figure 3.1. In this model, the shape of the feed radiation pattern is given in terms of the feed elevation angle (θ_f) and a power factor q, which in general is different for E- and H-planes, that is,

$$C_E(\theta_f) = \cos^{q_E}(\theta_f) \text{ E-plane pattern of the feed antenna,} \tag{3.3}$$

$$C_H(\theta_f) = \cos^{q_H}(\theta_f) \text{ H-plane pattern of the feed antenna} \tag{3.4}$$

Note that the back radiation of the feed pattern is assumed to be zero in these equations. The shape of the pattern is controlled by q_E and q_H, which can be determined by matching C_E and C_H to the measured pattern of the feed antenna. It should be noted that with this feed model, for any azimuth angle other than $\varphi_f = 0°$ and 90°, the field from the feed is approximated by interpolation.

As a simple example, first we consider a short dipole antenna, which is centered at the phase center point shown in Figure 3.1 and oriented along X_F. The feed principal planes (Feed P.P.1 and Feed P.P.2 in Figure 3.1) depend on the orientation of the feed antenna. With this setup, the E-plane is P.P.1 and the H-plane is P.P.2 which are given by

$$C_E(\theta_f) = \cos\theta_f, C_H(\theta_f) = 1, \text{ for an } x_f\text{-polarized short dipole.} \tag{3.5}$$

Besides the linearly polarized feeds, circularly polarized (CP) feed patterns can also be represented with this model [5], which is given by

$$E^F(r_f, \theta_f, \varphi_f) = A_0 e^{j\tau\varphi_f}\left[\hat{\theta} C_E(\theta_f) + \hat{\varphi} j\tau C_H(\theta_f)\right]\frac{e^{-jkr_f}}{r_f}, \text{ for circularly polarized.} \tag{3.6}$$

For the circular polarized feed patterns, the parameter τ determines the sense of polarization. For a left-handed (LH) CP feed $\tau = +1$ and for right-handed (RH) CP feed $\tau = -1$. Note that this equation represents a perfect circular polarized wave only in the main beam direction ($\theta_f = 0$) of the feed coordinate system, unless $C_E(\theta_f) = C_H(\theta_f)$. It is important to note here that for circular polarization, E- and H-planes cannot be defined, and C_E and C_H simply refer to the patterns in xz and yz planes, respectively.

3.1.2 Balanced Feed Models

In the dipole feed example, q_E and q_H are 1 and 0, respectively, and the pattern is asymmetric. For reflector feeds, the asymmetric pattern is not desirable in most cases, and rotationally symmetric patterns are often desired, that is,

$$C_E(\theta_f) = C_H(\theta_f) = C(\theta_f) = \cos^q(\theta_f). \tag{3.7}$$

Feed antennas that exhibit rotationally symmetric patterns are referred to as balanced feeds. With the $\cos^q$ model, this means that $q_E = q_H = q$,

$$E^F(r_f, \theta_f, \varphi_f) = A_0 \frac{e^{-jkr_f}}{r_f} C(\theta_f)\left[\hat{\theta}\cos\varphi_f - \hat{\varphi}\sin\varphi_f\right], \text{ for } x \text{ polarized,} \tag{3.8}$$

$$E^F(r_f, \theta_f, \varphi_f) = A_0 \frac{e^{-jkr_f}}{r_f} C(\theta_f)\left[\hat{\theta}\sin\varphi_f + \hat{\varphi}\cos\varphi_f\right], \text{ for } y \text{ polarized,} \tag{3.9}$$

$$E^F(r_f, \theta_f, \varphi_f) = A_0 \frac{e^{-jkr_f}}{r_f} C(\theta_f) e^{j\tau\varphi_f}\left[\hat{\theta} + \hat{\varphi} j\tau\right], \text{ for circularly polarized.} \tag{3.10}$$

For the balanced feed described in this section, the normalized power pattern of the feed antenna is given by

$$P^F(\theta_f, \varphi_f) = \begin{cases} \cos^{2q}(\theta_f) & \text{for } 0 \le \theta \le \pi/2 \\ 0 & \text{elsewhere} \end{cases}. \tag{3.11}$$

With this feed radiation model, the feed antenna directivity and pattern shape are described by a single parameter q. Note that the back radiation of the feed pattern is assumed to be zero in this equation. The radiation pattern for this feed model is shown in Figure 3.2 for a few different values of q, where it can be seen that as q increases, the beamwidth of the feed pattern decreases. Consequently, the antenna directivity increases as the value of q is increased. The directivity as a function of the q value is given in Figure 3.3.

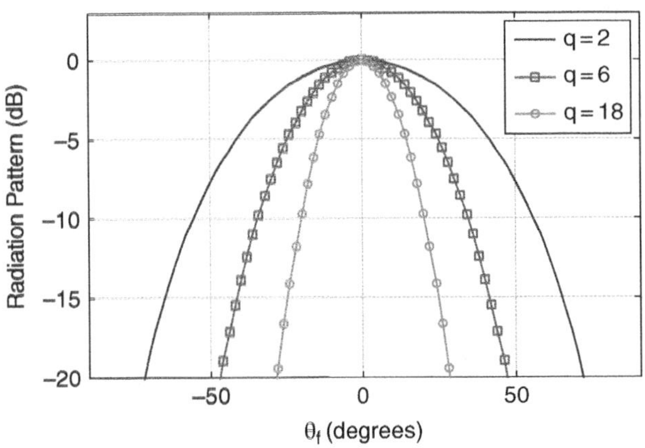

Figure 3.2 Radiation pattern of the $\cos^q$ feed pattern model for different values of q.

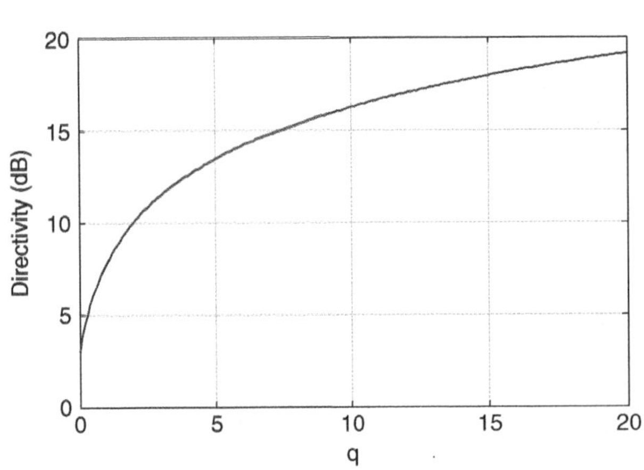

Figure 3.3 Directivity of the $\cos^q$ feed pattern model as a function of q.

3.2 Aperture Efficiency

In aperture type antennas such as reflectarrays, the physical aperture area A determines the theoretical maximum directivity of the antenna, that is,

$$D_{\max} = 4\pi \frac{A}{\lambda^2}. \tag{3.12}$$

Here λ is the wavelength at the design frequency. The reflectarray antenna gain can then be computed as

$$G = 4\pi \frac{A}{\lambda^2}\eta_{aperture}, \tag{3.13}$$

where $\eta_{aperture}$ is the aperture efficiency of the antenna. Similar to reflector antennas, many efficiency factors influence the overall aperture efficiency of reflectarray antennas, and the dominant terms are typically spillover (η_s) and illumination (η_i) efficiencies. In this section, we study the aperture efficiency as the product of these two terms,

$$\eta_{aperture} = \eta_s \eta_i. \tag{3.14}$$

It is important to note here that these two important efficiency factors are directly related to the antenna configuration parameters, hence they also serve as key factors for the reflectarray system design. Other efficiency factors are typically associated with the choice of reflectarray phasing elements such as material losses, losses due to phase error, mismatch, and polarization losses. Further discussion on these losses and their influence on aperture efficiency will be given when we discuss loss budget calculations in Chapter 4.

3.2.1 Spillover Efficiency

Spillover efficiency (η_s) is defined as the percentage of radiated power from the feed that is intercepted by the reflecting aperture relative to the total radiated power. Mathematically, this is given by

$$\eta_s = \frac{\iint_A \vec{P}(\vec{r}_f)\cdot d\vec{s}}{\iint_{Sphere} \vec{P}(\vec{r}_f)\cdot d\vec{s}} \tag{3.15}$$

where both integrals are fluxes of the Poynting vector, *P(r)*, evaluated over certain surfaces. The denominator is the total power radiated by the feed; therefore, the integral is performed over the entire surface area of a virtual spherical surface that encapsulates the feed antenna and is centered at the phase center of the feed. The integral of the numerator, the intercepted power, is evaluated over the aperture of the reflectarray antenna, as shown in Figure 3.4.

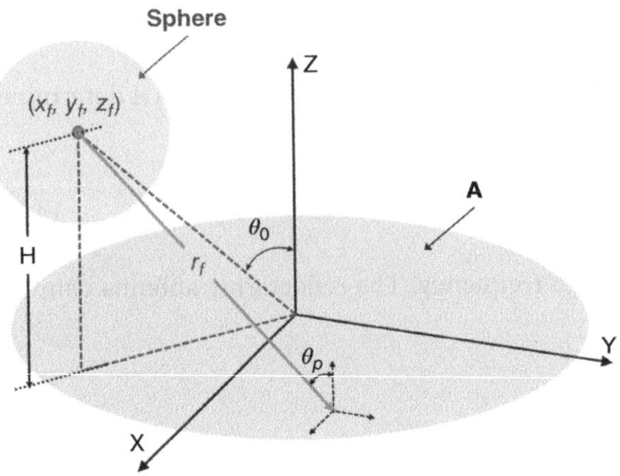

Figure 3.4 Geometrical setup of the feed and the reflectarray aperture.

Using the $\cos^q$ pattern model for a balanced feed, the Poynting vector of the feed antenna is defined as

$$\vec{P}(\vec{r}_f) = \hat{r}_f \frac{\cos^{2q}(\theta_f)}{r_f^2} \qquad \text{for } 0 \le \theta_f \le \pi/2. \tag{3.16}$$

The denominator in (3.15) can now be computed quite easily, and will have a simple closed form expression given by

$$\iint\limits_{Sphere} \vec{P}(\vec{r}_f).d\vec{s} = \int_0^{2\pi}\int_0^{\pi/2} \cos^{2q}(\theta_f)\sin\theta_f \, d\theta_f \, d\varphi_f = \frac{2\pi}{2q+1}. \tag{3.17}$$

The numerator in (3.15) is dependent on the feed position and aperture shape, and is typically computed numerically [6]. The geometrical setup of the system is shown in Figure 3.4.

3.2.2 Illumination Efficiency

Similar to spillover efficiency, the definition of illumination efficiency for reflectarrays is also extended from reflector antennas, which is mathematically given as

$$\eta_i = \frac{1}{A_a} \frac{\left| \iint\limits_A I(A')dA' \right|^2}{\iint\limits_A \left| I(A') \right|^2 dA'}. \tag{3.18}$$

Here I is the amplitude distribution over the reflectarray aperture, A. As discussed earlier, in the analysis outlined here it is assumed that the feed is purely in a certain polarization. The amplitude on the aperture is then a function of the feed (transmit) and reflectarray element (receive) patterns.

For reflectarrays, the radiation pattern of the element can also be modeled with a $\cos^q$ pattern, and to simplify the model, we assume that the element also has a rotationally symmetric pattern. To distinguish between the feed and element pattern power factor, we use q_e for the element pattern. The mathematical model of this element pattern is given by

$$U^E(\theta_p,\varphi_p)=\begin{cases}\cos^{2q_e}(\theta_p) & \text{for } 0\le\theta_p\le\pi/2 \\ 0 & \text{elsewhere}\end{cases}. \tag{3.19}$$

Here θ_p is the elevation angle in the local coordinate system for any point on the reflectarray aperture as shown in Figure 3.4. Typically, the values of q_e for reflectarray elements are much smaller than the values of q, in other words, the element pattern has a broader beamwidth. The nominal value for q_e is about 1. Note that this value of q_e corresponds to a directivity of 7.78 dB, which is typical for microstrip patches that are conventionally used for reflectarray phasing elements. Note that in some engineering designs, q_e is set to zero for simplicity.

The normalized amplitude at any point of the reflectarray aperture can now be expressed in a simple form, which is given by

$$I_n(A)=\frac{\cos^q(\theta_f)}{r_f}\cos^{q_e}(\theta_p). \tag{3.20}$$

Here the first term represents the amplitude on a certain point on the aperture coming from the feed antenna, thus the denominator r_f is added to account for the path length, since rays leaving the feed at the focal point spread out in all directions with a $1/r_f$ variation. It should be noted here that since the element receiving pattern is included in this calculation, the energy radiated from the aperture is smaller than the energy illuminated on the aperture.

With the amplitude distribution defined on the aperture, the illumination efficiency can be computed numerically. Note that when computing (3.20), the angles θ_f and θ_p are computed for every point on the aperture in terms of the feed coordinate system, Figure 3.1, and the local point coordinate system, Figure 3.4, respectively.

3.2.3 Effects of Aperture Shape on Efficiency

Similar to reflector antennas, the optimum aperture efficiency for a reflectarray is achieved when the feed pattern is matched to the reflectarray [3]. Consequently, the shape of the reflectarray aperture, the radiation pattern of the feed antenna, and the feed location play pivotal roles in yielding maximum aperture efficiency for the reflectarray system. Here we will compare the aperture efficiencies of a few typical reflectarray aperture configurations when a balanced feed with a rotationally symmetric pattern is used as the primary feed antenna [3].

A front-fed reflectarray configuration is considered here, where the feed antenna is placed at a distance of 18λ from the center of the aperture. Three different aperture shapes are studied here, where in each case, the power pattern of the feed (the value of q) is selected such that the reflectarray achieves its maximum aperture efficiency.

For the first case, we consider a 20λ × 30λ rectangular aperture for the reflectarray. Aperture efficiency as a function of q is given in Figure 3.5(a), where the optimum

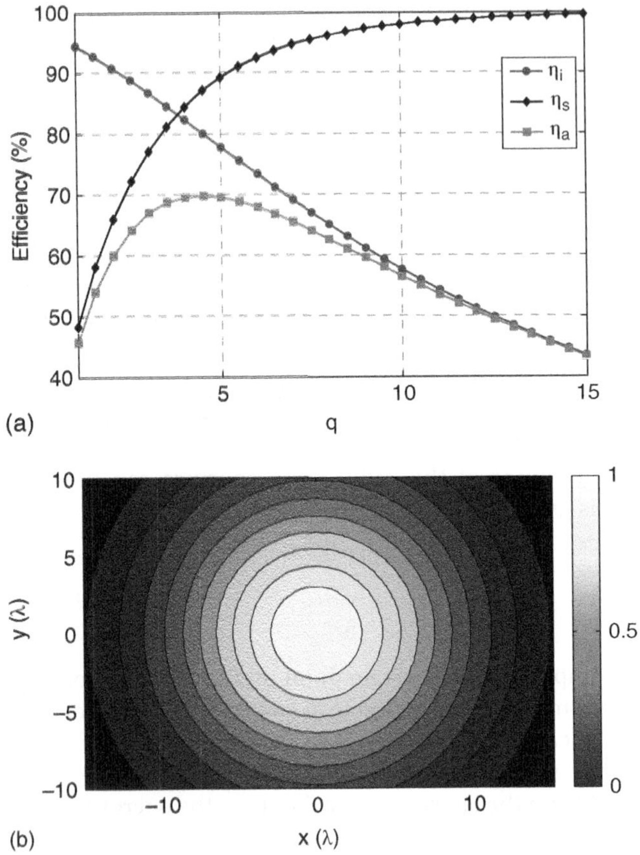

Figure 3.5 Aperture efficiency for a reflectarray with a rectangular aperture: (a) efficiency versus q, (b) normalized amplitude on the aperture for $q = 4.5$.

value is 4.5. Here the spillover and illumination efficiencies are computed using (3.15) and (3.18), respectively. The total aperture efficiency is the product of these two terms, which corresponds to an aperture efficiency of 69.76%. The normalized amplitude distribution on the aperture for this q value is also given in Figure 3.5(b).

It can be seen that for this aperture shape, even with an optimum feed pattern, the wider side of the reflectarray aperture is not properly illuminated, which leads to a low aperture efficiency. In general for these types of apertures, a feed antenna that can provide a pattern with different beamwidths in the two orthogonal directions proportional to the x/y ratio of the aperture, can significantly improve the aperture efficiency of the reflectarray system. However, such a reflectarray design will have a non-symmetric main beam that is broader in the direction of the smaller side of the aperture.

For the same system setup, now let us consider a square aperture with a side length of 20λ for the reflectarray. Similarly, aperture efficiency as a function of q is given in Figure 3.6(a). The optimum value of q is 6.5 in this case, which corresponds to an aperture efficiency of 75.35%. The normalized amplitude distribution on the aperture for

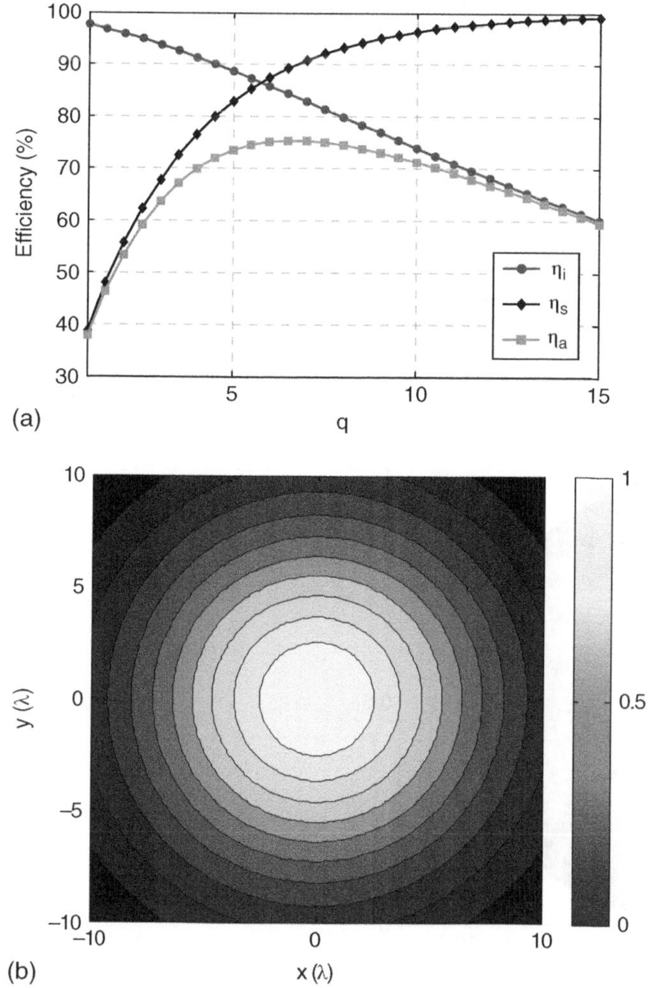

Figure 3.6 Aperture efficiency for a reflectarray with a square aperture: (a) efficiency versus q, (b) normalized amplitude on the aperture for $q = 6.5$.

this value of q is also given in Figure 3.6(b), where it can be seen that in comparison with the rectangular aperture studied earlier, a much better illumination is achieved on the aperture, which ultimately results in a better aperture efficiency. However, note that the corners of the square are still not properly illuminated. This is again due to the fact that the spherical wave front of the symmetric feed pattern cannot be matched to a square aperture shape.

The previous two cases clearly show the influence of aperture shape on the efficiency of reflectarray systems. For maximum efficiency, the shape of the aperture should directly correlate with the shape of the feed radiation pattern. Consequently, for a balanced feed pattern, the ideal aperture shape is circular. To verify this, here we study a reflectarray antenna with a circular aperture with a diameter of 20λ for the same system setup. For this aperture, the optimum value of q is 8.5, which corresponds to

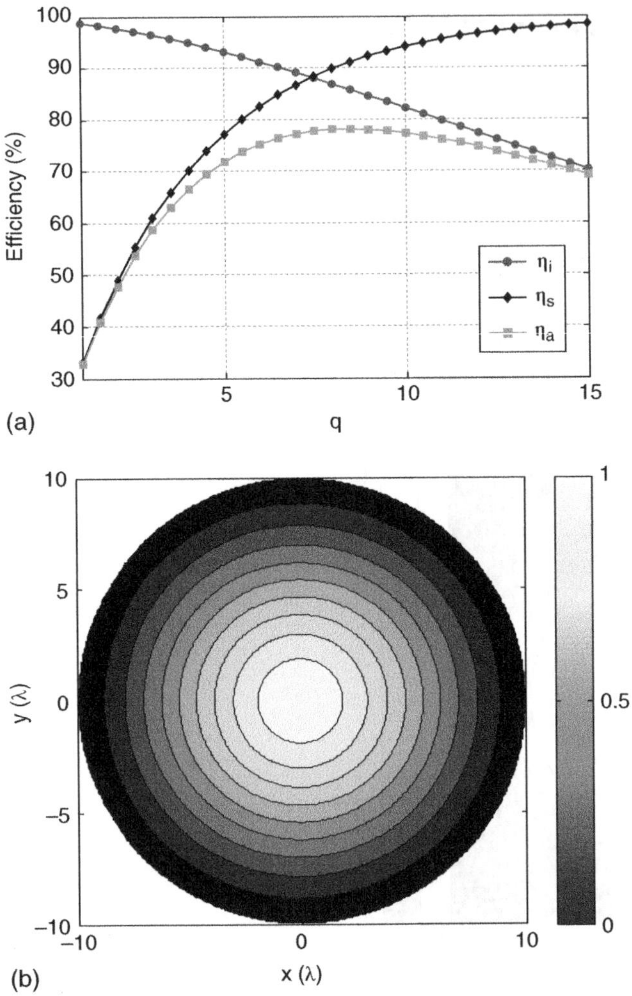

Figure 3.7 Aperture efficiency for a reflectarray with a circular aperture: (a) efficiency versus q, (b) normalized amplitude on the aperture for $q = 8.5$.

an aperture efficiency of 78.12%, and is notably higher than the previous two cases. Aperture efficiency as a function of q and the normalized amplitude distribution on the aperture for the optimum q value are both given in Figure 3.7.

The rotationally symmetric amplitude taper on the reflectarray aperture, Figure 3.7(b), also leads to a desirable radiation pattern for the reflectarray, and in general, a circular aperture is the optimum aperture shape for front-fed symmetric reflectarrays. For offset configurations, however, the asymmetry of the system results in a non-symmetric amplitude taper that degrades the efficiency. Elliptic aperture shapes may yield a better performance for offset configurations, however, for small feed offset angles, which is usually the case for reflectarrays, the circular aperture shape will yield good aperture efficiency.

3.2.4 Effects of Feed Location on Efficiency

In the previous aperture shape study, the feed location is fixed. In practical reflectarray designs, the feed location is an important system design parameter to design. For this setup, the two parameters in the design are the feed radiation pattern model, q, and the feed distance from the reflectarray aperture (z). Aperture efficiency as a function of both these parameters is given in Figure 3.8(a). It can be seen that for different feed patterns, the aperture efficiency can be maximized by selecting different feed positions. In general, the larger the q factor, the higher the feed position and the higher the aperture efficiency. Very broad feed patterns (small values of q) typically cannot yield high aperture efficiency. On the other hand, while very directive feed patterns (large values

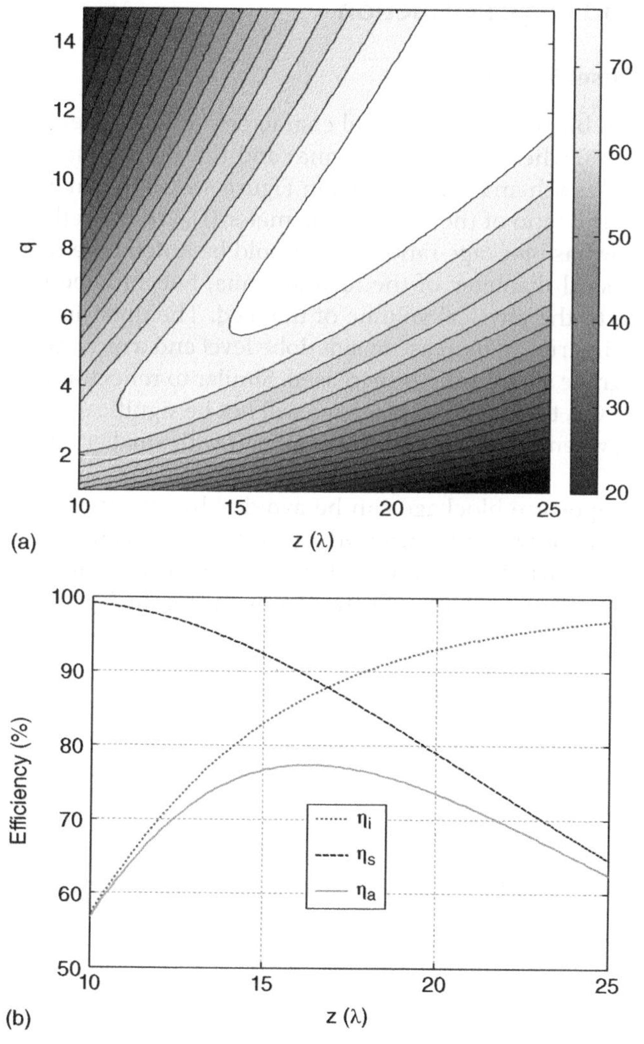

Figure 3.8 Efficiency for a reflectarray with a front-fed symmetric reflectarray: (a) aperture efficiency as a function of feed position and q, (b) efficiency as a function of feed position for $q = 6.5$.

of q) can attain high aperture efficiency, they usually are not considered a good choice for the reflectarray since the corresponding feed antennas typically have a large physical volume (resulting in high aperture blockage), and also reduce the relative gain improvement achieved by the reflector system. Values of q in the range of 6–8 are typically good choices.

For the same system, efficiency as a function of feed position is given for a feed with $q = 6.5$ in Figure 3.8(b). Note that illumination efficiency increases as the feed distance to aperture is increased, while the spillover efficiency decreases. Consequently, the optimum position is a nominal value in between, which is 16.3λ for this design and corresponds to an aperture efficiency of 77.34%.

3.3 Aperture Blockage and Edge Diffraction

3.3.1 Aperture Blockage and Offset Systems

Aperture blockage, mainly the feed blockage, is a physical characteristic that generally exists in reflectors and reflectarrays, where the feed antenna (and the support structure) blocks some of the radiation, as schematically shown in Figure 3.9(a). In general, aperture blockage is a function of the ratio of the blockage diameter (D_0) to the reflectarray diameter (D), which is known as blockage ratio [5]. It should be noted here that blockage diameter is not the physical diameter of the feed antenna, but an effective projected diameter that is related to the physical volume of the feed. The undesirable effects of aperture blockage are primarily an increase in side-lobe level and a reduction in antenna gain, which further degrade as this ratio is increased. Similar to reflectors, as long as the blockage ratio is below 0.2, the effects of blockage will not be significant [5], however, this is more challenging when one deals with reflectarrays with smaller aperture sizes.

The problems associated with aperture blockage can be avoided by offsetting the feed antenna, which is schematically shown in Figure 3.9(b). The basic characteristics of the offset system are similar to the symmetric case, and as such the offset system is generally preferred, especially for moderate size reflectarrays with nominal gain of

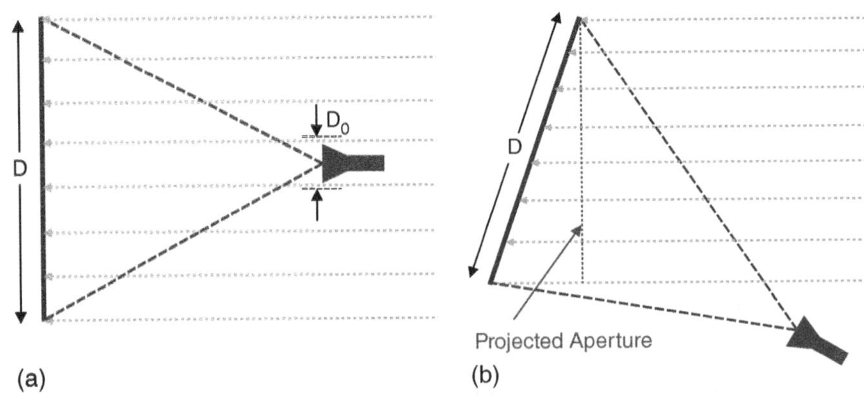

Figure 3.9 Aperture blockage in front-fed reflectarray systems: (a) symmetric, (b) offset.

about 30 dB. The offset system, however, exhibits a higher cross-polarization level due to the fact that the focal-length-to-diameter for offset systems is usually smaller [3]. In addition, for the same physical aperture diameter (D), the gain of the offset system is lower due to the smaller projected aperture effect. Nonetheless, elimination of feed blockage with the offset system is a significant advantage over the symmetric configuration.

As discussed earlier, efficiency analysis is a critical step in the system design. In this section, we will present some numerical studies on the aperture efficiency for offset reflectarray systems, which would give the reader a better understanding on the importance of efficiency analysis in system design. The basic formulations for calculating the aperture efficiency of reflectarray systems were outlined in the previous section and it was shown that the aperture efficiency of a reflectarray system is a function of the feed (position, orientation, and radiation pattern), reflectarray element pattern, and shape of the aperture. As discussed earlier, circular apertures are generally the optimum choice for reflectarrays, and thus selected for this study. A reflectarray with a diameter of 20λ is considered for the design here. In general, the element pattern shape has very little effect on the efficiency of the system [6]. Moreover, while the reflection characteristics of different element shapes are quite different, in most cases the element pattern is quite similar, as seen in Chapter 2. Thus, here we consider a $\cos(\theta)$ pattern model for the element.

Now let's study the aperture efficiency of an offset system configuration. For offset configurations, the offset angle is typically selected to minimize the feed blockage. Here we select a feed pattern with $q = 6.5$, and an offset angle of 20°, where the feed points to the geometrical center of the aperture. The aperture efficiency as a function of feed distance to reflectarray aperture is given in Figure 3.10(a), where it can be seen that the general trend of the efficiency factors is similar to the symmetric case. The optimum position of feed is at a distance of 15.2λ from the reflectarray aperture, yielding an aperture efficiency of 75.69%. It can be seen that with optimum feed placement, the offset system efficiency can achieve an aperture efficiency close to the symmetric case (77.34%), with the added advantage of minimized feed blockage.

It is important to note that the offset angle plays a pivotal role here. For the same feed pattern, the aperture efficiency as a function of both feed position and offset angle are given in Figure 3.10(b). It can be seen that for small values of offset angle (less than 20°), a high aperture efficiency can be achieved, however, as the offset angle increases beyond this, the system can no longer yield high efficiency. In addition, for reflectarrays, a large offset angle for the feed will result in elements being excited with very oblique angles, which will further degrade the performance of the system. As such, in general it is desirable to keep the offset angle to a minimum.

In the studies presented so far, the feed antenna was pointed to the geometrical center of the reflectarray aperture. While this is the optimum choice for symmetric designs, for offset systems the optimum beam pointing direction of the feed is not the geometrical center of the array. To study this, here we consider the same offset reflectarray system where the feed is placed with an offset angle of 20°, at a height of 15.2λ from the aperture. The beam pointing direction is then adjusted by displacing the feed along the x-axis in Figure 3.1. The aperture efficiency as a function of beam pointing position (X_0) of the feed is given in Figure 3.11. It can be seen that for offset systems, the optimum beam point for the feed is slightly below the geometrical center of the aperture. In this case, this point is −0.6λ, and with this setup the system will achieve an aperture

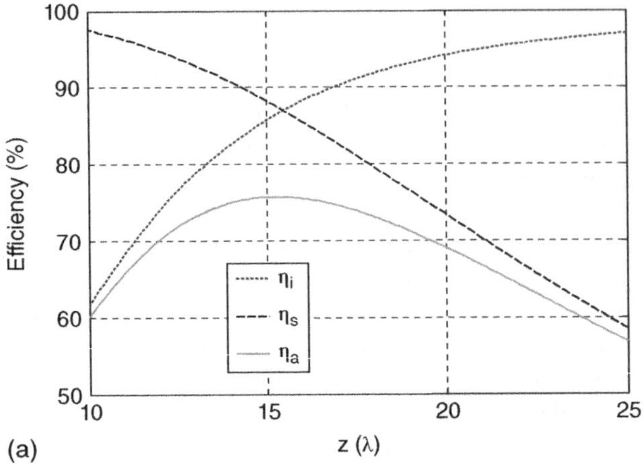

(a)

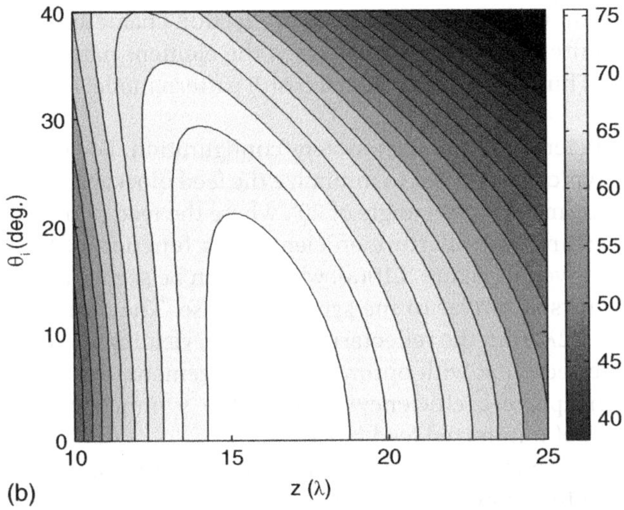

(b)

Figure 3.10 Efficiency for an offset reflectarray with $q = 6.5$: (a) efficiency as a function of feed position for $\theta_{\text{offset}} = 20°$, (b) aperture efficiency as a function of feed position and θ_{offset}.

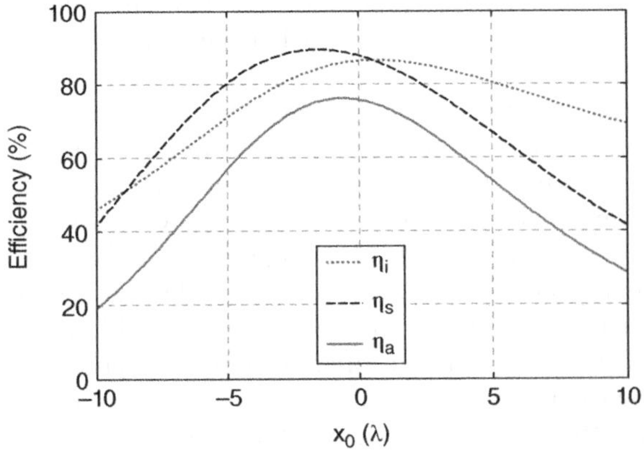

Figure 3.11 Efficiency for an offset reflectarray as a function of feed beam pointing position on the aperture.

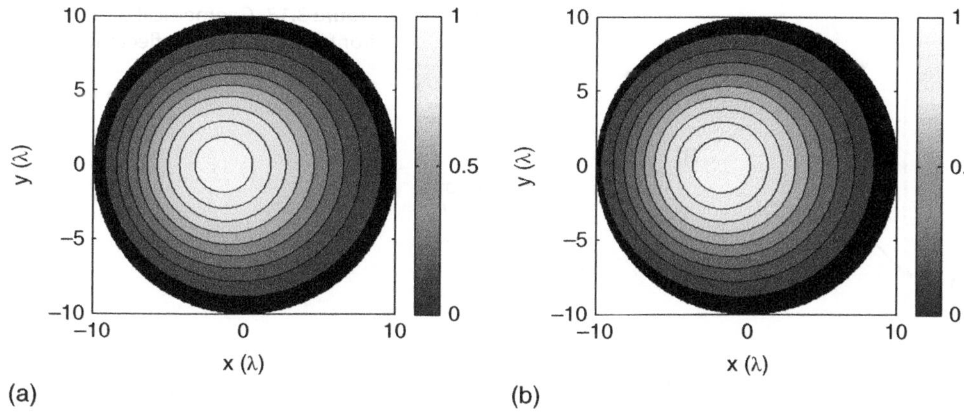

Figure 3.12 Amplitude distribution on the reflectarray aperture: (a) feed points to the geometrical center, (b) feed points to the optimum point.

efficiency of 76.1%. The amplitude distribution on the reflectarray when the feed points to the geometric center, and the optimum position are given in Figure 3.12(a) and (b), respectively. Note that with these optimized values (the offset angle, feed location, and beam pointing direction), the efficiency of the offset reflectarray is only 1% lower than the symmetric case.

3.3.2 Edge Taper and Edge Diffraction

An important consideration in designing a reflectarray system is the relative field strength at the edge of the reflectarray aperture with respect to the maximum, known as edge taper (ET). Similar to reflector antennas, the fields at the edges of the reflectarray aperture will inevitably exhibit diffraction that causes the rays to diverge. These edge diffracted fields degrade the radiation characteristics of the antenna since they no longer exhibit the ideal properties of a collimated wave. To minimize the effects of edge diffraction on the radiation performance of the antenna, the illumination level at the edges, that is, the ET, should be low.

For front-fed symmetric reflectors, ET is defined as the ratio of field intensity at the reflector rim to the intensity at its center and is typically given in decibels. The same formulation is adopted here for reflectarrays. The geometrical setup of a front-fed symmetric reflectarray is given in Figure 3.13. The two parameters that define this reflectarray system are the aperture diameter (D_{RA}) and the distance from the focal point to the aperture surface (H_F), which is similar to parabolic reflectors, where the diameter and focal length fully describe the shape of the parabola. The analogy between parabolic reflectors and reflectarrays will be discussed later on in Section 3.4.

For the reflectarray setup here, the half subtended angle is given by

$$\theta_0 = \tan^{-1}\left(\frac{D_{RA}}{2H_F}\right). \tag{3.21}$$

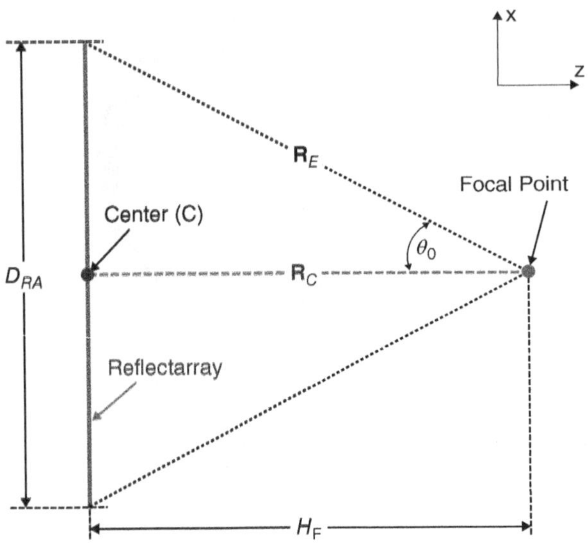

Figure 3.13 Geometrical setup of a front-fed symmetric reflectarray.

Within this subtended angle, the illumination varies from a maximum value at the center, to a minimum value at the edge. As before, we use the $\cos^q$ pattern model for a balanced feed, where the illumination at any point on the aperture can then be given by

$$I = \frac{\cos^q(\theta_f)}{r_f}. \tag{3.22}$$

Here r_f is the distance from the focal point to any point on the aperture, and θ_f is the angle between r_f and the surface normal (R_c). The illumination at the center of the reflectarray aperture can then be given as

$$I_{RAC} = \frac{\cos^q(\theta = 0^o)}{H_F} = \frac{1}{H_F}. \tag{3.23}$$

Similarly, the illumination at the edge of the reflector is given by

$$I_{RAE} = \frac{\cos^q(\theta = \theta_0)}{R_E}, \text{ where } \quad R_E = \frac{D_{RA}}{2\sin\theta_0}. \tag{3.24}$$

It should be noted here that, in this setup it is assumed that the system is rotationally symmetric. In other words, the reflectarray has a circular aperture, which as demonstrated earlier provides the highest aperture efficiency, and the feed points to the geometrical center. For the front-fed symmetric reflectarray system, the ET is then given as

$$ET_{RA} = 20\log_{10}\left(\frac{I_{RAE}}{I_{RAC}}\right) = 20\log_{10}\left(2\frac{H_F}{D_{RA}}\sin(\theta_0)\cos^q(\theta_0)\right). \tag{3.25}$$

It is worthwhile pointing out that this formula also allows one to select the desired feed power pattern, the value of q in this case, based on the desired value of ET. For reflectarrays this is given by

$$q = \log_{\cos(\theta_0)} \frac{D_{RA} 10^{\frac{ET_{RA}}{20}}}{2H_F \sin(\theta_0)}. \tag{3.26}$$

As a simple example here, we study a reflectarray with a circular aperture diameter of 20λ. In the first case, we fix the geometrical setup of the system, that is, $H_F/D_{RA} = 0.8$, and study ET as a function of feed pattern. This is given in Figure 3.14(a), where it can be seen that a monotonous decrease in ET is observed as the value of q increases. The typical or commonly used value for ET is about −10 dB, which for this system corresponds to a q value of 5.98. The illumination on the aperture for this q value is also given

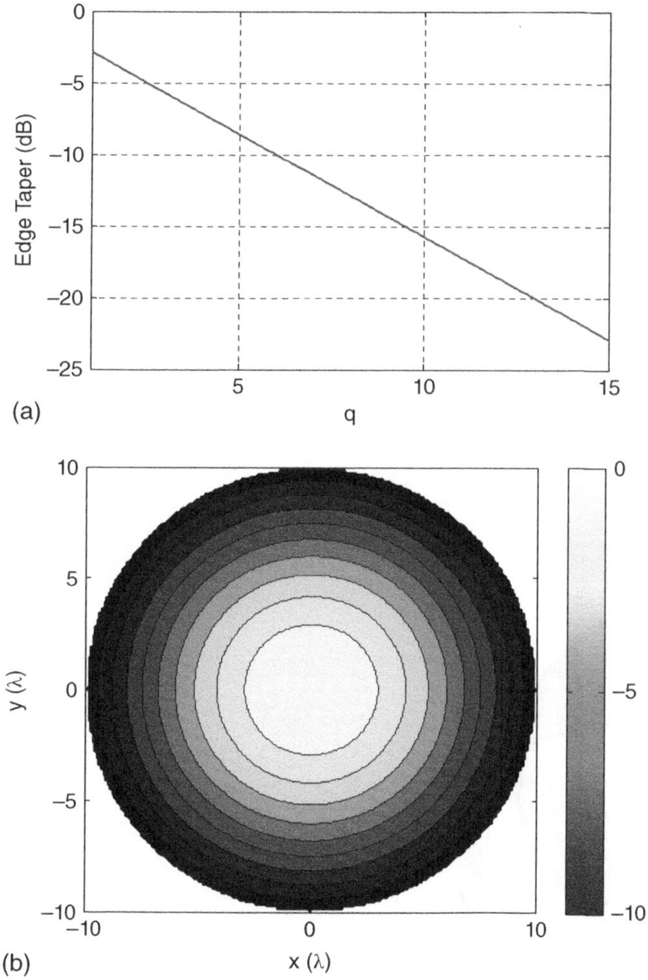

Figure 3.14 Aperture illumination and edge taper for a front-fed symmetric reflectarray: (a) ET versus q for $H_F/D = 0.8$, (b) normalized illumination on the aperture for $q = 5.98$ in dB scale.

in Figure 3.14(b). It is important to note that in this case, a feed antenna has to be designed that satisfies this pattern requirement.

For the second case, we study ET as a function of feed position (H_F) which is the practical approach when an available feed antenna has to be used for the system. We select the same aperture size of 20λ, and a feed antenna with a q value of 7.4, which corresponds to a directivity of 15 dB. Here we study H_F/D_{RA} in the range from 0.4 to 1. The results are given in Figure 3.15, where it can be seen that the ET changes quite rapidly for smaller values of H_F/D_{RA}. With this feed, for a 10 dB ET, the value of H_F/D_{RA} is 0.9. It is worthwhile to note that while the typical value of ET is about −10 dB, slightly lower value of edge illumination can yield better aperture efficiency.

Now let us study the offset feed configuration. The geometrical setup of an offset fed reflectarray is shown in Figure 3.16. As with reflectors [5], unfortunately the definition of ET that was introduced earlier for symmetric systems is ambiguous for offset configurations, since the taper on the reflectarray edge will be a function of φ_a. One approach to quantify the ET is to study it on the upper, lower, and side edges.

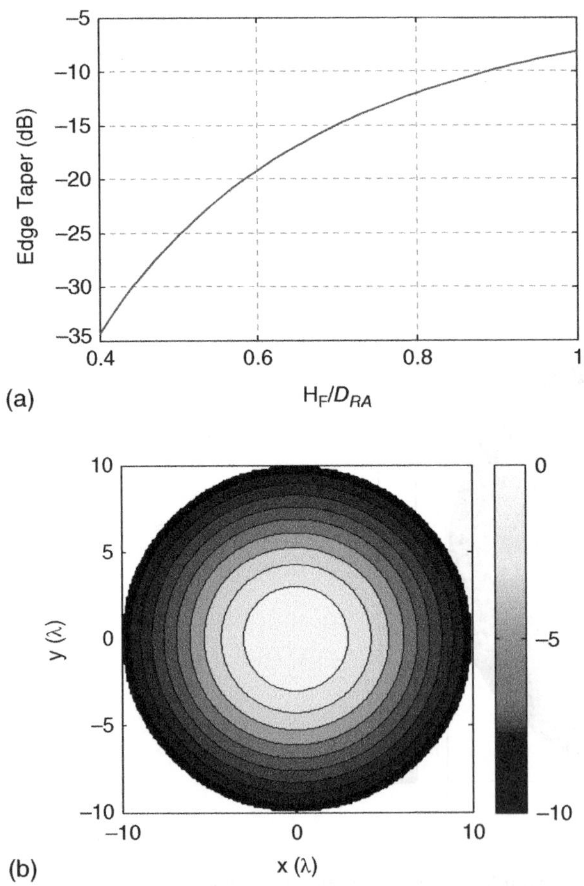

Figure 3.15 Aperture illumination and edge taper for a front-fed symmetric reflectarray: (a) ET versus H_F for a fixed feed pattern (q=7.4), (b) normalized illumination on the aperture for H_F/D_{RA}=0.9 in dB scale.

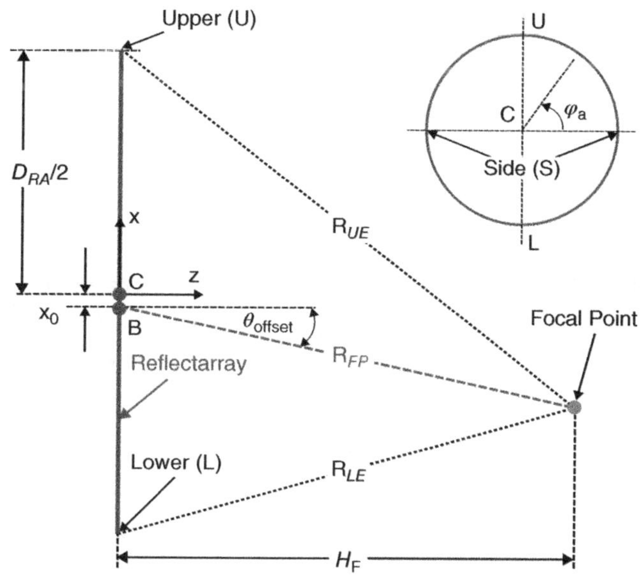

Figure 3.16 Geometrical setup of an offset reflectarray system.

Note that unlike the symmetric case, where the feed antenna points to the geometrical center of the circular aperture, the offset case may have its feed pointed to a different point as shown in Figure 3.16. In any case, the system can be properly defined by a unique set of parameters, namely D_{RA}, H_F, Δ_B, and θ_{offset} which are shown in Figure 3.20.

Similar to the symmetric case, the radial distance from the focal point is computed for several points on the aperture. Note that in the offset configuration, aperture efficiency can be maximized when the feed is pointing to a point slightly below the center [6], shown as B in Figure 3.16, therefore this geometrical setup is adopted with the equations given here.

$$R_{FP} = \frac{H_F}{\cos(\theta_{offset})}, \text{ focal point to B,} \tag{3.27}$$

$$R_C = \sqrt{\Delta_B^2 + R_{FP}^2 - 2R_{FP}\Delta_B \cos\left(\theta_{offset} + \frac{\pi}{2}\right)}, \text{ focal point to C,} \tag{3.28}$$

$$R_U = \sqrt{\left(\frac{D_{RA}}{2} + \Delta_B\right)^2 + R_{FP}^2 - 2R_{FP}\left(\frac{D_{RA}}{2} + \Delta_B\right)\cos\left(\theta_{offset} + \frac{\pi}{2}\right)}, \text{ focal point to upper edge,} \tag{3.29}$$

$$R_L = \sqrt{\left(\frac{D_{RA}}{2} - \Delta_B\right)^2 + R_{FP}^2 - 2R_{FP}\left(\frac{D_{RA}}{2} - \Delta_B\right)\cos\left(\frac{\pi}{2} - \theta_{offset}\right)}, \text{ focal point to lower edge,} \tag{3.30}$$

$$R_S = \sqrt{\left(\frac{D_{RA}}{2}\right)^2 + R_C^2}, \text{ focal point to side edges.} \tag{3.31}$$

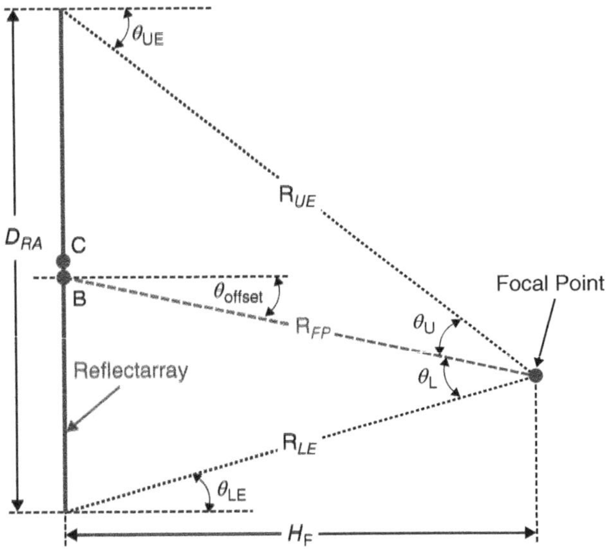

Figure 3.17 The side-view of an offset reflectarray system showing the edge and feed subtended angles.

In order to compute the illumination at the edge, it is also necessary to compute the angle at each point with respect to the feed focal point as shown in Figure 3.17. These angles are given by

$$\theta_{UE} = \cos^{-1}\left(\frac{H_F}{R_U}\right),\ \theta_U = \theta_{UE} - \theta_{offset},\ \text{for upper edge,} \tag{3.32}$$

$$\theta_{LE} = \cos^{-1}\left(\frac{H_F}{R_L}\right),\ \theta_L = \theta_{LE} + \theta_{offset},\ \text{for lower edge,} \tag{3.33}$$

$$\theta_S = \tan^{-1}\left(\frac{D_{RA}}{2R_C}\right),\ \text{for side edges.} \tag{3.34}$$

Note that for the offset configuration the subtended feed angle is $\theta_L + \theta_U$. The illumination at any of the desired edge points can now be computed using (3.13). As with the symmetric case, taper has to be computed with respect to the maximum illumination, however, in the offset configuration the position of maximum illumination is unknown, and must be determined. With the geometrical setup of Figure 3.16, the maximum illumination is along the x-axis, and typically at $-x_0 > x_{max} > -D/2$. In any case once the maximum illumination is computed, ET can easily be determined at the desired points on the edge of the reflectarray aperture using

$$ET_U = 20\log_{10}\left(\frac{I_{UE}}{I_{MAX}}\right) = 20\log_{10}\left(\frac{\cos^q(\theta_U)}{I_{MAX}R_U}\right),\ \text{for upper edge,} \tag{3.35}$$

$$ET_L = 20\log_{10}\left(\frac{I_{LE}}{I_{MAX}}\right) = 20\log_{10}\left(\frac{\cos^q(\theta_L)}{I_{MAX}R_L}\right),\ \text{for lower edge,} \tag{3.36}$$

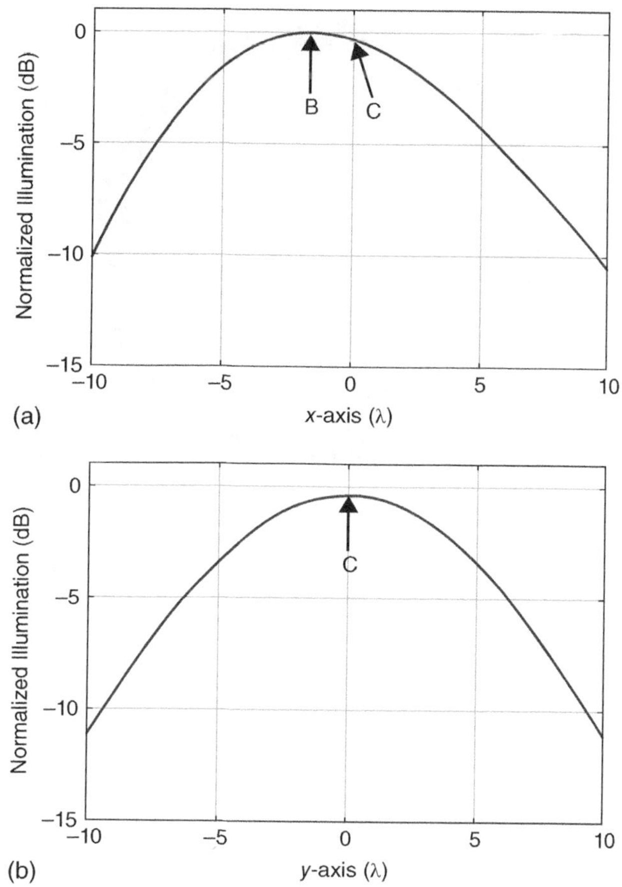

Figure 3.18 Illumination on the offset reflectarray aperture: (a) along *x*-axis, (b) along *y*-axis.

$$ET_S = 20\log_{10}\left(\frac{I_{SE}}{I_{MAX}}\right) = 20\log_{10}\left(\frac{\cos^q(\theta_S)}{I_{MAX}R_S}\right), \text{ for side edges.} \tag{3.37}$$

As an example, here we study an offset reflectarray system with a circular aperture diameter of 20λ, $H_F = 16\lambda$, $\theta_{offset} = 20°$, and $x_0 = \lambda$. The feed antenna has a q value of 7.4, which corresponds to a directivity of 15dB. The illumination along the x- and y-axes of the reflectarray aperture is given in Figure 3.18.

For this system, the maximum illumination occurs at $x = -1.8515\lambda$. As discussed earlier, in all cases the maximum illumination is at a point that is slightly lower than where the feed points to, that is, point B in Figure 3.16. The aperture illumination is also given in Figure 3.19. As discussed earlier, for an offset system ET is ambiguous. However, the formulation outlined in this section allows one to observe the taper on the aperture at certain edge points, and control the ET by proper feed design or placement. In any case optimum system design is generally conducted by means of an efficiency analysis as demonstrated earlier.

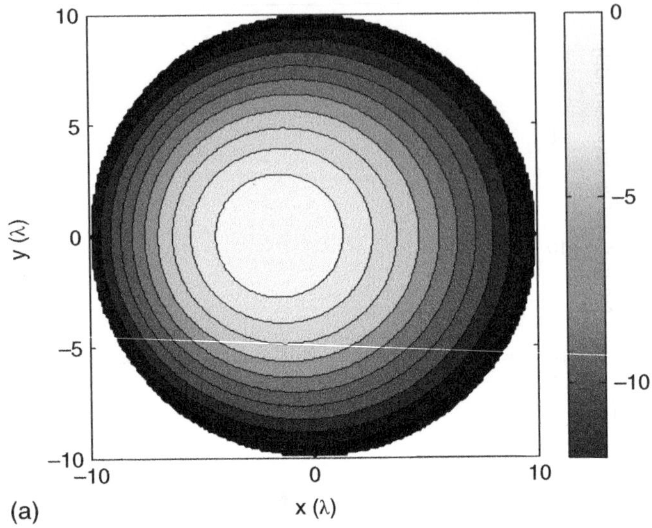

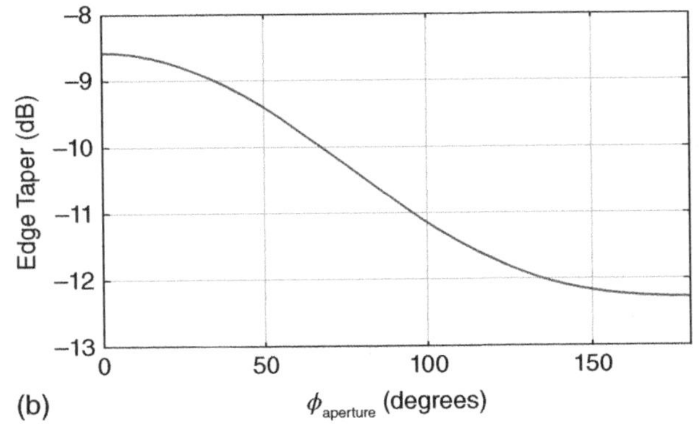

Figure 3.19 (a) Aperture illumination of an offset reflectarray. (b) ET as a function of polar angle on the reflectarray aperture.

3.4 The Analogy between a Reflectarray and a Parabolic Reflector

Parabolic reflector antennas have long served as the most suitable antenna for high-gain operation in both terrestrial and spaceborne applications [1], [2]. Reflectarray antennas are essentially built based on the concept of reflector antennas. Given the fact that reflector antenna technology is mature and well developed, it certainly is beneficial to take advantage of the developments that have been made in reflector design over generations. To this end one should be able to describe a reflectarray system that exactly mimics the parabolic reflector system. However, in many cases, the design parameters for the reflectarray don't correlate directly to the reflector design parameters. This is due to the fact that for each antenna, the design parameters are given in manner that best describes its own system. While this is less of an issue for axisymmetric designs, the offset configuration is more complex. In particular, many important reflector design parameters such as focal depth are typically misinterpreted, while other parameters such as feed tilt angle, and offset height are essentially nonexistent in offset reflectarrays [7].

In this section, we will first review the offset parabolic reflector design characteristics, and then define the analogous relationship between offset reflector and reflectarray systems. We then present the mathematical formulation that converts the design parameters of the parabolic reflector system to the reflectarray design parameters and vice versa. These equations simplify the configuration setup in offset reflectarrays, by using a coordinate system and a set of parameters that is well suited for design and analysis of offset reflectarrays, and can be advantageous for complex reflectarray configurations, such as offset dual-reflectarray systems [8], [9].

3.4.1 The Offset System Configurations

In an offset reflector configuration, many parameters exist that can be utilized to achieve the required design goals [5]. However, the most important reflector design parameters, which can fully describe a focused offset system, can be given in terms of four key parameters. The geometrical model of the offset parabolic reflector system is given in Figure 3.20(a), where the design parameters are:

- D = reflector diameter,
- F = focal length,
- H = offset height,
- ψ_B = feed tilt angle.

Note that the first three parameters completely describe the geometrical setup of the parabolic reflector; however, since in general the primary feed does not point to the reflector center (C), for a full system specification it is necessary to specify the feed tilt

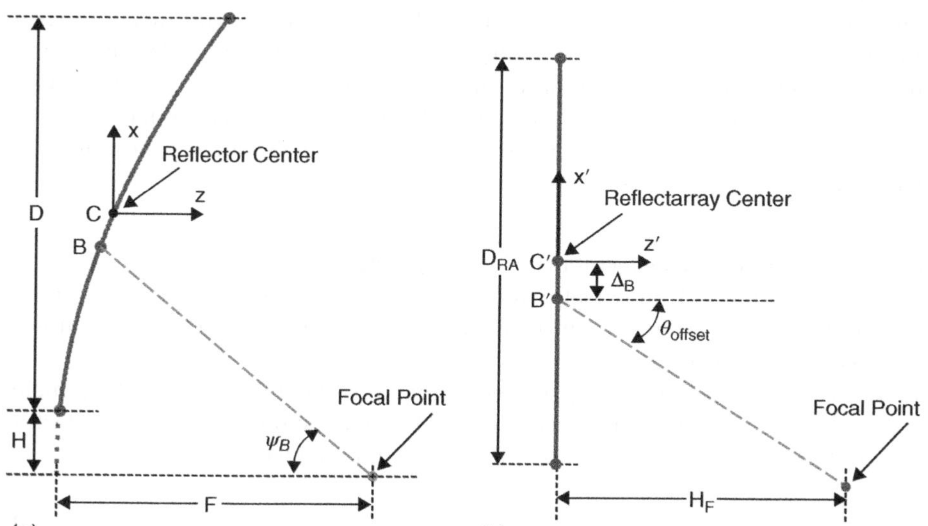

Figure 3.20 The geometrical models and key parameters of the offset systems: (a) parabolic reflector, (b) reflectarray.

angle. In the general case, the feed points to point B on the reflector surface, as shown in Figure 3.20(a).

As discussed in earlier sections, the offset reflectarray system can also be fully specified with four key parameters. The geometrical model of the offset reflectarray system is given in Figure 3.20(b), where the design parameters are:

- D_{RA} = reflectarray diameter,
- H_F = focal point to aperture distance,
- Δ_B = offset distance from reflectarray center,
- θ_{offset} = feed offset angle.

Similar to the reflector system, the primary feed points to a point on the reflectarray aperture (B′) which is not necessarily the geometrical center, where the optimal position for B′ is determined by efficiency analysis. Note that in order to distinguish the design parameters of the two systems, we use the non-prime coordinates for the reflector system, and the prime coordinates for the reflectarray system.

The two sets of parameters described here for these offset configurations make it possible to fully describe the antenna systems. In order to establish an analogy between these two systems, one must be able to derive the relationship between the parameters in the two sets. It should be noted here that in the reflector system, the main beam of the antenna points in the z-direction in the coordinate setup of Figure 3.20(a). For the reflectarray system, however, one has the option to control the beam direction by adding a progressive phase to the aperture. More discussion on this will be given in the following sections.

3.4.2 Analogous Offset Reflector

The key system design parameters for offset reflectors and reflectarrays were described in the previous sections. In general, in order to analyze the antennas one makes use of a coordinate system that best suits the geometrical setup, which are different for the reflector and reflectarray systems. In order to establish the analogy between these two systems, we will use the coordinate setup of the reflector antenna. It is implicit that once the relationship between the design parameters is determined, one can describe the systems in any desired coordinate system.

3.4.2.1 Transformation from Reflector to Reflectarray System

The geometrical models of the offset configurations in the reflector coordinate setup are shown in Figure 3.21. Here we consider the case where the reflector design parameters (F, D, H, ψ_B) are given.

First, we derive some additional parameters for the reflector using the four main parameters that would aid in our derivations. The angle subtended to the reflector lower tip, and the distance from focal point to this point are given as

$$\psi_L = 2\tan^{-1}\left(\frac{H}{2F}\right), \tag{3.38}$$

$$R_L = \frac{2F}{1+\cos\psi_L}. \tag{3.39}$$

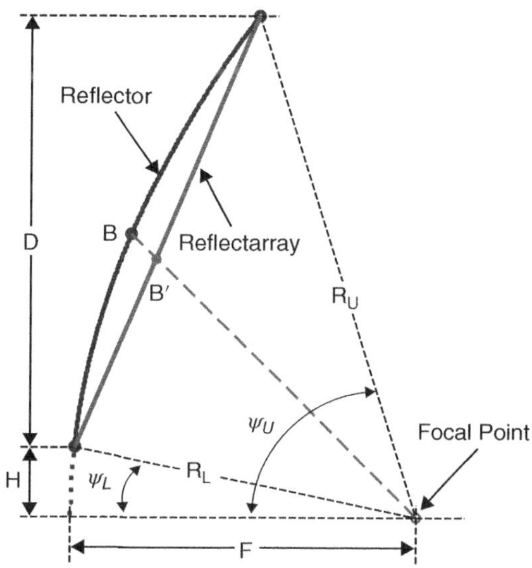

Figure 3.21 The analogous geometrical setup of the offset reflector and reflectarray systems.

Similarly, the angle subtended to the reflector upper tip, and the distance from focal point to this point are given by

$$\psi_U = 2\tan^{-1}\left(\frac{D+H}{2F}\right), \tag{3.40}$$

$$R_U = \frac{2F}{1+\cos\psi_U}. \tag{3.41}$$

Note that with this setup, the feed subtended angle, ψ_S, is identical in both systems and is given by

$$\psi_S = \psi_U - \psi_L. \tag{3.42}$$

With these parameters defined, now let us derive the reflectarray system parameters in this coordinate setup. The system setup is given in Figure 3.22, where the feed subtended angle is $\psi_S = \theta_U + \theta_L$. These angles can be given in terms of reflector parameters as

$$\theta_L = \psi_B - \psi_L, \text{and } \theta_U = \psi_U - \psi_B. \tag{3.43}$$

The diameter of the reflectarray antenna can now be calculated in terms of these parameters as

$$D_{RA} = \sqrt{R_U^2 + R_L^2 - 2R_U R_L \cos\psi_S}. \tag{3.44}$$

As discussed earlier, with the reflectarray antenna one can direct the main beam to any desired direction. However, since here we aim at describing analogous systems, the main beam for the reflectarray points in the same direction as the reflector, as shown in Figure 3.23. Note that in the current setup, the reflectarray aperture is tilted by a certain angle with respect to the reflector coordinates which is given by

$$\theta_{\text{tilt}} = \cos^{-1}\left(\frac{D}{D_{RA}}\right). \tag{3.45}$$

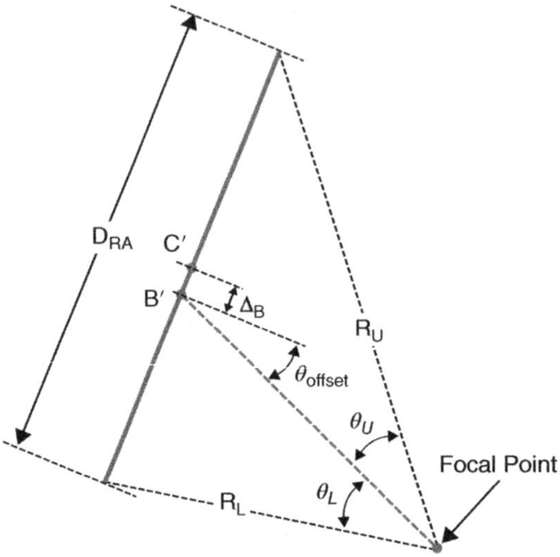

Figure 3.22 The geometrical setup of the offset reflectarray in the reflector coordinates.

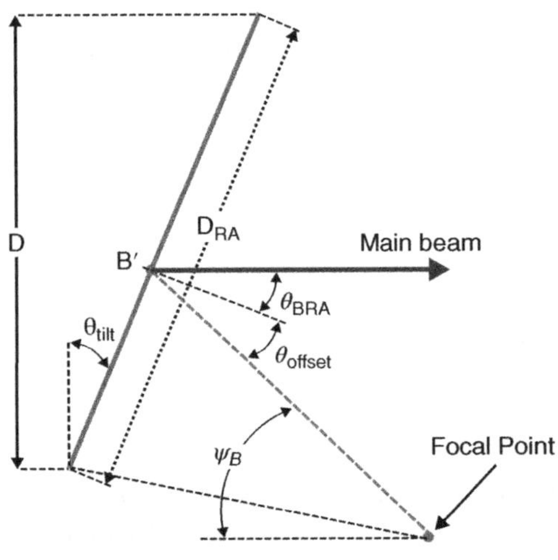

Figure 3.23 The main beam direction and geometrical setup of the offset reflectarray system.

To direct the reflectarray main beam in the same direction as the reflector, one must have $\theta_{\text{BRA}} = \theta_{\text{tilt}}$, where θ_{BRA} is the angle with respect to the normal on the reflectarray aperture. The feed offset angle for the reflectarray can then be given as

$$\theta_{\text{offset}} = \psi_B - \theta_{\text{BRA}}. \tag{3.46}$$

The distance from focal point to B′ is given as

$$R_{FP} = R_L \frac{\sin\left(\frac{\pi}{2} - \theta_L + \theta_{\text{offset}}\right)}{\sin\left(\frac{\pi}{2} - \theta_{\text{offset}}\right)}. \tag{3.47}$$

The focal point to reflectarray aperture distance, and the feed point offset distance from reflectarray center can now be given in terms of these parameters as

$$H_F = R_{FP}\cos(\theta_{\text{offset}}), \tag{3.48}$$

$$\Delta_B = \frac{1}{2}\left(\sqrt{R_{FP}^2 + R_U^2 - 2R_{FP}R_U\cos\theta_U} - \sqrt{R_{FP}^2 + R_L^2 - 2R_{FP}R_L\cos\theta_L}\right). \tag{3.49}$$

The equations given in this section can fully describe the analogous reflectarray system in terms of the reflector design parameters. In particular, the four parameters of the offset reflectarray are computed from equations (3.44), (3.46), (3.48), and (3.49), respectively.

3.4.2.2 Transformation from Reflectarray to Reflector System

Now let's consider the case where the reflectarray design parameters (H_F, D_{RA}, Δ_B, θ_{offset}) are given. The geometrical setup of the reflectarray is essentially identical to the previous case shown in Figure 3.22. Similarly, we derive the distance from focal point to the reflectarray upper and lower tips using

$$R_L = \sqrt{R_{FP}^2 + \left(\frac{D_{RA}}{2} - \Delta_B\right)^2 - 2R_{FP}\left(\frac{D_{RA}}{2} - \Delta_B\right)\cos\left(\frac{\pi}{2} - \theta_{\text{offset}}\right)}, \tag{3.50}$$

$$R_U = \sqrt{R_{FP}^2 + \left(\frac{D_{RA}}{2} + \Delta_B\right)^2 - 2R_{FP}\left(\frac{D_{RA}}{2} + \Delta_B\right)\cos\left(\frac{\pi}{2} + \theta_{\text{offset}}\right)}, \tag{3.51}$$

where R_{FP} is the distance from focal point to B′ and can be derived using (3.47). The feed angles subtended to the upper and lower tips are then given as

$$\theta_L = \cos^{-1}\left(\frac{\left(\frac{D_{RA}}{2} - \Delta_B\right)^2 - R_L^2 - R_{FP}^2}{-2R_{FP}R_L}\right), \tag{3.52}$$

$$\theta_U = \cos^{-1}\left(\frac{\left(\frac{D_{RA}}{2} + \Delta_B\right)^2 - R_U^2 - R_{FP}^2}{-2R_{FP}R_U}\right). \tag{3.53}$$

Using the geometrical formula for a paraboloid, the distance from focal point to the upper and lower edges can also be given using

$$R_L = \frac{2F}{1 + \cos\psi_L}, \tag{3.54}$$

$$R_U = \frac{2F}{1 + \cos(\psi_L + \psi_S)}, \tag{3.55}$$

where ψ_S is the total feed subtended angle as given in (3.41). The angle subtended to the reflector lower tip can be derived by solving these two equations simultaneously as

$$\psi_L = \text{Re}\left\{-i\ln\left(-\frac{R_L - R_U + 2i\sin\left(\frac{\psi_S}{2}\right)}{R_L - R_U e^{i\psi_S}}\right)\right\}. \tag{3.56}$$

The reflector design parameters can now be computed using

$$F = \frac{R_L}{2}\left(1+\cos(\psi_L)\right), \tag{3.57}$$

$$H = R_L \sin(\psi_L), \tag{3.58}$$

$$D = R_U \sin(\psi_L + \psi_S) - H, \tag{3.59}$$

$$\psi_B = \cos^{-1}\left(\frac{D}{D_{RA}}\right) + \theta_{\text{offset}}. \tag{3.60}$$

These equations make it possible to describe an analogous offset reflector system in terms of the reflectarray design parameters. One important note here is that in the reflectarray to reflector transform, the direction of the main beam of the reflectarray cannot be defined independently. In the setup outlined here, the direction of the reflectarray main beam is in the same direction as the reflector. However, as discussed earlier, the reflectarray can be designed to scan the beam in any desired direction, but in any case, all the important reflector characteristics such as focal depth, offset height, and feed tilt angle, will be identical. In other words, with the same configuration setup for both antennas (Figure 3.21), the reflectarray allows for more flexibility in the system design.

3.4.3 Example of Analogous Offset Systems

The formulations presented in the previous section allow one to accurately determine the important parameters of offset reflectarray systems [10], [11], in a fashion that has been well established for parabolic reflector antennas. Here, let us study the performance of an analogous parabolic reflector and reflectarray system. A Ka-band reflectarray antenna with an aperture diameter of 190 mm, designed for the center frequency of 32 GHz, is considered. We select a balanced feed antenna with a q value ($\cos^q\theta_f$ radiation pattern model) of 6.5. The optimal values of the reflectarray design parameters are determined based on efficiency analysis, and are summarized in Table 3.1. For the analogous reflector antenna, the design parameters are determined using the formulation presented in Section 3.4.2, and are also summarized in this table.

The F/D ratio for this system is 0.83. As discussed earlier, the main beam direction of the reflectarray (θ_{BRA}) is determined based on the parabolic reflector parameters, and should be equal to 19°. The 2D-model (cross-sectional view) of this parabolic reflector

Table 3.1 Design parameters for the offset reflectarray and reflector.

Reflectarray	D_{RA}	H_F	Δ_B	θ_{offset}
	190 mm	144.71 mm	6 mm	20°
Reflector	D	F	H	ψ_B
	179.64 mm	148.93 mm	12.78 mm	39°

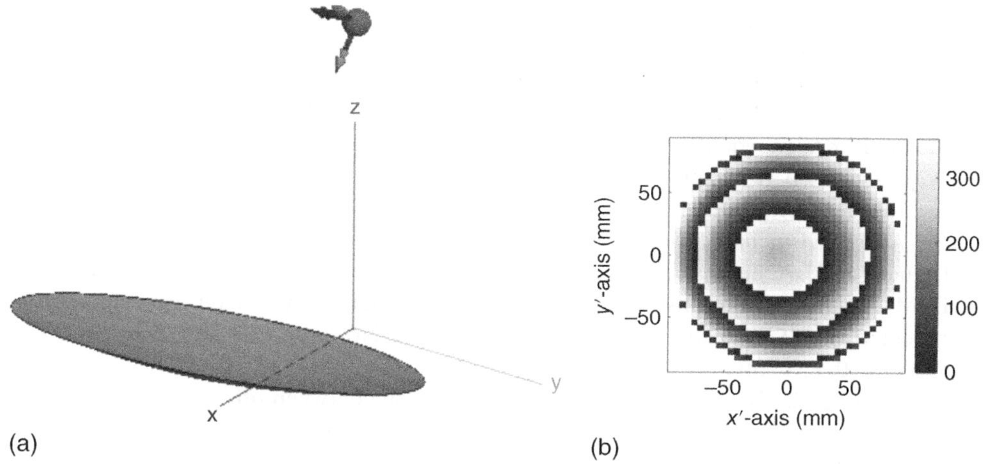

Figure 3.24 (a) The parabolic reflector antenna model in FEKO. (b) The element phase distribution on the analogous reflectarray antenna aperture.

and reflectarray system is similar to that shown in Figure 3.21. For the parabolic reflector antenna, the design is essentially complete, however, for the reflectarray antenna, the array nature of the system requires one to also specify the element spacing. Here we set the element spacing to 4.7 mm ($\sim\lambda/2$ at 32 GHz). This corresponds to 1184 elements on the aperture of the reflectarray. In this study, we are comparing the idealized performance of both systems, thus for the reflectarray we use ideal elements, that is, elements with no reflection loss ($|\Gamma| = 1$), and the exact (non-quantized value) required phase shift. Also for the parabolic reflector we use perfect electric conductor. The 3D model of the parabolic reflector antenna in FEKO [12], and the required element phase shift on the aperture of the reflectarray antenna are given in Figure 3.24. It is important to note that the radiation patterns of these two antennas are almost identical, as expected, and the interested readers are referred to [13].

References

1 C. J. Sletten, *Reflector and Lens Antennas: Analysis and Design Using Personal Computers*, Artech House Inc., 1988.

2 M. E. Cooley, and D. Davis, "Reflector antennas," in *Radar Handbook*, 3rd Edn, McGraw-Hill, 2008.

3 W. L. Stutzman and G. A. Thiele, *Antenna Theory and Design*, 3rd Edn, John Wiley & Sons Inc., 2012.

4 C. A. Balanis, *Antenna Theory: Analysis and Design*, 3rd Edn, John Wiley & Sons Inc., 2005.

5 Y. Rahmat-Samii, "Reflector Antennas," in *Antenna Handbook: Theory, Applications, and Design*, Y. T. Lo and S. W. Lee, Van Nostrand Reinhold, 1988.

6 A. Yu, F. Yang, A. Z. Elsherbeni, J. Huang, and Y. Rahmat-Samii, "Aperture efficiency analysis of reflectarray antennas," *Microwave and Optical Technology Letters*, Vol. 52, No. 2, pp. 364–372, Feb. 2010.

7 P. Nayeri, A. Z. Elsherbeni, and F. Yang, "The analogy between offset configurations of parabolic reflectors and reflectarrays," *IEEE Antennas and Propagation Society International Symposium*, Vancouver, Canada, July 2015.
8 E. Almajali, D. McNamara, J. Shaker, and M. R. Chaharmir, "Derivation and validation of the basic design equations for symmetric sub-reflectarrays," *IEEE Trans. Antennas Propag.*, Vol. 60, No. 5, pp. 2336–2346, May 2012.
9 C. Tienda, M. Arrebola, J. A. Encinar, G. Toso "Analysis of a dual-reflectarray antenna," *IET Microw. Antennas Propag.*, Vol. 5, No. 13, pp. 1636–1645, 2011.
10 J. Huang and J. A. Encinar, *Reflectarray Antennas*. New York, NY, USA: Wiley-IEEE, 2008.
11 D. M. Pozar, S. D. Targonski, and H. D. Syrigos, "Design of millimeter wave microstrip reflectarrays," *IEEE Trans. Antennas Propag.*, Vol. 45, No. 2, pp. 287–296, Feb. 1997.
12 FEKO v7.0, EM Software & Systems Inc., 2014.
13 P. Nayeri, A. Z. Elsherbeni, and F. Yang, "The analogy between offset configurations of parabolic reflectors and reflectarrays," *IEEE Antennas and Propagation Society International Symposium*, Vancouver, Canada, July 2015.

4

Radiation Analysis Techniques

The design and analysis procedure for reflectarray elements and the reflectarray system were outlined in detail in Chapters 2 and 3, respectively. The next task is to calculate the radiation characteristics of reflectarray antennas, such radiation pattern, directivity, gain, and so on. This chapter focuses on radiation analysis techniques for reflectarray antennas. As discussed in earlier chapters, reflectarrays combine the numerous advantages of both printed arrays and parabolic reflectors and create the new hybrid high-gain antenna [1]. In terms of analysis approaches for reflectarrays, it is implicit that they are essentially adapted from the analysis techniques developed for their predecessors over generations.

Various analysis approaches have been developed to calculate the radiation characteristics of the reflectarray. Most reflectarray analysis approaches approximate the elements on the reflectarray aperture as identical elements that form an array, and then array summation or far-field transformation of currents is used to calculate the radiation pattern of the reflectarray antenna [2], [3]. Accurate characterization of the reflection coefficients of the reflectarray elements holds a significant importance in the radiation analysis for these approaches. These coefficients are therefore usually obtained with a full-wave simulation technique, using an infinite-array approach that takes into account the mutual coupling between the elements as described in Chapter 2. Alternatively, the current distributions on the reflectarray elements (obtained from the full-wave simulations under the local-periodicity approximation) could be directly used, and the radiation pattern would then be calculated using far-field transformation of the currents for the total array [4]. The great advantage of these analysis methods is the fast computational time. However, it is implicit that due to the approximations in the analysis, some discrepancies with practical results may be observed. On the other hand, a full-wave simulation of the entire reflectarray can provide accurate results. The electrically large size of the reflectarray antenna aperture, combined with hundreds of elements with dimensions smaller than a wavelength, demands a significantly high computational time and large resources for full-wave analysis [5], [6]. While in many cases a full-wave simulation will be advantageous at the final stage of a design, fast computational approaches are still necessary tools for the initial design. This is in addition to several stages of optimization that may be required for a reflectarray antenna design [7], [8].

In this chapter, we first present two basic methods for analysis of the radiation performance of reflectarray antennas. The two different approaches, namely the array theory

Reflectarray Antennas: Theory, Designs, and Applications, First Edition.
Payam Nayeri, Fan Yang, and Atef Z. Elsherbeni.

method and the aperture field method, are described, and numerical results are then presented for various reflectarray configurations. The advantages and limitations for each method are discussed. Moreover, some important radiation analysis concepts such as directivity, gain, and loss budget calculations are described in detail. Next, full-wave simulation approaches are discussed, and comparison between the theoretical analysis techniques and full-wave simulations are also presented, which shows a close agreement.

4.1 Array Theory Approach: The Robust Analysis Technique

Analysis of electrically large antennas is a topic of great interest in antenna engineering, and in the broad sense it can be divided into analog and digital techniques. Continuous large antennas, such as the parabolic reflector, form the analog section and must be analyzed using integrals that are usually time consuming. Arrays on the other hand form the digital section and thus can be analyzed using fast and simple summations. The reflectarray antenna by nature possesses the basic properties of antenna arrays, and as such one should be able to use the array theory technique for analysis. However, given that this antenna is space-fed (in contrast to conventional arrays which have a feed network), the array theory analysis for reflectarray antennas demands some further considerations, which will be outlined in this section.

4.1.1 Idealized Feed and Element Patterns

In the array theory approach, the most primary step is to determine the element tapers. For the reflectarray antenna, the space-fed nature of the system implies that the taper is directly dependent on the position and radiation pattern of the feed antenna, and as such characterization of the feed is of paramount importance. The feed radiation pattern model was described in Chapter 3 and can be applied in the array theory analysis; however, polarization of the feed is not considered in the analysis here. We will provide more details about feed polarization later on in this chapter. For simplicity, the radiation pattern of the feed antenna is assumed to be rotationally symmetric and is given by

$$E^F(\theta_f,\varphi_f)=\begin{cases}\cos^q(\theta_f) & \text{for}\quad 0\le\theta_f\le\pi/2\\ 0 & \text{elsewhere}\end{cases}, \tag{4.1}$$

where similar to (3.10), θ_f is the feed elevation angle and q is the feed power pattern.

The simplified feed radiation pattern model described here makes it very easy to compute the illumination of each element on the reflectarray aperture. However, note that for the reflectarray antenna, accurate modeling of the array excitation also requires one to include the reflection coefficients of the element when computing the element taper. More discussion on this is given in the next section.

For an antenna array, the radiation pattern of the antenna is essentially the summation of the radiation pattern of all of its elements. As such another important step in the array theory analysis for reflectarrays is to characterize the radiation pattern of the elements. Here we use the same element pattern described in Section 3.2.2 for the element, that is, a rotationally symmetric pattern modeled with a cosine q function, which is given by

$$E^E(\theta_p,\varphi_p)=\begin{cases}\cos^{q_e}(\theta_p) & \text{for } 0\le\theta_p\le\pi/2\\ 0 & \text{elsewhere}\end{cases}, \tag{4.2}$$

where θ_p is the elevation angle in the local coordinate system for any element on the reflectarray aperture as shown in Figure 3.4.

4.1.2 Element Excitations and Reflectarray Radiation Pattern

The conventional array theory can be implemented to calculate the far-field radiation pattern of the reflectarray antenna. The radiation pattern of a 2D planar array with M × N elements can be calculated as

$$\begin{aligned}\vec{E}(\hat{u})&=\sum_{m=1}^{M}\sum_{n=1}^{N}\vec{A}_{mn}(\hat{u})\cdot\vec{I}(\vec{r}_{mn}),\\ \hat{u}&=\hat{x}\sin\theta\cos\varphi+\hat{y}\sin\theta\sin\varphi+\hat{z}\cos\theta,\end{aligned} \tag{4.3}$$

where A is the element pattern vector function, I is the element excitation vector function, and $\vec{r}_{mn}$ is the position vector of the mn^{th} element. The coordinate system for the reflectarray antenna is given in Figure 4.1.

To simplify calculations, usually one uses scalar functions in the analysis. For the element pattern function, A, a scalar approximation considers a cosine q_e model for each element with no azimuth dependence, that is,

$$A_{mn}(\theta,\varphi)\approx\cos^{q_e}(\theta)\cdot e^{jk(\vec{r}_{mn}\cdot\hat{u})}. \tag{4.4}$$

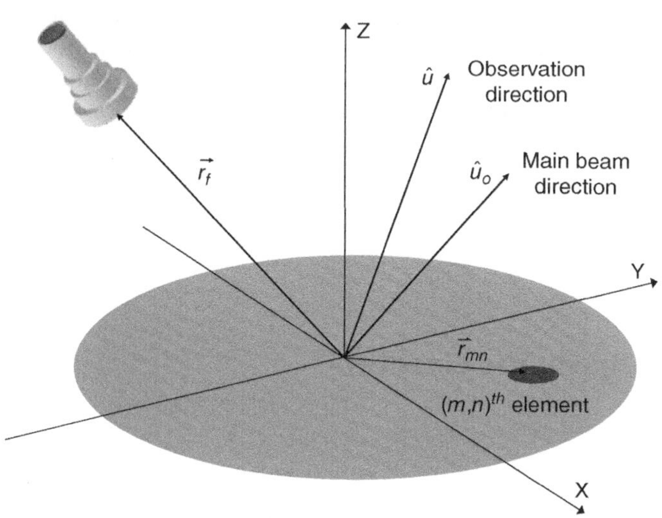

Figure 4.1 Coordinate system of the reflectarray antenna for array theory analysis.

The element excitation function $I(m,n)$, is determined by the incident field and element property. By approximating the feed horn pattern function using a cosine q model and taking into account the Euclidian distance between the feed horn and the element, the illumination of the aperture can be obtained. The element excitation can then be expressed as

$$I(m,n) \approx \frac{\text{Cos}^{q_f}\theta_f(m,n)}{|\vec{r}_{mn}-\vec{r}_f|} \cdot e^{-jk(|\vec{r}_{mn}-\vec{r}_f|)} \cdot |\Gamma_{mn}| e^{j\phi_{mn}}. \tag{4.5}$$

Here, θ_f is the spherical angle in the feed coordinate system and $\vec{r}_f$ is the position vector of the feed.

In addition, for each element this excitation can take into account the receive mode pattern, that is $|\Gamma_{mn}|$. This pattern is also modeled by a cosine q_e function based on the local element coordinates, that is,

$$|\Gamma_{mn}| = \text{Cos}^{q_e}\theta_e(m,n). \tag{4.6}$$

It should be noted that this definition of $|\Gamma_{mn}|$ is for the receive/transmit (R/T) model. In the scattering model analysis approach, this value is obtained directly from the unit-cell analysis.

The required phase delay of the mn^{th} element, ϕ_{mn}, is designed to set the main beam in the $\hat{u}_o$ direction as described in Chapter 2. It should be noted that since the infinite-array approach is often used for analysis of reflectarray elements, ϕ_{mn} does contain the mutual coupling effects under local periodicity approximation.

With these approximations, the radiation pattern can be simplified to the scalar form

$$E(\theta,\varphi) = \sum_{m=1}^{M}\sum_{n=1}^{N} \text{Cos}^{q_e}\theta \frac{\text{Cos}^{q_f}\theta_f(m,n)}{|\vec{r}_{mn}-\vec{r}_f|} e^{-jk(|\vec{r}_{mn}-\vec{r}_f|-\vec{r}_{mn}\cdot\hat{u})} \text{Cos}^{q_e}\theta_e(m,n) e^{j\phi_{mn}}. \tag{4.7}$$

The radiation pattern calculation method described here uses a conventional array summation technique. In general, the array theory formulation will yield good main beamwidth, beam direction, and general pattern shape; however, since the polarization of the feed horn and elements are not accounted for in the simplified cosine q model, the cross-polarization of the reflectarray antenna is not calculated in this procedure.

In summary, the advantages and disadvantages of this approach are:

Advantages:

- Simplicity of the formulation and program development
- Fast computational time

Limitations:

- Cross-polarization is not modeled

4.2 Aperture Field Approach: The Classical Analysis Technique

4.2.1 Complex Feed Patterns

Antenna feeds play a major role in the radiation performance and efficiency of the reflectarray system. The balanced feed $\cos^q$ pattern described in Chapter 3 is a popular model for the feed antenna pattern, however, more accurate analysis may require one to use the radiation pattern of the practical feed antenna. A variety of antenna configurations such as open-ended waveguides (OEWG), microstrip patch arrays, pyramidal and conical horns, hybrid-mode horns, and corrugated horns have been used for reflector feeds: see [9], [10], and [11]. Analytical solutions for the far-field radiation pattern of these feed antennas are available in literature [12] and can be used directly for the radiation analysis of the reflectarray. In this section, we will briefly review the analytical radiation pattern of corrugated circular aperture or dual mode horns.

A circularly symmetric feed radiation pattern typically provides the best aperture efficiency for reflectors. Practical realization of such pattern can be accomplished by corrugated or dual mode horn antennas. The pattern of these antennas is similar to the H-plane pattern of a TE_{11} mode radiation from an open-ended circular waveguide, which is mathematically given as

$$E(\theta_f)=\frac{\sqrt{(ka)^2-(1.841)^2}+ka\sin\theta_f}{\sqrt{(ka)^2-(1.841)^2}+ka}\cdot\frac{J_0(ka\sin\theta_f)-J_2(ka\sin\theta_f)}{1-\left(\dfrac{ka\sin\theta_f}{1.841}\right)^2}. \tag{4.8}$$

Here ka is the electrical circumference of the waveguide. With a fixed phase center, the complete radiation pattern model for the feed is then given by

$$E^F(r_f,\theta_f,\varphi_f)=A_0\left[\hat{\theta}E(\theta_f)\cos\varphi_f-\hat{\varphi}E(\theta_f)\sin\varphi_f\right]\frac{e^{-jkr_f}}{r_f},\ \text{for x polarized,} \tag{4.9}$$

$$E^F(r_f,\theta_f,\varphi_f)=A_0\left[\hat{\theta}E(\theta_f)\sin\varphi_f+\hat{\varphi}E(\theta_f)\cos\varphi_f\right]\frac{e^{-jkr_f}}{r_f},\ \text{for y polarized.} \tag{4.10}$$

where r_f is the vector from the feed phase center to any point on the reflectarray aperture.

4.2.2 Field Transformations from Feed to Aperture and Equivalent Surface Current

The next step in the analysis is to determine the field radiated by the feed at each reflectarray element in terms of the Cartesian components in the reflectarray coordinate system [13]. A flowchart of this procedure is given in Figure 4.2. First, the incident field is calculated at the center point of each reflectarray element using the appropriate feed pattern model. Then the field is transformed into Cartesian components using the following equation

$$\begin{bmatrix}E_x^F\\E_y^F\\E_z^F\end{bmatrix}=\begin{bmatrix}\sin\theta_f\cos\varphi_f & \cos\theta_f\cos\varphi_f & -\sin\varphi_f\\ \sin\theta_f\sin\varphi_f & \cos\theta_f\sin\varphi_f & \cos\varphi_f\\ \cos\theta_f & -\sin\theta_f & 0\end{bmatrix}\begin{bmatrix}0\\E_{\theta_f}^F\\E_{\varphi_f}^F\end{bmatrix}. \tag{4.11}$$

In this equation (θ_f, φ_f) are the spherical angles of the reflectarray elements in the feed coordinate system. Since the spherical angles change with position of the element, this

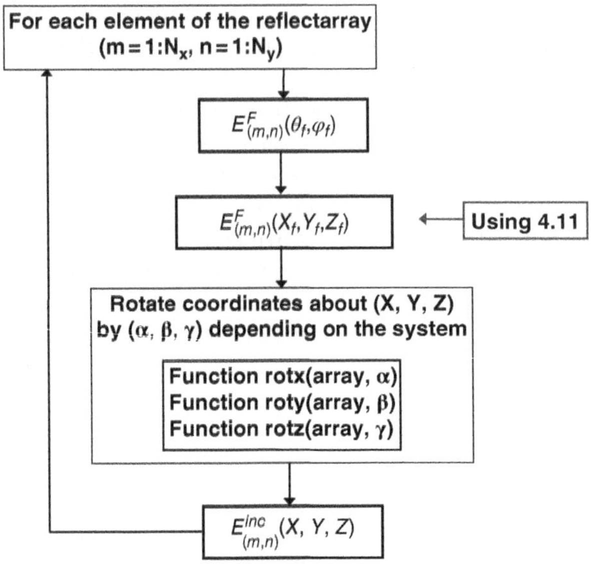

Figure 4.2 Flowchart of the transformations from feed to array coordinates.

transformation is carried out for each element of the reflectarray. It is worth emphasizing the superscript "F," which indicates the fields here are in the Cartesian coordinate of the feed antenna.

Transformation from the feed coordinate to the reflectarray coordinates can be accomplished using the rotation functions described in [14]. The three functions rotx(array, α), roty(array, β), and rotx(array, γ) rotate the Cartesian coordinates of the feed horn to the array coordinate system.

These transformations are repeated for every element on the reflectarray aperture. Here N_x and N_y are the total number of elements in x and y directions on the reflectarray aperture.

Once the incident electric fields on the aperture surface are computed, the reflected fields for every element in the array is then obtained using the generalized scattering matrix, which relates the Cartesian components of incident and reflected fields in the periodic cell, that is,

$$\begin{bmatrix} E_x^{ref}(m,n) \\ E_y^{ref}(m,n) \end{bmatrix} = \begin{bmatrix} \Gamma_{xx} & \Gamma_{xy} \\ \Gamma_{yx} & \Gamma_{yy} \end{bmatrix} \begin{bmatrix} E_x^{inc}(m,n) \\ E_y^{inc}(m,n) \end{bmatrix}. \tag{4.12}$$

The scattering matrix is usually obtained from a full-wave simulation of the reflectarray element with periodic boundary conditions, as discussed in Chapter 2. The phase shift produced for each element depends on the polarization of the feed and, also, takes into account the cross-polarized reflected fields.

In the aperture (reflected) field analysis, the Cartesian components of the tangential electric and magnetic currents on the reflectarray can be obtained using the equivalence principle, as expressed in (4.13):

$$J_{sx}(x,y) = H_y^{ref}(x,y), \quad J_{sy}(x,y) = -H_x^{ref}(x,y), \tag{4.13}$$

$$M_{sx}(x,y) = E_y^{ref}(x,y), \;\; M_{sy}(x,y) = -E_x^{ref}(x,y), \tag{4.14}$$

which give a complete description of the radiating currents of the aperture. Here the magnetic (*H*) fields are obtained using Maxwell's equations under local plane wave assumption [15], [16]. From the tangential currents on the reflectarray surface, the far-field radiation pattern can be calculated using the vector potentials, which will be described in the next section.

4.2.3 Near-Field to Far-Field Transforms and Reflectarray Radiation Pattern

The far-field electric and magnetic fields can be computed using the tangential currents on the aperture as follows:

$$\begin{aligned}
&E_r = 0 && H_r = 0 \\
&E_\theta = -\frac{jke^{-jkr}}{4\pi r}\left(L_\varphi + \eta N_\theta\right) && H_\theta = +\frac{jke^{-jkr}}{4\pi r}\left(N_\varphi - \frac{L_\theta}{\eta}\right) \\
&E_\varphi = +\frac{jke^{-jkr}}{4\pi r}\left(L_\theta - \eta N_\varphi\right) && H_\varphi = -\frac{jke^{-jkr}}{4\pi r}\left(N_\theta + \frac{L_\varphi}{\eta}\right)
\end{aligned} \tag{4.15}$$

Here *N* and *L* are functions that are evaluated directly from the tangential currents on the reflectarray surface and are given by

$$N_\theta = \iint_S \left(J_x \cos\theta \cos\varphi + J_y \cos\theta \sin\varphi\right) e^{jk(x'u+y'v)} ds' \tag{4.16}$$

$$N_\varphi = \iint_S \left(-J_x \sin\varphi + J_y \cos\varphi\right) e^{jk(x'u+y'v)} ds' \tag{4.17}$$

$$L_\theta = \iint_S \left(M_x \cos\theta \cos\varphi + M_y \cos\theta \sin\varphi\right) e^{jk(x'u+y'v)} ds' \tag{4.18}$$

$$L_\varphi = \iint_S \left(-M_x \sin\varphi + M_y \cos\varphi\right) e^{jk(x'u+y'v)} ds' \tag{4.19}$$

where u and v are the angular coordinates, ($u = \sin\theta \cos\varphi$, $v = \sin\theta \sin\varphi$). In general, the double integration must be extended to the entire plane $z = 0$, but it is always limited to the reflectarray surface, because it is assumed that the tangential currents are zero outside the reflectarray. In order to evaluate these integrals element by element, some change of parameters is required. The coordinates of the central point of each reflectarray element can be given by

$$x' = x'_e + md_x - \left(\frac{N_x - 1}{2}\right) d_x ; \; m = 0, 1, 2, \ldots, N_x - 1 \tag{4.20}$$

$$y' = y'_e + nd_y - \left(\frac{N_y - 1}{2}\right) d_y ; \; n = 0, 1, 2, \ldots, N_y - 1 \tag{4.21}$$

where N_x and N_y are the maximum number of elements in each direction, and d_x and d_y represent the periodicity of the reflectarray elements in the x and y directions, respectively. It should be noted that it is assumed that the tangential currents are constant on each element of the reflectarray and also x'_e and y'_e are limited to the unit-cell, that is, $\left(-\frac{d_x}{2} \le x'_e \le +\frac{d_x}{2}, -\frac{d_y}{2} \le y'_e \le +\frac{d_y}{2}\right)$.

To simplify the formulation, we define the constant $K_1 = e^{-j\frac{k}{2}\left[(N_x+1)ud_x+(N_y+1)vd_y\right]}$. Also, as mentioned earlier, since the tangential currents on each reflectarray element are assumed to be constant, they are pulled out of the integrals. The common terms which need to be evaluated for all the integrals can then be expressed in the form of sinc functions.

$$\int_{-d_x/2}^{d_x/2}\int_{-d_y/2}^{d_y/2} e^{jk\left(x'_e u + y'_e v\right)} ds'_e = d_x d_y \operatorname{sinc}\left(\frac{kud_x}{2}\right)\operatorname{sinc}\left(\frac{kvd_y}{2}\right). \tag{4.22}$$

The functions N and L can then be simplified to

$$N_\theta(\theta,\varphi) = K_1 d_x d_y \operatorname{sinc}\left(\frac{kud_x}{2}\right)\operatorname{sinc}\left(\frac{kvd_y}{2}\right) \cdot \sum_{m=1}^{N_x}\sum_{n=1}^{N_y}\left(J_x(x'_e,y'_e)\cos\theta\cos\varphi + J_y(x'_e,y'_e)\cos\theta\sin\varphi\right)e^{jk\left[mud_x+nvd_y\right]} \tag{4.23}$$

$$N_\varphi(\theta,\varphi) = K_1 d_x d_y \operatorname{sinc}\left(\frac{kud_x}{2}\right)\operatorname{sinc}\left(\frac{kvd_y}{2}\right) \cdot \sum_{m=1}^{N_x}\sum_{n=1}^{N_y}\left(-J_x(x'_e,y'_e)\sin\varphi + J_y(x'_e,y'_e)\cos\varphi\right)e^{jk\left[mud_x+nvd_y\right]} \tag{4.24}$$

$$L_\theta(\theta,\varphi) = K_1 d_x d_y \operatorname{sinc}\left(\frac{kud_x}{2}\right)\operatorname{sinc}\left(\frac{kvd_y}{2}\right) \cdot \sum_{m=1}^{N_x}\sum_{n=1}^{N_y}\left(M_x(x'_e,y'_e)\cos\theta\cos\varphi + M_y(x'_e,y'_e)\cos\theta\sin\varphi\right)e^{jk\left[mud_x+nvd_y\right]} \tag{4.25}$$

$$L_\varphi(\theta,\varphi) = K_1 d_x d_y \operatorname{sinc}\left(\frac{kud_x}{2}\right)\operatorname{sinc}\left(\frac{kvd_y}{2}\right) \cdot \sum_{m=1}^{N_x}\sum_{n=1}^{N_y}\left(-M_x(x'_e,y'_e)\sin\varphi + M_y(x'_e,y'_e)\cos\varphi\right)e^{jk\left[mud_x+nvd_y\right]}. \tag{4.26}$$

It should be noted that this summation is extended to a rectangular grid with N_x and N_y elements, but for circular or elliptical reflectarrays, the tangential fields for the elements outside the aperture boundary have to be set to zero.

The formulation presented here is the general aperture field analysis method, where the first principal of equivalence is used to obtain the currents on the reflectarray aperture and the corresponding far-field radiation pattern. It is worthwhile to mention that to simplify the calculations usually only the electric field components are used in the analysis, which corresponds to the second principle of equivalence. Alternatively, the magnetic field components may also be used to compute the far-field radiation pattern of the reflectarray antenna. More discussion on this topic will be given in a later section.

In summary, the advantages and disadvantages of the aperture field approach are:

Advantages:
- Accurate modeling of feed and element polarization.

Limitations:
- Complicated formulation and program development.
- Increased computational time.

4.3 Important Topics in Reflectarray Radiation Analysis

4.3.1 Principal Radiation Planes

The previous sections discuss calculating the radiation pattern of the antenna in the entire space. However, it is interesting to view the radiation pattern in the principal planes with respect to the main beam direction [17]. While for a broadside radiation the principal planes are the *xz*- and *yz*-planes, for off-broadside beams the principal plane positions do not lie on the spherical coordinate axis. Consider the coordinate system in Figure 4.3, which shows a main beam in an arbitrary direction of (θ_m, φ_m). It is necessary to choose the plane cuts that best capture the features of the 3D patterns.

For any arbitrary beam direction (θ_m, φ_m) it is possible to align the axis of the coordinate system with the main beam direction by defining two coordinate rotations. First a rotation of $\gamma = \varphi_m$ about the *z*-axis, and next a rotation of $\beta = \theta_m$ about the *y*-axis. By using these two rotations the main beam will be in the direction of z''-axis in the new coordinate system (x'', y'', z''). These coordinate rotations are shown pictorially in Figure 4.4.

In this new coordinate system, the principal planes (P.P.1, P.P.2) are simply defined as $x''z''$ and $y''z''$ planes, respectively. The 2D radiation patterns can then be plotted using the principal plane angles defined as α_1 and α_2 for P.P.1 and P.P.2. The principal planes and the corresponding angles are shown in Figure 4.5. The task here is to calculate the principal plane angles in terms of the initial coordinate system. As discussed earlier two coordinate rotations are used here. These transformations can be expressed in terms of matrix multiplications which transform the original coordinate system (x, y, z) to the new coordinate system (x'', y'', z'').

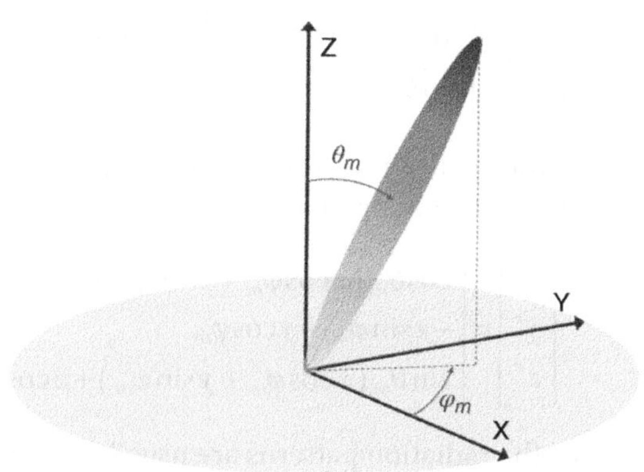

Figure 4.3 Far-field coordinate system for the reflectarray antenna.

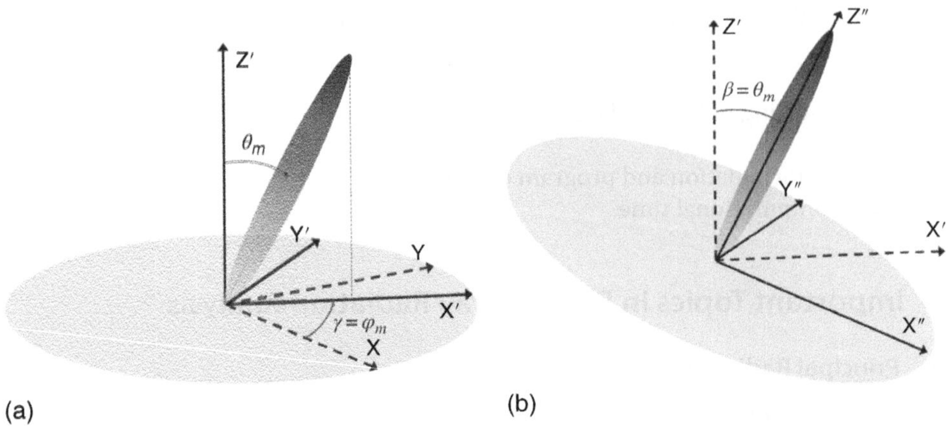

Figure 4.4 Coordinate rotations used for the reflectarray system: (a) rotation of $\gamma=\varphi_m$ about the z-axis with Z=Z′, (b) additional rotation of $\beta=\theta_m$ about the y'-axis with Y′=Y″.

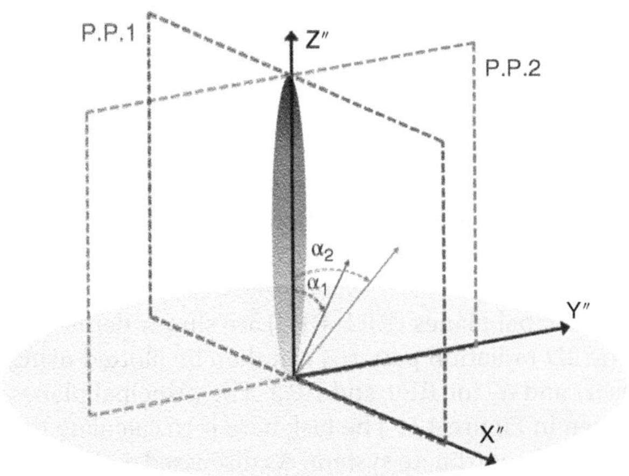

Figure 4.5 Principal planes of the reflectarray antenna.

$$\begin{bmatrix} x'' \\ y'' \\ z'' \end{bmatrix} = \begin{bmatrix} \cos\theta_m & 0 & -\sin\theta_m \\ 0 & 1 & 0 \\ \sin\theta_m & 0 & \cos\theta_m \end{bmatrix} \cdot \begin{bmatrix} \cos\varphi_m & \sin\varphi_m & 0 \\ -\sin\varphi_m & \cos\varphi_m & 0 \\ 0 & 0 & 1 \end{bmatrix} \cdot \begin{bmatrix} x \\ y \\ z \end{bmatrix}. \tag{4.27}$$

With this matrix transform a direct relation between the original and new coordinate systems can be obtained as

$$\begin{bmatrix} x'' \\ y'' \\ z'' \end{bmatrix} = \begin{bmatrix} \cos\theta_m\left(x\cos\varphi_m + y\sin\varphi_m\right) - z\sin\theta_m \\ -x\sin\varphi_m + y\cos\varphi_m \\ \sin\theta_m\left(x\cos\varphi_m + y\sin\varphi_m\right) + z\cos\theta_m \end{bmatrix}. \tag{4.28}$$

Since the radiation patterns are usually expressed in terms of the spherical or angular coordinates, we can use the formula

$$\begin{aligned} x &= u = \sin\theta\cos\varphi \\ y &= v = \sin\theta\sin\varphi \\ z &= w = \cos\theta \end{aligned} \tag{4.29}$$

to obtain the principal plane angles α_1 and α_2. From Figure 4.5 it can be seen that for P.P.1

$$\begin{aligned} x'' &= u'' = \sin\theta''\cos\varphi'' = \pm\sin\alpha_1 \\ y'' &= 0 \\ z &= w'' = \cos\theta'' = \cos\alpha_1. \end{aligned} \tag{4.30}$$

Similarly, for P.P.2

$$\begin{aligned} x'' &= 0 \\ y'' &= v'' = \sin\theta''\sin\varphi'' = \pm\sin\alpha_2 \\ z &= w'' = \cos\theta'' = \cos\alpha_2. \end{aligned} \tag{4.31}$$

Using (4.30) and (4.31) in (4.28) the spherical or angular coordinates corresponding to the principal plane angles can be calculated. For P.P.1 one obtains

$$\begin{aligned} u &= \cos\theta_m\cos\varphi_m\left(\cos\alpha_1\tan\theta_m \pm \sin\alpha_1\right) \\ v &= u\tan\varphi_m. \end{aligned} \tag{4.32}$$

Similarly, for P.P.2 one obtains

$$\begin{aligned} u &= \sin\theta_m\cos\varphi_m\left(\cos\alpha_2 \pm \sin\alpha_2\tan\varphi_m\sin\theta_m\right) \\ v &= u\tan\varphi_m \mp \frac{\sin\alpha_2}{\cos\varphi_m}. \end{aligned} \tag{4.33}$$

Equations (4.32) and (4.33) provide a complete description of the principal planes in terms of the angular coordinates (u,v). For any arbitrary beam direction P.P.1 is a straight line passing through the center, while P.P.2 is generally a curved line in the (u,v) plane. For example, the principal planes of a reflectarray antenna with a main beam in the direction of ($\theta_m = 26°$, $\varphi_m = 40°$) are given in Figure 4.6.

4.3.2 Co- and Cross-Polarized Patterns

In the previous sections (4.1 and 4.2), the radiation patterns of the reflectarray were obtained and expressed in terms of θ and φ components. However, in practice it is more common to use the co-polar and cross-polar components of the fields according to Ludwig's third definition [18].

For an x-polarized feed, the transformation from θ and φ to co-polar ($E_{co\text{-}pol}$) and cross-polar ($E_{x\text{-}pol}$) components are given by

$$\begin{bmatrix} E^x_{co-pol} \\ E^x_{x-pol} \end{bmatrix} = \begin{bmatrix} \cos\varphi & -\sin\varphi \\ -\sin\varphi & -\cos\varphi \end{bmatrix} \begin{bmatrix} E_\theta \\ E_\varphi \end{bmatrix}, \tag{4.34}$$

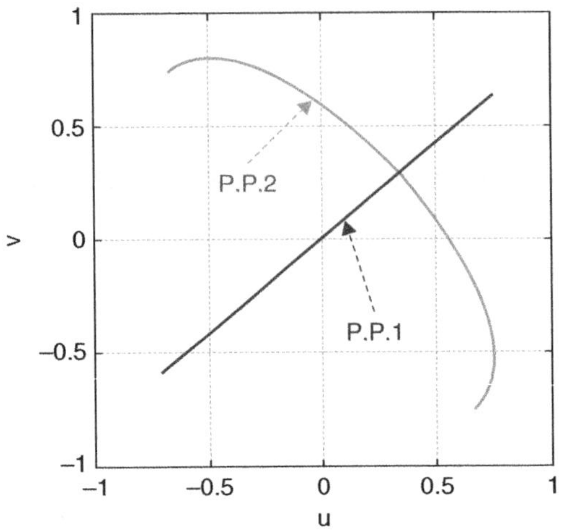

Figure 4.6 Principal planes of a reflectarray antenna in the angular coordinates with a main beam at the direction of ($\theta_m = 26°$, $\varphi_m = 40°$).

and for a y-polarized feed by

$$\begin{bmatrix} E^{y}_{co-pol} \\ E^{y}_{x-pol} \end{bmatrix} = \begin{bmatrix} \sin\varphi & \cos\varphi \\ \cos\varphi & -\sin\varphi \end{bmatrix} \begin{bmatrix} E_\theta \\ E_\varphi \end{bmatrix}. \tag{4.35}$$

For circular polarized feeds, the co- and cross-polar components are defined in terms of the sense of rotation where

$$E_{RHCP} = \frac{1}{\sqrt{2}}\left(E_\theta + jE_\varphi\right),\ E_{LHCP} = \frac{1}{\sqrt{2}}\left(E_\theta - jE_\varphi\right). \tag{4.36}$$

The equations provided in this section allow for an accurate calculation of the co- and cross-polar radiation patterns of the reflectarray antenna. This includes all possible sources of cross-polar radiation from the feed horn, the geometrical projections, and even the reflectarray phasing elements. For the latter case, the cross polarization of the phasing elements has to be obtained from the unit-cell simulations.

4.3.3 Antenna Directivity

Once the radiation pattern of the antenna is obtained from any of the methods described in the previous section, the antenna directivity can be obtained using

$$D_0 = \frac{4\pi\left|E\left(\theta_m, \varphi_m\right)\right|^2}{\int_0^{2\pi}\int_0^{\pi}\left|E(\theta,\varphi)\right|^2 \sin\theta\, d\theta\, d\varphi}. \tag{4.37}$$

Here, (θ_m, φ_m) is the direction of the main beam. The calculated directivity in (4.37) takes into account the illumination (taper efficiency) of the array, and the effect of projected aperture for off-broadside beams. Typically, the main challenge in directivity calculations is the evaluation of the denominator in (4.37), which is done numerically. It is important to point out here, that if the radiation patterns are computed in the angular coordinates (u, v) rather than the spherical coordinates (θ,φ), it will be necessary to use Jacobi transformation for the directivity calculations. The differential relation between angular and spherical coordinates is

$$u = \sin\theta\cos\varphi, \quad v = \sin\theta\sin\varphi$$
$$du = \frac{\partial u}{\partial\theta}d\theta + \frac{\partial u}{\partial\varphi}d\varphi, \; dv = \frac{\partial v}{\partial\theta}d\theta + \frac{\partial v}{\partial\varphi}d\varphi. \tag{4.38}$$

The matrix relation is then obtained as

$$\begin{bmatrix} d\theta \\ d\varphi \end{bmatrix} = \begin{bmatrix} \dfrac{\cos\varphi}{\cos\theta} & \dfrac{\sin\varphi}{\cos\theta} \\ -\dfrac{\sin\varphi}{\sin\theta} & \dfrac{\cos\varphi}{\sin\theta} \end{bmatrix} \begin{bmatrix} du \\ dv \end{bmatrix}. \tag{4.39}$$

The beam solid angle can then be expressed in terms of the angular coordinates as

$$d\Omega = \sin\theta\, d\theta\, d\varphi = \frac{1}{\cos\theta}\left(\cos\varphi du + \sin\varphi dv\right)\left(-\sin\varphi du + \cos\varphi dv\right) \tag{4.40}$$

From this equation, it can be observed that for the case of $\theta = 90°$ the beam solid angle shows a singular behavior. Two methods can be used to mitigate this problem where both approaches basically omit the singular points. The first method avoids the calculation of the integral at $\theta = 90°$. The second method removes this singularity in terms of the angular components. The beam solid angle expression in (4.40) is expressed completely in terms of the angular components, that is

$$\begin{aligned} d\Omega &= \frac{1}{\cos\theta}\left(\cos 2\varphi du dv + \frac{1}{2}\sin 2\varphi\left(dv^2 - du^2\right)\right) \\ &= \frac{1}{\sqrt{1-\left(u^2+v^2\right)}}\left(\frac{u^2-v^2}{u^2+v^2}du dv + \frac{uv}{u^2+v^2}\left(dv^2 - du^2\right)\right) \end{aligned} \tag{4.41}$$

In this method, the singularity is removed for the case where $u^2 + v^2 = 1$. It is worthwhile mentioning that since most reflectarrays are designed to generate a beam that is not far off from the broadside direction, the numerical singularity problem is in a region where radiated power is almost zero, hence has trivial effect on the value of the computed directivity.

4.3.4 Antenna Efficiency and Gain

Once the reflectarray directivity is calculated, the antenna gain can be obtained by taking into account the aperture efficiency. For an accurate calculation of the reflectarray antenna gain, it is imperative to determine the overall aperture efficiency of the antenna. Different kinds of efficiency factors are considered in reflectarray antennas, and in general these factors cause a deviation from ideal performance. Some of the most common factors are listed here:

- *Illumination efficiency* is a measure of the uniformity of the feed illumination, where we note that the maximum directivity of an aperture is obtained when the aperture is illuminated uniformly.
- *Spillover efficiency* is the percentage of radiated power from the feed that is intercepted by the reflecting aperture, as described in Chapter 3.
- *Phase efficiency* is a measure of the uniformity of the phase distribution on the reflectarray aperture. These include the pixelated (unit-cell) phase distribution on the reflectarray aperture as opposed to a smooth reflector aperture, quantization errors, and limited phase range of the reflectarray elements, and phase wraps on the reflectarray aperture.
- *Material loss* such as substrate and conductor loss in microstrip reflectarrays result in an overall reduction of the antenna gain and ultimately the reflectarray efficiency.
- *Polarization efficiency* is a measure of the power that is coupled into the cross-polarized radiated beam of the reflectarray antenna.
- *Blockage efficiency* is a measure of the blockage of aperture radiation by the feed assembly, which results in reduction of gain and efficiency in reflectarray antennas. As discussed earlier in Chapter 3, this loss can be completely avoided by using an offset system.
- *Feed efficiency* is a measure of the losses in the reflectarray feed system. For a horn antenna feed, these are typically Ohmic losses.

It is important to note that while all these efficiency factors are responsible for the total gain of the reflectarray antenna, as discussed earlier in Chapter 3, the major terms that influence the reflectarray antenna gain are the spillover and taper efficiency. The formulation presented in the previous section for calculating the radiation pattern accurately takes into account several of these factors. Namely illumination efficiency, phase efficiency, and polarization efficiency are already taken into account when one calculates the radiation pattern directivity. Blockage efficiency can also be included in the radiation pattern directivity [9]. The remainder of the efficiency factors, primarily dominated by spillover efficiency and material loss, influence the reflectarray antenna gain. Mathematically this is given as

$$G = \eta_g D. \tag{4.42}$$

Here D is the maximum directivity of the reflectarray antenna, which is defined as the maximum radiation intensity of the antenna over the average, and is calculated using (4.37), and η_g are the combined efficiency factors that are not included in the computed directivity.

The reflectarray antenna gain can also be computed from the far field radiation patterns while using the input power of the feed horn antenna as the reference. The antenna gain is defined as

$$G(\theta,\varphi) = 4\pi \frac{\text{intensity of the antenna in direction } (\theta,\varphi)}{\text{power incident at the antenna}} = 4\pi \frac{U(\theta,\varphi)}{P_{feed\ horn}} \tag{4.43}$$

The radiation intensity is defined as

$$U(\theta,\varphi)=\frac{1}{2}\mathrm{Re}\{E(\theta,\varphi)\times H^*(\theta,\varphi)\}.r^2\hat{r}. \tag{4.44}$$

The gain of the reflectarray antenna can then be evaluated as

$$G(\theta,\varphi)=4\pi\frac{|E(\theta,\varphi)|^2 r^2}{2\eta_0 P_{feed\ horn}}, \tag{4.45}$$

where η_0 is the intrinsic impedance of the free-space.

Assuming the losses in the feed horn antenna are negligible, the total power radiated by the feed horn antenna can be computed as

$$P_{feed\ horn}=\int_{\varphi=0}^{2\pi}\int_{\theta=0}^{\pi}\frac{|E^F(\theta,\varphi)|^2}{2\eta_0}r^2\sin\theta\, d\theta\, d\varphi. \tag{4.46}$$

In this equation, E^F is the radiation pattern of the horn antenna. For an x-polarized feed horn, the amplitude squared for the horn radiation pattern is

$$|E^{Fx}(\theta,\varphi)|^2=E^{Fx}(\theta,\varphi)(E^{Fx}(\theta,\varphi))^*=|A_0|^2(C^2{}_E(\theta)\cos^2\varphi+C^2{}_H(\theta)\sin^2\varphi). \tag{4.47}$$

Similarly, for a y-polarized feed horn, the amplitude squared for the horn radiation pattern is

$$|E^{Fy}(\theta,\varphi)|^2=E^{Fy}(\theta,\varphi)(E^{Fy}(\theta,\varphi))^*=|A_0|^2(C^2{}_E(\theta)\sin^2\varphi+C^2{}_H(\theta)\cos^2\varphi). \tag{4.48}$$

Assuming that $q_E=q_H=q$ in the cosine-q models used for the functions C_E and C_H, both can be expressed as

$$|E^{Fx}(\theta,\varphi)|^2=|E^{Fy}(\theta,\varphi)|^2=|A_0|^2\cos^{2q}\theta. \tag{4.49}$$

Substituting this in (4.46) and evaluating the integral, the total power radiated by a linearly polarized feed horn can be calculated as

$$P_{LP\ feed\ horn}=\frac{|A_0|^2\pi r^2}{\eta_0}\frac{2}{(2q+1)}. \tag{4.50}$$

Using (4.50) in (4.45), the gain of a linearly polarized reflectarray antenna is given as

$$G_{LPRA}(\theta,\varphi)=\frac{1}{|A_0|^2}(2q+1)|E(\theta,\varphi)|^2. \tag{4.51}$$

It is implicit that if the complex constant A_0 was taken into account in the calculation of the radiation patterns of the antenna, it will be cancelled out in the gain computations.

For a circularly polarized feed horn, the amplitude of the electric field power is given by

$$\left|E^{Fc}(\theta,\varphi)\right|^2 = E^{Fc}(\theta,\varphi)\left(E^{Fc}(\theta,\varphi)\right)^* = \frac{1}{2}|A_0|^2\left(C^2{}_E(\theta)+C^2{}_H(\theta)\right), \tag{4.52}$$

for both right- and left-hand polarizations. Considering that for an ideal circular polarized feed horn $q_E = q_H = q$, the total power radiated from the horn antenna can then be calculated as

$$P_{CP\,feed\,horn} = \frac{|A_0|^2\pi r^2}{\eta_0}\frac{2}{(2q+1)}. \tag{4.53}$$

Using (4.53) in (4.45), the gain of a circularly polarized reflectarray antenna is given as

$$G_{CPRA}(\theta,\varphi) = \frac{1}{|A_0|^2}(2q+1)|E(\theta,\varphi)|^2, \tag{4.54}$$

which as expected is equal to the gain of the linearly polarized antenna (4.51).

It is important to note that different methods for computation of reflectarray antenna gain are available in the literature and the interested reader is referred to [19] for a comparative study of these techniques. As a concluding remark, it should be noted that both formulas used for the gain calculations, that is (4.42) and (4.43), correctly take into account the losses due to aperture projection, illumination taper, and spillover. However, (4.43) is more general since in addition to the polarization, the effect of element losses on the antenna gain is also evaluated directly, given that they are included in the radiation pattern calculations.

4.3.5 Spectral Transforms and Computational Speedup

In the previous sections it was shown that both radiation analysis approaches, namely array theory and aperture field analysis, require evaluation of a double summation for far-field calculations [20]. This double summation can be replaced by a 2D inverse discrete Fourier transform (IDFT), defined as

$$f(p,q) = \frac{1}{N_x N_y}\sum_{m=0}^{N_x-1}\sum_{n=0}^{N_y-1} F(m,n)e^{j\frac{2mp\pi}{N_x}}e^{j\frac{2nq\pi}{N_y}}. \tag{4.55}$$

Here the spectral functions will be obtained in a discrete number of angular coordinates. These points in the (u,v) plane are defined by the Fourier transform as:

$$\begin{aligned} u &= \frac{2\pi}{N_x d_x k_0}p;\ \ p = 0,1,2,\ldots N_x - 1, \\ v &= \frac{2\pi}{N_y d_y k_0}q;\ \ q = 0,1,2,\ldots N_y - 1. \end{aligned} \tag{4.56}$$

The main advantage of using spectral functions in the calculations is a significant reduction of computational time, which is simply achieved by replacing the double summations with Fourier transforms. It should be noted that the definition of the propagating wave direction determines whether the discrete Fourier transform or IDFT is to be used for the transforms. In the formulation presented here, we followed the definition in [1], where the IDFT is used to replace the double summation in the spectral functions. It should be noted here that the condition which needs to be satisfied here is that the sets (m,n) and (p,q) must have a one on one relation. This condition can be satisfied by setting the number of far-field points that are calculated equal to the number of elements on the array. With these constraints defined, it is possible to replace the double summation with the 2D IDFT. By using the 2D inverse fast Fourier transform (2D-IFFT) algorithm available in Matlab [21], it is possible to calculate the radiation patterns of the reflectarrays very efficiently. Some important considerations about implementing the 2D-IFFT are listed next.

1) The 2D-IFFT routine described here cannot be implemented if the radiation pattern is being calculated in terms of the spherical coordinates (θ,φ). This is due to the fact that (θ,φ) don't satisfy the one on one condition with (m,n). However, it is trivial that once the pattern is calculated in the angular coordinates, it is possible to transform it to any other coordinate system.
2) The spectral functions are obtained in a discrete number of angular coordinates. It is possible to make this range symmetric by defining a shift in the variables. The range of (u,v) is typically defined as

$$-\frac{\lambda_0}{2d_x} \le u \le \frac{\lambda_0}{2d_x}, \quad -\frac{\lambda_0}{2d_y} \le v \le \frac{\lambda_0}{2d_y}, \tag{4.57}$$

It should be pointed out that if the periodic cell is $\lambda_o/2$, the radiation pattern will be computed in the whole visible range which is defined by the circle $u^2 + v^2 \le 1$. In addition, note that in this case the absolute value of the range of u and v will be smaller than 1, due to the $\frac{N_x-1}{N_x}, \frac{N_y-1}{N_y}$ factors in (4.56). For periodic cells smaller than $\lambda_o/2$, the radiation patterns will be computed in a larger range, while for periodic cells larger than $\lambda_o/2$, the radiation patterns will be computed in a smaller range [1].
3) The number of points in the (u,v) plane for radiation pattern calculations are equal to the number of elements on the array. For a higher resolution in pattern calculations, it is possible to increase the number of points by extending the grid on the reflectarray surface. This extended grid can be viewed as a virtual reflectarray surface where the amplitude of all the elements outside of the reflectarray is set to zero. With this definition of extended grid, it is possible to obtain high resolution radiation patterns for small size arrays.
4) The implementation of the 2D-IFFT routine in Matlab also requires some discussion. The FFT routine generally has only one zero frequency point. For the pattern calculations here this point is actually representing the center point in the (u,v) plane, which is (0,0). To satisfy this condition it is necessary to use an odd number of points for radiation pattern calculations. In addition, Matlab's 2D-IFFT command swaps the coordinates so that the zero-frequency point is located at the lower left corner.

After the 2D-IFFT command (**ifft2**) is executed, it is necessary to correct the obtained results by using the **fftshift** command, which corrects the quadrants so that the center of the (u,v) plane is placed in the center of the matrix.

4.4 Full-Wave Simulation Approaches

4.4.1 Constructed Aperture Currents Under Local-Periodicity Approximation

In this approach, the current distribution on the total reflectarray aperture is constructed from unit-cell analysis of individual elements under local-periodicity approximations. A common method for determining the currents is based on the spectral domain method of moments (SDMoM). Three different techniques have been proposed to calculate the radiation pattern of printed reflectarrays using this approach, which are briefly summarized here.

In the first technique, the radiation pattern is computed from the currents on the array elements with stationary phase evaluation of the spatial dyadic Green's function (DGF) assuming an infinite ground plane. Despite being slightly more accurate than the array theory and aperture field approaches, the primary drawback of this technique is that the finite size of the reflectarray aperture is not taken into account, and as such diffraction effects are not properly modeled with this approach.

The second technique computes the equivalent currents on a surface enclosing the entire reflectarray. The equivalent currents are determined through the Floquet space harmonics from the SDMoM formulation, where the contribution from each array element to the equivalent current is confined to its unit cell. To account for the finite physical size of the array, unit-cells with no array elements are typically placed at the edges. While the use of equivalent currents on a volume surrounding the reflectarray tends to be more accurate than the first technique, the approximation of the equivalent currents on the unit-cells using only the fundamental Floquet space harmonics, as well as the discontinuities in the equivalent currents due to truncations at the unit-cell borders, may result in some errors in computed radiation patterns.

The third technique combines the advantages of both approaches by using the equivalent currents approach with a continuous spectrum formulation. In this method, the equivalent currents are constructed on a surface enclosing the entire reflectarray, where the total field on the back side and edges of the reflectarray are assumed to be zero. The tangential fields at the plane of the array are then expanded in a spectrum of plane waves. Note that with this approach, the contribution from each array element over the entire surface is taken into account. This is due to the fact that the electric and magnetic fields at the reflectarray surface are related through the continuous plane wave spectrum and not through the plane wave relation of the fundamental Floquet harmonic as in the second technique. More discussions on the accuracy of these techniques, along with numerical and measured results will be given in the next section.

4.4.2 Complete Reflectarray Models

The analysis techniques presented thus far, analyze the radiation performance of the reflectarray antenna under local periodicity approximation. While this approximation is valid to a good degree, accurate analysis of the radiation pattern requires electromagnetic modeling of the full array. The challenge here is that the electrically large size of the reflectarray antenna aperture, combined with thousands of elements with dimensions smaller than a wavelength, demands a considerably high computational time and resources for full-wave analysis. Despite this drawback, the full-wave analysis approach is generally considered to be the most accurate technique for reflectarray analysis.

Many commercial electromagnetic simulation software can be used for analysis of reflectarray antennas. A challenge in geometry modeling of the reflectarray antenna is the variable dimension/rotation of the large number of elements. An efficient way to overcome this is to input the dimensions and geometrical locations of the elements into the software, by generating a data file containing this information in a format that is recognized by the software. The geometrical location of the reflectarray elements can best be specified by the center position of the unit-cell. Typically, for printed reflectarrays, only the x- and y-locations of each unit-cell center needs to be specified. As an example, for variable size patches, the coordinate information can be given in the first and second columns, while the next columns contain the width and length of the variable size patches.

A very common geometry file format for 2D structures is DXF. The overall file organization of a DXF file contains a header, classes, tables, blocks, entities, and objects sections, and can be used to generate almost any 2D geometry. Here we will briefly outline the format for generating DXF files for square patches, which are the most common type of reflectarray elements. First, the (x, y, z) coordinates of each vertex have to be computed and imported in the given format. While the computation of the vertex positions is straightforward, the vertex positioning has to follow the format order. The vertex positions have to be specified in a manner that creates a closed loop with a counter clockwise order. Namely, that is the top left, bottom left, bottom right, and top right vertices of the square. Using this approach, a full array containing hundreds of elements can be easily imported into any simulation software. However, note that these are geometry files, and once imported the electrical parameters, such as conductivities, have to be specified for full-wave simulations. An image of variable size square patches for a reflectarray antenna imported into Ansys HFSS is shown in Figure 4.7.

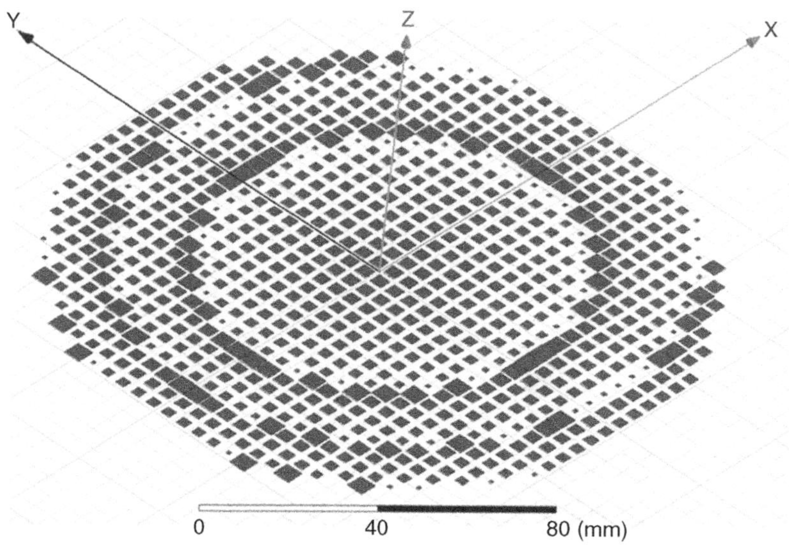

Figure 4.7 Geometry of reflectarray patch elements modeled in Ansys HFSS.

Once the element geometries have been modeled in the full-wave software, modeling the substrate and feed antenna is relatively straightforward.

Two different full-wave analyses have been proposed for reflectarrays that are referred to as transmit mode and receive mode analysis techniques. In the transmit mode analysis, the reflectarray aperture is illuminated by the feed antenna and the far-field radiation characteristics of the antenna is computed using the full-wave approach. In the receive mode analysis, the reflectarray is excited with a plane wave impinging on the aperture, and the field distribution at the focal region of the array (where the feed antenna phase center would be placed) is analyzed. Examples of full-wave reflectarray analysis techniques are presented in the next section.

4.5 Numerical Examples

4.5.1 Comparison of the Array Theory and Aperture Field Analysis Techniques

We consider Ka-band reflectarray designs with a circular aperture and a diameter of 17λ at the design frequency. The phasing elements, used in this study, are variable size square patches with a unit-cell periodicity of λ/2 at the design frequency of 32 GHz and are fabricated on a 20 mil Rogers 5880 substrate. The reflection phase response (S-curve) of the phasing elements obtained using the infinite array approach, is generated using Ansoft Designer, and is given in Figure 4.8.

Note that as discussed in Chapter 2, the reflection characteristics of the phasing elements are angle dependent; however, for this design, normal incidence can present good approximations for oblique incidence angles up to 30°. Thus, the reflectarrays here are designed based on the simulated reflection coefficients obtained with normal incidence. In addition. the element losses and cross-polarization are also negligible here, hence are ignored in the computations.

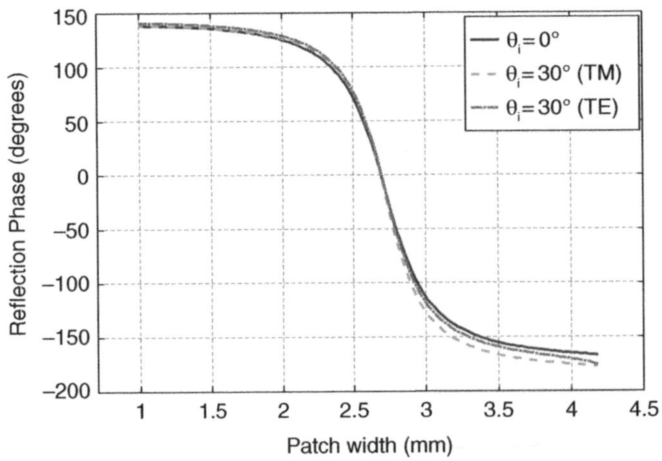

Figure 4.8 Reflection phase versus patch width for the reflectarray elements.

4.5.1.1 Example 1: Reflectarray Antenna with a Broadside Beam

We first consider an axisymmetric design. The reflectarray phasing elements are designed to generate a beam in the broadside direction. An x-polarized prime-focus feed horn is positioned with an F/D ratio of 0.735. For the horn model used in this study, the power q of the feed radiation pattern is 6.5 at 32 GHz. As discussed previously, in the array theory calculations, the polarization of the feed horn is not modeled. The aperture taper and the ideal phase requirement for the reflectarray elements are given in Figure 4.9.

For this system, the spillover and illumination efficiency are 93.01% and 81.98%, respectively. This corresponds to an aperture efficiency of 76.25%. In the next stage, the dimensions of the patch elements are selected from the S-curve in Figure 4.8 to match the required phase distribution on the aperture. The mask of the reflectarray antenna and the obtained reflection phase of the elements are given in Figure 4.10. It can be seen

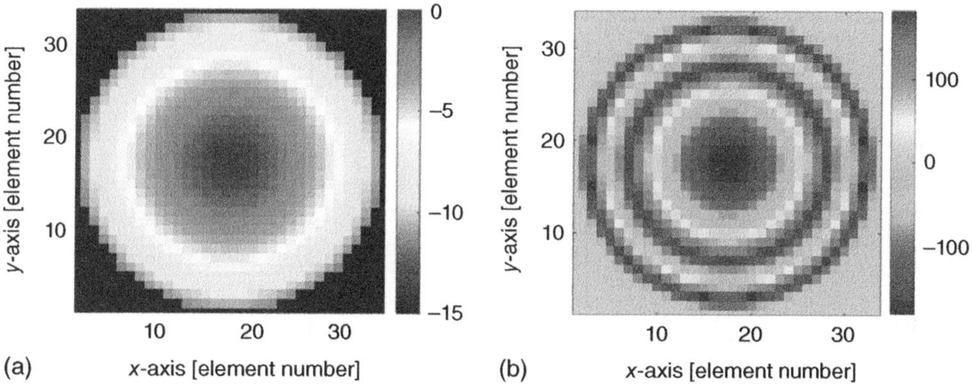

Figure 4.9 Reflectarray design with a broadside beam: (a) Aperture taper, (b) Phase requirement on the aperture.

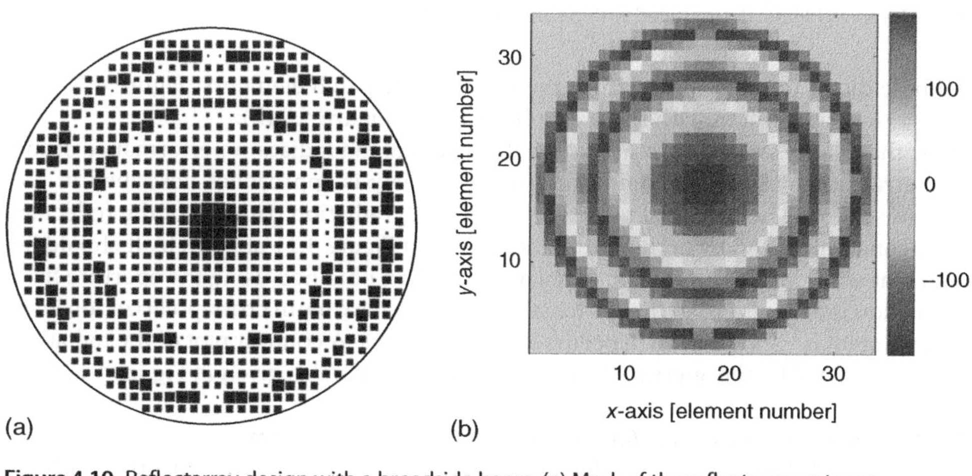

Figure 4.10 Reflectarray design with a broadside beam: (a) Mask of the reflectarray antenna, (b) Reflection phase of the elements on the reflectarray aperture.

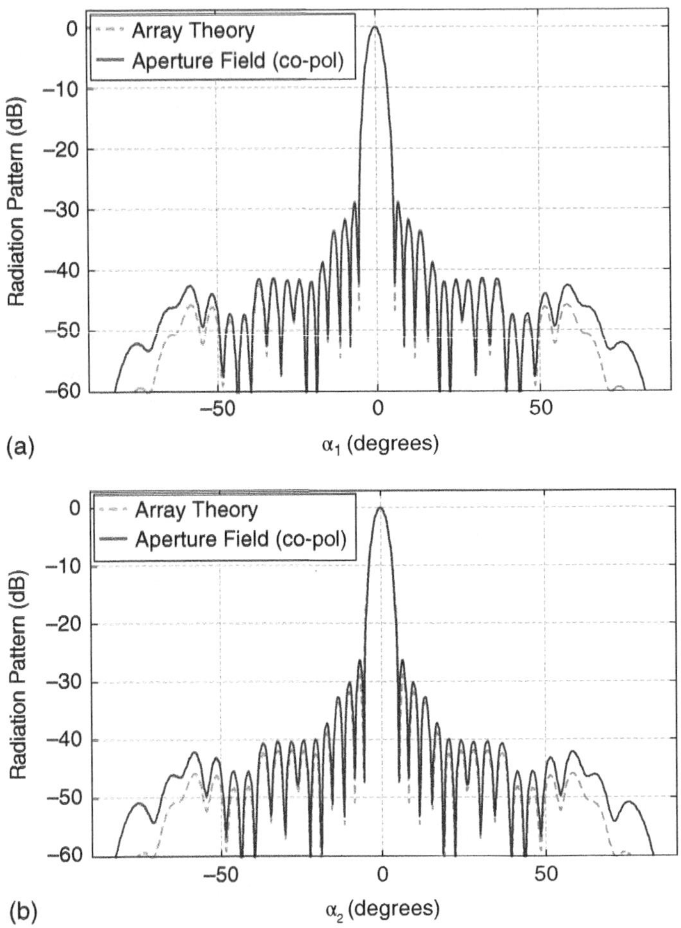

Figure 4.11 Radiation pattern of the reflectarray with a broadside beam: (a) *xz*-plane, (b) *yz*-plane. *Source:* Nayeri 2013 [20]. Reproduced with permission of IEEE.

the phase distribution obtained from the variable size patch elements (Figure 4.10 (b)) shows a close agreement with the ideal phase, that is Figure 4.9 (b).

Now that all the system parameters are determined, the radiation performance of the reflectarray antenna can be computed using the analysis approaches described in the previous section. The principal plane (P.P.1 and P.P.2) radiation patterns of the reflectarray antenna, calculated by both methods are given in Figure 4.11 at 32 GHz. It should be noted that with this design the cross-polarized pattern obtained using the aperture field formulation is almost zero in the principal planes. The maximum cross-polarization level for this system is −36.1 dB, which occurs in the 45° planes. A contour plot of the radiation patterns obtained by the aperture field analysis approach is also given in Figure 4.12.

4.5.1.2 Example 2: Reflectarray Antenna with an Off-Broadside Beam

For the second design, we consider an offset-fed reflectarray with an off-broadside beam. The phasing elements are designed to generate a beam in the direction of $(\theta, \varphi) = (25°, 0°)$. The offset feed horn position is $X_{feed} = -45.9\,\text{mm}$, $Y_{feed} = 0\,\text{mm}$,

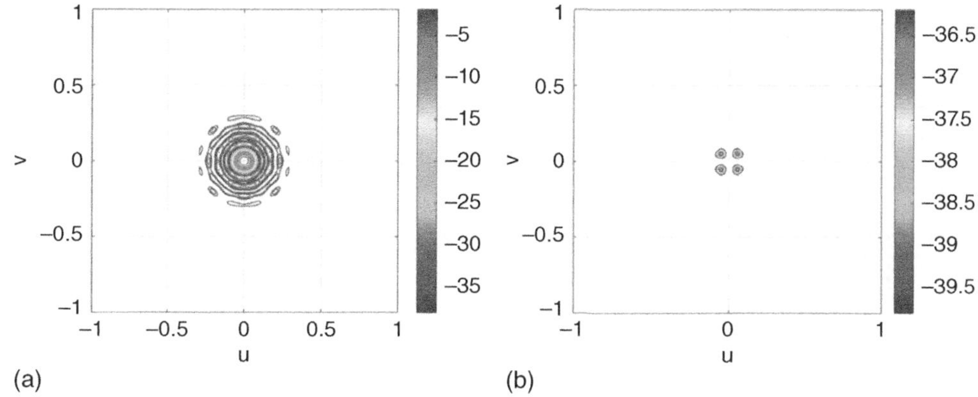

Figure 4.12 Radiation pattern of the reflectarray with a broadside beam: (a) co-pol, (b) cross-pol.

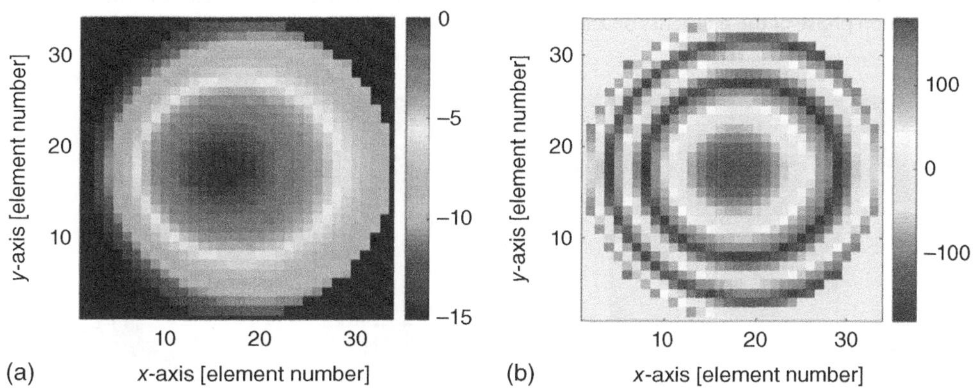

Figure 4.13 Reflectarray design with an off-broadside beam: (a) Aperture taper, (b) Phase requirement on the aperture.

Z_{feed} = 98.4 mm. The feed horn is LHCP and the power q of the feed radiation pattern is 6.5 at 32 GHz. The aperture taper and the ideal phase requirement for the reflectarray elements are given in Figure 4.13. For this system, the spillover and illumination efficiency are 92.19 and 77.92%, respectively. This corresponds to an aperture efficiency of 71.83%. Similar to the previous design, the dimensions of the patch elements can now be selected from the S-curve in Figure 4.8 to match the required phase distribution on the aperture. The mask of the reflectarray antenna and the obtained reflection phase of the elements are given in Figure 4.14. Similarly, it can be seen the phase distribution obtained from the variable size patch elements show a close agreement with the ideal phase, that is, Figure 4.13(b).

As discussed earlier, in the aperture field approach the polarization of the feed is modeled in the analysis. As such it would be interesting to observe the field distribution on the aperture for both incident and reflected fields. The incident electric fields on the reflectarray aperture are given in Figure 4.15. To generate a collimated beam, the reflected phase, after element phase compensation, should be flat along the y-axis and progressive in the direction of the scanned beam (x-axis). The phases of the

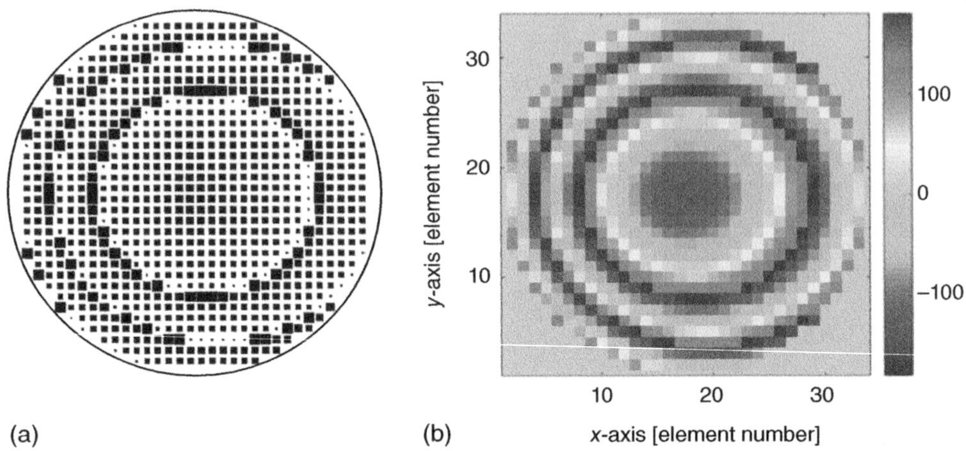

Figure 4.14 Reflectarray design with an off-broadside beam: (a) Mask of the reflectarray antenna, (b) Reflection phase of the elements on the reflectarray aperture.

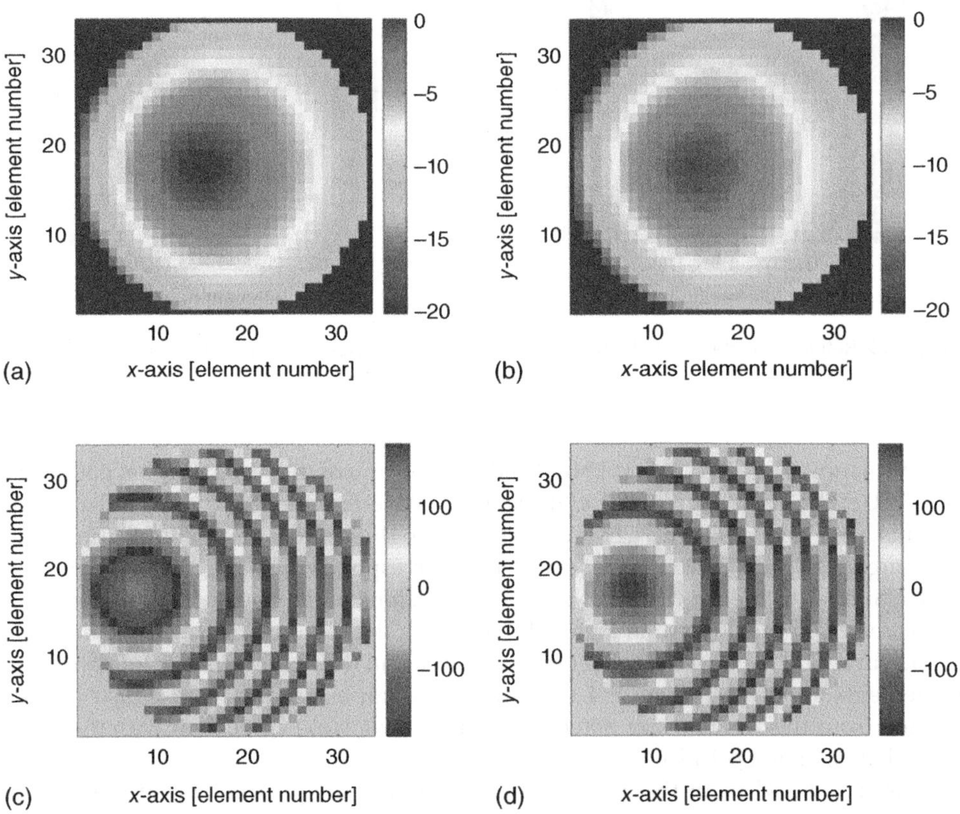

Figure 4.15 Incident electric fields on the reflectarray aperture: (a) $|E_x|$ (dB), (b) $|E_y|$ (dB), (c) phase of E_x, (d) phase of E_y.

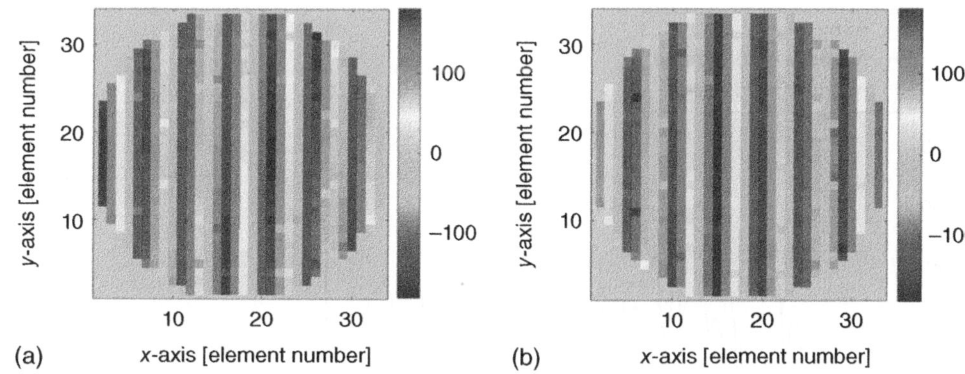

Figure 4.16 Reflected electric fields on the reflectarray aperture: (a) phase of E_x, (b) phase of E_y.

reflected electric fields are shown in Figure 4.16 where indeed such a phase distribution can be observed. It should be noted that the reflectarray elements used in the design here have a reflection coefficient magnitude close to unity, thus the magnitude of the reflected field remains almost the same as that of the incident field, and are not shown here for brevity.

The radiation patterns of this offset reflectarray antenna is also computed using both analysis approaches. The radiation patterns in the principal planes at 32 GHz are given in Figure 4.17, where again note that there is a very good agreement between both analysis techniques. For this offset system, the cross-polarized radiation pattern is also observed in both principal planes, where the maximum cross-polarization level is −30.0 dB.

4.5.1.3 Comparison of Calculated Directivity versus Frequency

From the results given in Figure 4.11 and Figure 4.17 for the two reflectarray systems, it can be seen that the calculated radiation pattern obtained by both methods are in close agreement with each other. In particular, the main beam direction, beamwidth, and general pattern shape are almost identical, however, a slight difference is observed in the side-lobe regions.

The antenna directivity is a suitable measure to compare the calculated radiation performance of these methods. In order to accurately model the reflectarray directivity versus frequency, the frequency behavior of the feed horn pattern and the element reflection characteristics are implemented into this calculation routine. For the phasing elements, the frequency behavior of the reflection phase is obtained across the band from full-wave simulations. For the horn model used in this study, the power q of the feed radiation pattern varies almost linearly from 5 at 30 GHz to 8.3 at 34 GHz according to the measured data.

For the two reflectarray systems studied here, the directivity versus frequency is given in Figure 4.18. It can be seen that the computed directivity versus frequency obtained by both methods also shows a close agreement. It should be noted here that, as expected, the off-broadside system shows a lower directivity. At the center frequency of 32 GHz, the difference in computed directivity is less than 0.1 dB for both designs. It is worth pointing out that a possible reason for the slightly lower directivity obtained with the

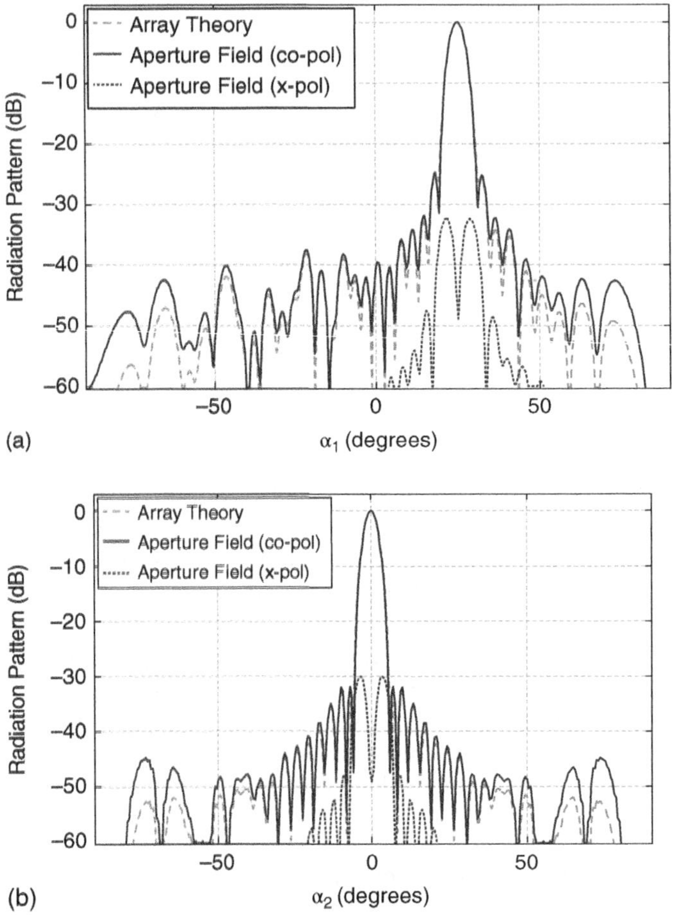

Figure 4.17 Radiation pattern of the reflectarray with an off-broadside beam: (a) P.P.1, (b) P.P.2. *Source:* Nayeri 2013 [20]. Reproduced with permission of IEEE.

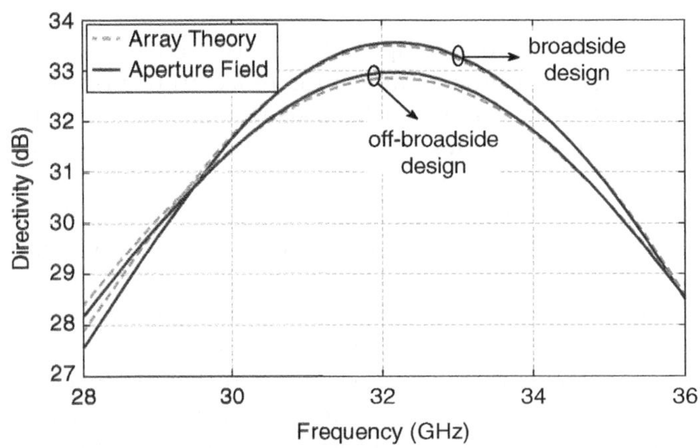

Figure 4.18 Directivity versus frequency for the two reflectarray systems. *Source:* Nayeri 2013 [20]. Reproduced with permission of IEEE.

array theory approach is the effect of the element receive mode pattern. In the aperture field method, the reflection coefficients of the elements are obtained under normal incidence approximation. In other word, an omni-directional element receive pattern is assumed in the aperture field analysis.

4.5.2 Consideration in the Array Theory Technique: Element Pattern Effect

In the array theory formulation, the radiation pattern of the reflectarray antenna is computed using (4.7), which takes into account the element receive (R_x) and transmit (T_x) patterns. As discussed earlier, the elements are modeled with a cosine q_e radiation pattern, where in many cases $q_e = 1$. Meanwhile an isotropic element pattern is also used for the elements of large reflectarrays. Therefore, it is interesting to observe the effects of element pattern on the radiation pattern of the reflectarray antenna.

To study this effect, we consider the same off-broadside system studied in the previous section. The radiation pattern in the principal plane is given in Figure 4.19 for four different cases. From these results, it can be seen that although the radiation patterns show a close agreement around the area of the main beam, a noticeable difference in the pattern shape is observed in the off-main beam areas. In particular, the element transmit pattern plays an important role here, where in comparison almost 10 dB difference is observed in the minor lobes. Similar results were observed in the other principle plane. In addition, note that while the difference is quite small, when the element transmit mode pattern is ignored, the antenna beam width is slightly wider, which would also correspond to a lower computed directivity. This study demonstrates the importance of the element pattern shape in the radiation analysis of reflectarray antennas. While for the square type patch elements studied here the element pattern was a simple cosine model, accurately determining the element pattern shape is essential before one computes the radiation pattern of a reflectarray antenna.

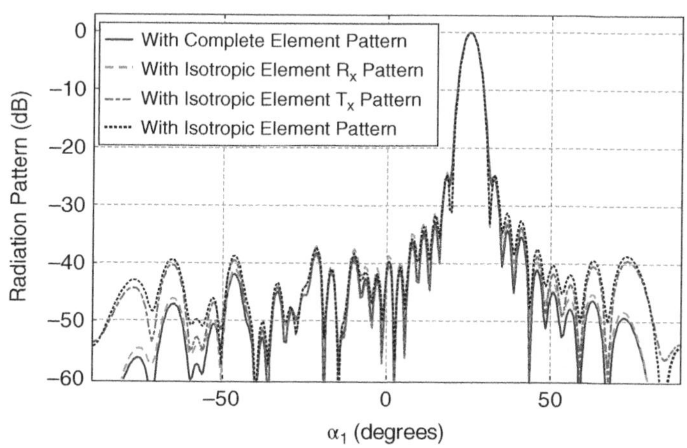

Figure 4.19 Radiation pattern of the reflectarray antenna and the effect of element pattern in P.P.1.

4.5.3 Consideration in the Aperture Field Technique: Variations of Equivalence Principle

The aperture field approach is based on the principle of equivalence. It was discussed earlier that the radiation pattern of the antenna can be obtained using any of the three variations of equivalence principle, that is, both electric and magnetic currents, only electric currents, and only magnetic currents. To study this numerically, we consider the same off-broadside reflectarray system. The radiation pattern in the principal plane is given in Figure 4.20 for the three cases. Similar results were observed in the other principle plane. These results show that as expected the co-polarized radiation pattern obtained by these three methods are almost identical, hence either approach can be applied. On the other hand, some difference is observed in the cross-polarized radiation patterns. In general, using both electric and magnetic fields on the aperture yields a better representation of the problem, and as such it is expected that using the first equivalence of principle is the most accurate analysis approach. Comparison with full-wave simulations will be presented in the next section, which further validates this discussion.

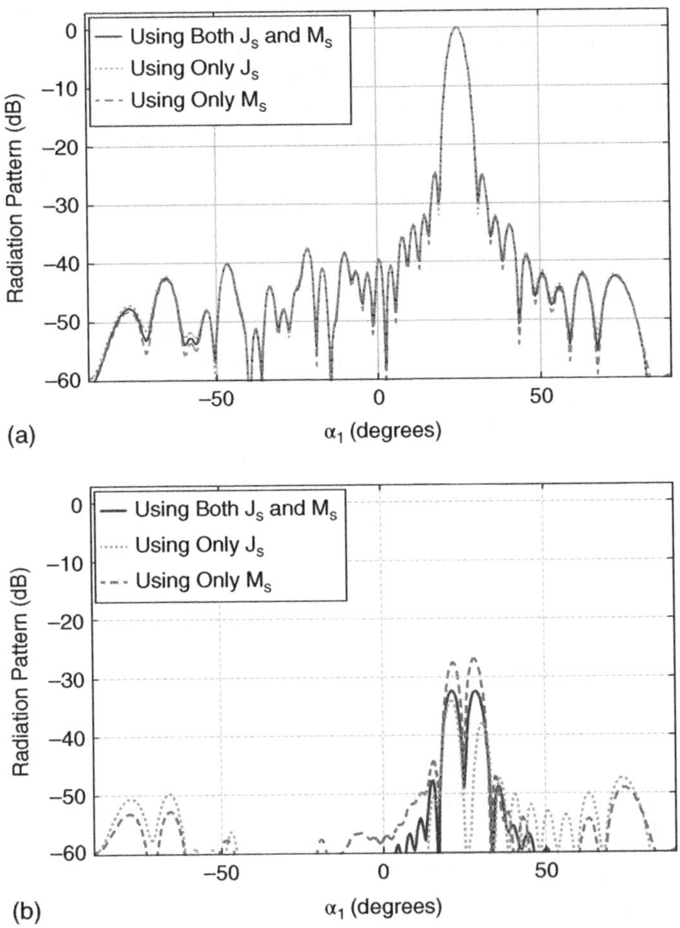

Figure 4.20 Radiation pattern of the reflectarray antenna and the effect of aperture currents in equivalence principle: (a) P.P.1, (b) P.P.1 zoomed view.

4.5.4 Comparisons with Full-Wave Technique

As discussed earlier, full-wave techniques provide the most accurate solution to the reflectarray radiation problem. In this section, we compare the accuracy of the reflectarray radiation pattern computed using earlier approaches by comparing them with full-wave simulations. Note that for this comparative study, full-wave simulations will be more advantageous than measured results, since measurement results are susceptible to both fabrication and measurement errors.

Due to the limited computational resources, we consider a smaller Ka-band reflectarray with a circular aperture and a diameter of 14.5λ at the design frequency of 32 GHz. The feed is positioned at $X_{feed} = -45.90$ mm, $Y_{feed} = 0$ mm, $Z_{feed} = 98.44$ mm based on the coordinate system. The elements phases are designed to generate a beam in the direction of $(\theta, \varphi) = (25°, 0°)$. Similarly, for the reflectarray phasing elements, the variable size square patches are selected from the S-curve data in Figure 2.23. The 609-element reflectarray antenna is modeled using the commercial electromagnetic software FEKO. For the excitation of the reflectarray, a point source feed model with a $\cos^{6.5}\theta$ radiation pattern is used. The advantage of using a point source rather than a feed horn here is that a point source model does not have a blockage aperture, which makes it more suitable for comparison purpose, since blockage is typically not modeled in the classical methods. For this design 568,435 unknown basis functions have to be calculated for the FEKO method of moments (MoM) solution. Considering the large number of unknowns, the multilevel fast multi-pole method (MLFMM) solver in FEKO was selected for this simulation. The geometry of the reflectarray antenna modeled in FEKO and the simulated 3D radiation pattern are shown in Figure 4.21. The full-wave simulation here take into account all approximations in reflectarray element design and mutual

(a) (b)

Figure 4.21 Reflectarray antenna simulated using FEKO, (a) top view of the reflectarray, (b) 3D radiation pattern of the reflectarray. *Source:* Nayeri 2013 [20]. Reproduced with permission of IEEE.

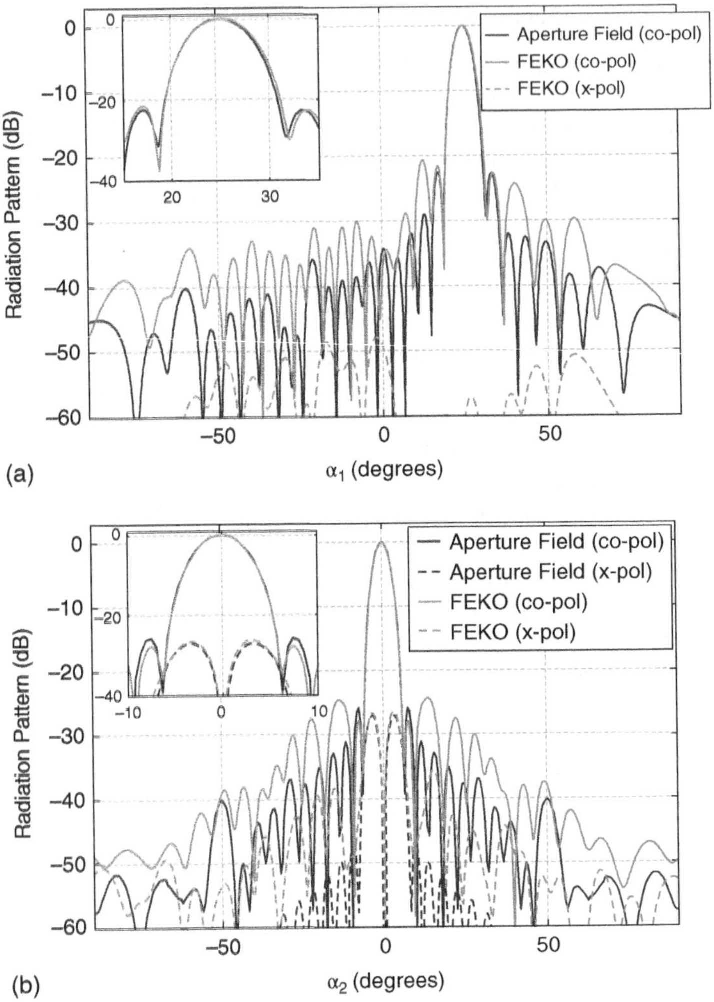

Figure 4.22 Radiation patterns of reflectarray antenna: (a) P.P.1, (b) P.P.2. *Source:* Nayeri 2013 [20]. Reproduced with permission of IEEE.

coupling, as well as the edge diffraction effects. Therefore, comparing these simulation results with the results obtained using classical approaches can provide a good measure in terms of the accuracy of these approaches.

The principal plane (P.P.1 and P.P.2) radiation patterns of the reflectarray antenna calculated using the aperture field method and the full-wave simulation are given in Figure 4.22. The computed directivity and gain of the reflectarray antenna using the aperture field approach is 30.48 dB and 29.68 dB, respectively. To accurately model the cross-polarization of the reflectarray antenna in the aperture field analysis, the incident fields on the reflectarray aperture obtained from the full-wave simulations were used in these calculations. It should be noted that the calculated co-polarized radiation pattern of the reflectarray antenna obtained using the array theory approach was similar to the aperture field results, and is not shown in this section for the sake of brevity.

Comparison of the results given here show that the aperture field analysis approach accurately calculate the general pattern shape, main beam direction, beam-width, and the side-lobe and cross-polarization level in the main beam area. Outside the main beam area, however, some discrepancies are observed between these results. These are primarily due to element design approximations and edge diffraction effects that are not taken into account in the analysis approaches.

In particular, comparison between the ideal phase shift and the phase shift obtained by the reflectarray elements, Figure 4.23, indicates that while the reflectarray aperture does indeed generate a phase shift to collimate the beam, this difference in phase shift is the primary reason for the discrepancies in the radiation pattern. To better observe the difference between the ideal phase shift and the phase shift obtained by the reflectarray elements, the phase distribution along the x-axis ($y = 0$) is also given in Figure 4.24.

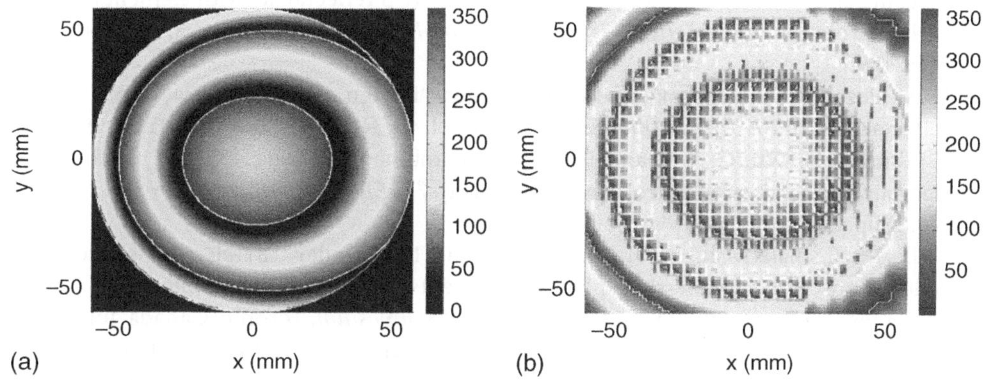

Figure 4.23 Phase shift on the reflectarray aperture, (a) ideal phase shift, (b) phase shift obtained by the reflectarray elements. *Source:* Nayeri 2013 [20]. Reproduced with permission of IEEE.

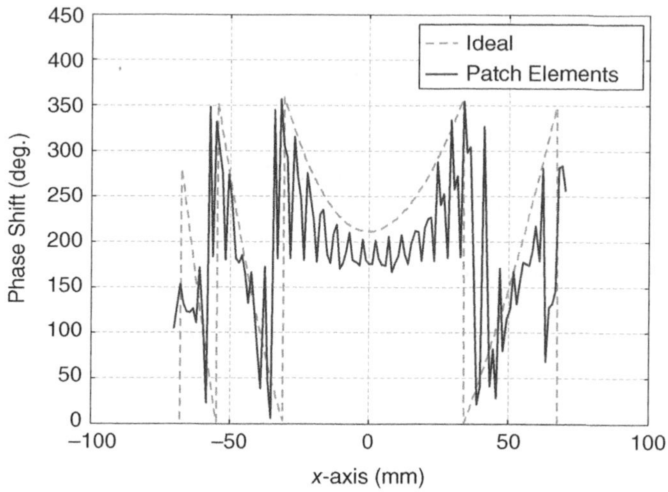

Figure 4.24 The progressive phase shift on the reflectarray aperture.

In conclusion, while it is implicit that a full-wave simulation will provide accurate results, the main disadvantage of full-wave simulations is the high computational time and resources required. In total, the full-wave simulation here required 29.56 GB of memory with a CPU time of 26.97 hours on an 8 core 2.66 GHz Intel(R) Xeon(R) E5430 computer. In comparison, the CPU time for the array theory and aperture field calculations were 1.78 s and 46.4 s on the same computer using a single core, respectively. It is worth pointing out that, in general, the classical approaches have limited accuracy and when accurate radiation pattern computation is required, full-wave simulation of the reflectarray system is necessary.

References

1 J. Huang and J. A. Encinar, *Reflectarray Antennas*. New York, NY, USA: Wiley-IEEE, 2008.
2 J. Huang and R. J. Pogorzelski, "A Ka-band microstrip reflectarray with elements having variable rotation angles," *IEEE Trans. Antennas Propag.*, Vol. 46, No. 5, pp. 650–656, May. 1998.
3 D. M. Pozar, S. D. Targonski, and H. D. Syrigos, "Design of millimeter wave microstrip reflectarrays," *IEEE Trans. Antennas Propag.*, Vol. 45, No. 2, pp. 287–296, Feb. 1997.
4 Y. Zhuang, K. Wu, C. Wu, and J. Litva, "A combined full-wave CG-FFT method for rigorous analysis of large microstrip antenna arrays," *IEEE Trans. Antennas Propag.*, Vol. 44, No. 1, pp. 102–109, Jan. 1996.
5 Y. Li, M. E. Bialkowski, K. H. Sayidmarie, and N.V. Shuley, "81-element single-layer reflectarray with double-ring phasing elements for wideband applications," *IEEE Antennas and Propagation Society International Symposium*, Toronto, Canada, July 2010.
6 H. Rajagopalan, S. Xu, and Y. Rahmat-Samii, "Graphical visualization of reflectarray reflection phase response," *IEEE Antennas and Propagation Society International Symposium*, Washington, USA, July 2011.
7 J. A. Encinar and J. A. Zornoza, "Three-layer printed reflectarrays for contoured beam space applications," *IEEE Trans. Antennas Propag.*, Vol. 52, No. 5, pp. 1138–1148, May 2004.
8 P. Nayeri, F. Yang, and A. Z. Elsherbeni, "Single-feed multi-beam reflectarray antennas," *IEEE Antennas and Propagation Society International Symposium*, Toronto, Canada, July 2010.
9 Y. Rahmat-Samii, "Reflector Antennas," in *Antenna Handbook: Theory, Applications, and Design*, Y. T. Lo and S. W. Lee, Van Nostrand Reinhold, 1988.
10 C. J. Sletten, *Reflector and Lens Antennas: Analysis and Design Using Personal Computers*, Artech House Inc., 1988.
11 M. E. Cooley, and D. Davis, "Reflector antennas," in *Radar Handbook*, 3rd Edn, McGraw-Hill, 2008.
12 A. D. Olver, P. J. B. Clarricoats, A. A. Kishk, and L. Shafai, *Microwave Horns and Feeds*, The Institution of Electrical Engineers, 1994.
13 Y. Rahmat-Samii, "Useful coordinate transformations for antenna applications," *IEEE Trans. Antennas Propag.*, pp. 571–574, July 1979.
14 L. Diaz and T. Milligan, *Antenna Engineering Using Physical Optics*, Norwood, MA: Artech House, 1996.

15 W. L. Stutzman and G. A. Thiele, *Antenna Theory and Design*, 3rd Edn, Hoboken, NJ: John Wiley & Sons Inc., 2012.

16 C. A. Balanis, *Antenna Theory: Analysis and Design*, 3rd Edn, Hoboken, NJ: John Wiley & Sons Inc., 2005.

17 A. Yu, "Microstrip reflectarray antennas: Modeling, design and measurement," Ph.D. dissertation, Dept. Elect. Eng., University of Mississippi, Oxford, MS, 2010.

18 A. Ludwig, "The definition of cross polarization," *IEEE Trans. Antennas Propag.*, Vol. 21, No. 1, pp. 116–119, Jan. 1973.

19 B. Devireddy, A. Yu, F. Yang, and A. Z. Elsherbeni, "Gain and bandwidth limitations of reflectarrays," *ACES Journal*, Vol. 26, No. 2, pp. 170–178, Feb. 2011.

20 P. Nayeri, A. Z. Elsherbeni, and F. Yang, "Radiation analysis approaches for reflectarray antennas," *IEEE Antennas Propag. Magazine*, Vol. 55, No. 1, pp. 127–134, Feb. 2013.

21 Matlab, The Mathworks Inc., Natick, MA.

5

Bandwidth of Reflectarray Antennas

As discussed in the earlier chapters, the reflectarray combines the numerous advantages of both antenna arrays and reflectors and creates a low-profile, low-mass, and low-cost antenna with a high gain. Despite the numerous advantages this hybrid configuration can offer, the reflectarray antenna suffers from a major limitation: an inherently narrow bandwidth. The bandwidth of reflectarray antennas depends on its element, aperture size, focal length, and so on, and typically does not exceed beyond 10%. In this chapter, we will study the fundamental physics that determine the bandwidth of reflectarray antennas.

5.1 Bandwidth Constraints in Reflectarray Antennas

5.1.1 Frequency Behavior of Element Phase Error

While reflectarrays mimic the traditional reflector antennas [1], there is an inherent fundamental difference between the two when it comes to system bandwidth [2]–[5]. A reflector antenna is essentially frequency-independent, that is, it has an infinite bandwidth, and in practice the primary feed antenna is responsible for limiting the bandwidth of the reflector. On the other hand, a reflectarray suffers from bandwidth limitations due to the physical nature of its phasing elements and also the flat nature of its aperture. Both factors introduce phase errors as a function of frequency, which limit the bandwidth of reflectarray antennas.

For microstrip reflectarrays, the element itself has a bandwidth of typically only a few percent, ultimately resulting in a narrow bandwidth for the reflectarray system. Techniques such as using thick substrates, stacked patches, or multi-resonant elements can improve the bandwidth of the elements and the reflectarray antenna, and bandwidths in excess of 20% have been reported.

Regardless of the choice of phasing elements used for the design, reflectarray elements are usually selected to satisfy the required phase at the design frequency. As the frequency changes, the required phase in the array will change, and the reflectarray element phase will also change. If these two phase changes do not match with each other, which is usually the case, it will introduce phase errors in the array, resulting in pattern deterioration, gain reduction, and ultimately, bandwidth limitations of the antenna.

Reflectarray Antennas: Theory, Designs, and Applications, First Edition.
Payam Nayeri, Fan Yang, and Atef Z. Elsherbeni.

It's worthwhile pointing out that the phase difference between reflectarray elements is important here rather than the reflection phase of an individual element, since in a reflectarray design, the phase difference between the elements is important for obtaining the desired radiation pattern. Consequently, defining an element bandwidth for reflectarray phasing elements is not practical without considering the relative phase requirements of all the reflectarray elements. It is, however, possible to study the phase error by considering the frequency behavior of the reflectarray elements for certain relative phase requirements.

Without loss of generality, we first consider a reflectarray with a main beam at the broadside direction ($\bar{r}_i.\hat{r}_o = 0$). The geometrical center of the reflectarray surface (point 1) is selected as a phase reference, and the center point of an arbitrarily element (point 2) is selected on the reflectarray surface to explain the phase relation. The phase difference between these two points at the center frequency (f_0) satisfies the following relation

$$\psi_2(f_0)-\psi_1(f_0)=k_0\cdot(R_2-R_1). \tag{5.1}$$

As the frequency changes, the phase difference will also change. An ideal phase relation is

$$\psi_2(f)-\psi_1(f)=k\cdot(R_2-R_1). \tag{5.2}$$

Combining (5.1) with (5.2), one can obtain

$$\frac{\psi_2(f)-\psi_1(f)}{f}=\frac{\psi_2(f_0)-\psi_1(f_0)}{f_0}=\frac{2\pi}{c}\times(R_2-R_1). \tag{5.3}$$

It is clear from these equations that the phase difference ($\Delta\psi(f)=\psi_2(f)-\psi_1(f)$) should be a function of frequency. When the frequency increases, the phase difference should also increase.

In practice, this ideal frequency behavior of the phase difference cannot be satisfied, and phase errors will occur with frequency, which reduces the reflectarray gain and narrows the reflectarray bandwidth. To study this effect quantitatively, a phase error term is defined as follows:

$$PhaseError(f)=\{\psi_2(f)-\psi_1(f)\}-k\cdot(R_2-R_1). \tag{5.4}$$

In this equation, the first term represents the frequency behavior of the elements, while the second term is the required spatial phase delay. In total, this equation represents the phase error of elements relative to a certain phase requirement. Thus, it provides a good measure to compare different element designs: the smaller the phase error, the wider the reflectarray bandwidth. It should be pointed out that in this phase error formula, the quantized phase of the reflectarray elements will result in a non-zero phase error at the center frequency (f_0), however, quantization phase error is usually small in comparison.

To quantitatively observe the phase errors introduced on the aperture of a reflectarray antenna, we consider a case where two elements on the aperture are required to have a 90° relative phase difference at the center design frequency of 32 GHz:

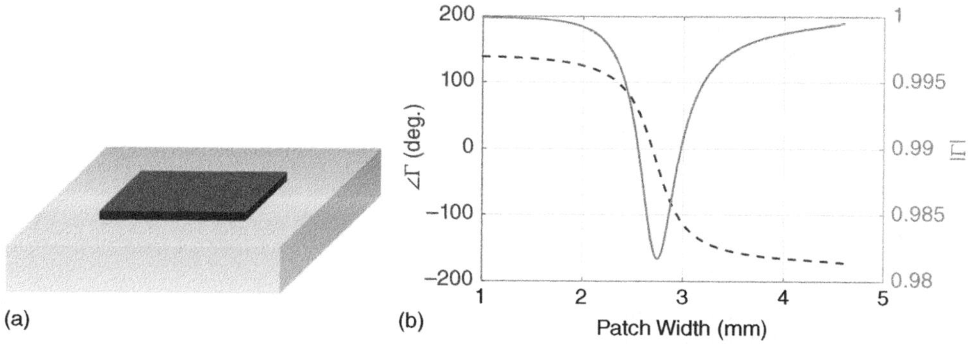

Figure 5.1 A Ka-band reflectarray variable-size square patch element: (a) geometry, (b) reflection coefficients at 32 GHz.

$$\psi_2(f_0)-\psi_1(f_0)=k_0\cdot(R_2-R_1)=90°. \tag{5.5}$$

The element chosen for this study is a variable-size square patch, which was studied in detail in Chapter 2. The patch is fabricated on a 20 mil Rogers Duroid 5880 laminate where the unit-cell size is $4.7\times 4.7\text{mm}^2$. The reflection coefficients as a function of patch width are given in Figure 5.1. Using this element data, for a 0° reflection phase, the patch width should be 2.69 mm; for a 90° reflection phase, it corresponds to a patch size of 2.41 mm. Figure 5.2(a) shows how the reflection phases of these 0° and 90° elements vary with frequency. As discussed previously, it is the frequency behavior of the phase difference that determines the reflectarray bandwidth. Thus, the phase error curve, as defined in (5.4), is also given in Figure 5.2(b). At the center frequency 32 GHz, the phase error is zero. As the frequency deviates from the center frequency, the phase error increases. At 30 GHz and 34 GHz, the phase error is −41.02 and 35.21°, respectively.

5.1.2 Frequency Behavior of Spatial Phase Delay

The bandwidth limitation imposed by the radiating element is most significant for small or moderate size reflectarrays, and bandwidth of large reflectarrays is primarily governed by a second factor, namely the differential spatial delay. For the space-fed system of the reflectarray antenna, the different path lengths from the feed phase center to each element on the aperture result in differential spatial delays for the elements. In order to illustrate this, we look at the cross-sectional view of a generic front-fed axisymmetric reflectarray antenna as shown in Figure 5.3. The rays emanating from the phase center of the feed antenna expand spherically. We show two rays traveling the paths R_1 and R_2, where the former travels the distance from the feed phase center to the reflectarray aperture center, and the latter travels from the same starting point to the reflectarray edge. ΔR is the path difference between the two rays. Depending on the system setup, ΔR can be several wavelengths at the center design frequency. Mathematically this can be written as

$$\Delta R=(N+n)\lambda, \tag{5.6}$$

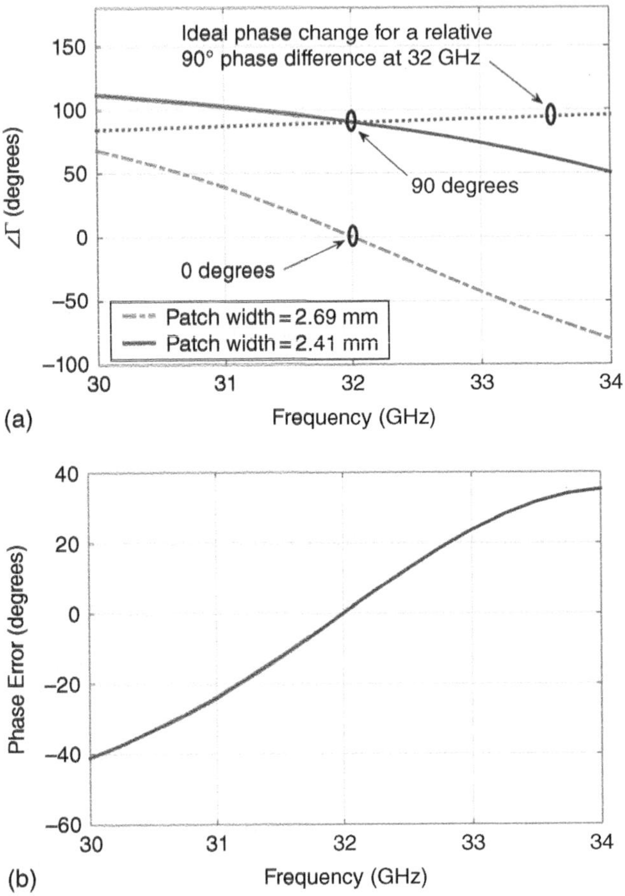

Figure 5.2 (a) Reflection phase as a function of frequency for two patch widths. (b) Phase error as a function of frequency.

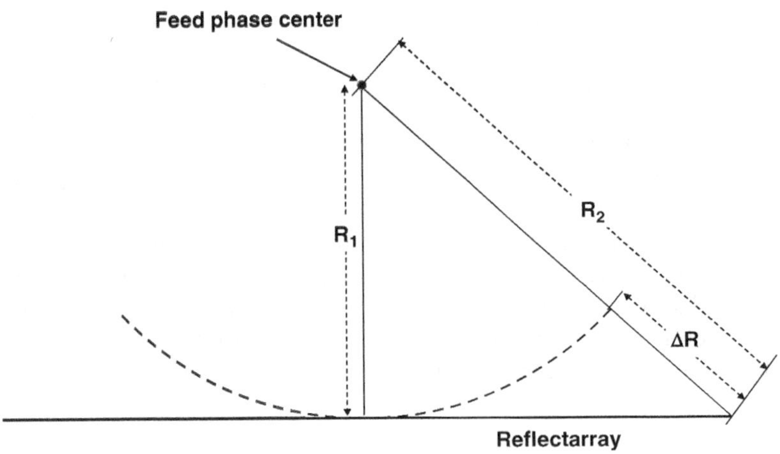

Figure 5.3 Cross sectional view of a reflectarray system and path delays.

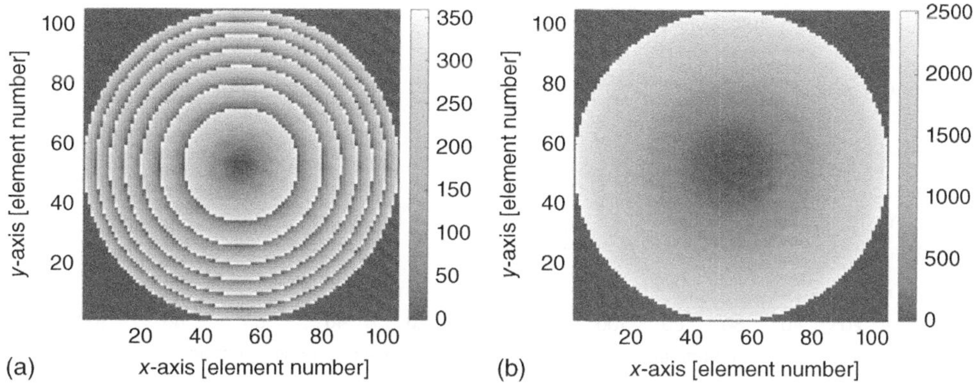

Figure 5.4 Phase distribution on a reflectarray antenna aperture in degrees: (a) truncated to one phase cycle, (b) no truncation.

where λ is the wavelength at the center frequency, N is an integer, and n is a fractional number.

In a reflectarray antenna, typically the phasing element is designed to compensate for the $n\lambda$ term, by providing a phase shift between 0 and 2π. To further illustrate this, we consider an axisymmetric reflectarray with a circular aperture and a diameter of 50 wavelengths. The feed is placed at a distance of 40 wavelengths from the center of the aperture and is pointing to the geometrical center. The phase distributions on the aperture are given for two cases in Figure 5.4. In the first case, the phases are truncated to provide a phase shift only between 0 and 2π. In the second case, the true physical difference is considered to determine the phase without wrapping, where it can be seen that this corresponds to several phase cycles for the majority of elements.

To understand the frequency behavior of the spatial phase delay, we consider an off-center frequency with the wavenumber given as

$$k_1 = \frac{2\pi}{\lambda_0 + \Delta\lambda}. \tag{5.7}$$

When the 2π phase truncation was implemented, as shown in Figure 5.4(a), the phase compensation was designed for each element at the center design frequency. This, however, results in a phase error term at off-center frequencies that is mathematically given by

$$Phase\ Error = \left(k_1 - k_0\right)\Delta R. \tag{5.8}$$

Depending on the design, this phase error could be very large portions of a full 2π phase cycle, and is the primary bandwidth limitation factor for large reflectarray antennas or moderate size reflectarrays with small f/D ratios. For the 50 wavelength reflectarray studied here the phase wraps on the aperture and the phase error at an off-center frequency of $\Delta\lambda = -0.1\lambda$ is given in Figure 5.5(a) and (b), respectively. Note that for this design, six complete phase wraps are observed on the reflectarray aperture. It can be seen that as a result of frequency change a phase error is generated on the aperture which moves outward from the center of the reflectarray aperture. In other

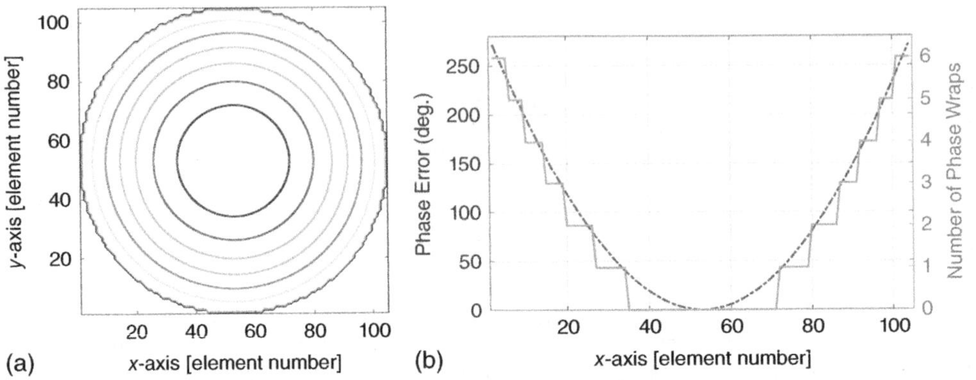

Figure 5.5 (a) Phase wraps on the aperture of the reflectarray. (b) Phase errors on the aperture of the reflectarray along the center *x*-axis at an off-center frequency with $\Delta\lambda = -0.1\lambda_0$. A 180° out of phase is observed when $N = 4$.

words, the maximum phase error will be observed at the edges of the aperture. The phase error plot clearly shows that as frequency deviates further from the center design frequency, the phase errors on the aperture increase significantly. Further, note that toward the edges of the reflectarray aperture the phase error approaches 180°, in other words the elements are completely out of phase which results in severe degradation of the array performance.

In order to reduce the errors associated with this differential spatial delay for flat reflectarrays, the integer N has to be reduced. Ideally, the error could be completely removed if true-time-delay lines can be used; that is, no phase wrap. However, as seen earlier, in a reflectarray this could correspond to several phase cycles, making it impractical. A more practical solution is to use elements that can provide more than a 2π cycle to alleviate this phase error, such as elements with longer delay lines. The main issues for such designs would be the available real estate in the lattice and losses in the delay lines that make the element design quite complicated.

5.1.3 Aperture Phase Error and Reflectarray Bandwidth Limitations

To get a better understanding on the bandwidth performance of a reflectarray antenna, it is necessary to observe the phase error performance for all the elements with respect to a common reference point. Such an approach takes into account the relative phase errors of the elements as well as the effects of spatial phase delay. Theoretically, the relative phase requirement on an array surface can be calculated with respect to any element in the array. However, since in most reflectarray designs the feed is pointing to the geometrical center of the array, we consider this point as the reference for phase calculations. The general formula for an ideal phase relation is defined as

$$\psi_i(f) - \psi_{ref}(f) = k \cdot \left\{ R_i - R_{ref} - \left(\bar{r}_i \cdot \hat{r}_o - \bar{r}_{ref} \cdot \hat{r}_o \right) \right\}. \tag{5.9}$$

Since the center element is used as the phase reference, $\left| \bar{r}_{ref} \right| = 0$. The relative phase error for the ith element in the array with respect to the center of the array can then be calculated as

$$Phase\,Error_i(f)=\{\psi_i(f)-\psi_{ref}(f)\}-k\cdot\{R_i-R_{ref}-\vec{r}_i\cdot\hat{r}_o\}. \tag{5.10}$$

This definition of phase error on the array surface takes into account the relative phase requirements for every element of the array for a general array geometry and beam direction. To further expand this phase error formulation for a practical case, the feed illumination must also be taken into account. The reason is that the excitation of each element will determine the overall effect of the phase error of that individual element on the reflectarray performance. The illumination of a reflectarray antenna is calculated using the normalized radiation pattern of the feed horn simulated as a $Cos^q(\theta)$ function, which is the widely adopted model to approximate practical feed patterns, as discussed in Chapter 2. The weighted phase error (*WPE*) is then defined as the product of the phase error and the normalized array illumination, that is,

$$WPE_i(f)=Phase\,Error_i(f)\cdot Normalized\,Illumination_i(f). \tag{5.11}$$

To quantitatively analyze the performance of phase errors, we consider a practical example of a Ka-band reflectarray. A circular aperture reflectarray with a diameter of 159 mm is designed for the operating frequency of 32 GHz to generate a beam in the direction of $(\theta, \varphi)=(25°, 0°)$. The phase center of the feed horn is placed at $X=-45.9$ mm, $Y=0$ mm, $Z=98.4$ mm based on the coordinate system shown in Figure 5.6.

For the horn model used in this study, the power q of the feed radiation pattern varies linearly from 5 at 30 GHz to 8.3 at 34 GHz. From the system design specifications, the location of all the array elements, the required element phase shift, and the illumination on the array surface can be calculated. The patch dimensions are then determined from the unit-cell simulations at the center design frequency of 32 GHz. From these patch size dimensions, the frequency behavior of the elements reflection phase is obtained across the frequency band. The WPE on the array surface can then calculated at any specified frequency using (5.11), which is plotted in Figure 5.7 for two off-center frequencies.

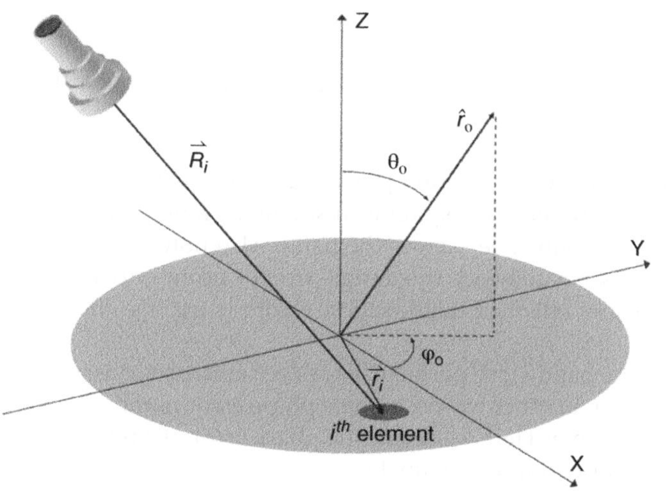

Figure 5.6 The coordinate system of the reflectarray antenna.

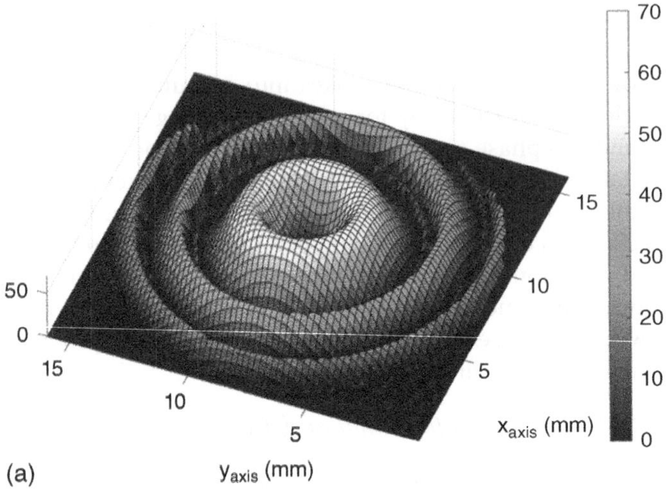

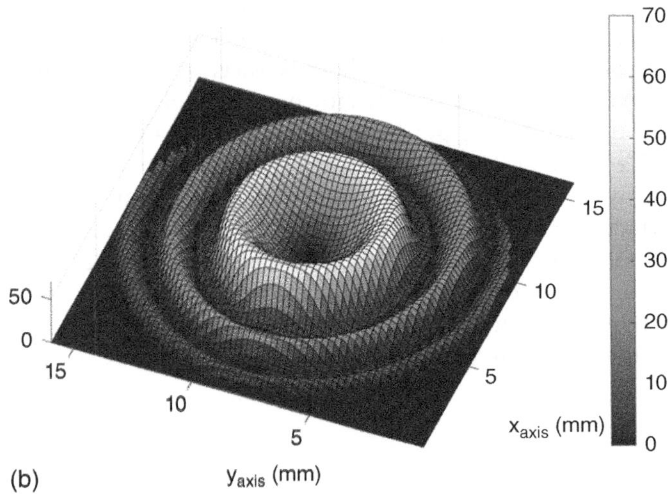

Figure 5.7 Phase errors on the reflectarray aperture at off-center frequencies: (a) 30 GHz, (b) 34 GHz.

The phase error plots on the reflectarray aperture provide a clear visual understanding of the bandwidth limitations in reflectarray systems. It should be noted that similar results were observed at different frequencies across the band. Also note that in these results, the quantization errors associated with the design are not taken into account. Although as mentioned earlier, the effect of quantization errors is not significant in these designs.

To further aid in quantifying the bandwidth performance of reflectarrays, it is beneficial to study the system phase error, in other words, a single phase error number for any reflectarray systems at every frequency. Here we define the phase error of a reflectarray antenna system using two different definitions, namely

$$Average\ Phase\ Error(f) = \frac{1}{N}\sum_{i=1}^{N} Phase\ Error_i(f), \tag{5.12}$$

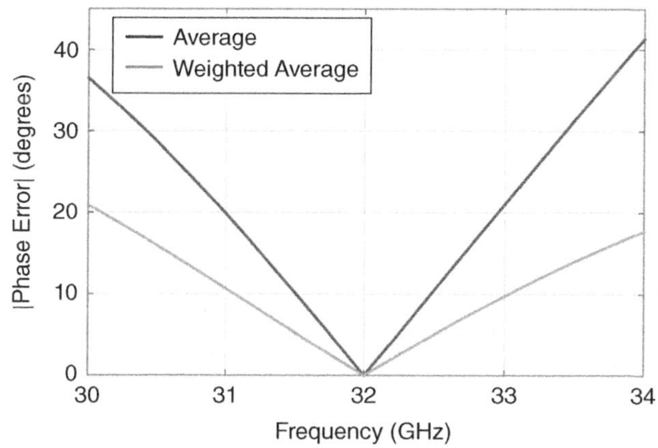

Figure 5.8 Average phase error of the reflectarray antenna as a function of frequency.

$$Average\ Weighted\ Phase\ Error(f) = \frac{1}{N}\sum_{i=1}^{N} WPE_i(f). \tag{5.13}$$

Here, N is the number of phasing elements of the reflectarray antenna. Note that in comparison between the two equations, the latter also includes the amplitude taper of the reflectarray aperture. These phase error terms as a function of frequency are given in Figure 5.8 for the previous reflectarray system, where it can be seen that at off-center frequencies phase errors will be generated on the aperture, and grow rapidly as we move away from the center frequency, ultimately resulting in bandwidth limitations of the reflectarray antenna. It is important to note that, while there is no direct correlation between these phase error terms and the gain bandwidth of a reflectarray, the formulation provided in this section makes it possible to quantify the bandwidth performance of different reflectarray elements and systems using basic element reflection coefficient data.

5.2 Reflectarray Element Bandwidth

5.2.1 Physics of Element Bandwidth Constraints

Various designs of broadband elements have been demonstrated over the years [6]–[15]. As discussed earlier, regardless of the choice of phase tuning approaches, reflectarray elements exhibit some form of phase error as a function of frequency, which ultimately results in bandwidth reduction of the reflectarray system. Depending on the phase tuning approaches. however, the physics behind the bandwidth limitation is different. In the delay-line approach, impedance matching between the transmission-line and the patch is the key factor, since it ensures that the power is received or transmitted from the patch to the delay-line and vice versa, thus controlling the element bandwidth. It is important to point out that while in principle, a good matching across the band will provide the appropriate phasing as a function of frequency, the resonances of the combined patch and line structure will result in somewhat non-linear phase curves versus stub length.

On the other hand, for the variable-size approach, element bandwidth cannot be defined in terms of input impedance bandwidth. In fact, a solid definition of element bandwidth is not available for this approach. The classic single-layer variable-size patch has a highly non-linear reflection phase curve as a function of patch width. These phase curves typically show a very steep slope near resonance and slow variations elsewhere. As a result, the reflection phase is highly sensitive to the frequency variations near resonance, which results in a significant phase change as a function of frequency, limiting the bandwidth of the reflectarray system. The general rule of thumb in designing variable-size elements with good bandwidth performance is to achieve phase curves that have a flat slope as a function of patch size and also show remain parallel to each other at different frequencies. This is typically achieved by using thick substrates and multi-resonance configurations, such as stacked patches or multiple rings.

The element rotation technique, which is popularly used for circularly polarized reflectarrays, has different physics that limit the operation bandwidth. In this approach, the reflected phase of the co-polarized field is dependent on the angle of rotation, and remains constant as a function of the frequency. Hence, a major concern is to maintain a good co-polarized field level over a broad frequency band. Meanwhile, since the spatial phase delay is frequency-dependent and the reflection phase is frequency-independent, according to (5.2), phase error occurs as the frequency changes, which limits the bandwidth of reflectarray systems, especially for large-size reflectarrays.

5.2.2 Parametric Studies on Element Bandwidth

In the previous section the basic physics of bandwidth limitations were outlined for all three phase tuning approaches. The variable-size element, however, is the most widely adopted method for reflectarray antennas, therefore in this section we will present comparative studies for several element topologies.

In this study, the element bandwidth is defined by

$$BW = \frac{f_{u\max} - f_{l\max}}{(f_{u\max} + f_{l\max})/2}. \tag{5.14}$$

To illustrate this, in Figure 5.9(a) we show an example of phase curves obtained at different frequencies for a reflectarray element, which is designed at the center frequency of 13.5 GHz. By calculating the phase differences between other frequencies and center frequency, the results were obtained and shown in Figure 5.9(b). Those two frequencies at which the maximum absolute value of phase difference is 45° are referred to as upper and lower frequencies limits, indicated by $f_{u\max}$ and $f_{l\max}$, respectively. Once the upper and lower frequencies are obtained, the element bandwidth is calculated from (5.14).

Essentially, in this approach the element bandwidth is defined by the maximum phase difference, which in this case is considered to be 45°. This bandwidth definition is a measure on how fast the element phase varies with frequency. If an element phase varies quickly with frequency, a narrow element bandwidth is obtained; otherwise, a broad element bandwidth is obtained. It is worthwhile pointing out that this element bandwidth is not the same as the reflectarray bandwidth. Nevertheless, a broad element bandwidth usually leads to a broad reflectarray bandwidth.

Using this definition, in this section we study the bandwidth performance of three representative variable-size elements; namely, the square patch, square loop, and cross-dipole, which are shown in Figure 5.10.

For each element geometry, the design variable "*a*" is used to change the reflected phase of the element. The square loop and cross-dipole element have a secondary design variable "*w*," which slightly affect the element performance and are typically used to tune

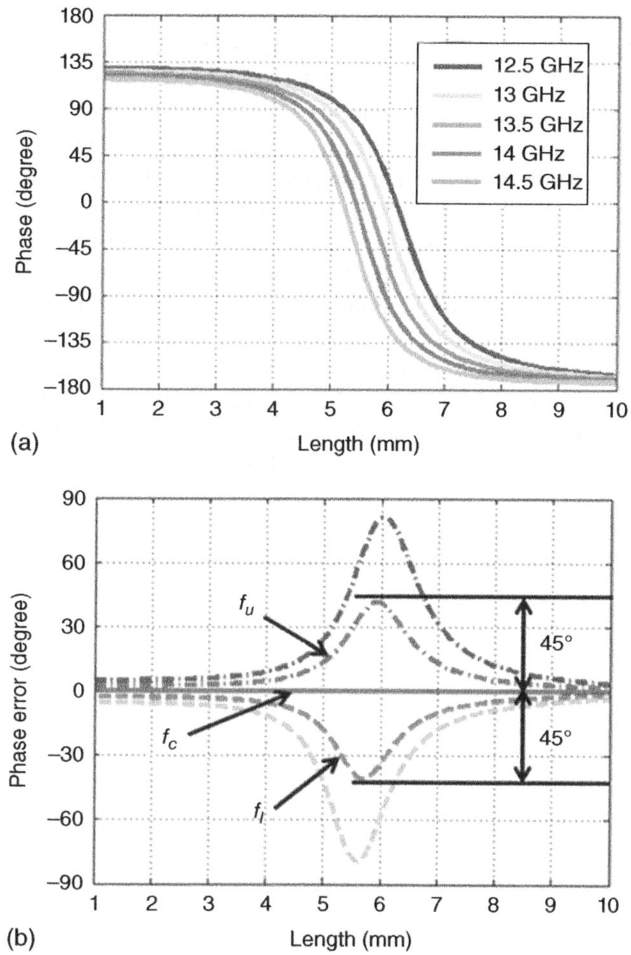

Figure 5.9 (a) Phase curves obtained at different frequencies. (b) The definition for the upper and lower frequency limits in the definition of element bandwidth.

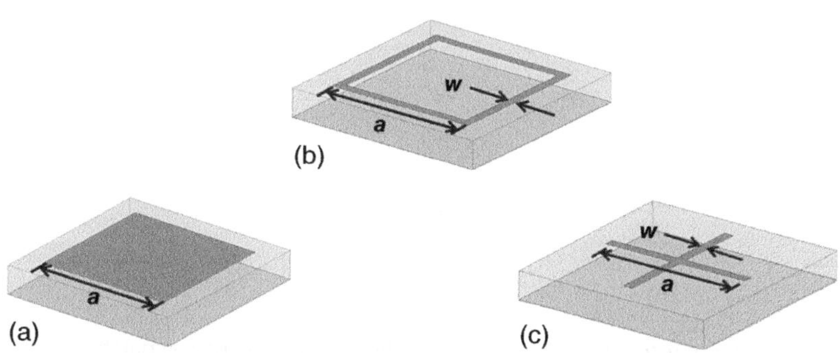

Figure 5.10 Geometrical models of variable-size elements: (a) square patch, (b) square loop, and (c) cross-dipole.

the element. All the three elements studied here are single-resonant, and can only provide a single-phase cycle with a phase tuning range less than 2π. The elements are designed at the center frequency of 13.5 GHz. The unit-cell size is 11.2 mm, and a substrate with a thickness of 1.58 mm and a dielectric constant of 2.55 is used for all elements.

Figure 5.11 shows the phase curves of these three different elements at 13.5 GHz. The resonant dimensions, which correspond to a zero reflection phase, are around 5.65, 4.6, and 7.1 mm for patch, loop and cross-dipole elements, respectively. Note that in these studies w was set to 0.5 mm for both square loop and cross-dipole elements. The current distributions for the resonant dimensions of each element are also depicted in Figure 5.12, which shows that these three elements geometries resonate at the fundamental modes of patch, loop and dipole antennas.

The element phase curves of these three elements at different frequencies are given in Figure 5.13. It can be seen that the patch element has the smoothest curve but with the smallest phase range [12]. The phase differences are also obtained with respect to the center frequency and are given in Figure 5.14, which show that the loop and

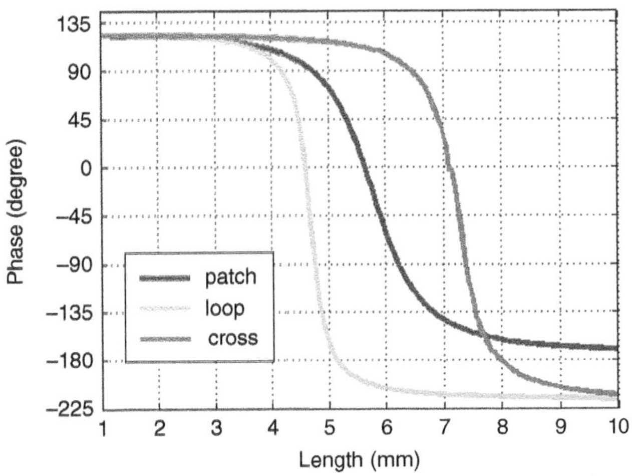

Figure 5.11 Phase curves obtained at 13.5 GHz for three different elements.

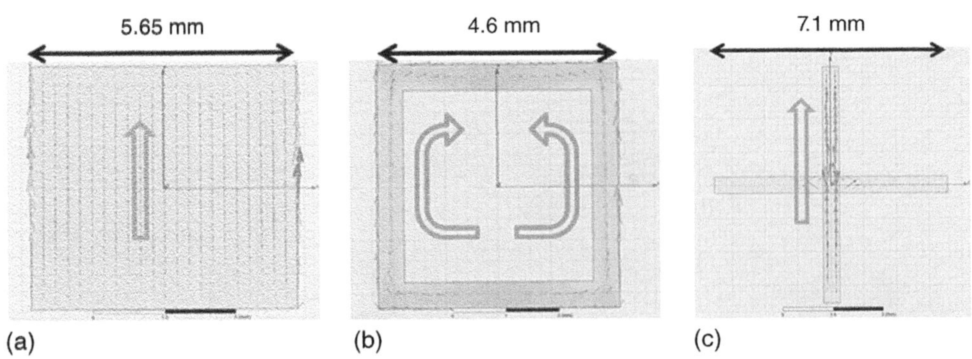

Figure 5.12 Current distributions at resonance dimensions for reflectarray elements: (a) patch, (b) loop, and (c) cross-dipole.

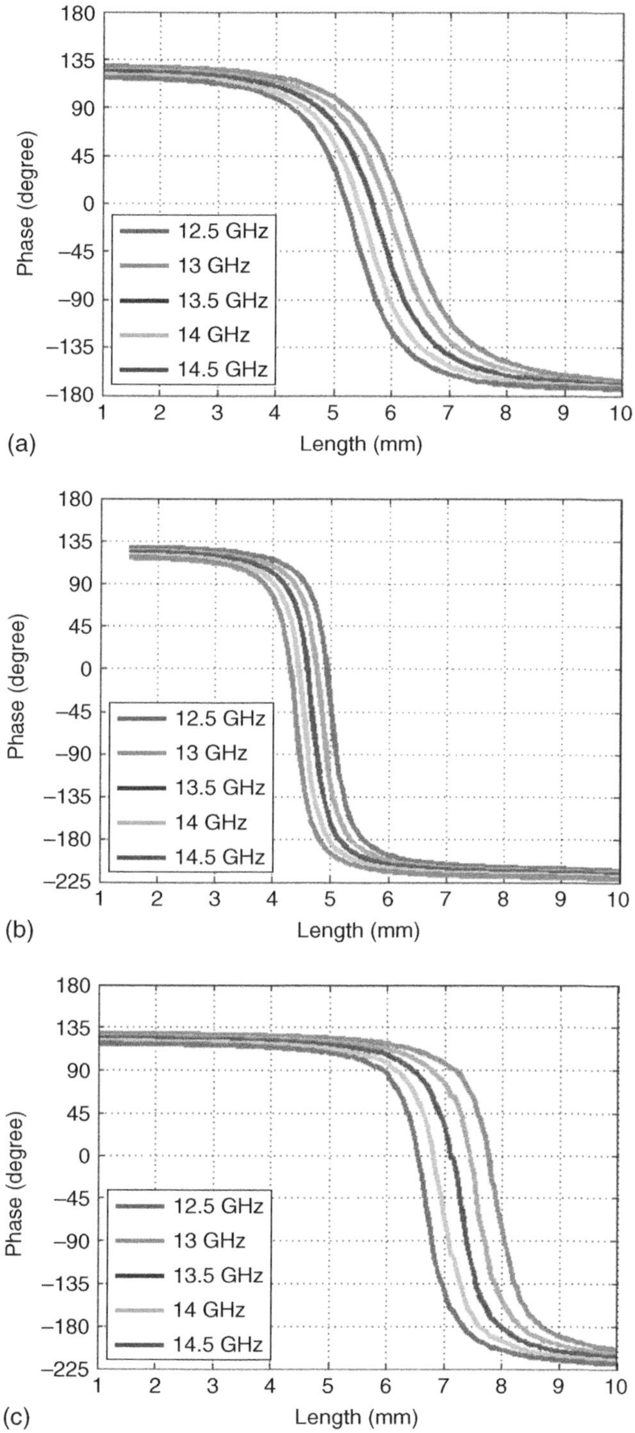

Figure 5.13 Reflection phase curves of reflectarray elements at different frequencies: (a) patch, (b) loop, and (c) cross-dipole.

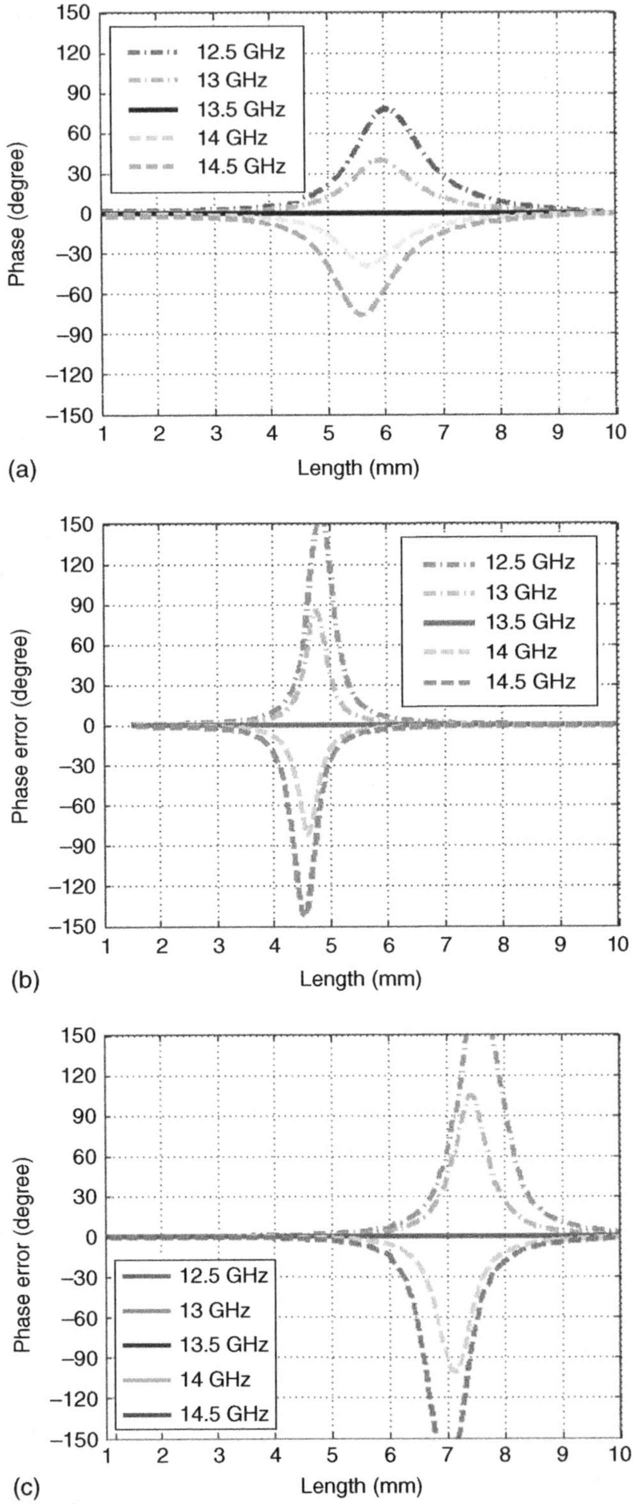

Figure 5.14 Phase errors compared to the one obtained at center frequency for: (a) patch (b) loop (c) cross-dipole.

Table 5.1 Summary of reflectarray element performances.

Element	Patch	Loop	Cross-dipole
Phase range @ 13.5 GHz	293.0°	339.7°	335.6°
Element bandwidth	7.6%	5.3%	4.8%

cross-dipole elements have much higher phase differences than the patch. The elements phase range at 13.5 GHz and element bandwidth as defined in (5.14) are summarized in Table 5.1, which shows that in terms of phase range loop outperforms the others, and in terms of bandwidth, the patch element provides the best performance.

In the next stage, we study the effect of the substrate thickness on element performance. When the substrate thickness is increased, usually the reflection phase slope shows a more linear response at the cost of a reduced phase range, thereby essentially increasing the element bandwidth and decreasing the quantization error in fabrication. Theoretically, it could be explained by Q factor theory in patch antennas. It is well known that when the substrate thickness increases, the Q factor decreases, thus the element bandwidth is increased. In order to show the effects of the substrate thickness, the three elements are investigated with three different substrate thicknesses; namely 0.79, 1.58, and 3.18 mm.

The phase curves obtained at 13.5 GHz for the patch elements with different substrate thicknesses are given in Figure 5.15, where it can be seen that as the substrate thickness increases, the phase curves become more linear and the phase range is greatly reduced.

In order to study the element bandwidth, phase curves obtained at different frequencies are given in Figure 5.16. Phase differences are also computed and depicted in Figure 5.17. It is seen that for the thinnest laminate thickness of 0.79 mm, the phase error could reach around 90° when the frequency changes only by 0.5 GHz. On the other hand, for the thickest laminate thickness of 3.18 mm, a frequency shift of 1 GHz only results in a phase error around 30°. The phase ranges at 13.5 GHz and the element

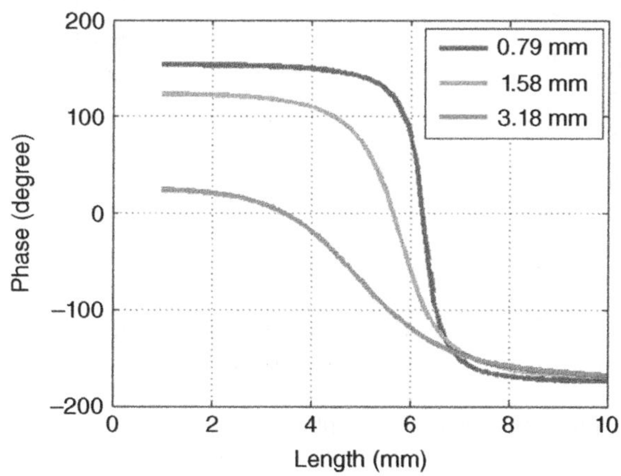

Figure 5.15 Comparison of phase curves obtained at 13.5 GHz for patch elements with different substrate thicknesses.

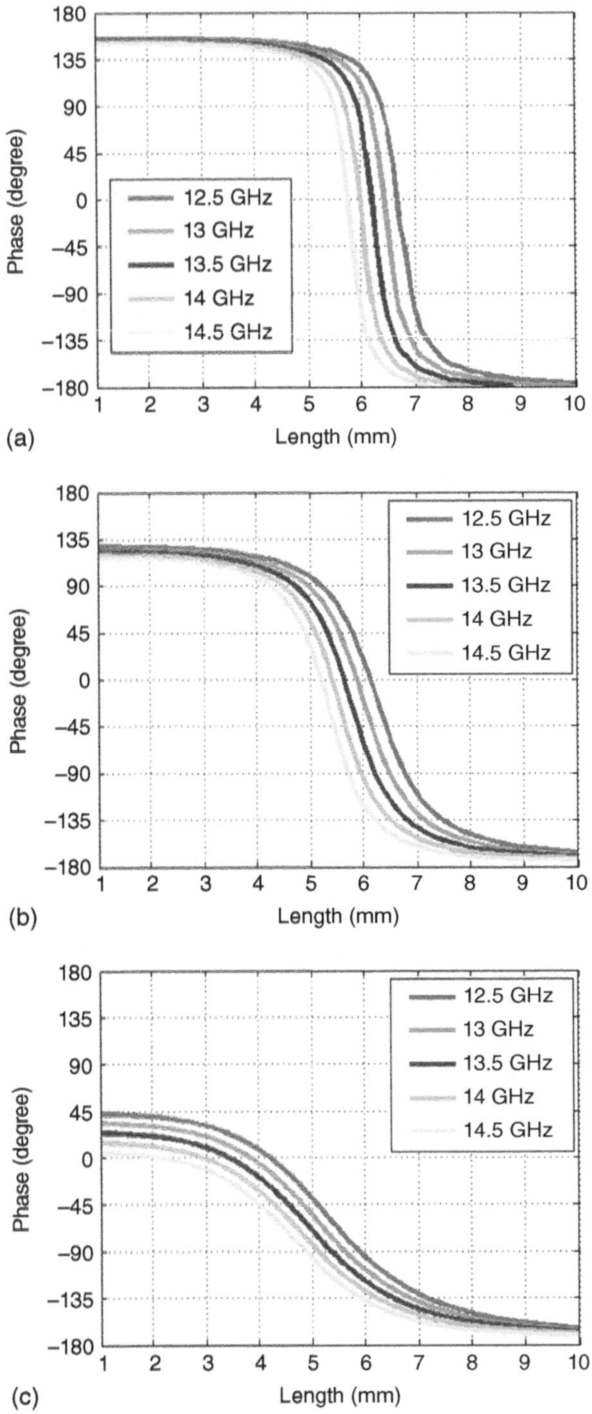

Figure 5.16 Comparison of phase curves for the patch elements with different substrate thicknesses: (a) t = 0.79 mm, (b) t = 1.58 mm, and (c) t = 3.18 mm.

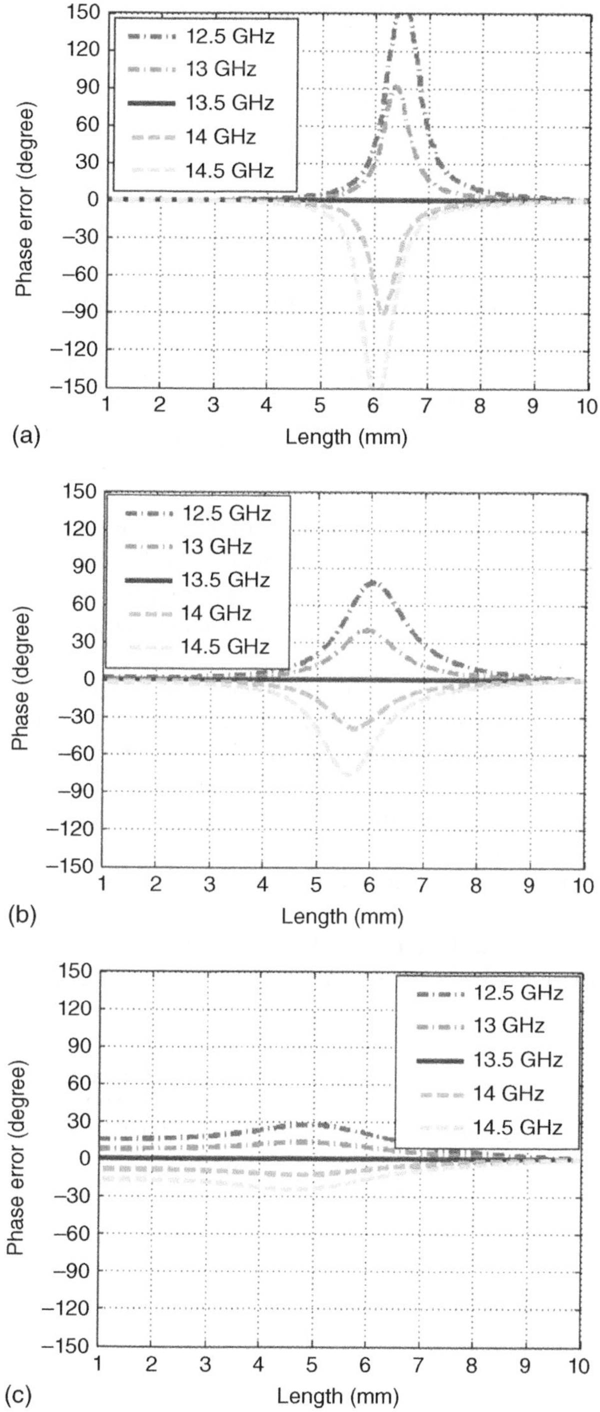

Figure 5.17 Comparison of phase differences for patch elements with different substrate thicknesses: (a) t = 0.79 mm, (b) t = 1.58 mm, and (c) t = 3.18 mm.

Table 5.2 Summary of phase ranges at 13.5 GHz and element bandwidth for the patch elements with different substrate thicknesses.

Element	t=0.79 mm	t=1.58 mm	t=3.18 mm
Phase range (@13.5 GHz)	327.6°	293.0°	191.9°
Element bandwidth	5.1%	7.6%	14.7%

bandwidth for the patch elements with various substrate thicknesses are summarized in Table 5.2.

Similar analysis is performed to understand the element characteristics of the square loop and the cross-dipole for different substrate thicknesses. These results are given in Figure 5.18 to Figure 5.23, and summarized in Tables 5.3 and 5.4. The observations are the same as the patch case: the thicker the substrate, the smaller the phase range and the wider the bandwidth.

The results presented here clearly show that in all element geometries using a thick substrate is advantageous in terms of element bandwidth. However, as discussed in earlier chapters, a phase tuning range below 300° typically results in poor performance of the array, hence appropriate compromises have to be made when selecting the substrate thickness.

As discussed earlier, the loop and cross-dipole elements offer a second degree of freedom for the element design. The width of the trace in both element geometries can be used to fine-tune the element performance. Here we study the effect of this design parameter on the performance of these elements by considering three different trace widths of 0.5, 1, and 1.5 mm. Similar analysis are conducted as before, and the results are given in Figure 5.24 to Figure 5.27, as well as Tables 5.5 and 5.6.

The results presented here show that the change trace width slightly affects the bandwidth and phase range performance of both the loop and cross-dipole elements. In the case of the loop, as the trace width increases, the bandwidth also increases by a small amount. Similar observations are made with the cross-dipole element, and in general

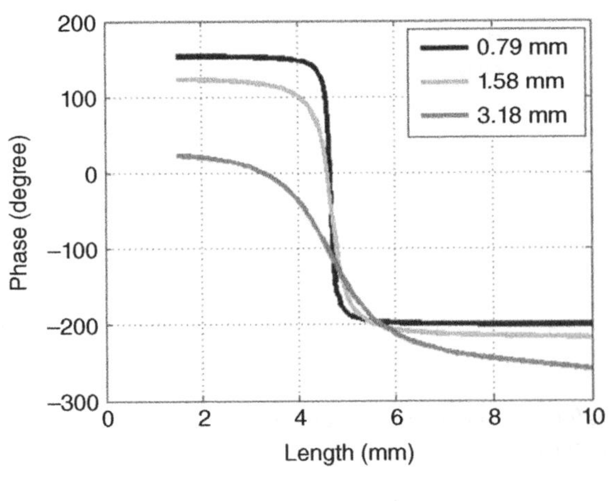

Figure 5.18 Comparison of phase curves obtained at 13.5 GHz for the square loop elements with different substrate thicknesses.

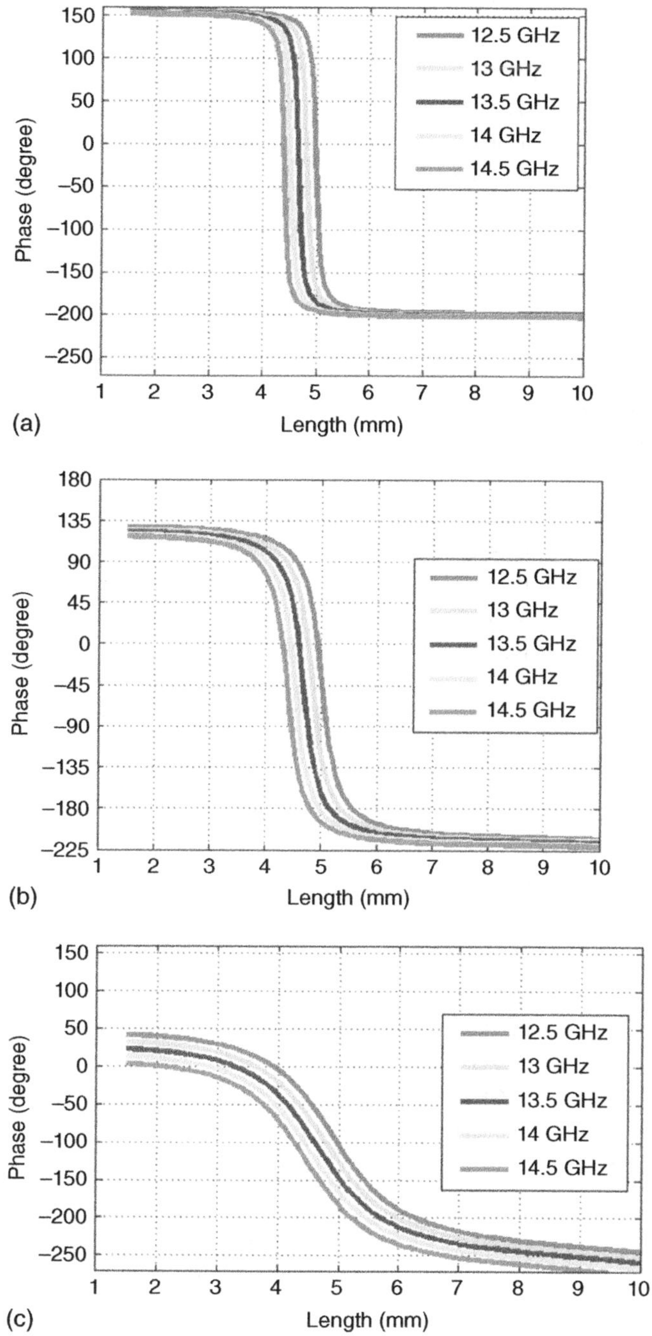

Figure 5.19 Comparison of phase curves for the square loop elements with different substrate thicknesses: (a) t = 0.79 mm, (b) t = 1.58 mm, and (c) t = 3.18 mm.

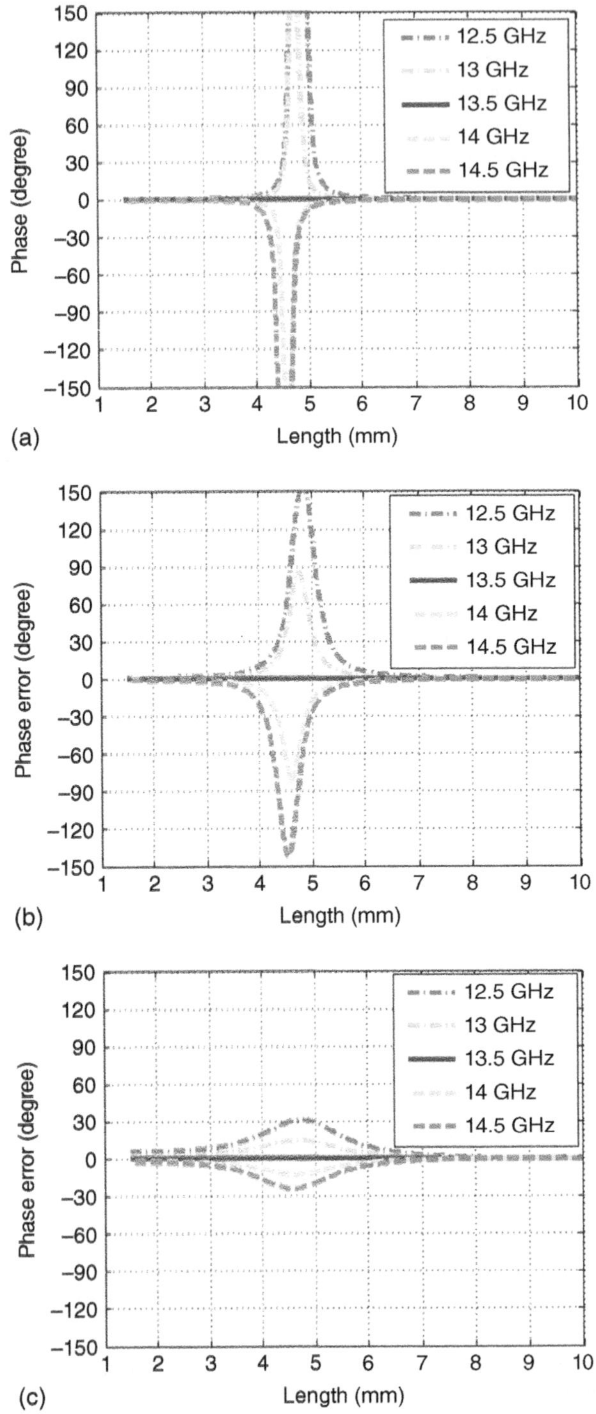

Figure 5.20 Comparison of phase differences for the square loop elements with different substrate thicknesses: (a) t = 0.79 mm, (b) t = 1.58 mm, and (c) t = 3.18 mm.

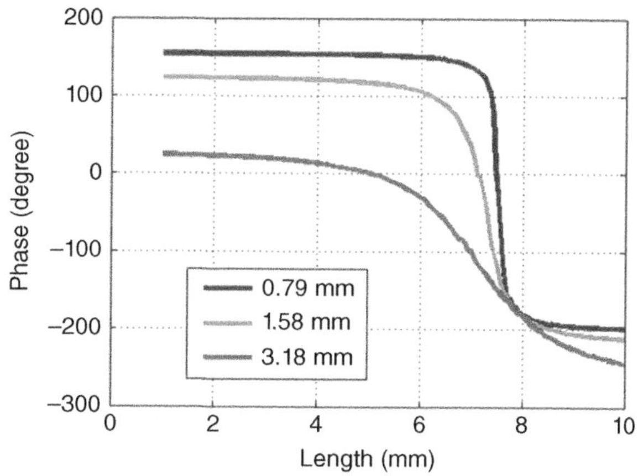

Figure 5.21 Comparison of phase curves obtained at 13.5 GHz for the cross-dipole elements with different substrate thicknesses.

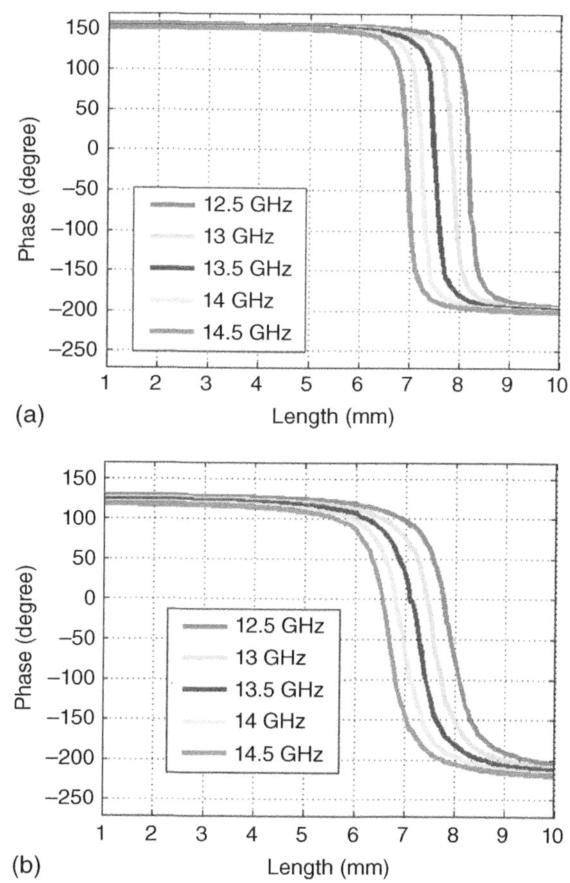

Figure 5.22 Comparison of phase curves for the cross-dipole elements with different substrate thicknesses: (a) t = 0.79 mm, (b) t = 1.58 mm, and (c) t = 3.18 mm.

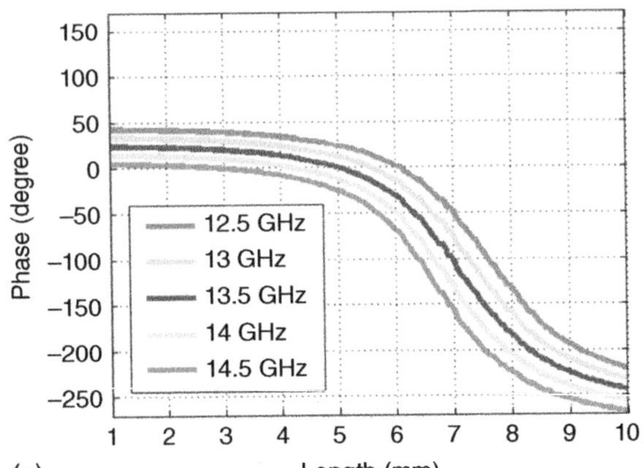

(c)

Figure 5.22 (Continued)

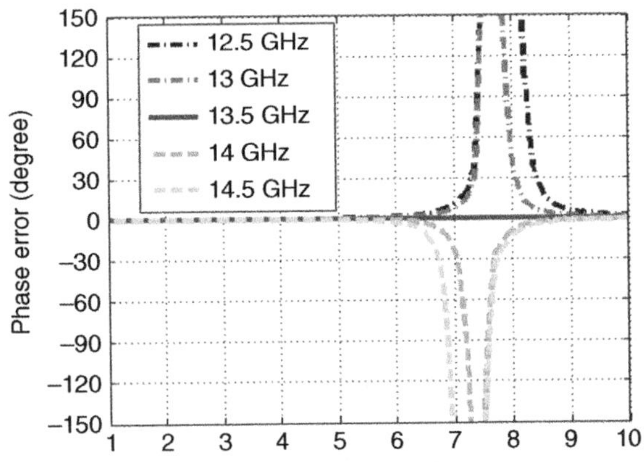

(a)

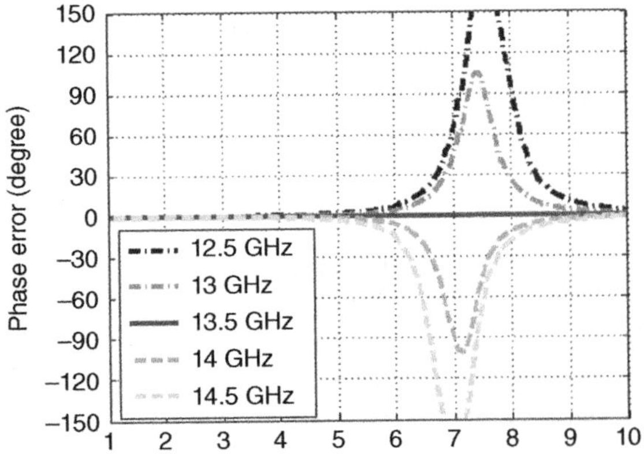

(b)

Figure 5.23 Comparison of phase differences for the cross-dipole elements with different substrate thicknesses: (a) t = 0.79 mm, (b) t = 1.58 mm, and (c) t = 3.18 mm.

Figure 5.23 (Continued)

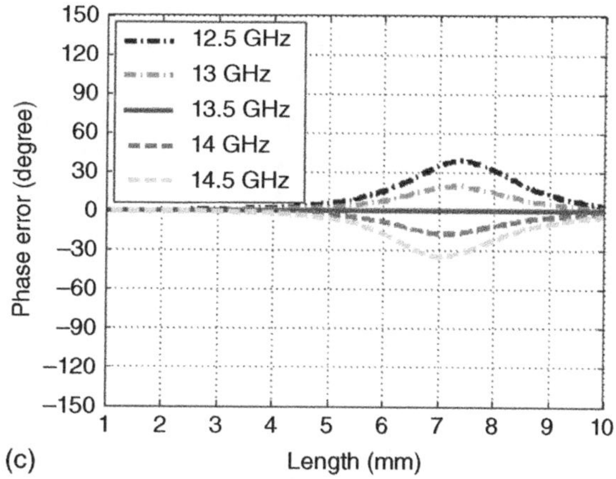

(c)

Table 5.3 Summary of phase ranges at 13.5 GHz and element bandwidth for the square loop elements with different substrate thicknesses.

Element	t=0.79 mm	t=1.58 mm	t=3.18 mm
Phase range (@13.5 GHz)	353.7°	339.7°	281.1°
Element bandwidth	3.9%	5.3%	11.9%

Table 5.4 Summary of phase ranges at 13.5 GHz and element bandwidth for patch elements with different substrate thickness.

Element	t=0.79 mm	t=1.58 mm	t=3.18 mm
Phase range (@13.5 GHz)	352.2°	335.6°	268.2°
Element bandwidth	3.7%	4.8%	10.4%

for these types of configurations a larger trace width is advantageous to achieve a better performance.

5.3 Reflectarray System Bandwidth

5.3.1 Effect of Aperture Size on Reflectarray Bandwidth

After the study on the bandwidth of reflectarray elements, now we focus on the bandwidth performance of reflectarray antennas. As discussed earlier, for large reflectarray antennas, the primary factor limiting the bandwidth is the effect of spatial phase delay. To quantitatively demonstrate the effect of aperture size on reflectarray bandwidth, here we consider four cases with aperture sizes of 13, 25, 50, and 100λ at the center design

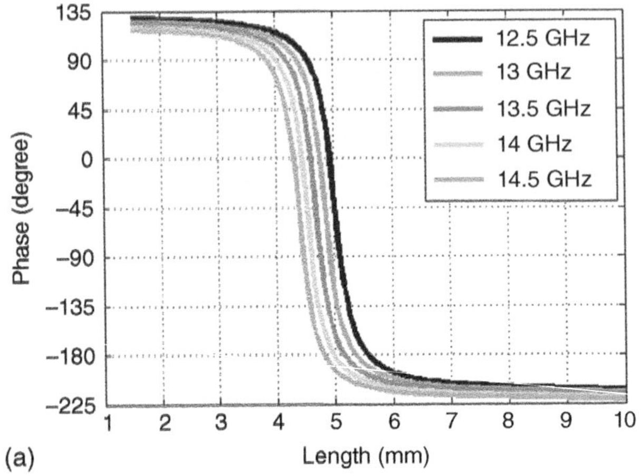

(a)

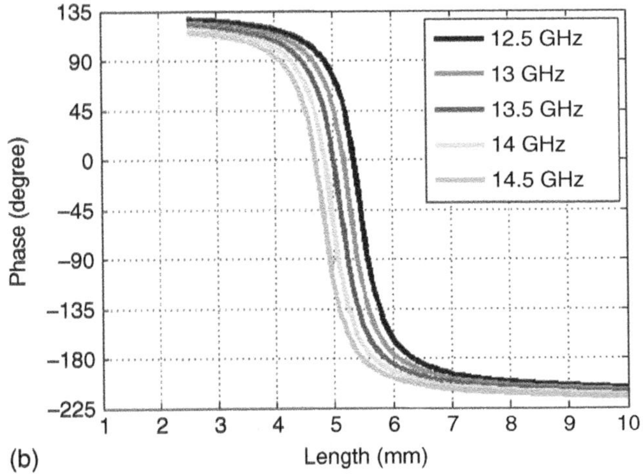

(b)

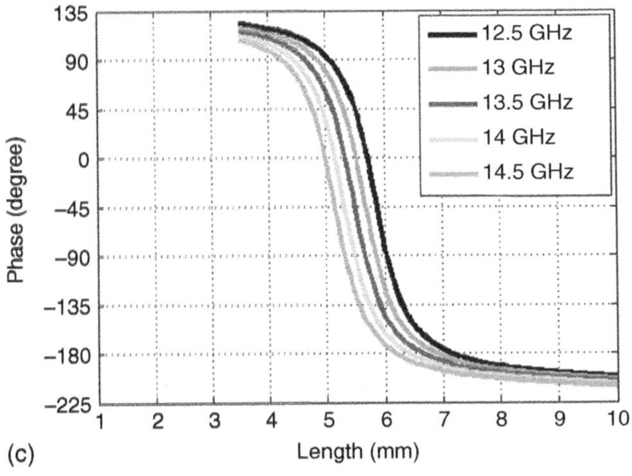

(c)

Figure 5.24 Comparison of phase curves for the square loop elements with trace width: (a) $w = 0.5$ mm, (b) $w = 1$ mm, and (c) $w = 1.5$ mm.

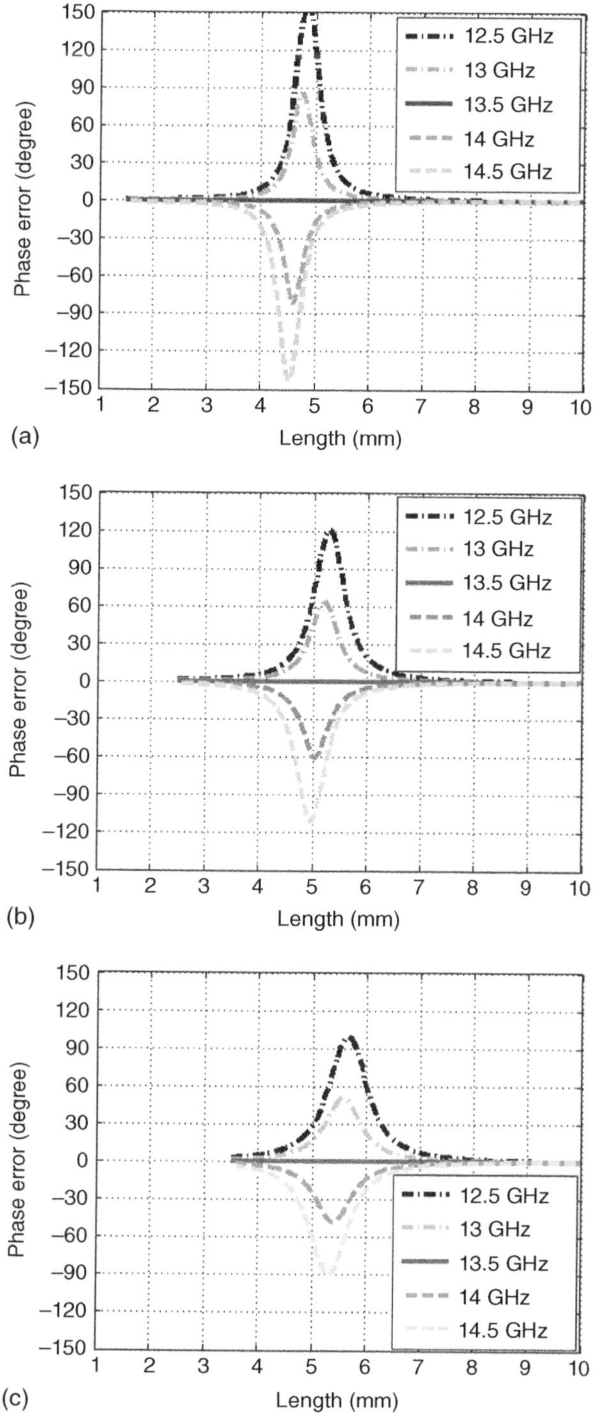

Figure 5.25 Comparison of phase differences for the square loop elements with trace width: (a) w = 0.5 mm, (b) w = 1 mm, and (c) w = 1.5 mm.

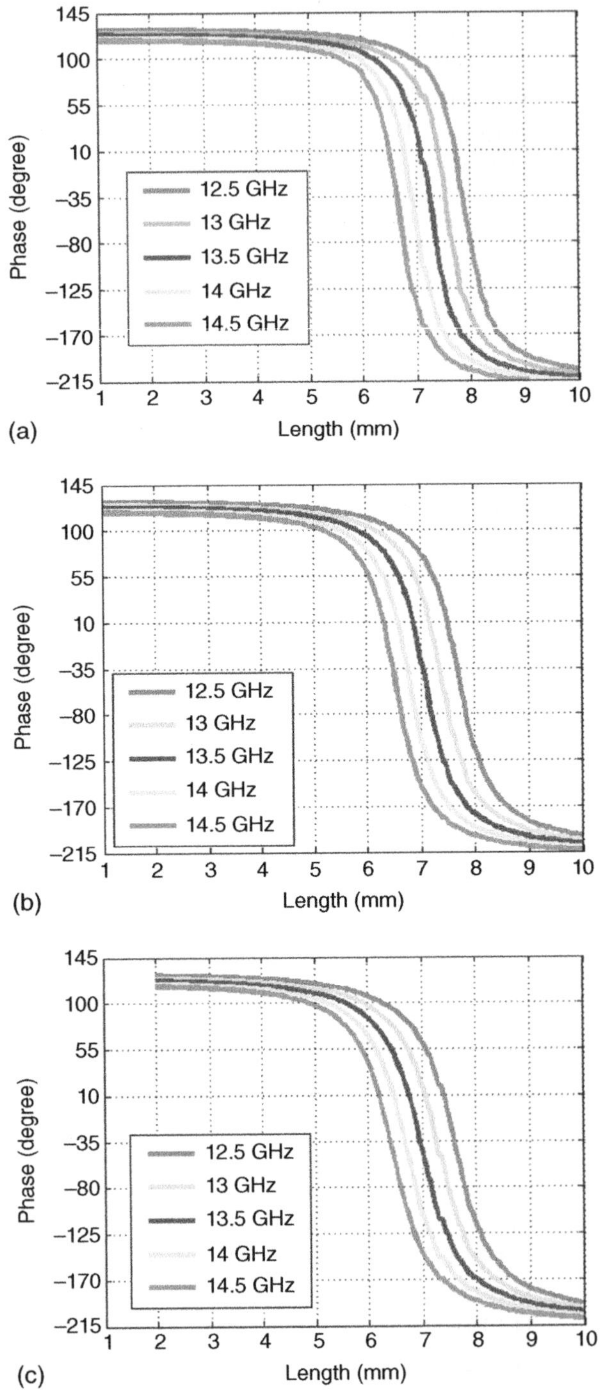

Figure 5.26 Comparison of phase curves for the cross-dipole elements with trace width: (a) $w = 0.5$ mm, (b) $w = 1$ mm, and (c) $w = 1.5$ mm.

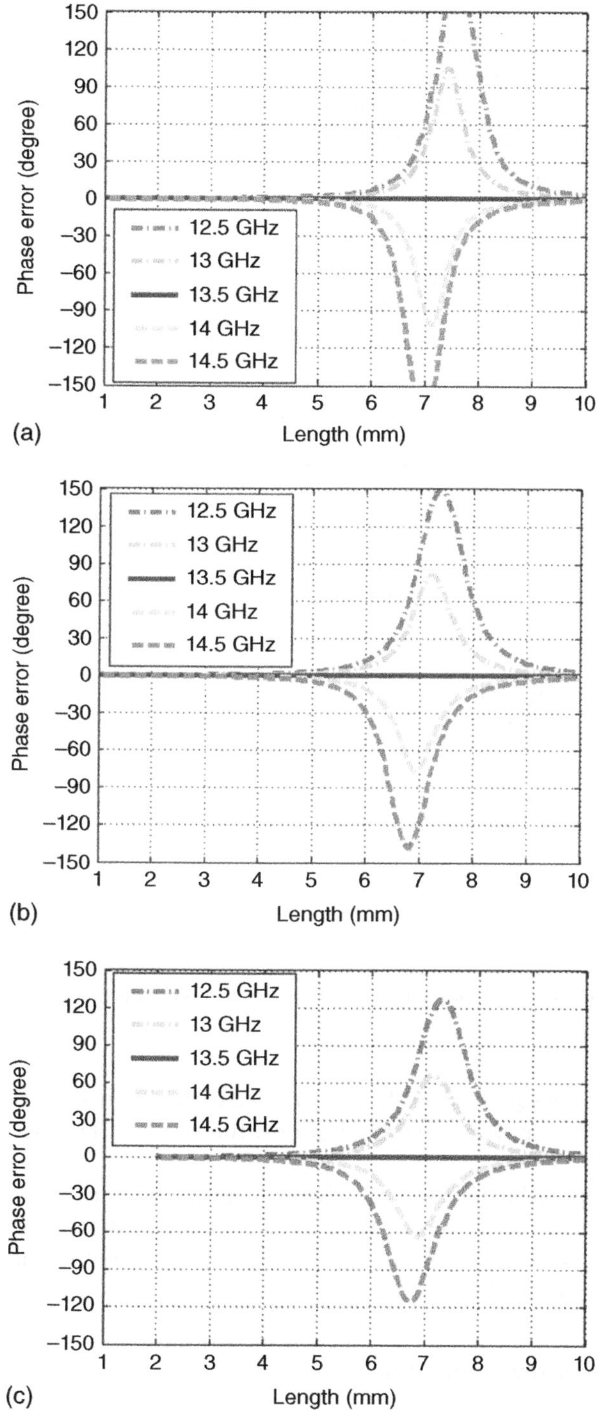

Figure 5.27 Comparison of phase differences for the cross-dipole elements with trace width: (a) $w = 0.5$ mm, (b) $w = 1$ mm, and (c) $w = 1.5$ mm.

Table 5.5 Summary of phase ranges at 13.5 GHz and element bandwidth for the square loop elements with different trace widths.

Width (w)	0.5 mm	1 mm	1.5 mm
Phase range @13.5 GHz	339.7°	332.5°	327.0°
Element bandwidth	5.3%	6.0%	6.7%

Table 5.6 Summary of phase ranges at 13.5 GHz and element bandwidth for the cross-dipole elements with different trace widths.

Width (w)	0.5 mm	1 mm	1.5 mm
Phase range @13.5 GHz	335.6°	330.1°	324.3°
Element bandwidth	4.8%	5.3%	5.8%

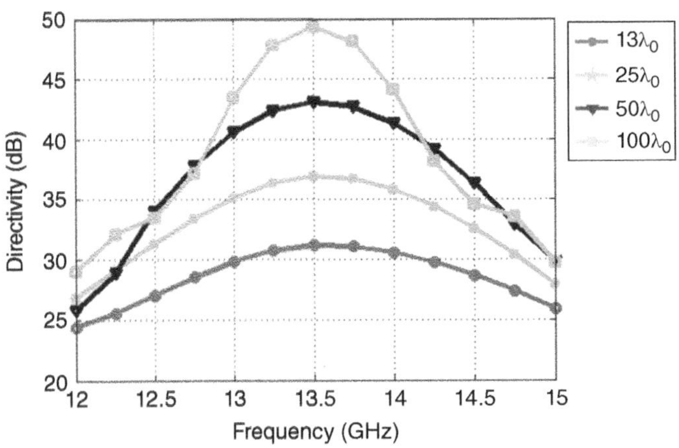

Figure 5.28 Directivity of the reflectarray antenna as a function of frequency for different aperture sizes.

frequency of 13.5 GHz. We consider the square loop element studied earlier in this chapter with a substrate thickness of 1.58 mm. For all configurations, the system f/D is set to 0.735.

The reflectarray directivity is computed using the array theory approach described in Chapter 4 and is given in Figure 5.28. It can be seen that as the aperture size increases, the antenna directivity also increases; however, this is at the expense of a significant reduction in directivity bandwidth of the antenna.

5.3.2 Effects of Element on Reflectarray Bandwidth

Although a quantitative relation between the element bandwidth and the reflectarray bandwidth is not available, a basic rule of thumb in reflectarray system design exists, which is if an element shows good bandwidth performance, it would translate to wide

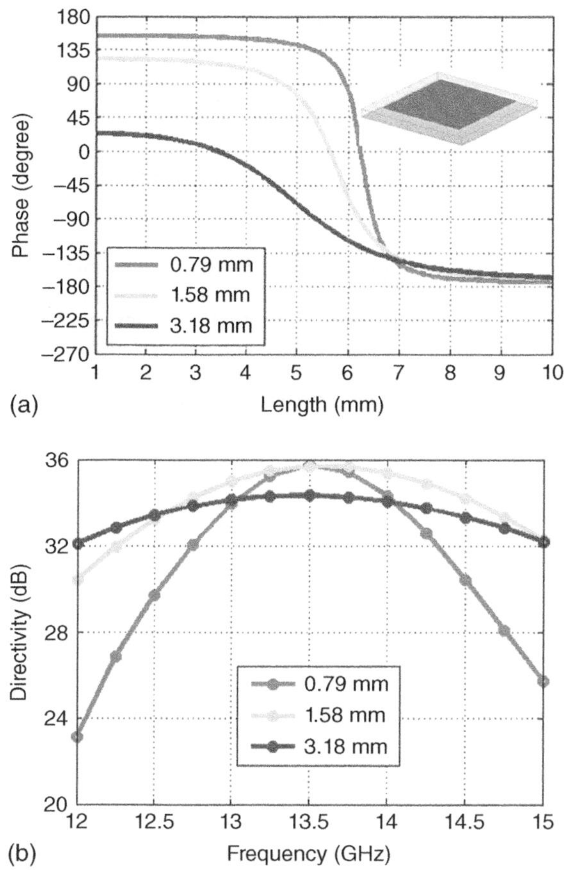

Figure 5.29 Performance of a variable-size square patch reflectarray using different substrate thickness: (a) Phase curves as a function of patch length, (b) directivity bandwidth of the reflectarray antenna with various thicknesses.

bandwidth performance for the reflectarray system. To quantify the effects of elements on the bandwidth of the reflectarray antenna, in this section we will study a reflectarray system using different types of elements. The reflectarray is designed for the center frequency of 13.5 GHz and has a diameter of 516 mm, with an f/D of 0.735.

First, we consider the patch element and study the bandwidth performance of the reflectarray using different substrate thicknesses. Phase curves as a function of patch length and reflectarray directivity bandwidth are given in Figure 5.29. The directivity bandwidth results given here clearly show that as the substrate thickness increases, the reflectarray bandwidth increases. Note that, however, for the very thick substrate (3.18 mm), the significantly reduced phase range of 192° results in a reduction of directivity by about 1.5 dB. As mentioned earlier, increasing the substrate thickness to a point where the element phase range drops below the 300° range is not practical. A summary of these results is given in Table 5.7. It is worthwhile pointing out that the reflectarray bandwidth here is defined by 1 dB directivity drop.

Table 5.7 Summary of directivity and bandwidth for reflectarray using the square patches.

Substrate thickness	0.79 mm	1.58 mm	3.18 mm
Directivity @13.5GHz (dB)	35.7	35.7	34.2
Bandwidth	5.5%	10.6%	15.0%

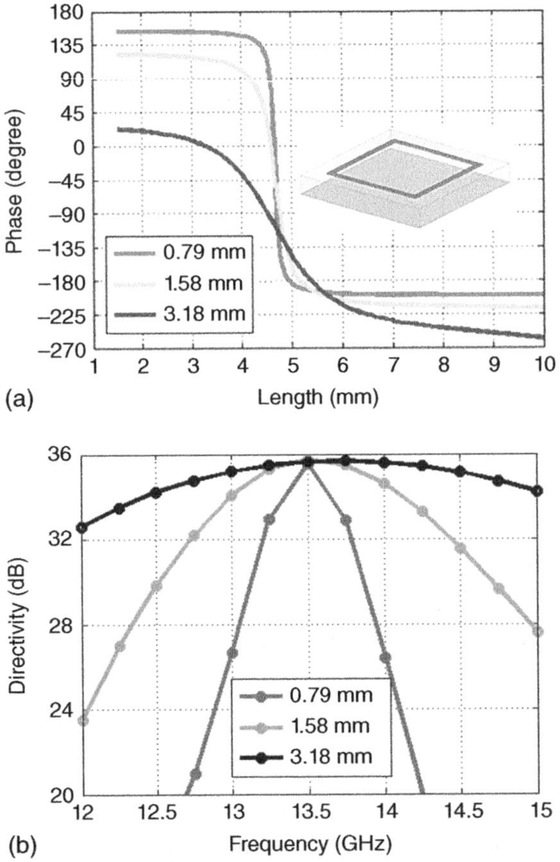

Figure 5.30 Performance of a variable-size square loop reflectarray using different substrate thicknesses: (a) phase curves as a function of loop length, (b) directivity bandwidth of the reflectarray antenna with various thicknesses.

Similar studies are also conducted for reflectarrays consisted of the square loop and cross-dipole elements, and results are given in Figure 5.30 and Figure 5.31, and summarized in Tables 5.8 and 5.9.

The results presented here clearly show a distinct trend that bandwidth will improve by using thicker substrates regardless of the element geometry. One thing that needs to be pointed out is that for the cross-dipole configuration when the substrate thickness is 1.58 mm, the directivity is highest. The reason is that there are two main factors controlling the element phase errors at the design frequency, namely, the phase range and the quantization error, which ultimately influence the peak directivity. When the

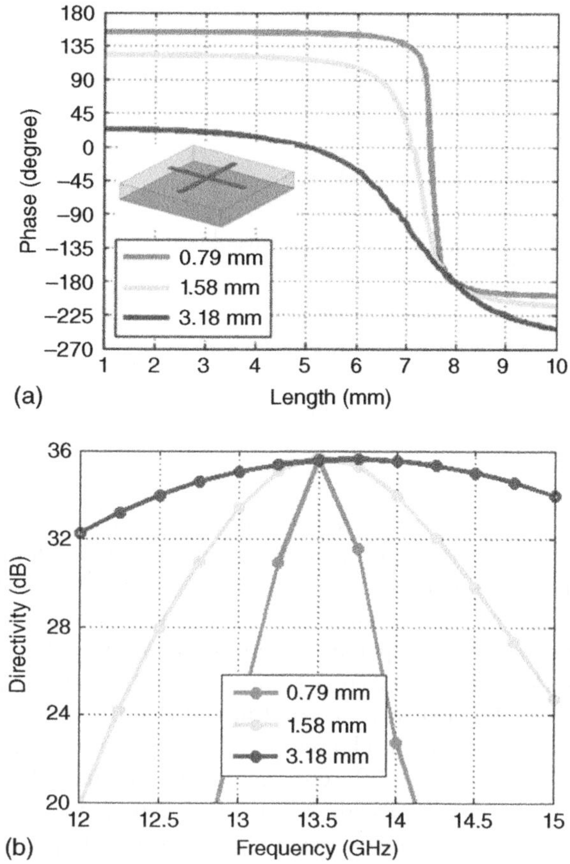

Figure 5.31 Performance of a variable-size cross-dipole reflectarray using different substrate thicknesses: (a) phase curves as a function of dipole length, (b) directivity bandwidth of the reflectarray antenna with various thicknesses.

Table 5.8 Summary of directivity and bandwidth for reflectarray using the square loops.

Substrate thickness	0.79 mm	1.58 mm	3.18 mm
Directivity @13.5GHz (dB)	35.5	35.7	35.6
Bandwidth	0.2%	6.2%	14.8%

Table 5.9 Summary of directivity and bandwidth for reflectarray using the cross-dipoles.

Substrate thickness	0.79 mm	1.58 mm	3.18 mm
Directivity @13.5GHz (dB)	35.6	35.7	35.6
Bandwidth	0.2%	4.5%	14.0%

substrate thickness differs, both phase range and quantization error vary, which changes the phase errors in two opposite ways. When the substrate thickness is 1.58 mm, it allows a smaller quantization error compared to thinner substrate, while a larger phase range compared to the thicker one; hence it has the smallest total phase error and consequently the largest peak directivity.

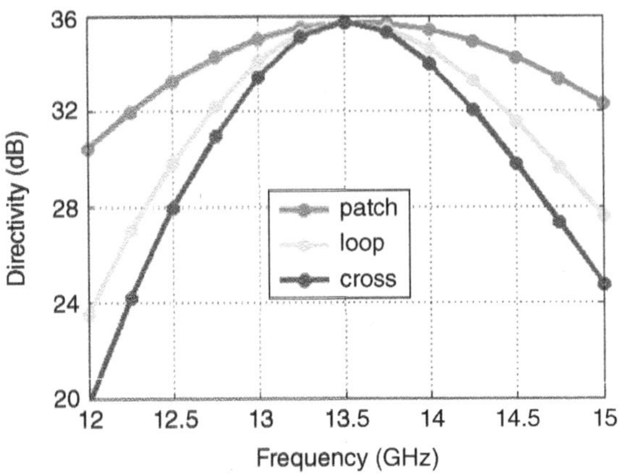

Figure 5.32 Directivity as a function of frequency for reflectarrays using different elements.

As a final remark on this comparative study, it is interesting to compare the performance of these reflectarrays using the three different element shapes. A substrate thickness of 1.58 mm is used for all three designs and the aperture size is 25λ at the center design frequency of 13.5 GHz. The directivity bandwidths are given in Figure 5.32 where it can be seen that the patch element performs best. The 1-dB directivity bandwidth for these reflectarrays are 10.6, 6.2, and 4.5% for the patch, square loop, and cross-dipole, respectively. It is worthwhile pointing out that the reflectarray bandwidth results here has the same trend as the element bandwidth results in Table 5.1.

In summary, different methodologies have been proposed to design reflectarray antennas with a wide bandwidth. Typically, a thick substrate has to be utilized and multiple resonances need to be introduced to provide a good phase tuning range with a linear slope. Discussions on wide band reflectarray antennas will be given in Chapter 7.

References

1 C. J. Sletten, *Reflector and Lens Antennas: Analysis and Design Using Personal Computers*, Artech House Inc., 1988.
2 J. Huang and J. A. Encinar, *Reflectarray Antennas*. New York, NY, USA: Wiley-IEEE, 2008.
3 D. M. Pozar, S. D. Targonski, and H. D. Syrigos, "Design of millimeter wave microstrip reflectarrays," *IEEE Trans. Antennas Propag.*, Vol. 45, No. 2, pp. 287–296, Feb. 1997.
4 J. Huang, "Bandwidth study of microstrip reflectarray and a novel phased reflectarray concept," presented at the *IEEE Antennas and Propagation Society International Symposium, California*, USA, June 1995.
5 D. M. Pozar, "Bandwidth of reflectarrays," *Electron. Lett.*, Vol. 39, No. 21, Oct. 2003.
6 J. A. Encinar, "Design of two-layer printed reflectarrays using patches of variable size," *IEEE Trans. Antennas Propag.*, Vol. 49, No. 10, pp. 1403–1410, Oct. 2001.
7 N. Misran, R. Cahill, and V. Fusco, "Reflection phase response of microstrip stacked ring elements," *Electron. Lett.*, Volume 38, No. 8, pp. 356–357, April 2002.
8 J. A. Encinar and J. A. Zornoza, "Broadband design of three-layer printed reflectarrays," *IEEE Trans. Antennas Propag.*, Vol. 51, No. 7, pp. 1662–1664, July 2003.

9 M. E. Bialkowski and K. H. Sayidmarie, "Investigations into phase characteristics of a single-layer reflectarray employing patch or ring elements of variable size," *IEEE Trans. Antennas Propag.*, Vol. 56, No. 11, pp. 3366–3372, Nov. 2008.

10 M. Bozzi, S. Germani, and L. Perregrini, "Performance comparison of different element shapes used in printed reflectarrays," in Proc. *IEEE Antennas Wireless Propag. Lett.*, 2003, Vol. 2, pp. 219–222.

11 N. Misran, R. Cahill, and V. Fusco, "Design optimization of ring elements for broadband reflectarray antennas," *Microwaves, Antennas and Propagation, IEE Proceedings*, Vol. 150, No. 6, pp. 440–444, Dec. 2003.

12 K. H. Sayidmarie and M. E. Bialkowski, "Investigations into unit cells offering an increased phasing range for single-layer printed reflectarrays", *Microwave and Optical Technology Letters*, Vol. 50, No. 4, pp. 1028–1032, Apr 2008.

13 P. Nayeri, M. E. Bialkowski, and K. H. Sayidmarie, "Design of unit cells of double circular rings for a single-layer microstrip reflectarray," *20th Asia Pacific Microwave Conference*, Hong Kong, China, Dec. 2008.

14 M. R. Chaharmir, J. Shaker, M. Cuhaci, and A. Ittipiboon, "A broadband reflectarray antenna with double square rings," *Microw. Opt. Technol. Lett.*, Vol. 48, No. 7, pp. 1317–1320, Jul. 2006.

15 J. Shaker, M. R. Chaharmir, and J. Ethier, *Reflectarray Antennas, Analysis, Design, Fabrication, and Measurement*, Artech House, 2013.

6

Reflectarray Design Examples

In the previous chapters, a systematic design and analysis procedure for reflectarray antennas was outlined in detail. In this chapter, we present several design examples of reflectarray antennas that can serve as a valuable reference for the readers. The first example is a comprehensive design of Ku-band reflectarray, outlining all steps in design, analysis, fabrication, and testing. Moreover, several representative design examples, including a Ka-band circularly polarized reflectarray antenna and a comparative study of Ku-band reflectarrays using different element shapes, are also provided in a concise fashion.

6.1 A Ku-band Reflectarray Antenna: A Step-by-Step Design Example

6.1.1 Feed Antenna Characteristics

As discussed earlier, the feed antenna radiation characteristics are one of the most influential measures in the design. Similar to reflector antennas [1]–[3], typically the engineer has two options for the feed design. Either a specific feed antenna is to be used for the design, in which case the system parameters need to be selected appropriately; or a feed antenna can be designed that achieves optimum radiation characteristics for a desired system. While the latter option generally provides the engineer with several degrees of freedom in designing the reflectarray, low-cost prototyping typically limits the engineer to the former option.

For this Ku-band design, we select a commercially available pyramidal horn antenna (LB 62–15) from A-INFO [4]. The aperture opening of the horn is $50 \times 35\,\text{mm}^2$, and the antenna has a gain of 15.8 dBi at the center design frequency of 14.25 GHz. A photo of the horn antenna is given in Figure 6.1.

To analyze the radiation characteristics, the antenna is modeled in the commercial EM solver, ANSYS HFSS [5]. The simulation model is shown in Figure 6.2. One of the notable advantages of the full-wave simulation here is that in addition to calculating the far-field radiation performance, it can also provide the electromagnetic field quantities inside the aperture of the horn which gives one a better understanding of the antenna radiation characteristics. The magnitudes of the total electric fields in E- and H-planes

Reflectarray Antennas: Theory, Designs, and Applications, First Edition.
Payam Nayeri, Fan Yang, and Atef Z. Elsherbeni.

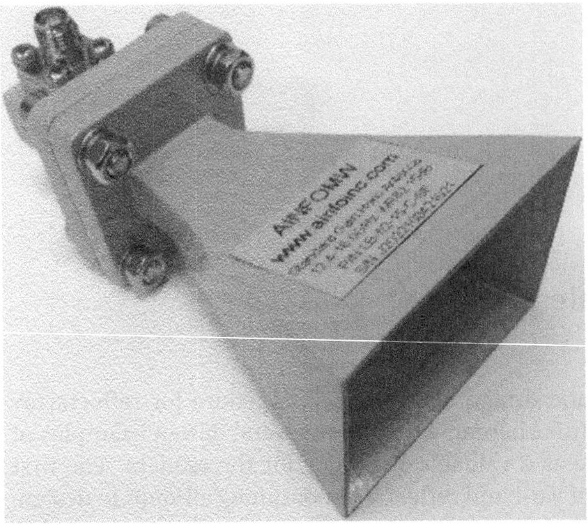

Figure 6.1 The LB 62–15 pyramidal horn antenna from A-INFO.

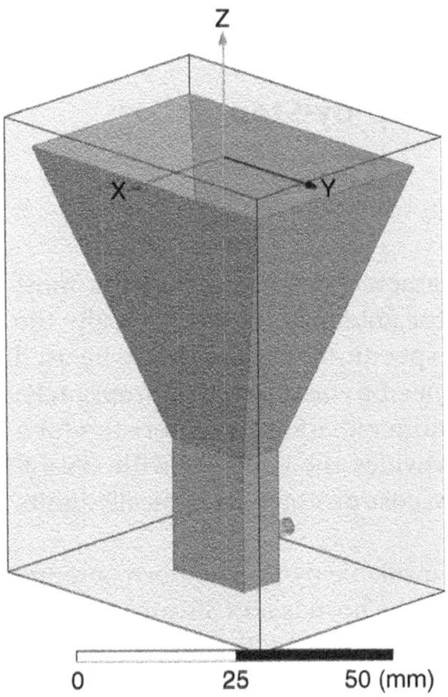

Figure 6.2 The feed horn antenna model in ANSYS HFSS.

at 14.25 GHz are shown in Figure 6.3. Note that in this setup, E- and H-planes correspond to *xz*- and *yz*-planes, respectively. It can be seen that the field distributions inside the horn antenna clearly show the transition from guided to free space waves.

The horn is a classic aperture antenna, and as such, its radiation pattern is a manifest of the field distribution on its aperture. The tangential electric fields on the aperture are given in Figure 6.4, where it can be seen that, as expected for a pyramidal horn, the co-polarized (E_x) amplitude taper is different in the two orthogonal directions. Due to the difference in both size and taper of the rectangular shaped aperture, the radiation pattern of a pyramidal horn will exhibit some asymmetry [6], [7].

As shown in Figure 6.5, the horn radiates a directive beam, which is quite suitable for illuminating a reflectarray aperture. The simulated peak gain of the antenna is 15.74 dBi, which is very close to the value reported in the data sheet [4]. The half-power beam-width in both E- and H-planes is about 30 degrees, and the side-lobe level is −14 dB. As discussed earlier, the pattern of the feed antenna is slightly asymmetric; and while for a reflectarray a feed antenna with a symmetric radiation pattern is usually more desirable, the difference in this feed horn is quite small. In fact, the 2D radiation patterns given in Figure 6.6 show that the patterns in both planes match up very closely up to about ±30°, which is quite sufficient for a reflectarray antenna design.

As discussed in Chapter 3, it is desirable to use a radiation pattern model for the feed antenna in order to effectively design and analyze the reflectarray system. For this design, we use the cosine-q radiation pattern model for the feed antenna. The simulated radiation pattern of the antenna is therefore matched to a cosine-q model in order to

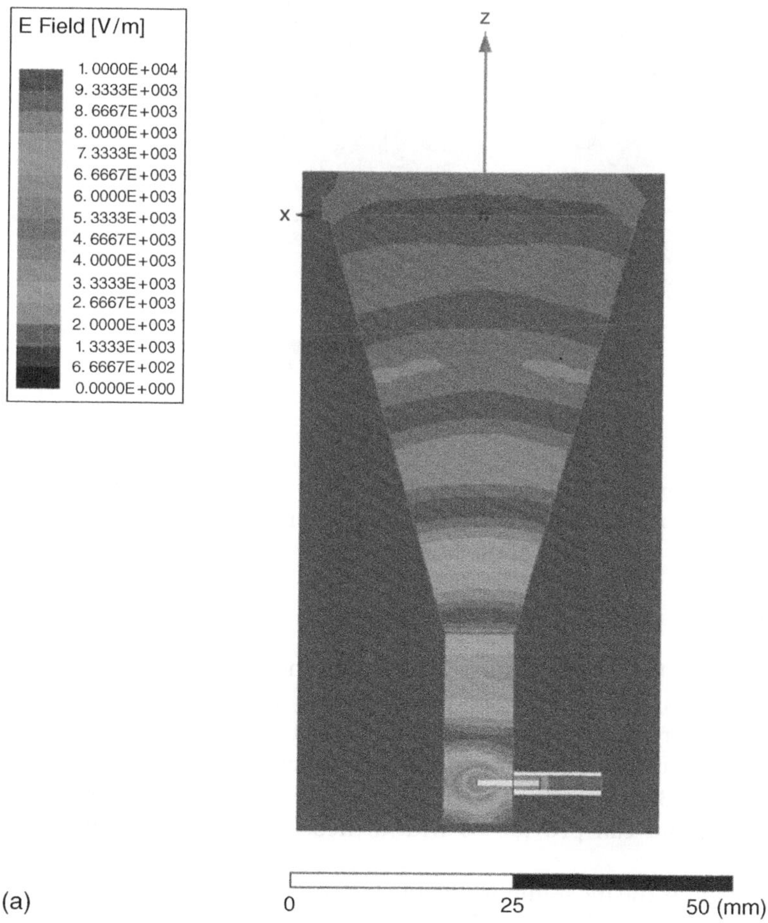

(a)

Figure 6.3 The electric field magnitude inside the horn antenna at 14.25 GHz: (a) E-plane, (b) H-plane.

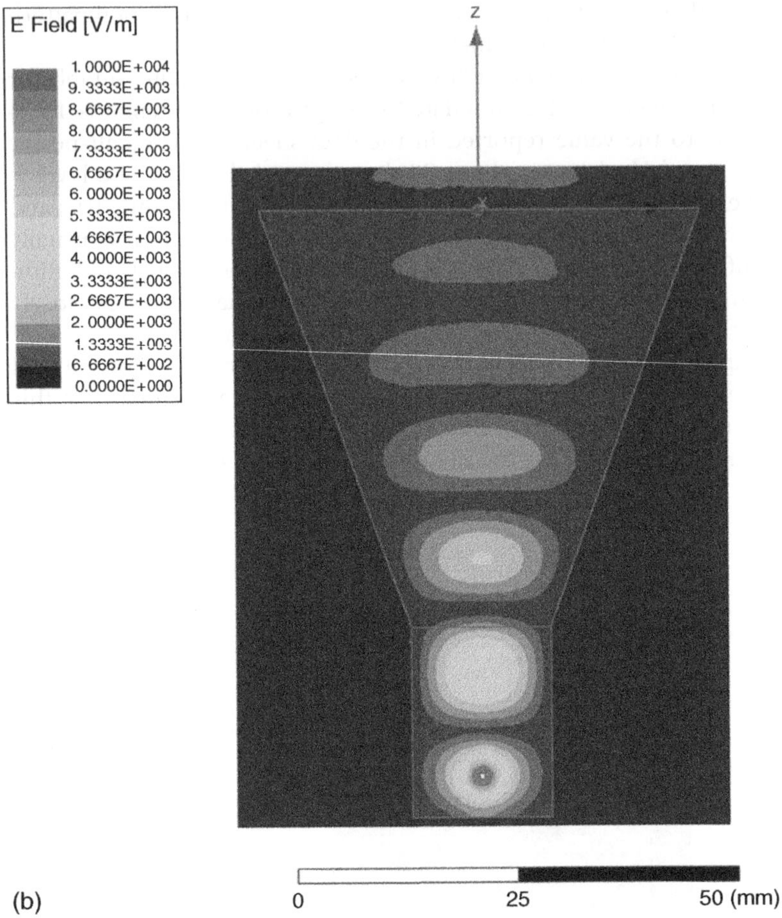

(b)

Figure 6.3 (Continued)

determine the value of q. For this horn the value of q is 10.5, which matches the simulated patterns very well up to about ±30°. Within this angular range, the phase center of this horn is 10.16 mm below the surface of the horn aperture.

6.1.2 Reflectarray System Design

With the feed antenna selected, the next stage is to design the reflectarray system. As discussed in earlier chapters, the antenna gain relates directly to the aperture size, and as such selecting the appropriate aperture size is typically the first step in system design. In order to achieve a moderately high gain with this reflectarray, we select a circular aperture with a diameter of 360 mm, which is about 17 wavelengths at the center design frequency of 14.25 GHz, and can provide us with a theoretical maximum aperture directivity of 34.6 dBi.

Feed antenna placement is the most important and influential step in the reflectarray system design. To avoid feed blockage, we design an offset system with a feed offset angle of 15°, where the feed points to the geometrical center of the array. The optimum

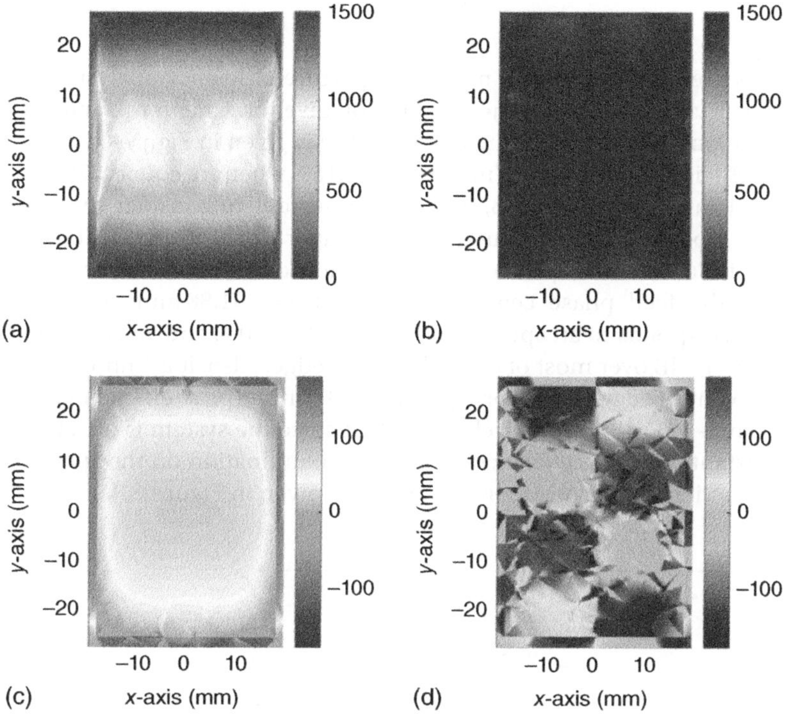

Figure 6.4 Electric fields on the aperture of the horn antenna: (a) $|E_x|$ (V/m), (b) $|E_y|$ (V/m), (c) phase of E_x (degrees), and (d) phase of E_y (degrees).

Figure 6.5 A 3D pattern of the horn antenna.

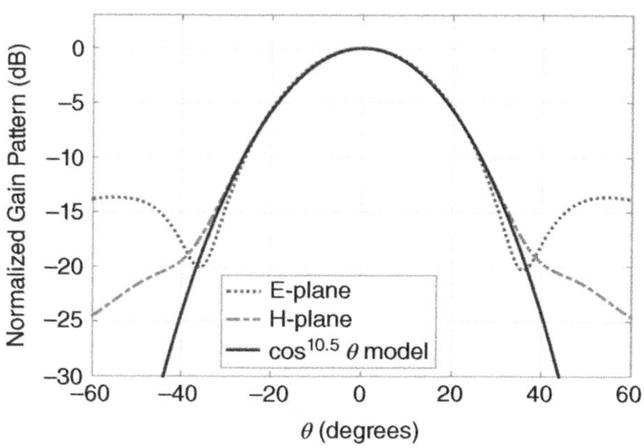

Figure 6.6 Normalized and cosine-*q* model of the radiation patterns of the horn antenna.

position of the feed is then determined by means of aperture efficiency analysis. The reflectarray efficiency is computed using the formulation given in Chapter 3, and efficiency as a function of focal point to aperture distance (H_F) is given in Figure 6.7.

The peak aperture efficiency for this system is about 78.15% when H_F equals 370 mm. However, as discussed earlier in Chapter 3, it is also important to consider the edge taper on the aperture. Typically, a slight reduction of peak aperture efficiency will result in a better edge taper for a reflectarray. After some parametric studies, we select H_F to be 342.9 mm, that is the feed phase center is placed at $x = -91.88$ mm, $y = 0$, and $z = 342.90$ mm. This corresponds to an aperture efficiency of 78.10%, and an edge taper which is almost below −10 dB over most of the reflectarray edge. Also, it is important to point out that the maximum edge tapers for H_F equal to 370 mm (peak efficiency) and 342.9 mm are −8.57 and −9.65 dB, respectively. Note that since the system is offset, the edge taper is not constant over the reflectarray edge. The illumination on the aperture and edge taper as a function of aperture azimuth angle are given in Figure 6.8.

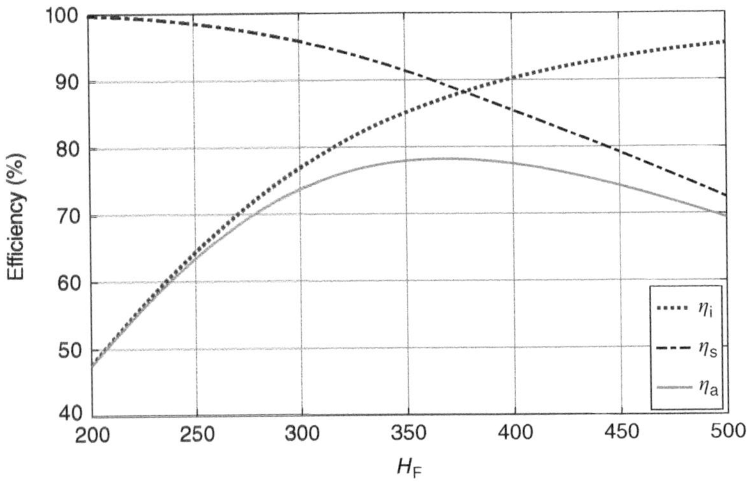

Figure 6.7 Reflectarray aperture efficiency as a function of focal point to aperture distance.

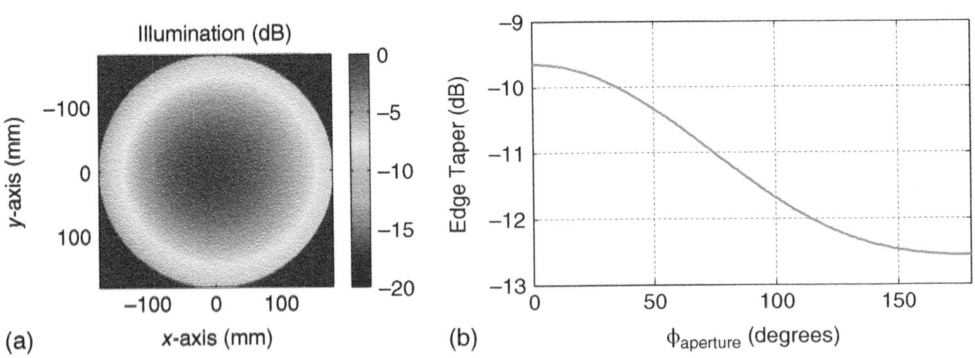

Figure 6.8 (a) Illumination on the reflectarray aperture. (b) Edge taper as a function of aperture azimuth angle.

Table 6.1 Design parameters for the offset reflectarray and reflector.

Reflectarray	D_{RA}	H_F	Δ_B	θ_{offset}
	360 mm	342.9 mm	0 mm	15°
Reflector	D	F	H	ψ_B
	150.3 mm	343.44 mm	−12.45 mm	28.32°

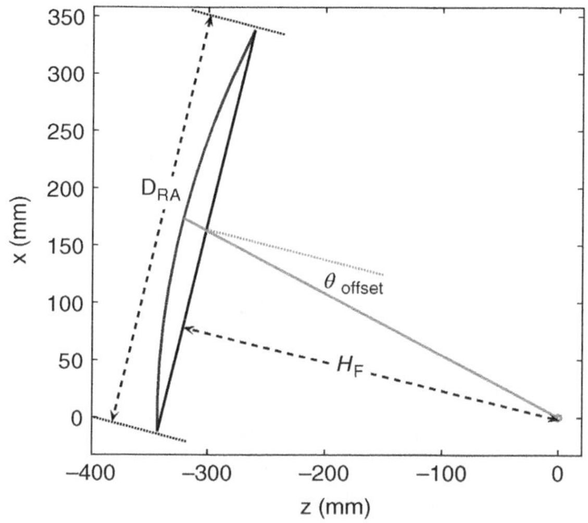

Figure 6.9 Geometrical setup of the reflectarray and reflector in the reflector coordinate system.

As discussed in earlier chapters, the reflectarray systems designed in this section, along with the analogous reflector, can each be completely defined with four design parameters. These parameters are given in Table 6.1. Note that with this setup the reflector offset height is a negative number, which would cause some blockage for such a reflector system. However, one of the advantages of reflectarrays over reflectors is that the direction of the main beam can be further controlled by adding a progressive phase on the aperture. In this design, we select a main-beam direction of 15° for the reflectarray, which will ensure that no feed blockage will be observed for the system and will also minimize beam squint for the reflectarray [8]. The geometrical setup of the reflectarray and the analogous reflector antenna is given in Figure 6.9.

6.1.3 Reflectarray Element Design

A critical step in designing the reflectarray antenna aperture, is designing appropriate phasing elements. These elements need to provide a phase shift in order to compensate for the spatial delay of the feed antenna, as well as the necessary progressive phase to generate a collimated beam in the required direction. This reflectarray is designed to focus the main beam 15° off broadside. Using the formulation presented in Chapter 2, and the selected position for the feed antenna, the required phase shift on the aperture can be computed. An important consideration here is the size of the unit-cells. Here we select a square lattice with a side dimension of 10 mm, which is 0.475 half-wavelength at the center design frequency. With a diameter of 360 mm for the circular aperture of the reflectarray, this corresponds to 936 elements on the aperture. An image of the grid layout for the array and the required phase shift on the aperture are both given in Figure 6.10.

Now that we have determined the required phase shift on the aperture, we need to design a phasing element that can provide this phase distribution on the aperture. For this reflectarray we select the variable size technique and use a classic square patch element. A 31 mil Taconic TLX8 laminate ($\varepsilon_r = 2.55$, tan $\delta = 0.0018$) is selected for the substrate. The thickness is about 0.06 wavelengths in the dielectric, which should provide a good phase tuning range for the element. The element unit-cell is modeled in Ansoft Designer [9], using the periodic boundary conditions. A snap shot of the simulation model is given in Figure 6.11.

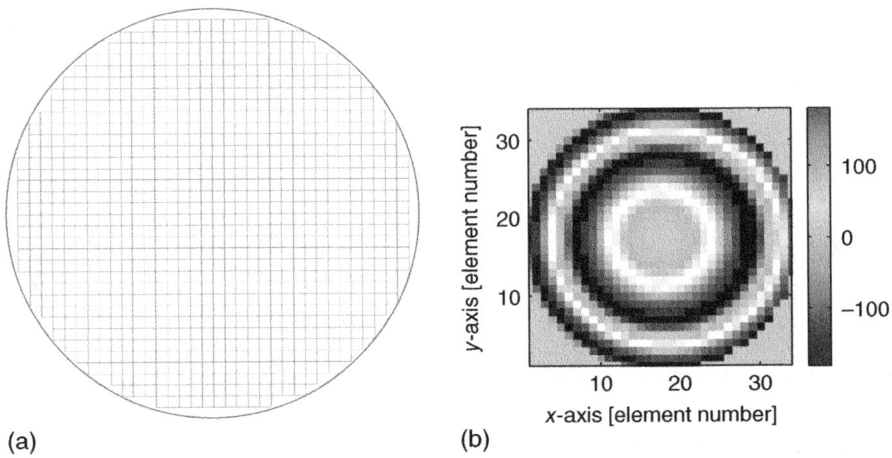

Figure 6.10 (a) Grid layout of the array. (b) Required phase shift on the reflectarray aperture.

Figure 6.11 Reflectarray element simulation model.

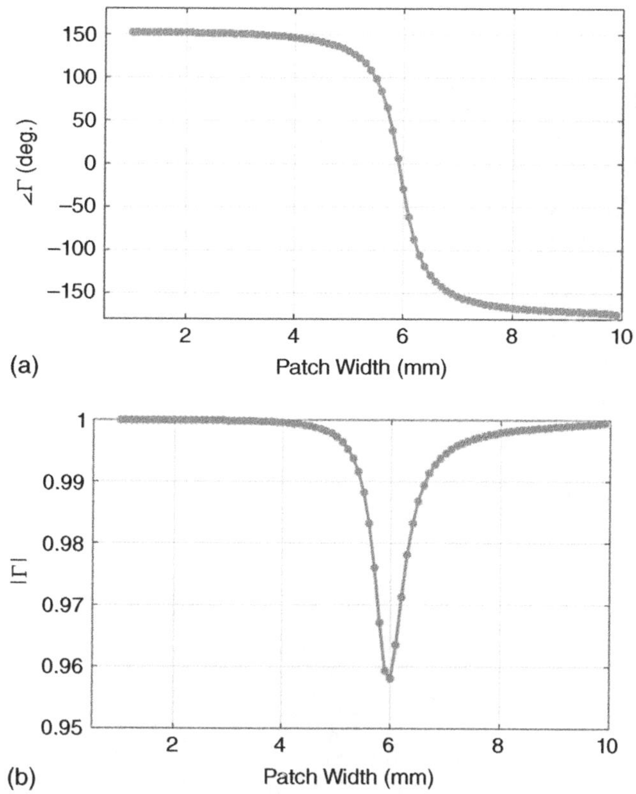

Figure 6.12 Reflection coefficients of the element at 14.25 GHz: (a) phase, (b) magnitude.

The reflection coefficients of the element as a function of patch width are given in Figure 6.12. This element provides a phase tuning range of 326°, which is quite sufficient for a reflectarray antenna. In addition, these elements provide a reflection magnitude that is close to unity, thus almost all the power impinging on the reflectarray aperture will be reflected. Note that due to fabrication considerations, the minimum and maximum sizes of the patches are selected to be 0.5 and 9.9 mm, respectively. While as discussed earlier, the phase variation obtained with these elements is non-linear, with a fabrication precision of 0.05 mm (about 2 mils), the maximum phase quantization errors will be about 15°, which is acceptable for this relatively large array.

The phasing elements designed here provide the means to realize the desired reflection phase on the aperture of the reflectarray antenna. Using the data in Figure 6.12(a), we select patch widths that provide reflection phases that match the ideal required phase distribution of Figure 6.10(b), to the best extent possible. The mask model of the reflectarray antenna is shown in Figure 6.13. It should be noted here that the elements are designed under normal incidence approximation, however, as discussed earlier; this approximation is quite valid for microstrip elements with thin substrates.

Due to fabrication tolerances in the design, and the limited phase range of the elements that are less than the full phase cycle, the reflection phase of the reflectarray elements will not exactly match that of Figure 6.10(b). However, in our design, these

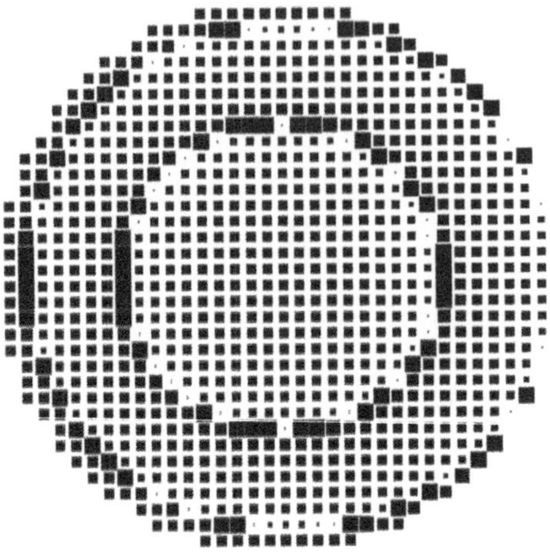

Figure 6.13 Mask plot of the reflectarray antenna.

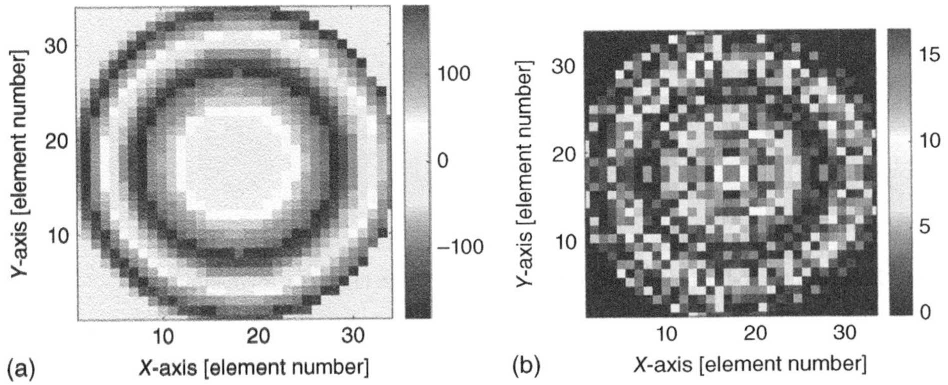

Figure 6.14 (a) Quantized phase distribution on the aperture. (b) Quantization phase errors on the aperture.

errors are relatively small. The quantized phase distribution on the aperture, as well as the relative phase errors are given in Figure 6.14, where it can be seen that the aperture phase distribution obtained by these elements is very close to the ideal case.

6.1.4 Radiation Analysis

With the reflectarray mask designed in the previous section, we have completed all the design stages of the reflectarray antenna. The next important task is to verify the design by analyzing the radiation characteristics of the antenna. In this section, we compute the radiation pattern of the antenna using two approaches: the array theory approach and the full-wave simulation of the entire reflectarray. We start with the robust array theory analysis technique described in Chapter 4. This method of analysis is very fast and can help us get a good idea about the radiation characteristics of the antenna. Two-dimensional plane cuts of the radiation pattern of the reflectarray

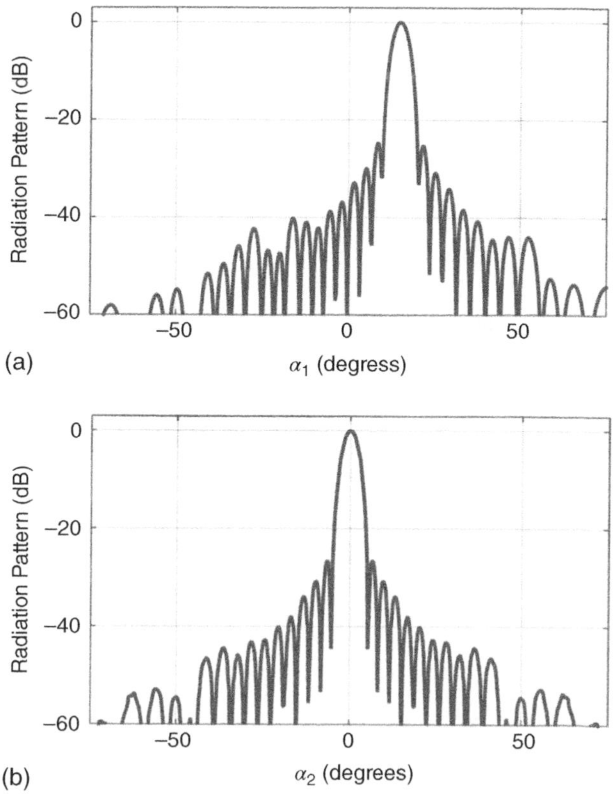

Figure 6.15 Radiation pattern of the reflectarray antenna in the principal planes: (a) P.P.1, (b) P.P.2.

in its principal planes at 14.25 GHz are given in Figure 6.15, where a high gain pencil beam radiation pattern is clearly observed. Also note that the beam is correctly scanned to 15° in the elevation plane (P.P.1). Note that these radiation patterns are obtained using ideal elements.

In the next stage of the analysis, we now compute the radiation pattern of the antenna using the realistic phase shift, which are obtained by the patch elements. Note that in this case, limited phase range of the patch elements, along with the quantization effects of the elements phase, leads to some errors in the antenna performance. Comparison between the radiation patterns of the antenna using ideal elements and practical patch elements is given in Figure 6.16, where it can be seen that while the phase errors give rise to some degradation in antenna pattern, the effects are almost negligible. The radiation pattern of the antenna in the *uv*-plane is also given in Figure 6.17, where it can be seen that the reflectarray shows a very directive pattern in the entire 3D space.

In the final stage of radiation analysis, we compute the radiation pattern of the reflectarray antenna with a full-wave electromagnetic simulation software. This method of analysis is highly accurate since it takes into account all approximations in the earlier approaches, however, the cost is a significantly high computational time and large memory requirements. For this simulation, the total computational time was 60 hours

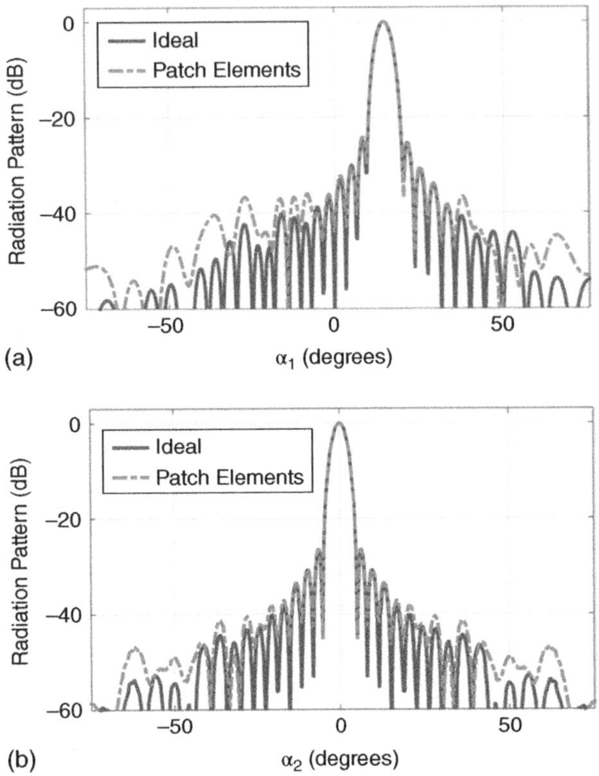

Figure 6.16 Comparison of the radiation pattern of the reflectarray antenna at 14.25 GHz using ideal and patch elements: (a) P.P.1, (b) P.P.2.

and 40 minutes, on an Intel Xeon E5–2680 processor, and the simulation utilized 141.1 GB of memory. Nonetheless, if resources are available, one generally would use this approach before prototyping the reflectarray. For our design, we simulated the antenna in ANSYS HFSS. Images of the simulation setup are given in Figure 6.18. The 3D radiation pattern of the reflectarray antenna is given in Figure 6.19, where it can be seen that a pencil beam radiation pattern is obtained with this reflectarray antenna as expected.

It is also interesting to compare the radiation pattern obtained by the full-wave simulation with the array theory approach. These are given in Figure 6.20, where it can be seen that a good agreement is observed between these approaches, in particular in the area of the main beam and the first side lobes. Outside the main-beam area generally the full-wave approach is considered to be more accurate. It is also important to note that the directivity computed using the array theory approach and the one obtained using HFSS are 33.49 dB and 32.21 dB, respectively. This 1.28 dB discrepancy in directivity is partially attributed to the approximations in the array theory and aperture efficiency analysis as described in Chapters 3 and 4.

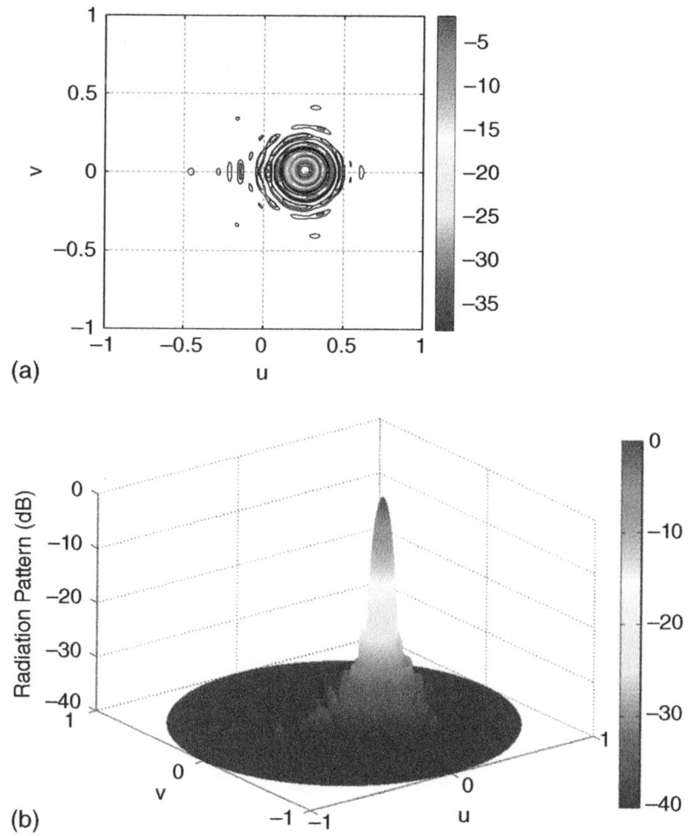

Figure 6.17 Radiation pattern of the reflectarray at 14.25 GHz in the *uv*-plane: (a) contour plot, (b) 3D pattern.

6.1.5 Fabrication and Measurements

In the final stage of the design process, the reflectarray antenna is fabricated and tested. Measurements are conducted using a planar NSI near-field scanner. A photo of the test setup is given in Figure 6.21. The waveguide probe is an OEWG WR62, which operates from 12.4 to 18 GHz. The distance between the waveguide probe and the reflectarray aperture is 24 inches. The sampling plane dimension is 30.0×30.0 inch^2 and the sample spacing is set to 0.25 inch. The average amplitude taper at the edge of the sampling plane is below 20 dB. Note that with the planar near-field system, far-field results are only valid within a certain observation angle, which is determined based on the size of the sampling plane, and the distance between the antenna under test (AUT) and the sampling plane. For this test setup, the maximum far-field angle was set to ±45°.

The sampled near-field electric fields from the reflectarray antenna are given in Figure 6.22. Note that the horn antenna is x-polarized, hence the y-component of the sampled electric field has a significantly smaller magnitude. The sampled

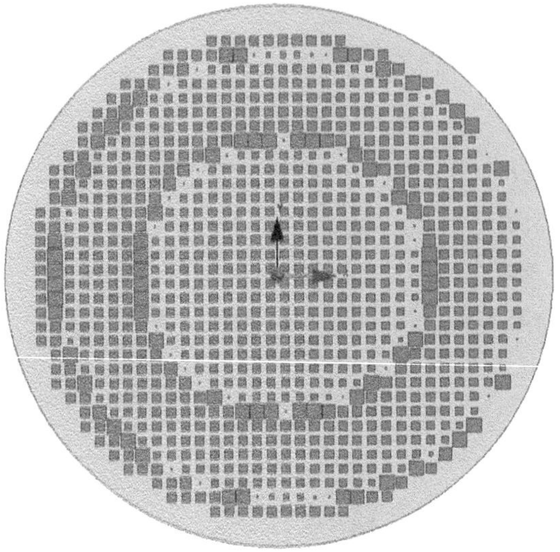

(a)

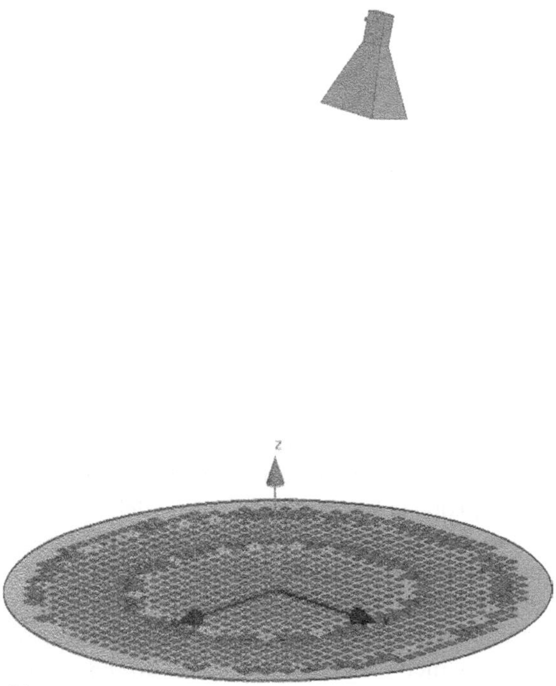

(b)

Figure 6.18 Images of the reflectarray antenna in ANSYS HFSS: (a) top view without the feed, (b) 3D view with the feed horn.

x-component of the electric field shows a strong field at the center of the sampling plane as expected. More importantly note that in the area of the strong illumination, the phases of these fields also form parallel lines indicating a progressive phase shift to generate a titled beam.

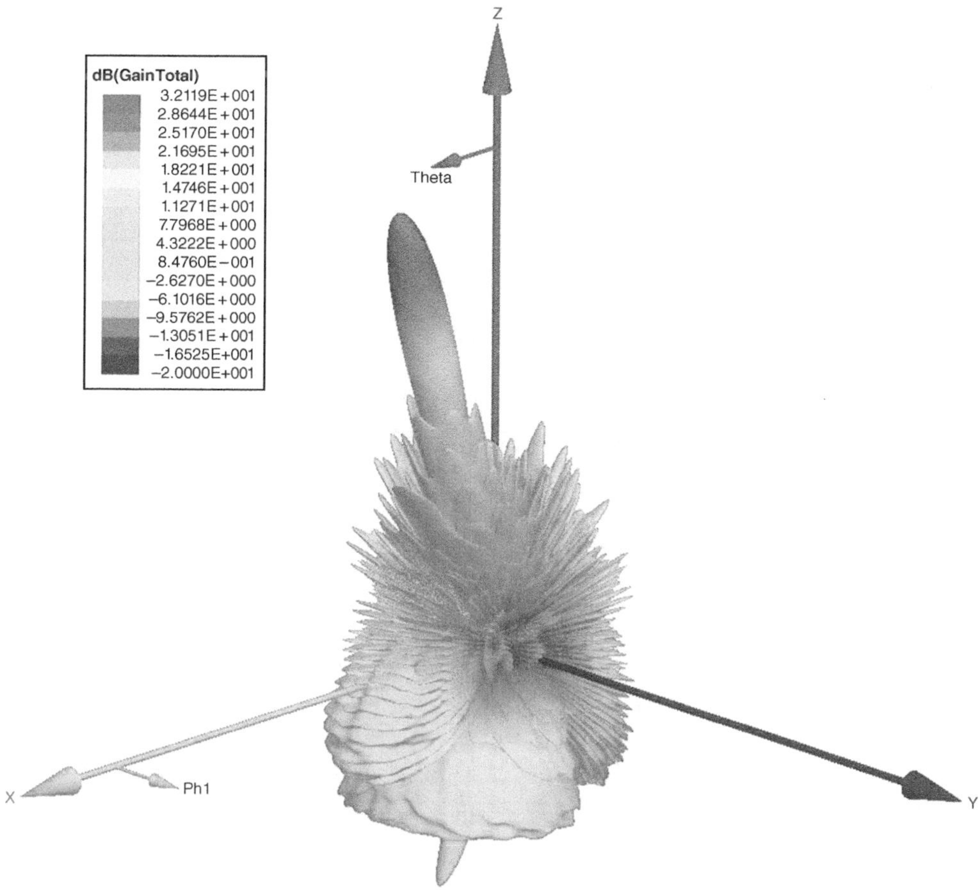

Figure 6.19 A 3D radiation pattern of the reflectarray antenna at 14.25 GHz obtained using ANSYS HFSS.

From these sampled near-fields, the far-field radiation pattern of the reflectarray antenna is obtained, which is given in Figure 6.23. It can be seen that these results match closely with the simulated results in Figure 6.17.

The measured radiation patterns obtained here are also compared with the results of the full-wave simulations in Figure 6.24, where it can be seen that a good agreement is observed. It is worthwhile to note that the discrepancies outside the area of the main beam is attributed to the fact that measured results obtained using planar near-field systems tend to be less accurate outside a nominal 60° angular distance from the main beam [10]. Nonetheless, a very good agreement is observed in the main-beam area. Also note that cross-polarization is below 33 dB in the two principal planes.

The measured gain and aperture efficiency of the antenna as a function of frequency are also given in Figure 6.25. The peak gain and aperture efficiency of this reflectarray are 32.28 dB and 59.8%, which are obtained at 14.1 GHz. At the center design frequency of 14.25 GHz, the measured gain is 32.06 dB, which is quite close

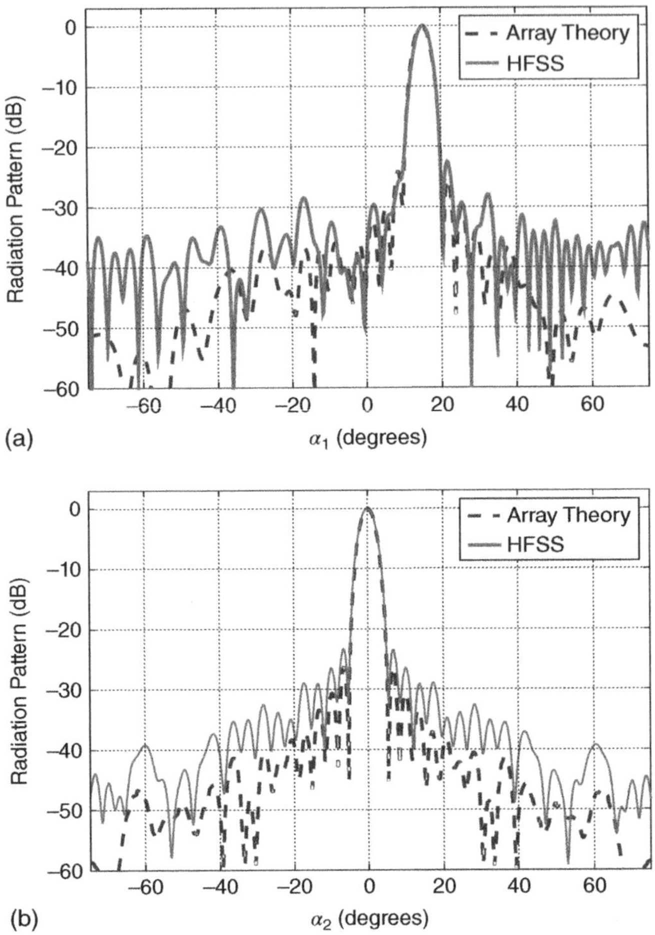

Figure 6.20 Comparison of the radiation patterns of the reflectarray antenna at 14.25 GHz obtained using two different approaches: (a) P.P.1, (b) P.P.2.

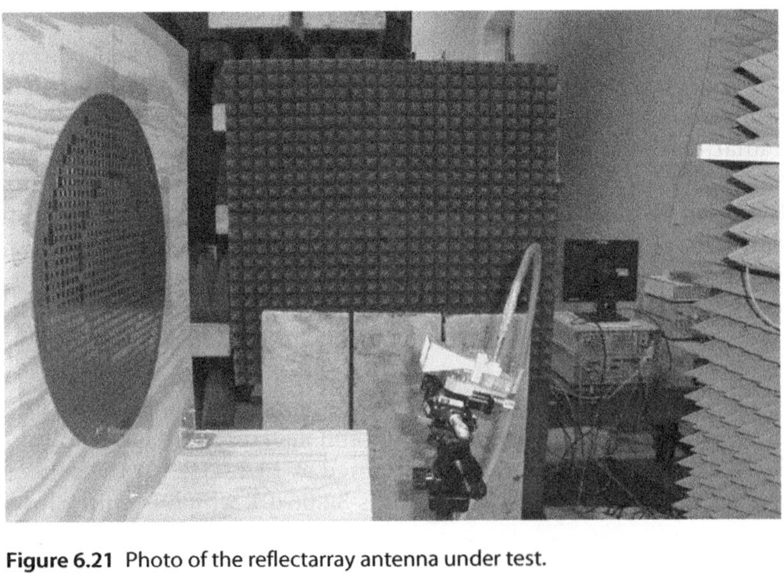

Figure 6.21 Photo of the reflectarray antenna under test.

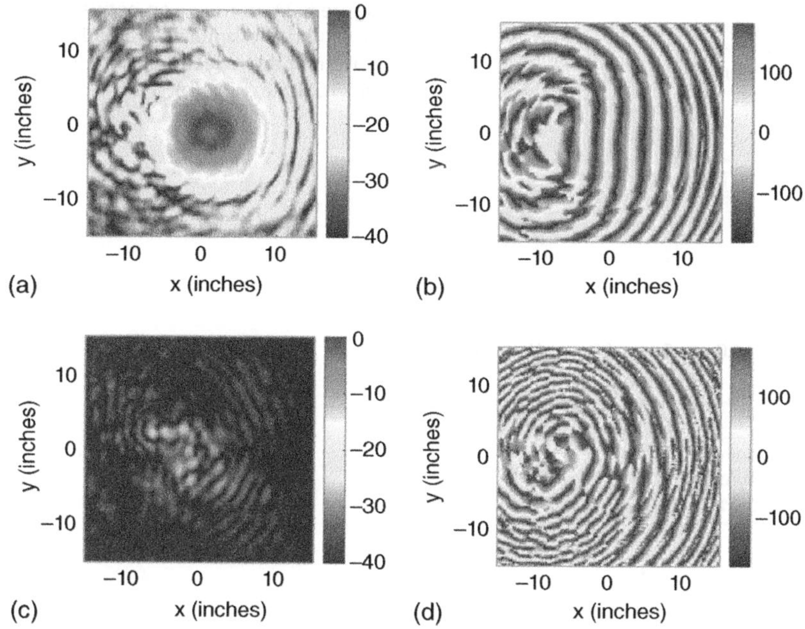

Figure 6.22 Electric fields sampled by the near-field scanner at 14.25 GHz: (a) $|E_x|$ in dB, (b) phase of E_x in degrees, (c) $|E_y|$ in dB, (d) phase of E_y in degrees.

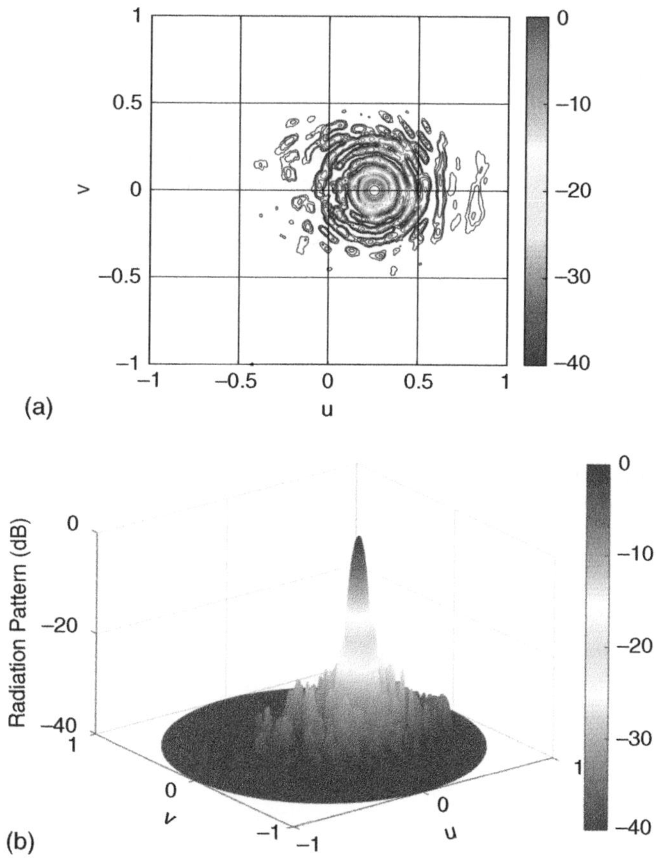

Figure 6.23 Measured radiation pattern of the reflectarray at 14.25 GHz in the *uv*-plane: (a) contour plot, (b) 3D pattern.

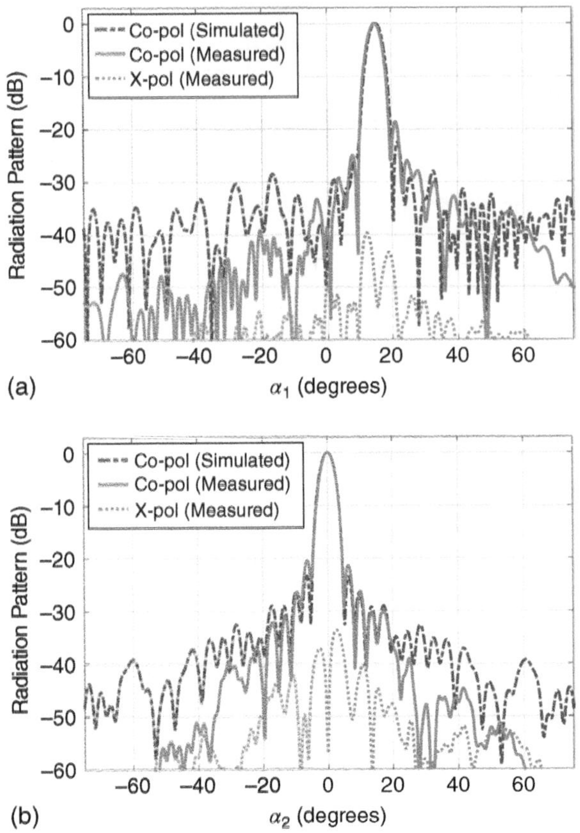

Figure 6.24 Comparison between the measured and simulated radiation pattern of the reflectarray antenna at 14.25 GHz: (a) P.P.1, (b) P.P.2.

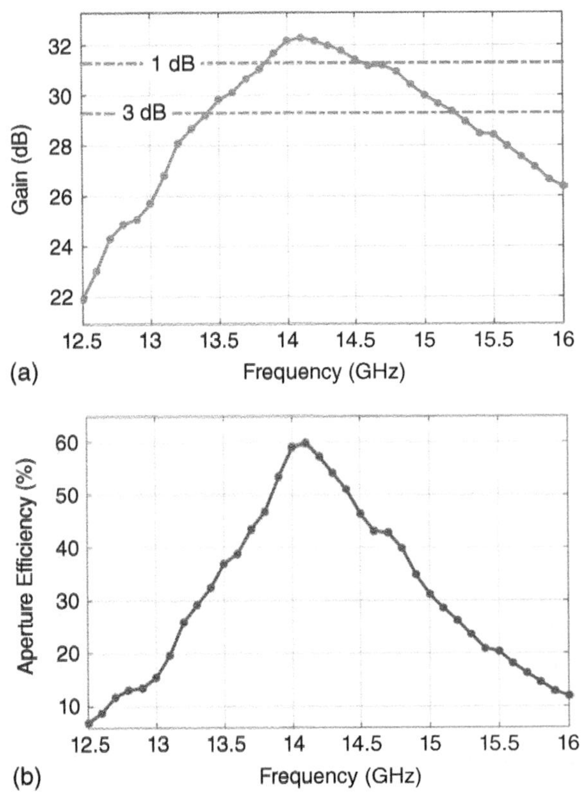

Figure 6.25 Measured performance of the reflectarray antenna as a function of frequency: (a) gain, (b) aperture efficiency.

to the value of 32.12 dB obtained using full-wave simulations. The 1 dB gain bandwidth of the reflectarray (13.84–14.55 GHz) is about 5%, while the 3-dB gain bandwidth 13.42–15.21 GHz is about 12.5%. The relatively narrow bandwidth of this reflectarray antenna is primarily attributed to the thin substrate used for the reflectarray elements.

6.2 A Circularly Polarized Reflectarray Antenna using an Element Rotation Technique

In this section, we present a design for a circularly polarized reflectarray antenna using the single split ring element described in Chapter 2. The feed antenna is a circularly polarized corrugated conical horn antenna. The antenna has two ports allowing for both right-hand and left-hand circular polarizations, where the septum polarizer switch controls the polarization of the antenna [11]. For this design, the polarization of the feed antenna is set to be left-handed. A photo of this horn antenna is given in Figure 6.26.

As discussed earlier, the common approach to characterize the radiation performance of a feed horn antenna is to use a cosine-q radiation pattern model, where the value of the parameter q is determined from the horn measurements. The measured radiation pattern of the horn antenna at 32 GHz is given in Figure 6.27, where it is also compared with a cosine-q model with $q = 6.5$. It can be seen that, in both planes, the cosine-q model gives a good representation of the horn pattern.

The geometrical model of the single split ring (SSR) element labeled with design parameters is shown in Figure 6.28. The element is designed for 32 GHz operation. A 31 mil Rogers 5880 laminate is used for the element. The design parameters are $R_{in} = 1.6$ mm, $R_{out} = 1.8$ mm, and $\theta_s = 118°$.

The magnitudes of the reflected co-polarized (LHCP) and the cross-polarized (RHCP) components, with a normal incident plane wave source, are depicted in Figure 6.29(a). The reflection phase of the co-polarized component as a function of rotation angle was

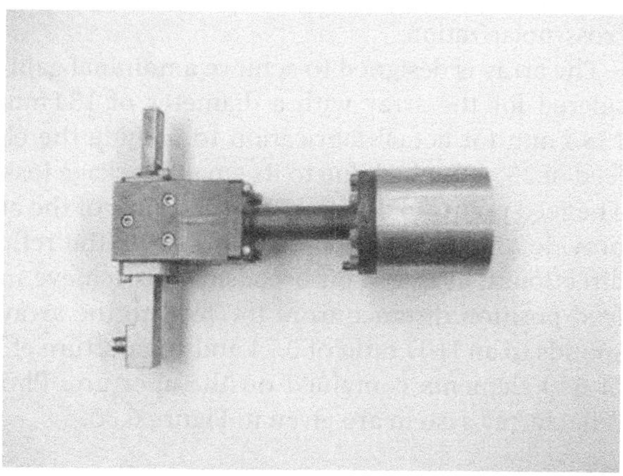

Figure 6.26 Photograph of the Ka-band feed horn antenna.

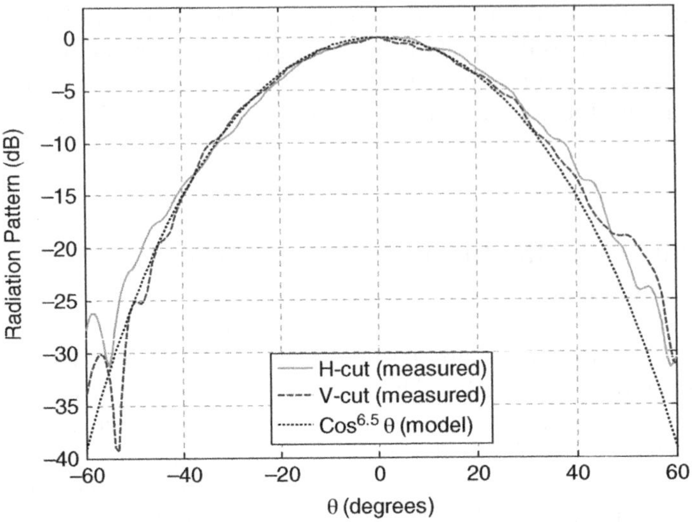

Figure 6.27 Measured far-field patterns of the Ka-band circularly polarized horn antenna.

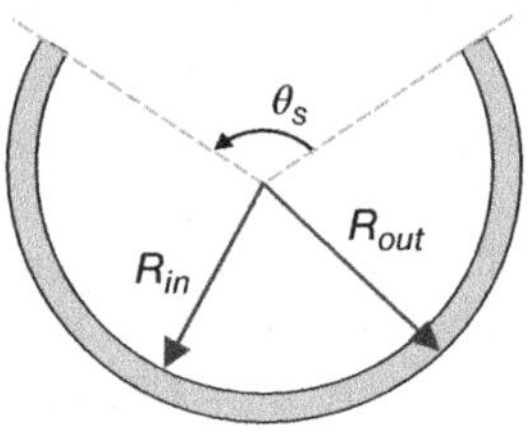

Figure 6.28 The SRR element along with the design parameters.

given in Chapter 2, and thus not repeated here. As discussed earlier, the cross-polarization level may increase to a large value when the incident angle θ grows up to 50° [12], as shown in Figure 6.29(b). When a prototype is built, the number of elements under an extreme incident angle is usually limited due to the system configuration. Hence this effect will be more or less mitigated. However, one may still foresee a relatively high cross-polarization.

The array is designed to achieve a nominal gain of 30 dB. A circular aperture is considered for the array with a diameter of 134 mm. This value has been extended to 138.2 mm for actual fabrication to include the elements on the boundary. An offset feed of 25° is applied due to its small blockage loss and convenience for measurement. The feed points to the geometrical center of the array, and the elements are rotated to provide a progressive phase that directs the reflectarray main beam in the specular direction, that is, 25° off broadside. To achieve maximum aperture efficiency, H, the feed position distance from the feed to the array plane is set to 98 mm. This corresponds to an H/D ratio of 0.74 and an aperture efficiency of 75%. The array has a total of 640 elements contained on the aperture. Photos of the array prototype and the reflectarray system are given in Figure 6.30.

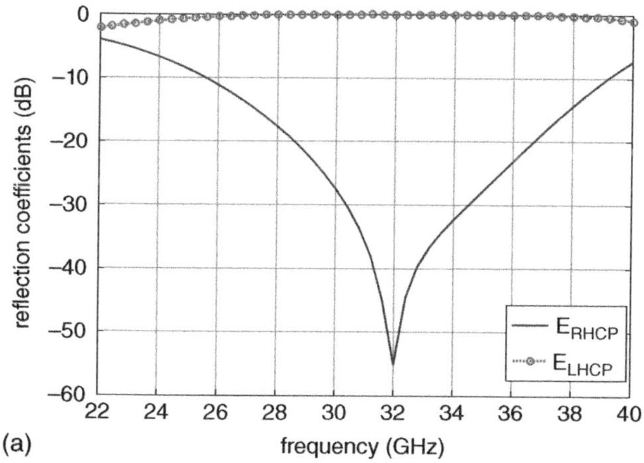

(a)

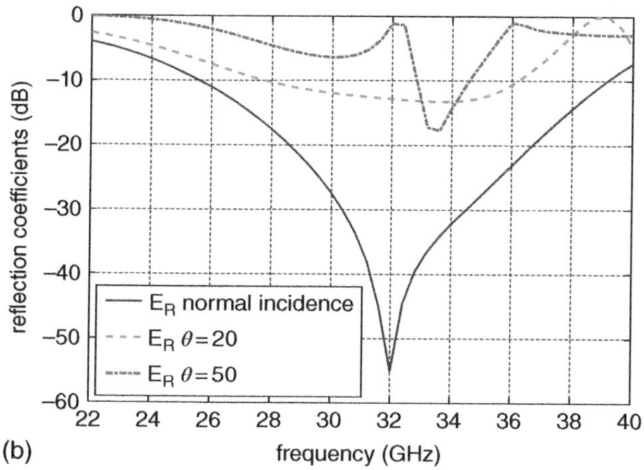

(b)

Figure 6.29 Magnitudes of co- and cross-polarized element reflection coefficients as a function of frequency. (b) Effects of incident angle on the reflection coefficient magnitudes as a function of frequency.

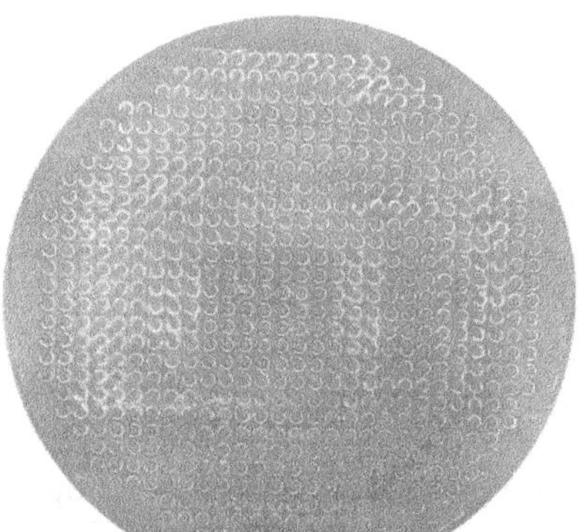
(a)

(b)

Figure 6.30 Fabricated circularly polarized reflectarray antenna: (a) array prototype, (b) reflectarray system.

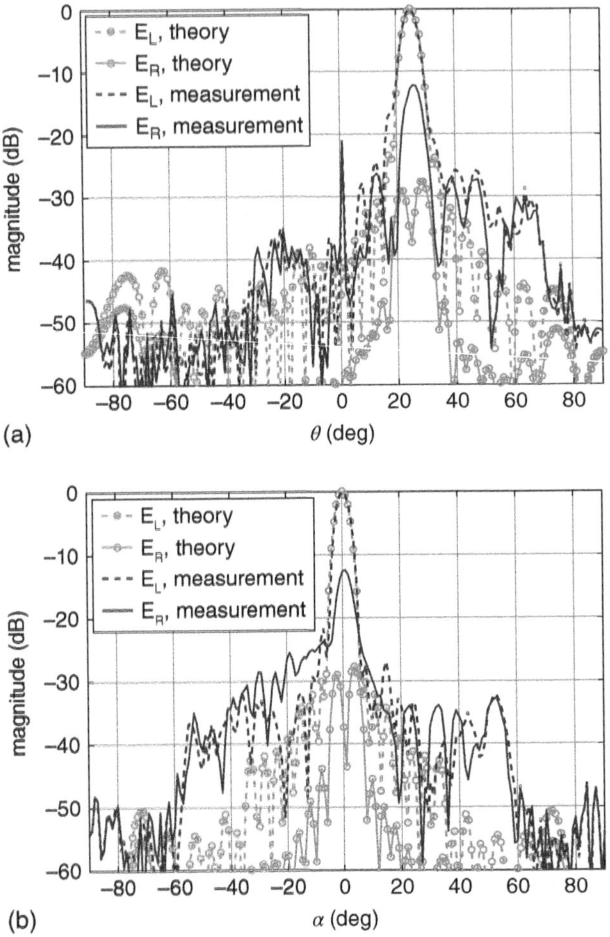

Figure 6.31 Comparison between measured and simulated radiation patterns of the reflectarray antenna at 32 GHz: (a) P.P.1, (b) P.P.2.

The radiation pattern of the antenna is computed using the aperture field method, and the antenna is tested using the planar near-field measurement system. Comparisons between the measured and simulated results are given in Figure 6.31. The results agree well with each other in the main lobe region. Deviations in the side lobes below −20 dB are mainly due to the diffraction of the incidence at the edge of the finite ground plane. The measured main beam is found at 24.2°. The slight deviation to the designed value of 25° could be due to the error of alignment in the measurement setup. The side-lobe levels are below −20 dB in both planes. The 3 dB beam widths are 4.6° and 5.2° in two principal planes, respectively. A relatively large cross-polarization is observed which is primarily attributed to the high cross-polarization of the feed antenna.

6.3 Bandwidth Comparison of Reflectarray Designs using Different Elements

As discussed in the previous chapters, different phasing elements have been proposed for reflectarray antennas. While they all provide the necessary phase shift needed to collimate the beam, they do exhibit different characteristics. In this section, we will compare the effect of different geometries on the reflectarray performance, in particular, the bandwidth characteristic.

To quantify some of these differences, in this section, three different reflectarray antennas, using square patch, cross dipole, and square loop elements, are compared. The geometrical models of the three different elements are depicted in Figure 6.32. The system setup for the measurement is also given in Figure 6.33. The horn antenna

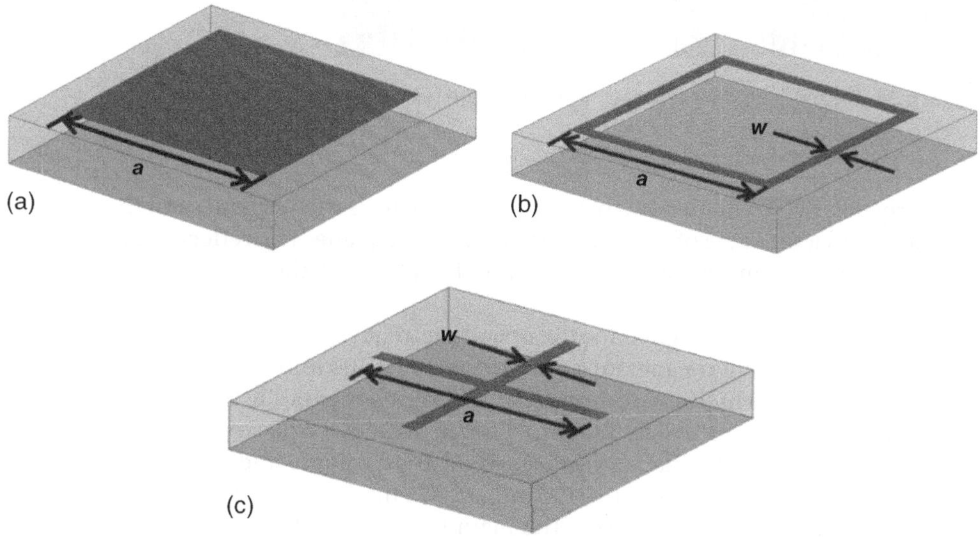

Figure 6.32 Geometrical models of variable size elements: (a) square patch, (b) square loop, and (c) cross dipole.

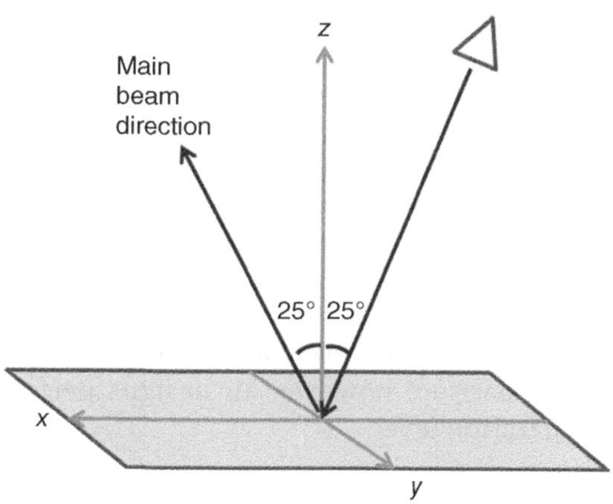

Figure 6.33 Coordinate setup for the reflectarray test.

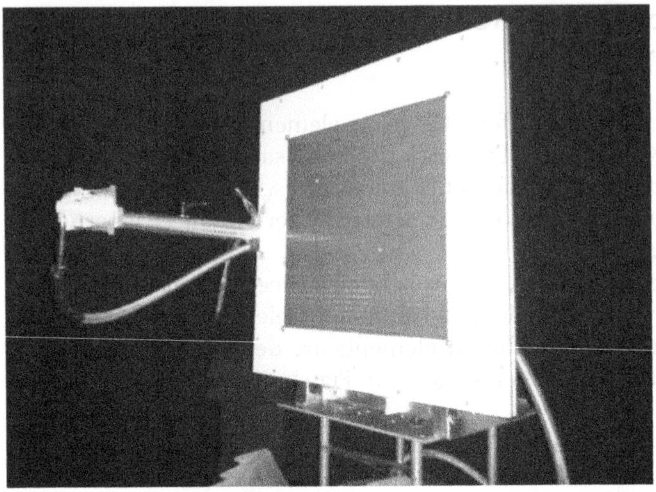

Figure 6.34 The measurement setup for far-field pattern measurement.

is linearly polarized, where note that here, the H-plane is the *xz*-plane, and the E-plane is a plane with a tilt angle of 25° with respect to the *yz*-plane. The reflectarray antennas were measured in an anechoic chamber, and a photo of the test setup is given in Figure 6.34.

The reflectarrays are designed for the operating frequency of 13.5 GHz. All reflectarrays have the same square aperture measuring 400 × 400 mm^2. The element spacing is set to 11.2 mm, which corresponds to 35 × 35 elements on the aperture of the reflectarray. A 1.58 mm thick Taconic TLX-8, with $\varepsilon_r = 2.55$ and $tan\ \delta = 0.0019$ is used for all designs. The planar surfaces lie in the *xy* plane and is illuminated by a linear-polarized horn located at (−175 mm, 0, 375 mm). The phase on the reflectarray apertures are designed to direct the main beam to $\theta = 25°$, $\varphi = 0°$ with respect to the coordinate system shown in Figure 6.33 [13]. Photos of the three reflectarray prototypes are given in Figure 6.35.

Comparison of the calculated and measured radiation patterns in E-plane and H-plane are presented in Figure 6.36 and Figure 6.37, respectively. While the agreement between measured and simulated results is quite reasonable for all designs, note that in comparison, the cross dipole shows the largest discrepancy. This is primarily attributed to the limited accuracy of the fabrication process, and the highly non-linear slope of the reflection phase curve for cross dipoles. In other words, square patches or loops generally perform better than cross dipoles, if sufficient fabrication precision cannot be realized for the given design.

The measured co- and cross-polarization performance in both planes are also given in Figure 6.38 and Figure 6.39, respectively. It can be seen that in all cases, the cross-polarization levels are below −25 dB in E-planes, and even less in H-planes. In comparison, however, all designs perform very similar in terms of the cross-polarization level.

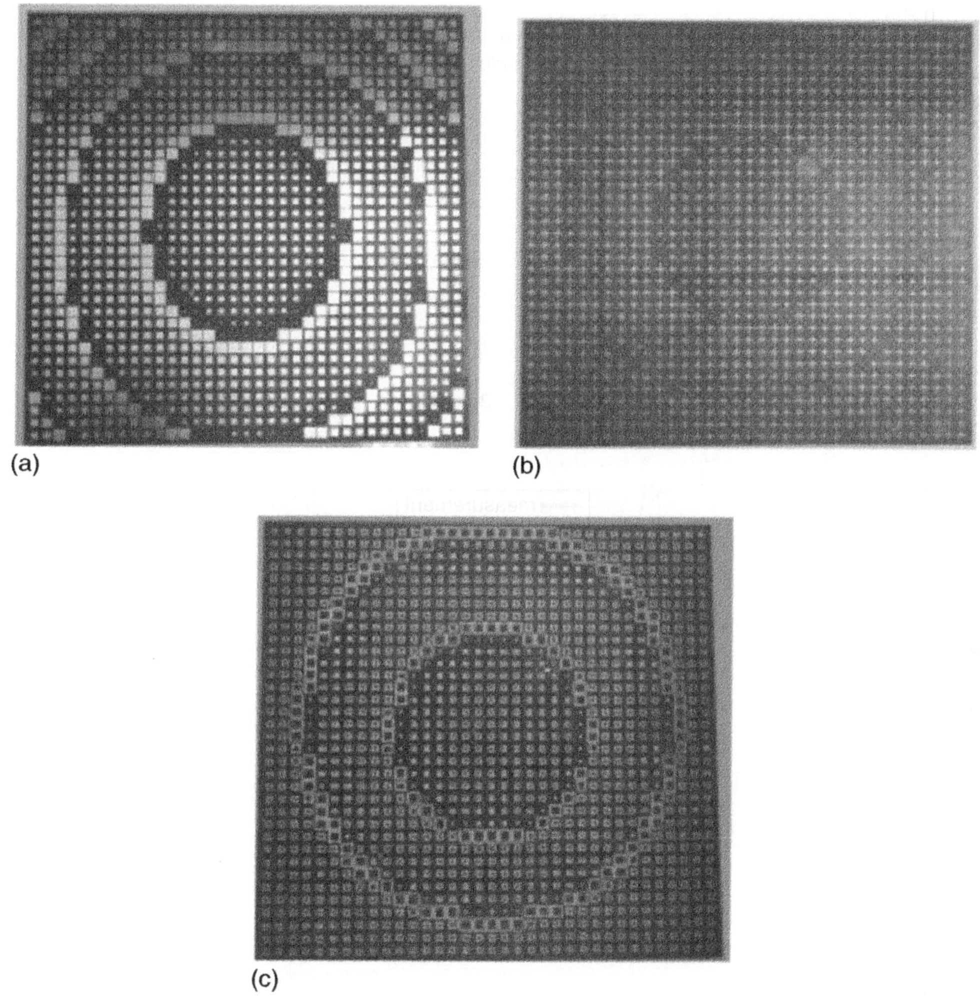

Figure 6.35 Photographs of fabricated reflectarray prototypes using variable size elements: (a) square patch, (b) cross dipole, and (c) square loop.

It is very interesting to observe the bandwidth performance of these reflectarrays. Measured gain and efficiency as a function of frequency are given in Figure 6.40, where the results clearly demonstrate that the reflectarray using square patch elements yields the best performance in terms of bandwidth. This result agrees well with the discussions in Chapter 5.

In summary, it appears that in comparison among these three types of elements, despite the fact that the square patch element has the smallest design freedom (only one design parameter, which is the patch length), it outperforms both other elements in terms of both fabrication sensitivity and bandwidth.

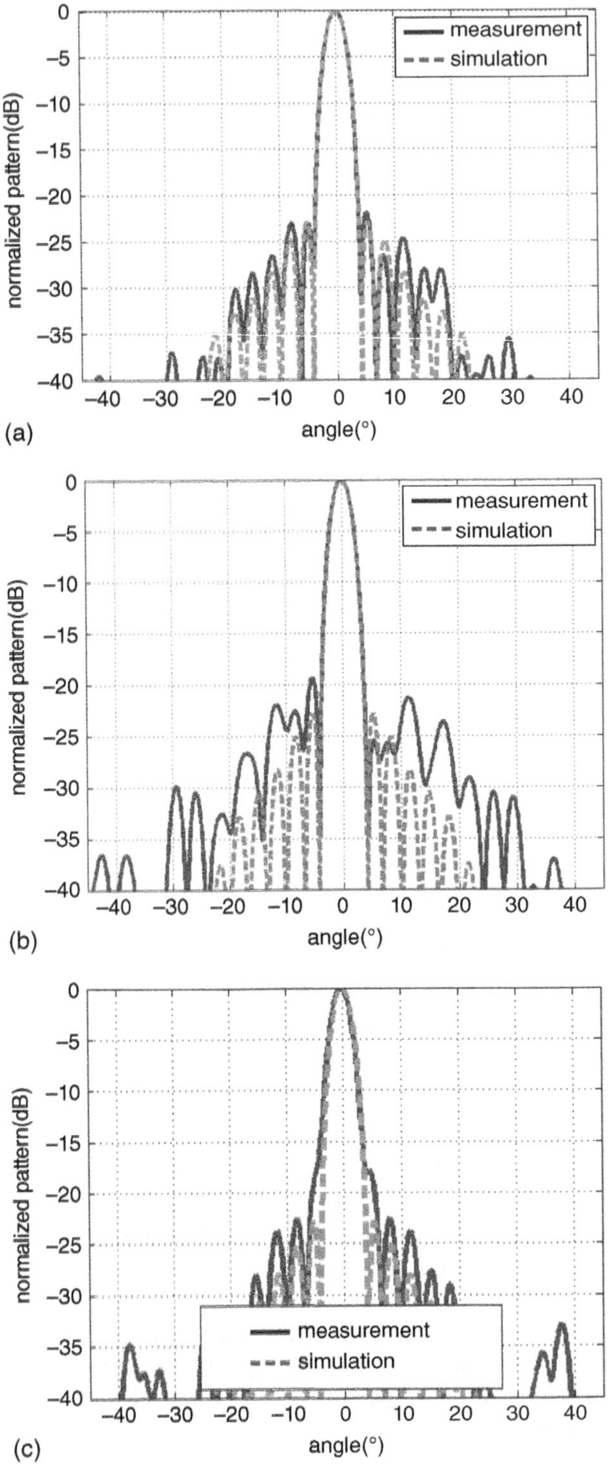

Figure 6.36 Comparison of the measured and simulated E-plane patterns at 13.5 GHz for reflectarrays using different elements: (a) square patch, (b) cross dipole, and (c) square loop.

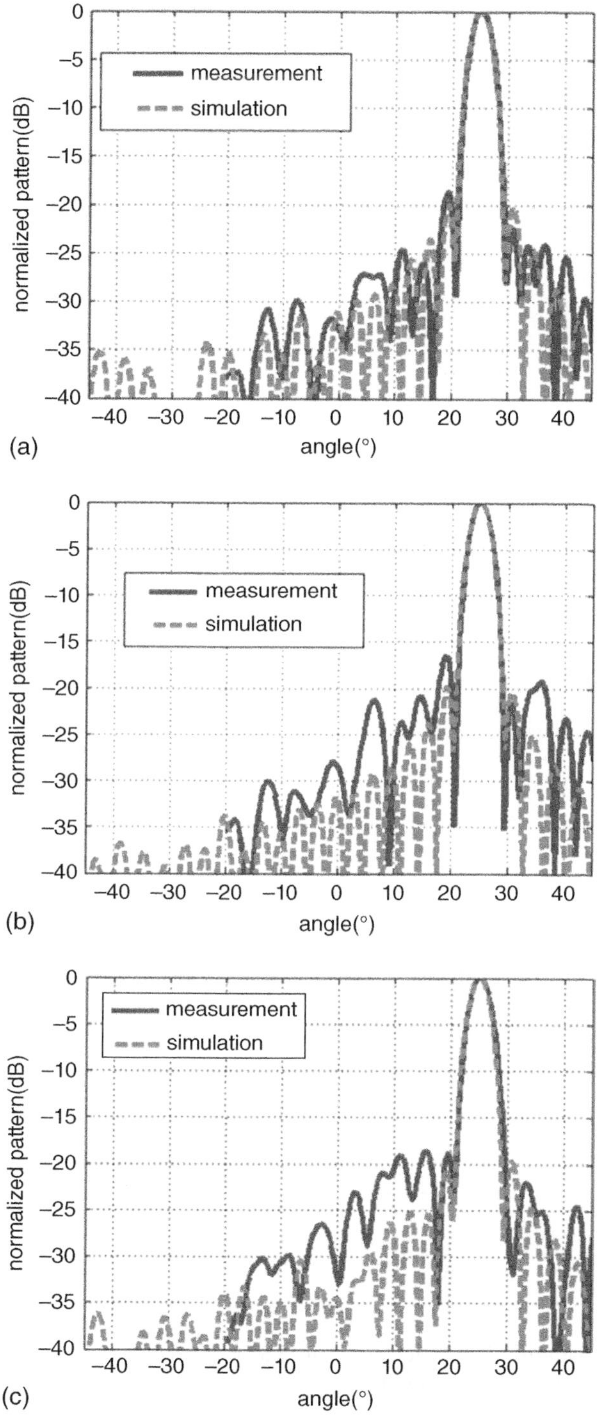

Figure 6.37 Comparison of the measured and simulated H-plane patterns at 13.5 GHz for reflectarrays using different elements: (a) square patch, (b) cross dipole, and (c) square loop.

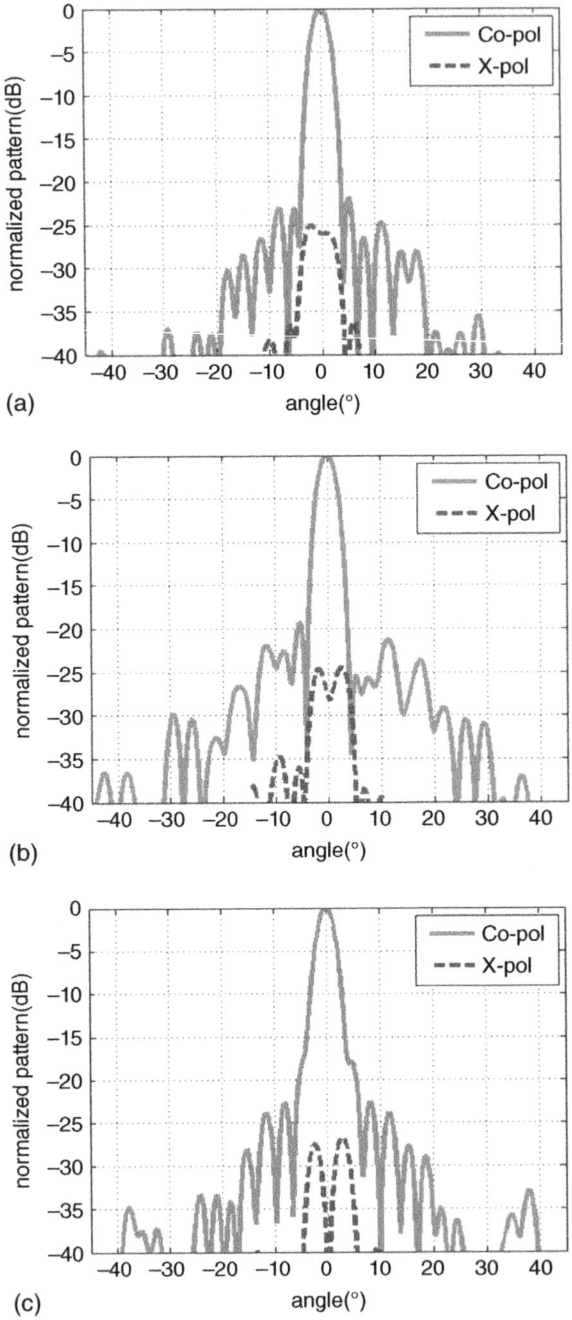

Figure 6.38 Measured E-plane radiation patterns of the reflectarrays at 13.5 GHz with different elements: (a) square patch, (b) cross dipole, and (c) square loop.

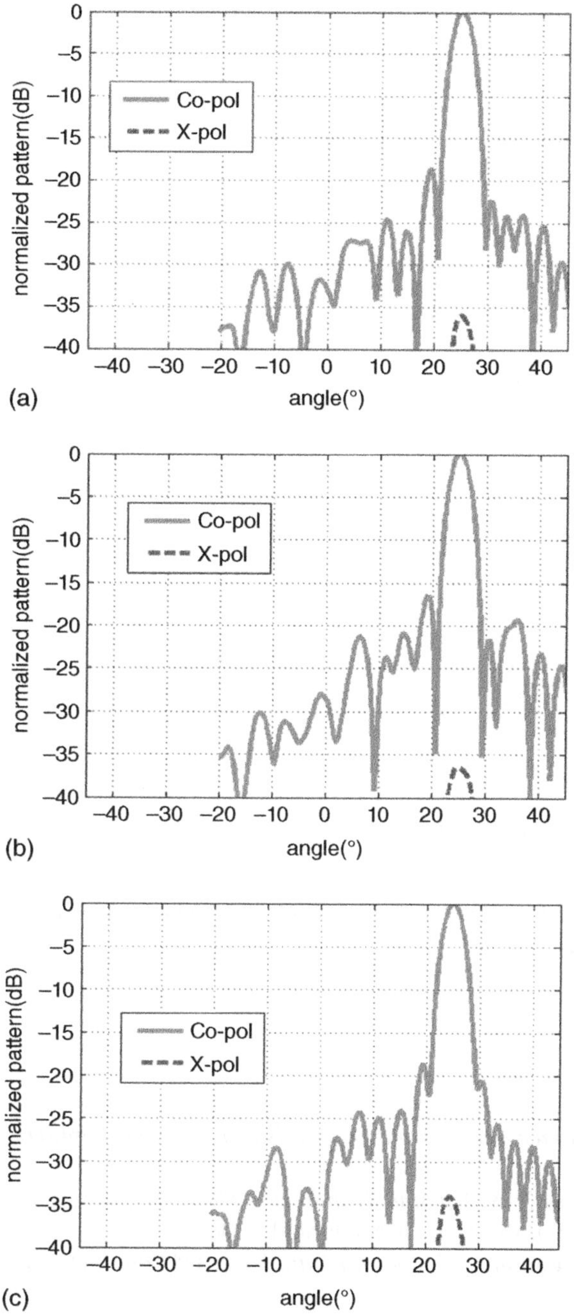

Figure 6.39 Measured H-plane radiation patterns of the reflectarrays at 13.5 GHz with different elements: (a) square patch, (b) cross dipole, and (c) square loop.

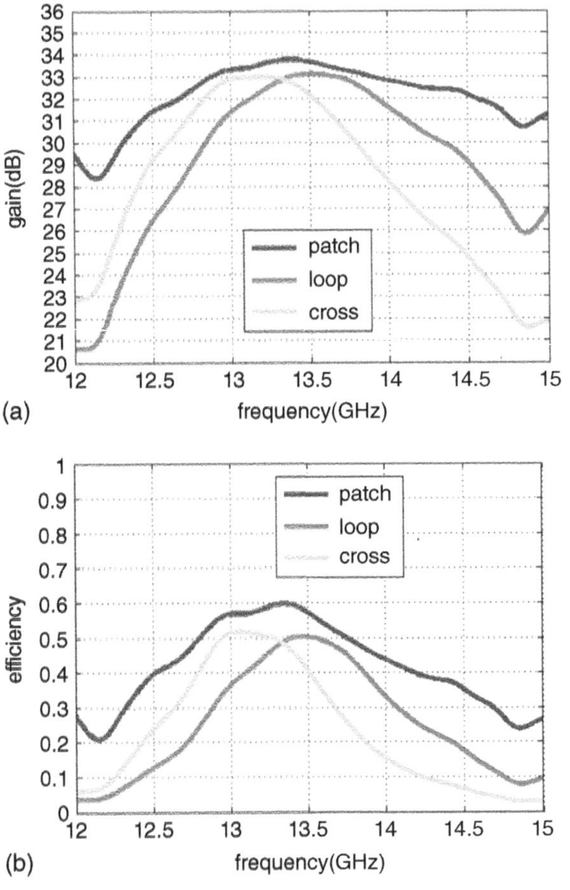

Figure 6.40 The measured performance of reflectarrays consisting of different element shapes as a function of frequency: (a) gain, (b) efficiency.

References

1 C. J. Sletten, *Reflector and Lens Antennas: Analysis and Design Using Personal Computers*, Artech House Inc., 1988.

2 Y. Rahmat-Samii, "Reflector antennas," in *Antenna Handbook: Theory, Applications, and Design*, Y. T. Lo and S. W. Lee (eds), Van Nostrand Reinhold, 1988.

3 M. E. Cooley, and D. Davis, "Reflector antennas," in *Radar Handbook*, 3rd Edn, McGraw-Hill, 2008.

4 A-Info. Website. Data sheet available at: www.ainfoinc.com/en/pro_pdf/new_products/antenna/Standard%20Gain%20Horn%20Antenna/tr_LB-62-15.pdf (accessed July 2017).

5 ANSYS HFSS, ANSYS Inc., 2014. Website available at www.ansys.com/ (accessed July 2017).

6 W. L. Stutzman and G. A. Thiele, *Antenna Theory and Design*, 3rd Edn, Hoboken, NJ: John Wiley & Sons, Inc., 2012.

7 C. A. Balanis, *Antenna Theory: Analysis and Design*, 3rd Edn, Hoboken, NJ: John Wiley & Sons, Inc., 2005.

8 D. M. Pozar, S. D. Targonski, and H. D. Syrigos, "Design of millimeter wave microstrip reflectarrays," *IEEE Trans. Antennas Propag.*, Vol. 45, No. 2, pp. 287–296, Feb. 1997.

9 Ansoft Designer, ANSYS Inc., 2014.

10 NSI MI website. Website. [Online] www.nearfield.com/products/NearFieldSystems.aspx (accessed July 2017).

11 M. J. Franco, "A high-performance dual-mode feed horn for parabolic reflectors with a stepped-septum polarizer in a circular waveguide," *IEEE Antennas and Propag. Magazine*, Vol. 53, No. 3, pp. 142–146, 2011.

12 A. Yu, *Microstrip Reflectarray Antennas: Modeling, Design and Measurement*, PhD dissertation, Department of Electrical Engineering, University of Mississippi, Oxford, MS, 2010.

13 Y. Mao, *Broadband Approaches and Beam-Scanning Techniques for Reflectarray Antennas*, PhD dissertation, Department of Electrical Engineering, University of Mississippi, Oxford, MS, 2014.

7

Broadband and Multiband Reflectarray Antennas

The reflectarray is inherently a narrowband antenna [1], [2]. With a typical design, the bandwidth of a reflectarray antenna is only a few percent, depending on the element selection, the size of its aperture, and f/D ratio. This basic limitation has stimulated engineers to find solutions to improve the bandwidth of reflectarrays and various broadband techniques have been developed over the years. Moreover, techniques have been proposed to provide coverage to two or more widely separated frequencies. In this chapter, we will present the design methodologies that have been developed for broadband and multiband reflectarrays.

7.1 Broadband Reflectarray Design Topologies

7.1.1 Multilayer Multi-Resonance Elements

As discussed in Chapter 2, a simple and practical way to obtain a progressive phase shift of the reflected field is by varying the resonant length of the patches [3], [4]. This technique allows simple manufacturing, and it produces lower cross-polarization levels than the stubs of different lengths attached to the radiating patches. To design a reflectarray, the phase of the reflected wave should have a progressive variation over the whole surface, essentially implying that a full phase cycle should be used for the reflection coefficient. The phase range obtained by varying the resonant length of a patch depends on the substrate thickness, that is, the smaller the thickness, the greater the phase range. For a thickness smaller than one-tenth of a wavelength, a phase range larger than 300 can be obtained, which allows for a practical design [5]. However, due to the narrowband behavior of microstrip patches, the reflection phase versus length is highly nonlinear, and shows a high slope near resonance and very slow variations near the extremes. The reflection phase is also very sensitive to frequency variations near resonance. A schematic view of variable size patch elements for reflectarrays and the reflection phase response of the elements as reported in [5] is given in Figure 7.1. As a consequence, the phase distribution on the surface of a reflectarray with patches of different sizes changes with frequency and reflectarrays are limited to narrowband applications.

It is clear that a smoother phase variation can provide a wider bandwidth. This can be achieved by increasing the dielectric substrate thickness, but in that case, the total phase range will be reduced, making the elements impractical. A solution to this

Reflectarray Antennas: Theory, Designs, and Applications, First Edition.
Payam Nayeri, Fan Yang, and Atef Z. Elsherbeni.

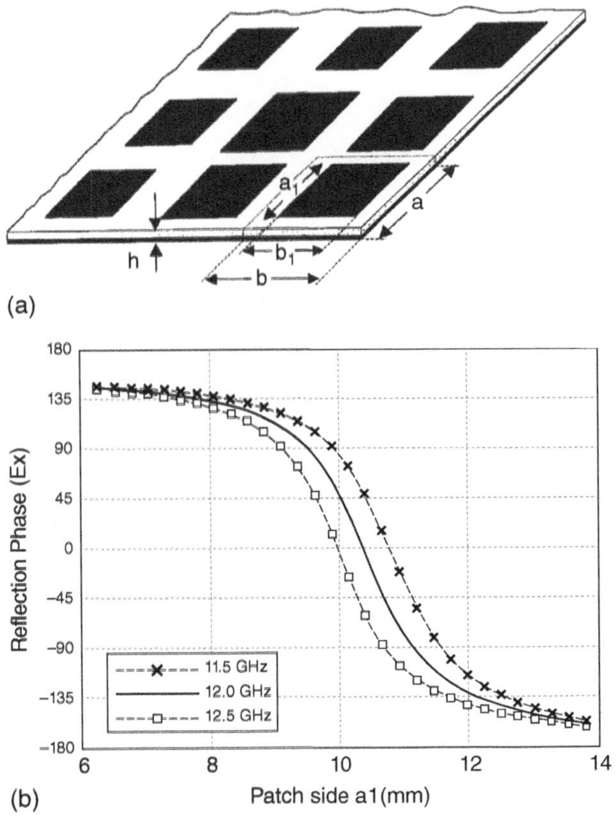

Figure 7.1 (a) Single-layer reflectarray with patches of variable size. (b) Reflection phase as a function of patch size. Reprinted with permission from [5].

problem is by stacking two or more patches; hence, a multi-resonant behavior is obtained and the phase range can be several phase cycles. Then, the dielectric substrate thickness of each layer can be increased to obtain a smoother and more linear phase variation, alleviating the reflectarray bandwidth limitation. A schematic view of a double-layer variable size patch elements for reflectarrays and the reflection phase response of the elements as reported in [5] are given in Figure 7.2. Note that the reflection phase shows a significantly smoother slope and a larger phase range in comparison with the single-layer design in Figure 7.1.

To demonstrate the broadband behavior of this element, a circular two-layer reflectarray of 406 mm diameter was designed and built using ROHACELL 51 as separators for the center operating frequency of 11.95 GHz. The arrays were cut from a cooper clad Kevlar skin using a cut plotter and the rest of the material removed. After manufacturing, tolerance errors of 0.2 mm were measured. A photo of the fabricated prototype along with the measured radiation patterns across the band is given in Figure 7.3. Note that the pattern is fairly stable across the entire band, demonstrating a good bandwidth response. The antenna has a peak gain of 31 dBi and gain variations less than 1.5 dB across 11–13 GHz.

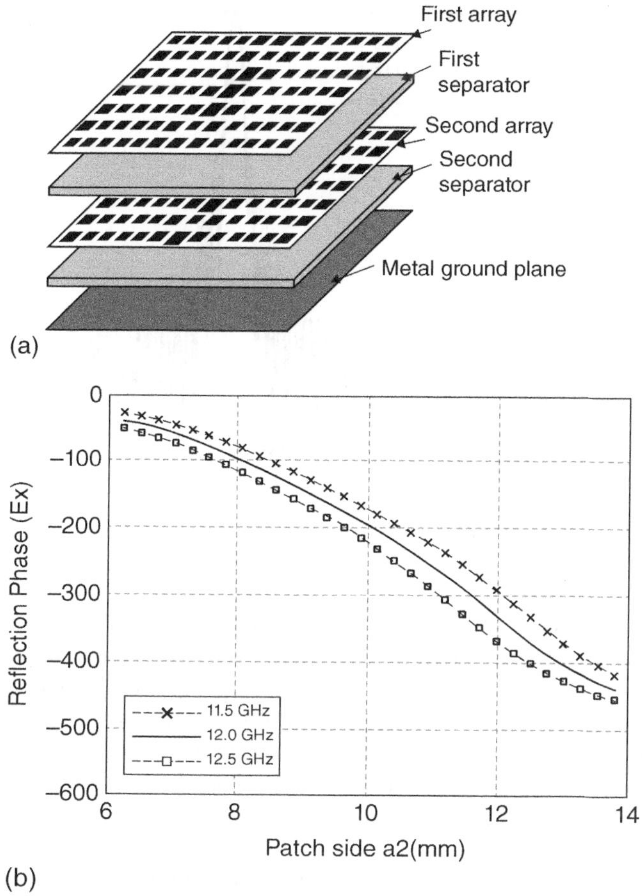

Figure 7.2 (a) Double-layer reflectarray with patches of variable size. (b) Reflection phase as a function of patch size. Reprinted with permission from [5].

In summary, by using stacked arrays of patches, a smooth phase variation with a phase range greater than 360 can be achieved, which allows improved bandwidth and reduced sensitivity to manufacturing tolerances. Several other designs of multilayer reflectarrays have been developed over the years [6]–[9], demonstrating the unique advantages of this broadband methodology.

7.1.2 Single-Layer Multi-Resonance Elements

The multilayer element design introduced in the previous section can indeed increase the bandwidth of reflectarray antennas; however, it is quite desirable to realize broadband designs with a single-layer configuration. The reason is that it is less expensive to manufacture and also it does not face problems associated with the precise layer alignment that occurs in its multilayer counterpart [10]. Note that the precise alignment of multilayers can be a serious problem in large size inflatable reflectarrays [11].

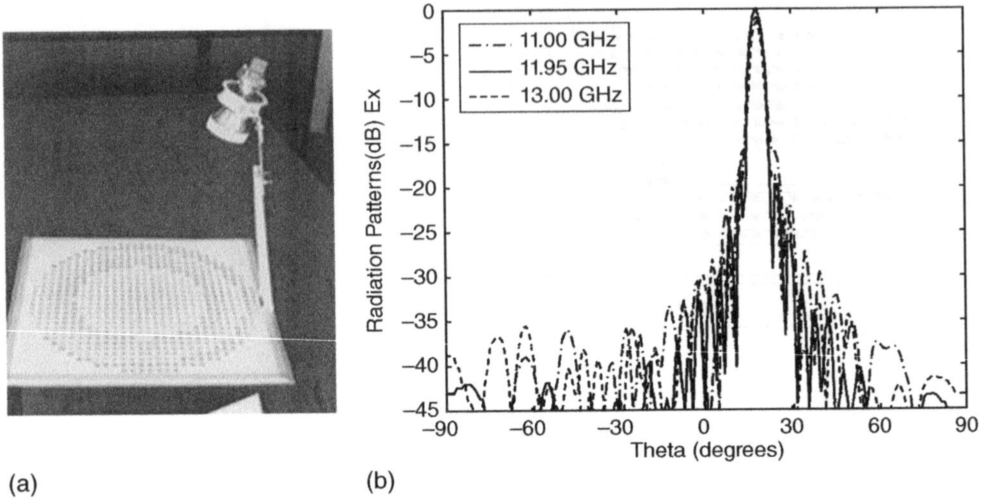

Figure 7.3 (a) A two-layer reflectarray prototype. (b) Measured radiation pattern of the antenna across the band. Reprinted with permission from [5].

As mentioned earlier, the difficulty with the typical single-layer microstrip reflectarrays is that a slower phase gradient is accompanied by a reduced phase range, making it impractical. The multilayer approach introduces multiple resonances, which was the key in increasing the phase range, allowing one to utilize thicker substrates and ultimately improve the bandwidth. Hence if one can introduce multiple resonances with a single-layer configuration, it could also be possible to use thicker substrates and improve the bandwidth. A detailed study on this design methodology was reported in [10], which is summarized here.

Multiple resonance elements in the form of multiple rings of square or circular shapes were studied for broadband single-layer reflectarrays, and the phase characteristics of a unit cell containing these uniplanar elements were compared with those of a unit cell containing stacked two-layer patches. The geometrical models of these multi-resonance elements are given in Figure 7.4.

Both elements have several design variables that can be utilized to achieve an almost linear phase response as a function of ring size. For the double-square ring, the size of the inner ring was set to 0.9 the size of the outer one. The width of the two rings was set to 0.025 of the outer ring size. Two different substrates were studied. The first substrate had a thickness of 1.57 mm and a dielectric constant of 3.2. The second substrate had a thickness of 3.175 mm and a dielectric constant of 2.2. The reflection phase response of this element is given in Figure 7.5(a), where it can be seen that the phase range obtained for the double-square ring is approaching 700°. The use of substrate with larger thickness and smaller dielectric constant results in lower slope and more linear response. This is at the expense of slightly reduced phase range of about 650°.

Similar trends were observed with double-circular rings on the same substrates. For this case, the inner ring radius was 0.8 of the outer ring radius while all the rings have a width of 0.08 of the outer radius. Similarly, the obtained phase range for the thinner substrate is approximately 700° while that for the thicker one is about 650°. It is also important to note that in comparison with the square-ring element, the slopes of the

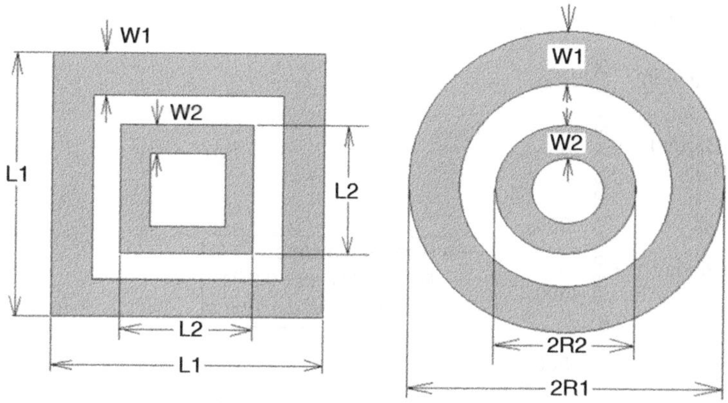

Figure 7.4 Geometries of double-square and double-circular rings. Reprinted with permission from [10].

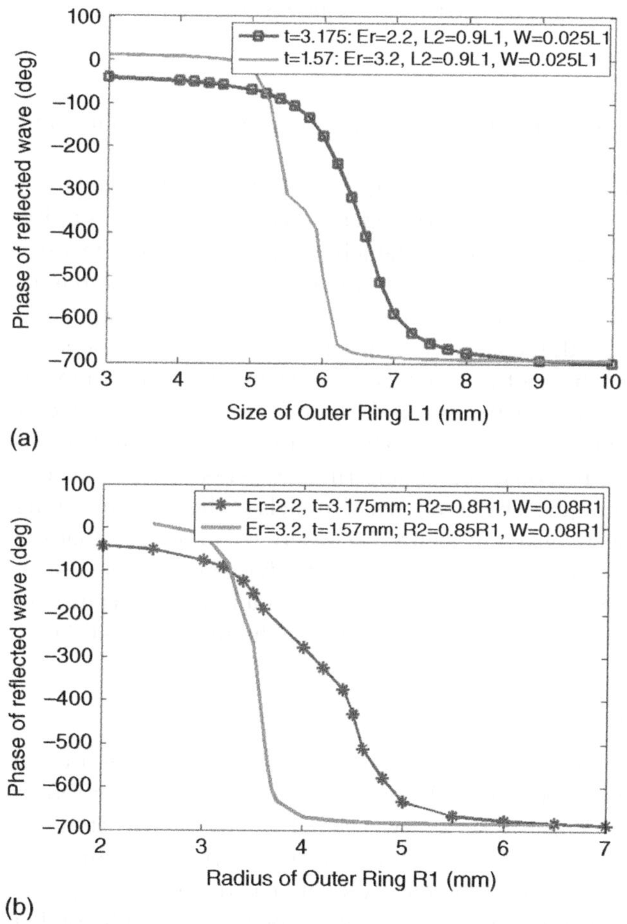

Figure 7.5 Phase response of the double-ring elements as a function of outer ring size: (a) double-square, (b) double-circle. Reprinted with permission from [10].

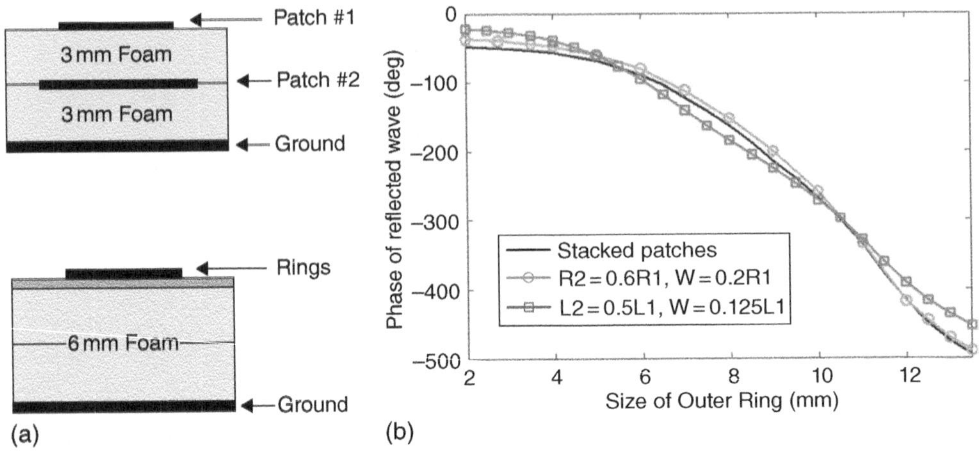

Figure 7.6 (a) Configurations of composite substrates. (b) Phase responses of double-circular rings and double-square rings compared to that of the stacked two-layer square patches. Reprinted with permission from [10].

circular rings are lower, which is also a trend that is observed with square and circular patch elements.

In designing double-ring elements, one has the choice of the sizes of the two rings, their aspect ratio, and their widths. For dual-polarized operation the aspect ratio is unity, and thus, the designer has three degrees of freedom, namely, the relative size of the inner ring and the widths of the two rings. Some of these parameters can be adjusted to linearize the response. The procedure is similar to staggering two resonance responses, where the size ratio is responsible for the separation between the resonance frequencies, and the ring widths mainly control the widths of the two responses or the local slope of the phase response. Furthermore, thicker substrate and lower dielectric constant facilitate an easier achievement of linear phase response. These parameters were utilized in [10] to demonstrate a linear phase response for double-ring elements, which was also compared with the multilayer technique. The geometrical models of the elements along with the reflection phase response are given in Figure 7.6, where it can be seen that the single-layer double-ring elements perform almost identical to the double-layer design. This single-layer multi-resonance approach has received considerable attention and several reflectarray designs of these elements have been reported in literature [12]–[16].

7.1.3 Sub-Wavelength Elements

Traditionally, the reflectarray phasing elements are designed with unit-cell sizes around half a wavelength. Advances on metamaterials [17], however, revealed that similar reflection phase response can also be realized using sub-wavelength elements. A patch element placed in a lattice with half-wavelength spacing can be self-resonant. In other words, it can exhibit resonance as a single element. A single electrically small patch, however, will no longer exhibit resonance and typically cannot be used as a radiator. For reflectarray designs, as long as a sufficient phase range is obtained with the element, smaller unit-cell periodicities can be used for the element. This is due to the fact that the adjacent elements in the periodic lattice of the reflectarray give rise to a coupled resonance behavior. This coupling mechanism between adjacent elements provides a useful resonance that can be utilized to phase the aperture of the reflectarray [18]. The design approach is basically identical to the traditional method, which makes it rather simple to implement. The only difference is that the smaller lattice size results in a larger number of elements.

The first mention of sub-wavelength elements for reflectarrays was given in [19], using the term "artificial impedance surface," and it was shown that these elements can provide a wider bandwidth than traditional half-wavelength elements. While some gain loss was observed, primarily due to a reduction of the phase range of sub-wavelength elements, a substantial improvement in reflectarray gain bandwidth was observed. To gain an understanding in the operating mechanism of sub-wavelength elements, we compare the reflection properties of reflectarray phasing elements with periodicities smaller than half-wavelength. The variable size square patch is used, and the only difference is the selection of element periodicity. In particular, $\lambda/2$, $\lambda/3$, and $\lambda/4$ are selected as the periodicities for this study. Although smaller periodicities such as $\lambda/10$ can be selected in the analysis and design, it will bring difficulties to the fabrication tolerance and increase the fabrication cost, which will be further discussed later on in this section. A 20 mil Rogers 5880 substrate ($\varepsilon_r = 2.2$) is used here for this study. The reflection coefficients versus patch size at the design frequency (32 GHz) for normal incidence are given in Figure 7.7.

Comparison between the reflection magnitudes of the elements shows that as the unit-cell size decreases, the resonance moves toward smaller patch sizes, indicating a stronger coupling effect between adjacent patches. Furthermore, note that the element losses also decrease. While the reflection coefficients for all the elements in this study are larger than 98%, mainly due to the low-loss properties of the substrate, sub-wavelength elements can potentially reduce the element losses [21], [22]. As pointed out earlier, comparison between the reflection phases of the elements also shows that as the unit-cell size decreases the phase range is reduced and the slope inclination becomes more vertical, essentially increasing the quantization error in the fabrication process. In general, this is one of the major challenges in sub-wavelength designs.

With respect to the phase reduction behavior of the sub-wavelength elements, it is important to note the effect of the gap size between the patches. In the unit-cell analysis here, a periodic boundary condition is being used to account for the coupling between the unit-cell elements. The coupling between the elements is a function of the patch size and the gap between the patches. For a fixed unit-cell size, the coupling between the patch elements increases by reducing the spacing between them. However, if the size of the unit cell is reduced, a closer spacing between the patch elements is required to achieve the same level of coupling for larger unit cell sizes. Consequently, this would mean that smaller unit-cell sizes with the same gap size would have a weaker coupling between the elements, which would reduce the phase range versus patch size for these elements. More discussion on the phase range of sub-wavelength elements and the effect of gap size between the patches will be given in a subsequent section.

Finally note that the results presented here are under normal incidence approximation. In general, the reflection characteristics are angle dependent and oblique incidence needs to be considered. However, as mentioned in earlier chapters, it has been shown that normal incidence can present good approximations for certain elements

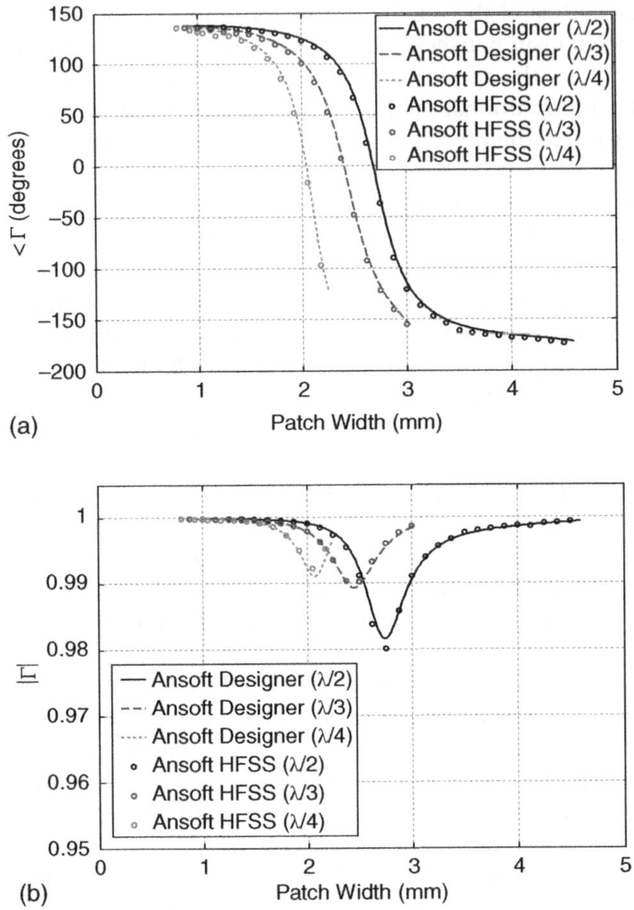

Figure 7.7 Reflection coefficients versus patch size at the center frequency 32 GHz for λ/2, λ/3, and λ/4 unit-cells with normal incidence: (a) phase, (b) magnitude. Reprinted with permission from [20].

with incidence angles up to 30°. Moreover, a study on the properties of sub-wavelength elements was given in [23], which showed that sub-wavelength elements provide a reduced angular sensitivity over their resonant counterparts, making the normal incidence approximation a more valid choice in these designs.

The element studies presented so far gives us a basic understanding of the element design methodology, however, fail to provide a solid explanation on bandwidth characteristics. From an element analysis viewpoint, operating off-self-resonance is considered to be the reason for the broadband behavior of sub-wavelength reflectarray elements. As discussed earlier in Chapter 5 the key to understanding the bandwidth behavior of reflectarray elements is the phase analysis as a function of frequency, thus, here we analyze the frequency response of half-wavelength and sub-wavelength reflectarray elements. Consider a case where two elements on the aperture are required to have a 90° relative phase difference. Using the element data in Figure 7.7, for a zero-degree reflection phase, the patch size will be 2.69, 2.41, and 2.04 mm for λ/2, λ/3, and λ/4 designs. When a 90° reflection phase is required, the patch size should be 2.41, 2.1, and 1.81 mm for λ/2, λ/3, and λ/4 designs.

Figure 7.8(a) shows how the reflection phases of these 0° and 90° elements vary with frequency. It is observed that λ/2, λ/3, and λ/4 designs have different frequency behaviors for the element reflection phase. When the frequency changes, the closely spaced elements show a smaller phase variation with frequency. As discussed previously, it is the frequency behavior of the phase difference that determines the reflectarray bandwidth. Thus, the phase differences between these 0° and 90° elements are calculated, and the phase error curves of the λ/2, λ/3, and λ/4 designs are obtained and plotted in Figure 7.8(b). It is clear from this figure that the closely spaced elements have a smaller phase error over the frequency than the half-wavelength elements. Similar studies have

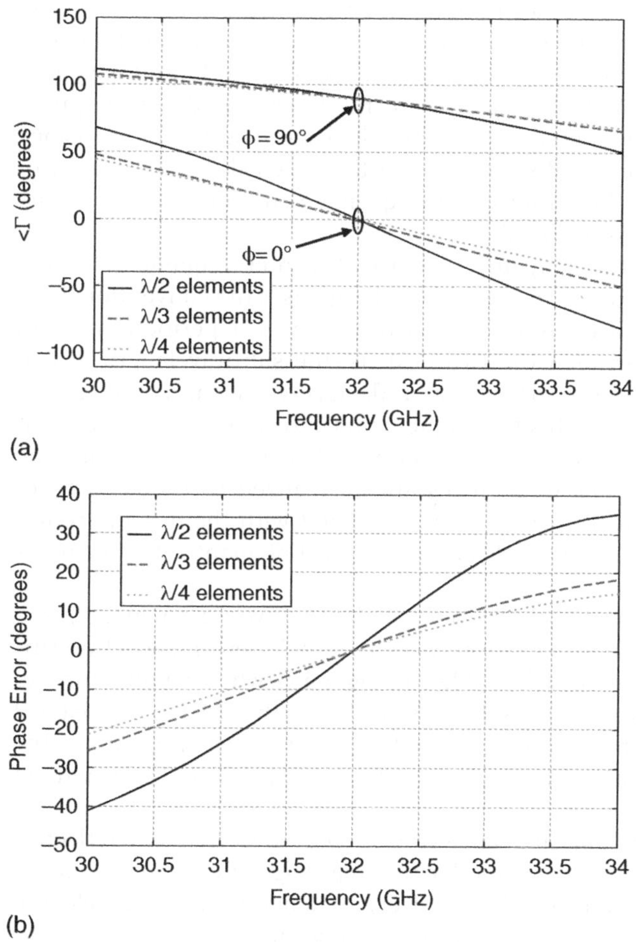

Figure 7.8 (a) Reflection phases versus frequency for 0° and 90° elements for λ/2, λ/3, and λ/4 unit-cells. (b) Phase errors for a 90° relative phase difference versus frequency for λ/2, λ/3, and λ/4 unit-cells. Reprinted with permission from [20].

been performed on the reflectarray elements with different relative phase requirements, and the same observation is obtained. As such this study confirms that these sub-wavelength elements have the potential to increase the reflectarray bandwidth.

7.1.4 Reflectarrays Employing Single-Layer and Double-Layer Sub-Wavelength Elements

As discussed in Chapter 5, to further enhance our understanding of the bandwidth performance of a reflectarray antenna, it is necessary to observe the phase error performance for all the elements with respect to a certain reference point. To quantitatively analyze the performance of the sub-wavelength phasing elements, we consider a practical example of a Ka-band reflectarray. A circular aperture reflectarray with a diameter of 159 mm is designed for the operating frequency of 32 GHz to generate a beam in the direction of $(\theta, \varphi) = (25°, 0°)$. The array geometrical center is placed at the center of the coordinate system and the offset feed horn phase center is located at $X = -45.9$ mm, $Y = 0$ mm, $Z = 98.4$ mm. For the horn model used in this study, the power q of the feed radiation pattern varies linearly from 5 at 30 GHz to 8.3 at 34 GHz. From the system design specifications, the location of all the array elements, the required element phase shift, and the illumination on the array surface can be calculated. The patch dimensions are then determined from the unit-cell simulations at the center design frequency of 32 GHz. From these patch size dimensions, the frequency behavior of the elements reflection phase is obtained across the band.

The weighted phase errors on the array surface at two off-center frequencies (30 and 34 GHz) are given in Figure 7.9 for two designs. One design uses the conventional $\lambda/2$ elements, while the other reflectarray is designed with $\lambda/4$ unit-cell periodicity. It can be seen that, in comparison between the two designs, the sub-wavelength design shows a significantly smaller phase error over the reflectarray aperture, demonstrating the bandwidth improvement of this technique. Similar results were observed at different frequencies across the band. It should be noted that in these results, the aperture phase errors at off-center frequencies are normalized to the center frequency. In other words, the quantization errors associated with the design are not taken into account. Although as mentioned earlier, the effect of quantization errors is not significant in these designs.

Experimental results for sub-wavelength Ka-band reflectarrays were given in [20], [22]. The authors noted that the minimum gap size between the patch elements was the critical factor in controlling the maximum achievable phasing range, where a minimum gap size of 0.1 mm was dictated by the size of the smallest drill bit in the LPKF ProtoMat S62 milling machine. By enforcing this fabrication limit in the unit-cell simulations, the phasing range of the elements was reduced to 310°, 290°, and 247° for $\lambda/2$, $\lambda/3$, and $\lambda/4$ elements, respectively.

As mentioned earlier, typically a phasing range around 300° is required for a reflectarray design. This study showed that the achievable phasing range of elements, which is directly related to fabrication tolerance of the gap sizes, particularly for sub-wavelength elements, should be viewed as the lower limit in selecting the lattice size. Since the phase range of the elements is decreased, phase errors lead to a reduction of antenna gain and an increase in the side-lobe levels. Note that, however, the antenna bandwidth is still mainly determined by the frequency behavior of the phasing elements.

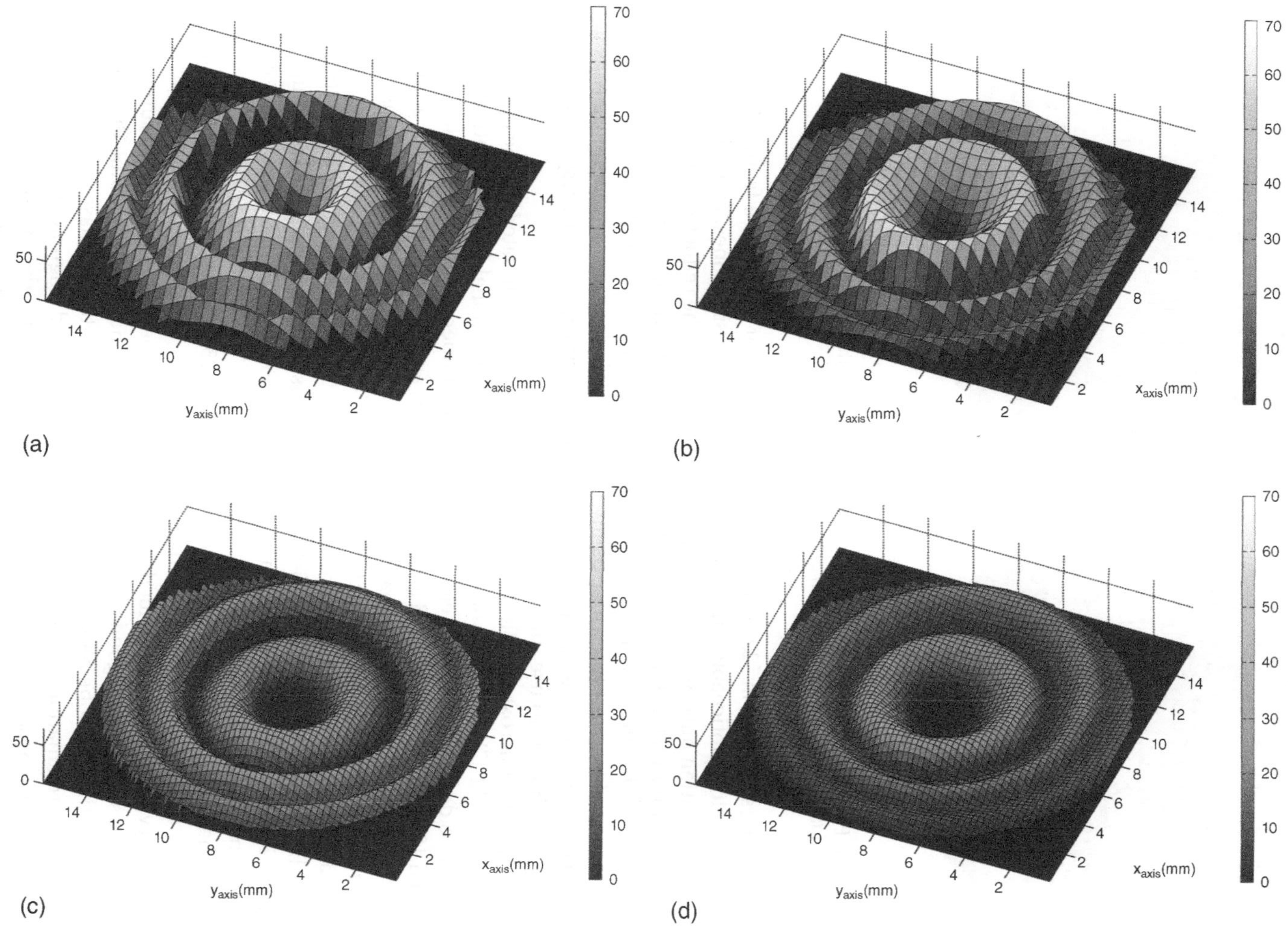

Figure 7.9 Absolute value of weighted phase error on the reflectarray aperture: (a) $\lambda/2$ elements – 30 GHz, (b) $\lambda/2$ elements – 34 GHz, (c) $\lambda/4$ elements – 30 GHz, (d) $\lambda/4$ elements – 34 GHz. Reprinted with permission from [20].

(a) (b)

Figure 7.10 Photographs of the fabricated arrays: (a) λ/2 array with 848 square patches, (b) λ/3 array with 1941 square patches. Reprinted with permission from [20].

The authors' studies in [22] showed that a reduction of lattice size from λ/2 to λ/3 does not result in noticeable gain reduction or increase in side-lobes, and is best suited for demonstration purposes. Note that for the λ/4 design, the reduction of phase range resulted in almost 0.5 dB loss in antenna gain and 6 dB increase in side-lobe level relative to the λ/2 design. It is important to point out that such performance degradation maybe acceptable given the wider bandwidth of the design, but in general there is a tradeoff between the reflectarray gain and radiation performance with respect to bandwidth if one chooses to use sub-wavelength elements.

Two Ka-band microstrip reflectarrays were designed and fabricated for the operating frequency of 32 GHz. One is designed using the conventional λ/2 elements and the other one is designed using λ/3 elements. The total number of square patch elements is 848 and 1941 for the λ/2 and λ/3 arrays, respectively, and the element phase is adjusted to scan the main beam 25° off broadside to minimize feed blockage. Photographs of the fabricated reflectarrays are shown in Figure 7.10.

The antennas performances were measured over the frequency range from 30.0 to 34.0 GHz using a NSI planar near-field system. Photos of the arrays in the test setup are given in Figure 7.11. The vertical plane radiation patterns at 32 GHz for both antennas are shown in Figure 7.12.

Note that the main beam is correctly scanned to 25° and the side-lobe levels are below −22 dB for both antennas. The simulated results here were obtained using the array theory formulation described in Chapter 4. It can be seen that the measured and simulated results show a good agreement in the main lobe; however, some discrepancies exist in the side-lobe region. These are primarily due to the phase errors (fabrication and approximations in the unit-cell simulations) and alignment errors. Similar results were also observed in the horizontal plane. At 32 GHz the measured half-power beam width for the λ/2 array is 4.5° and 4.6° for vertical and horizontal planes, respectively, and the measured gain is 30.99 dB, which corresponds to an overall aperture efficiency of 44.0%. For the λ/3 array the measured half-power beam width at 32 GHz is

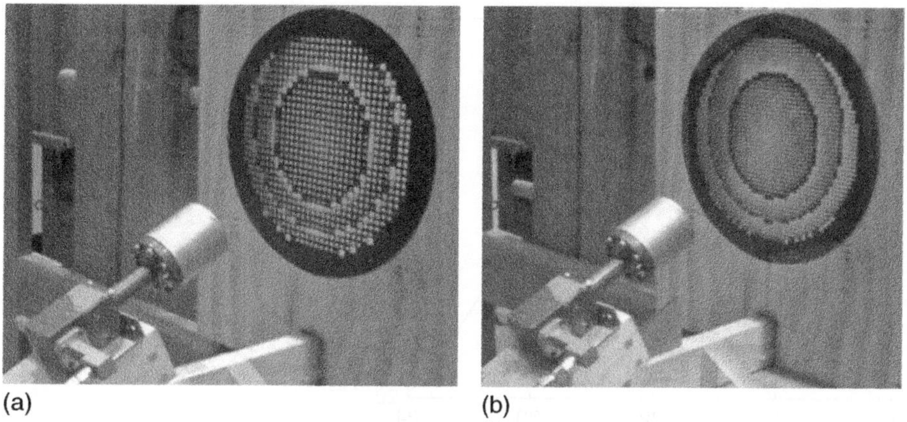

Figure 7.11 Photographs of the fabricated arrays: (a) λ/2 array with 848 square patches, (b) λ/3 array with 1941 square patches. Reprinted with permission from [20].

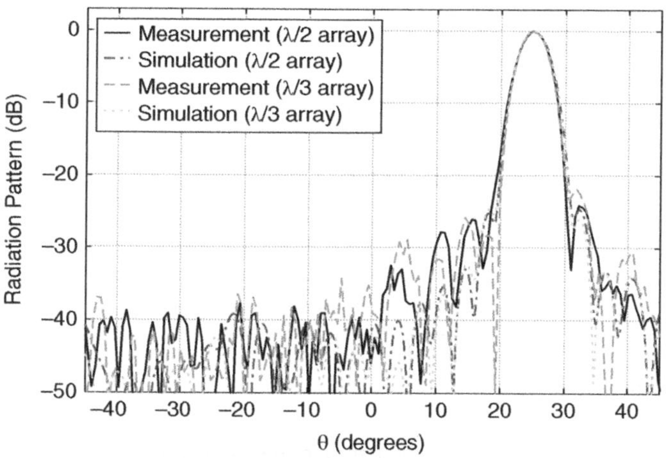

Figure 7.12 Measured and simulated radiation patterns of the reflectarray antennas at 32 GHz in the vertical plane. Reprinted with permission from [20].

4.4° and 4.6° for vertical and horizontal planes, respectively, and the measured gain is 30.95 dB, which corresponds to an overall aperture efficiency of 43.6%.

The measured gain and aperture efficiency of the antennas versus frequency are given in Figure 7.13. To observe the bandwidth improvement of the sub-wavelength element technique, a direct comparison between the measured gains of the two prototypes is also given in Figure 7.14. The 1-dB gain bandwidth of the λ/2 array is 8.0% and this antenna achieves its max gain of 31.34 dB at 31.8 GHz. The measured results also show that the λ/3 array achieves a considerable bandwidth improvement, where the 1 dB gain bandwidth has been increased to 10.9%. Similarly, this antenna achieves its max gain of 31.22 dB at 31.8 GHz. The measured radiation patterns show a similar performance across the entire band except for a slight increase in side-lobe level at the extreme

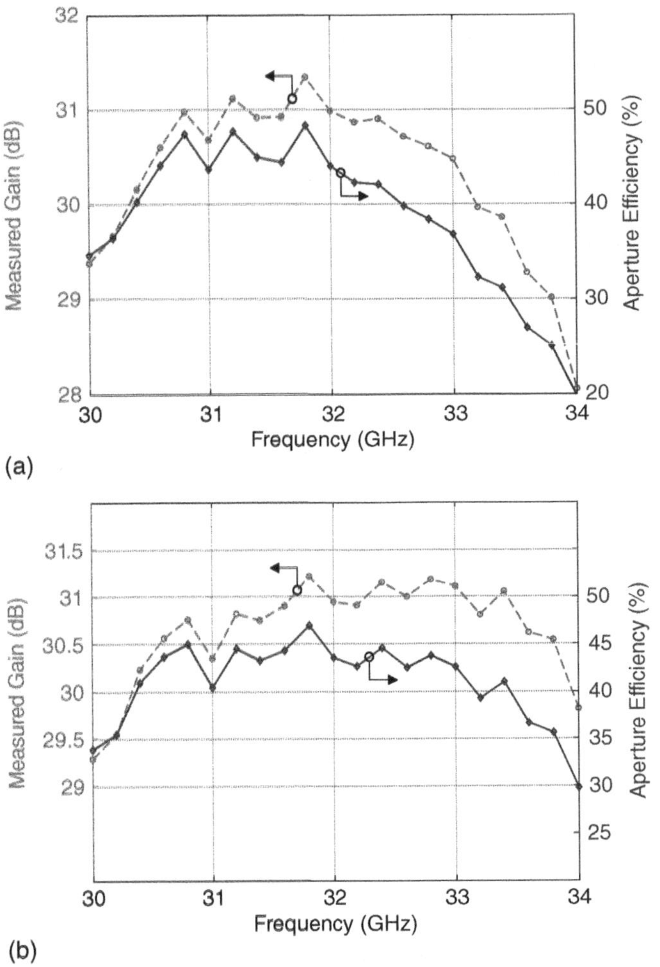

Figure 7.13 Measured gain and aperture efficiency versus frequency for the prototypes: (a) $\lambda/2$ array, (b) $\lambda/3$ array. Reprinted with permission from [20].

frequencies. It should be noted that the measured gain of the feed horn is 14.2 dB at 32 GHz and the variation across the 30 to 34 GHz band is less than 0.5 dB.

It is worthwhile to point out that as discussed earlier, the bandwidth of a reflectarray antenna is limited by two different factors, the bandwidth of the element and the bandwidth limitation by spatial phase delay. The sub-wavelength broadband technique studied here improves the reflectarray element bandwidth. Consequently, this broadband technique is applicable to small to moderate size reflectarray antennas. As the size of the aperture increases, the broadband effect of the sub-wavelength elements would be less effective due to the effect of the spatial phase delay that will become dominant. Numerical results given in [22] showed that for this Ka-band reflectarray going from $\lambda/2$ elements to $\lambda/3$ elements, the percentage of bandwidth improvement will be reduced as the aperture diameter is increased.

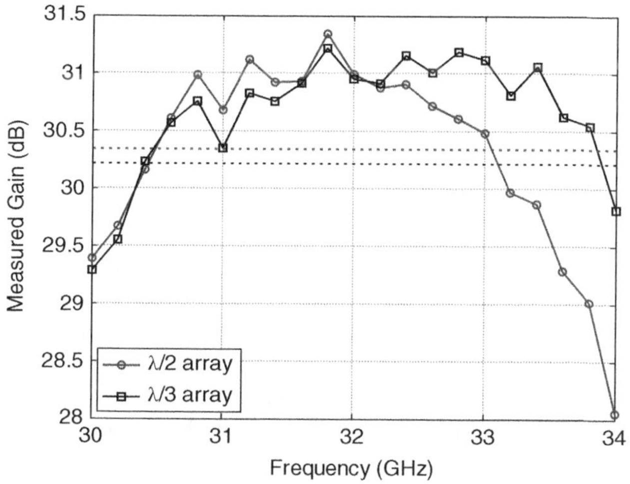

Figure 7.14 Comparison between the measured gain versus frequency for the λ/2 and λ/3 array. Reprinted with permission from [20].

The results presented here clearly show that the sub-wavelength technique is a unique and simple approach for broadband reflectarrays. Although there is no theoretical limit on using smaller unit-cells in reflectarray designs, the drawback of this approach primarily arises from the fabrication tolerance of the patches, which become a critical factor at high frequencies and therefore require higher fabrication precisions that increase the cost. In particular, since most of the reflection phase range occurs for patches with very thin gaps, the fabrication tolerance of the gap sizes controls the smallest unit-cell dimensions that can be etched reliably [24].

The effect of the minimum gap size on the phase range of sub-wavelength elements was studied in [25]. First, we consider the single-layer configuration, where the phase range of variable size square patch elements on a 20 mil Rogers Duroid 5880 substrate is given for various lattice sizes at 32 GHz. It can be seen from Table 7.1 that for most single-layer designs with a very small ideal gap size (0.001 mm) between the elements, a sufficient phase range (typically around 300°) can be achieved. However, practical fabrication tolerance limits the achievable phase range of elements. The minimum gap size of 0.1 mm used in practical designs is based on the typical low-cost fabrication capabilities of most milling machines. For these Ka-band elements, reducing the unit-cell size from λ/2 to λ/6 results in an almost 200° loss of phase range. Furthermore, the limited fabrication tolerance of the patch sizes will also increase the phase quantization

Table 7.1 Phase range of reflectarray elements at 32 GHz.

Unit-cell size design	λ/2	λ/3	λ/4	λ/6
Single-layer (0.001 mm gap)	312.6°	304.8°	294.9°	261.8°
Single-layer (0.1 mm gap)	310.0°	292.5°	258.3°	100.3°
Double-layer (0.1 mm gap)	583.9°	323.0°	277.0°	214.5°

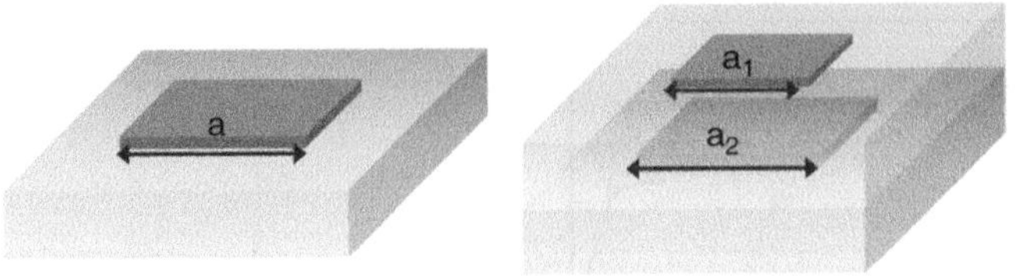

Figure 7.15 Geometry of single-layer and double-layer reflectarray phasing elements. Reprinted with permission from [25].

errors within this phase range. As a result, reflectarrays designed using single-layer sub-wavelength elements will show significant phase errors that degrade the antenna performance. In other words, due to the practical fabrication constrains, the bandwidth improvement achieved by means of sub-wavelength elements would be accompanied by a reduction in antenna gain, which is undesirable.

Another broadband technique that was introduced earlier in this chapter and has received considerable attention is by the use of stacked patches. The geometrical model of single and double-layer reflectarray elements is given in Figure 7.15. The multi-resonance design can increase the phase range of the elements and improve the slope linearity of the phase curve. By integrating multi-resonance elements with sub-wavelength unit-cells, one can take advantage of both broadband techniques. In addition, the phase range of the elements can be improved with the same fabrication tolerance. Comparison between the phase ranges of single- and double-layer designs is also given in Table 7.1. Note that for the double-layer configuration a ratio of $a_1 = 0.7a_2$ was selected for the patch sizes. It is observed that the double-layer designs can increase the achievable phase range of the elements as expected and shows promise as a practical low-cost approach for broadband reflectarrays.

To quantitatively analyze the performance of the double-layer sub-wavelength phasing elements, the authors studied the same reflectarray system described earlier and analyzed the weighted phase error on the array surface at two off-center frequencies. These results are shown in Figure 7.16 for two double-layer designs. One design uses $\lambda/2$ elements, while the other reflectarray is designed with $\lambda/4$ unit-cell periodicity. Comparing these results, it can be seen that as expected the double-layer designs show a significantly lower phase error on the aperture, which translates directly into a broader bandwidth for the reflectarray. The average phase error on the reflectarray was also computed using the formulation given in Chapter 5 and is shown in Figure 7.17. Note that although there is no direct correlation between reflectarray gain bandwidth and these absolute phase error values, the significantly lower phase errors of the double-layer designs clearly demonstrate that this hybrid approach, that is, a reflectarray designed with double-layer sub-wavelength elements, will provide the best bandwidth performance.

To further verify the broadband behavior of double-layer sub-wavelength designs, the gain bandwidth of the reflectarray antennas is computed using the aperture field formulation described in Chapter 4. The calculated gain here includes the spillover losses and the element losses in the reflectarray. In addition, the frequency behavior of the feed horn pattern and elements reflection characteristics are implemented into this calculation routine. Due to the limited phase range of Ka-band elements with a periodicity of

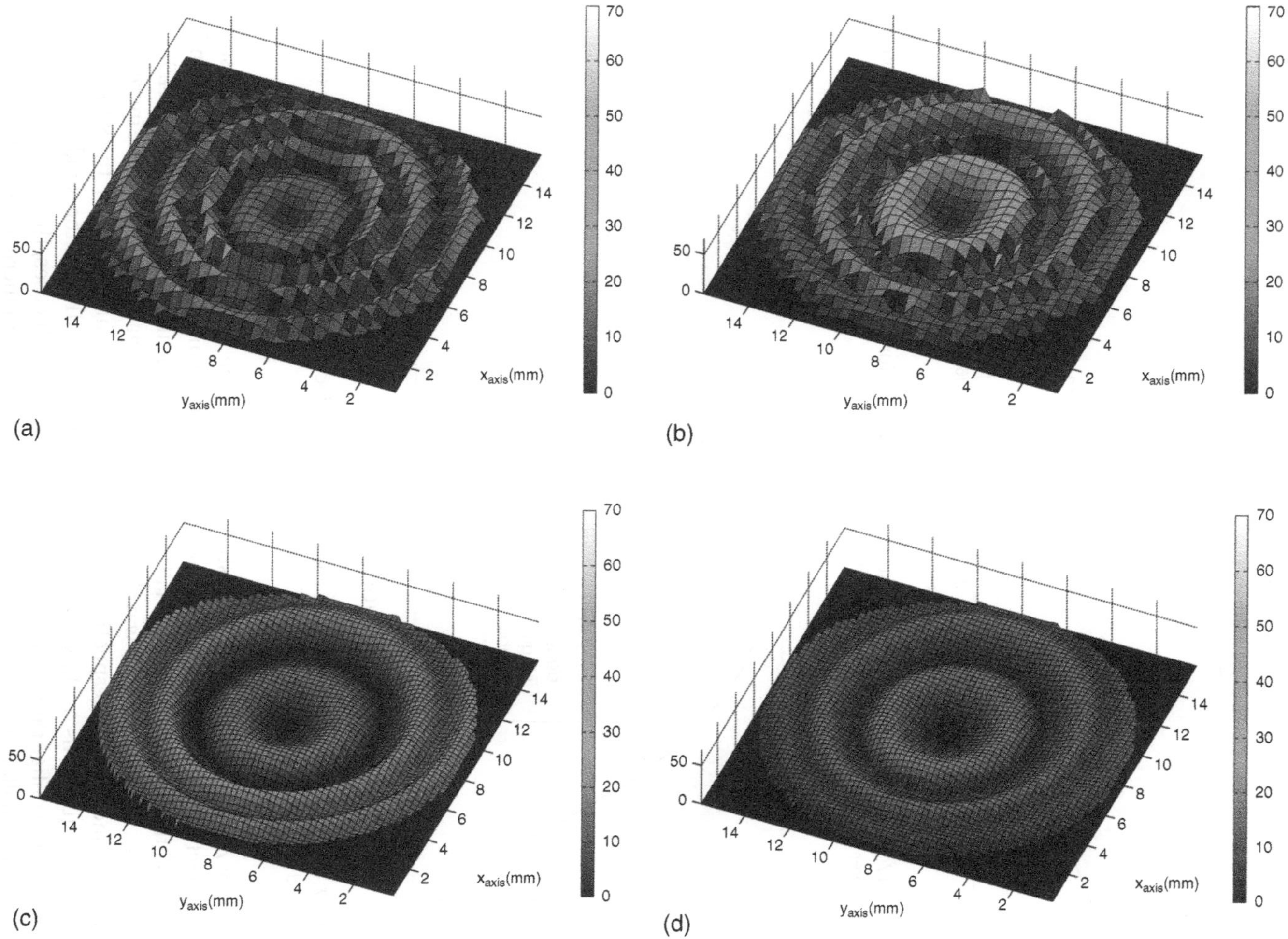

Figure 7.16 Absolute value of weighted phase error on the reflectarray aperture for double-layer designs: (a) $\lambda/2$ elements – 30 GHz, (b) $\lambda/2$ elements – 34 GHz, (c) $\lambda/4$ elements – 30 GHz, (d) $\lambda/4$ elements – 34 GHz. Reprinted with permission from [20].

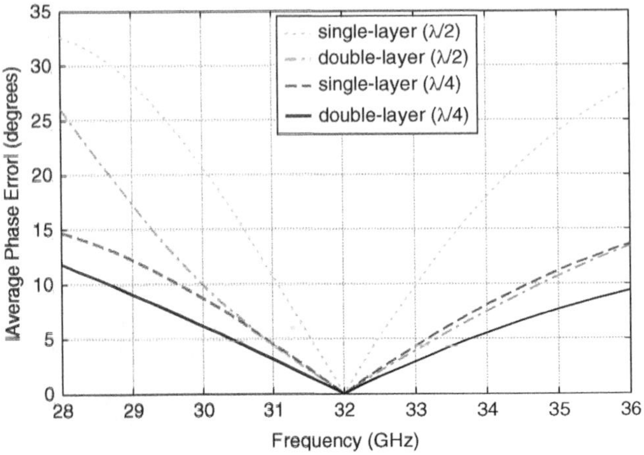

Figure 7.17 Average phase error of the reflectarray antenna as a function of frequency for single and double-layer elements. Reprinted with permission from [25].

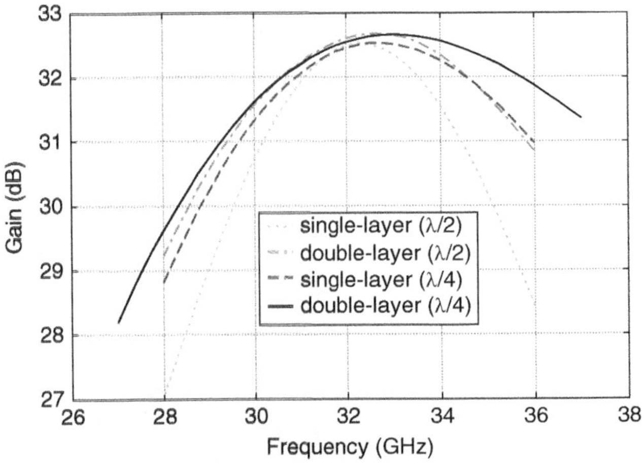

Figure 7.18 Gain vs frequency for single and double-layer reflectarrays using λ/2 and λ/4 elements. Reprinted with permission from [25].

λ/6, only λ/2 and λ/4 elements are studied here for the reflectarray gain comparison. The reflectarrays' gain versus frequency is given in Figure 7.18. The 1 dB gain bandwidths of the reflectarrays for half-wave element spacing are 10.55% and 14.98%, respectively. By using quarter-wavelength elements, these gain bandwidths are increased to 15.45% and 19.13%, respectively. The reflectarray antenna gains for the sub-wavelength configurations at 32 GHz are 32.45 dB and 32.55 dB, for the single-layer and double-layer configurations, respectively. Note that in both cases, the double-layer elements have smaller material loss, resulting in an overall gain increase of about 0.1 dB.

In summary, while the sub-wavelength technique is very effective for bandwidth improvement in small reflectarrays, combined with the multilayer approach, this technique was capable of increasing the 1 dB gain bandwidth of a fairly moderate size conventional single-layer $\lambda/2$ reflectarray (30.61–34.02 GHz) by more than 80% with a double-layer $\lambda/4$ reflectarray (30.07–36.43 GHz) configuration.

As a final remark on this unique broadband technique, we should note that a variety of different element geometries such as square loop, dogbone, extended dogbone, and fragmented elements [26], have also been proposed for sub-wavelength designs, which further show the effectiveness of this method for broadband reflectarrays [26]–[30].

7.1.5 Broadband Design Methods for Large Reflectarrays

The techniques described so far improve the bandwidth characteristics of the reflectarray elements. As mentioned earlier, for large reflectarrays, bandwidth is primarily limited by the different path lengths; therefore, most of the techniques introduced earlier will become less effective for large broadband reflectarrays. In this section, we will briefly introduce two techniques that are specifically designed for large reflectarrays.

One solution to reduce the errors associated with differential spatial phase delay is to implement true delay lines in order to compensate the real phase delay that can be several cycles of 2π. While this can be implemented using stubs with several wavelengths, the challenge is typically the real estate available on the lattice [9]. Rectangular patches with aperture-coupled lines [31] can produce a true-time delay [32], where the phase of the reflected wave is directly proportional to the length of the line. An image of a true-time delay reflectarray element as well as simulated and measured results showing the reflection phase curve as a function of line length are given in Figure 7.19. The range in phase delay can be enlarged by using a high dielectric constant for the line substrate, and also by bending the line to increase the physical length, and phase ranges up to five cycles have been reported [9]. One issue with these designs are the dissipative losses which increase slightly with the line length. By proper selection of materials, however, these losses can be reduced to less than 0.5 dB.

Another interesting approach to compensate for the differential spatial delay in large reflectarrays is by using a multi-panel configuration [33], [34]. Essentially this configuration is made of several flat reflectarray panels arranged to approximate a parabolic surface. Different configurations, such as a simple one-dimensional arrangement and more complex circumferential structures similar to an umbrella, have been proposed. Essentially each planar facet is a reflectarray that is designed to introduce the required phase shift to simulate the real parabolic surface. Images of different multi-panel reflectarray configurations are shown in Figure 7.20. Although these multi-panel (or multi-facet) configurations are more complex than a single flat reflectarray, they still offer the possibility of being folded and deployed for large antenna apertures [9].

7.2 Phase Synthesis for Broadband Operation

As discussed in the previous sections, a great deal of effort has been made to address the bandwidth limitations of reflectarrays with a focus on element performance improvement. By fine tuning the element design parameters, phase curves obtained at different frequencies can be almost parallel with each other, ultimately resulting in a broadband performance for the array. Most wideband elements, however, have certain undesirable

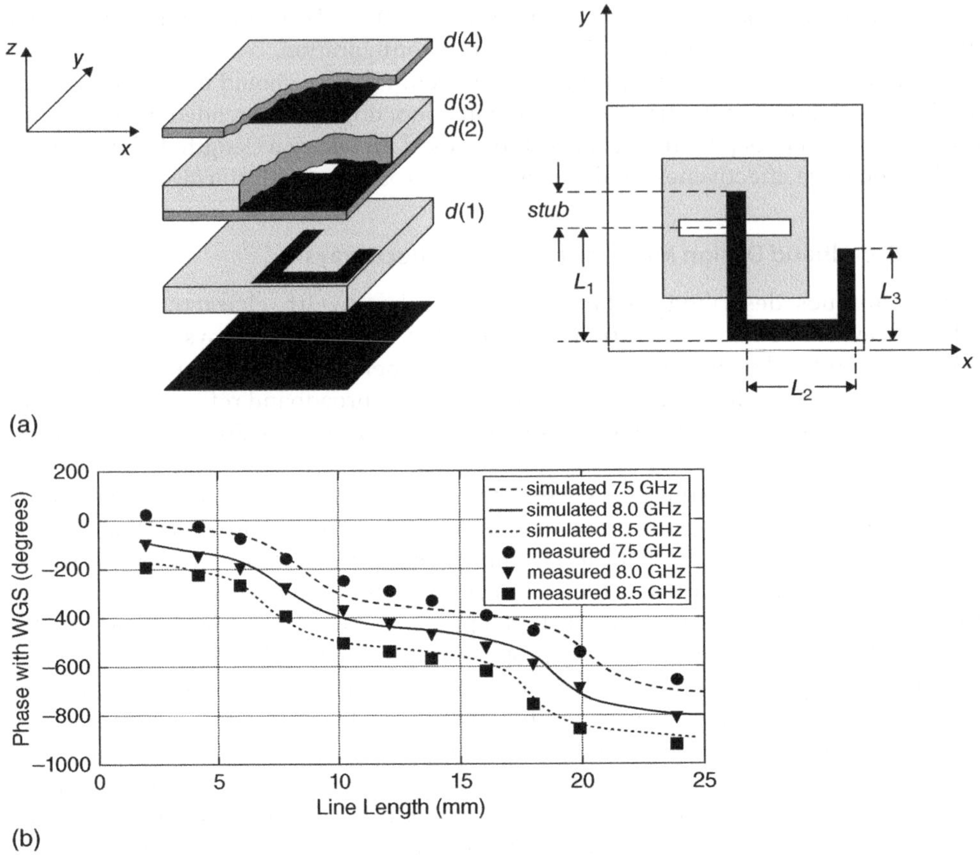

Figure 7.19 True-time delay reflectarray elements: (a) Aperture couple element. (b) Simulated and measured phases for the reflectarray element. Reprinted with permission from [32].

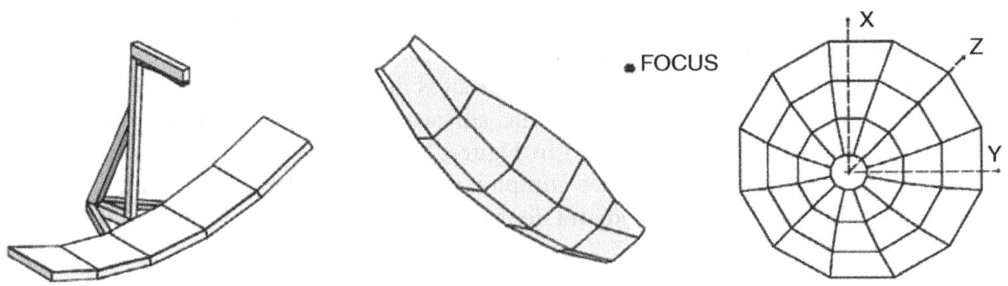

Figure 7.20 Multi-panel reflectarray configurations. Reprinted with permission from [9].

characteristics, such as a thicker substrate, which is heavier and increases the fabrication cost. The thick substrates also result in a larger phase difference between normal and oblique incident illuminations, complicating the design process by necessitating data tables for all oblique angles and polarizations.

An alternate approach for a broadband design is to synthesize the element phase over a frequency band for broadband operation. In other words, instead of designing the reflectarray at the center frequency, one can minimize the summed phase error (the difference between the desired and realized phases) at multiple frequencies simultaneously [7], [35], [36]. Moreover, the desired far-field patterns on a prescribed frequency band can directly be used in the synthesis technique to improve the frequency behavior [37]–[39]. These approaches, however, usually require a large number of available element phases, which is often achieved by tuning several geometrical parameters of the elements.

In order to better understand the principle of the phase synthesis approaches, two commonly used approaches are briefly introduced here. The first one, referred to as the single-frequency approach, is the most basic and widely used one, where the optimal performance in terms of gain is targeted at a single design frequency. The goal is to minimize the phase error of each element only at the design frequency f_0. Note that the required phase for each element is computed as described in Chapter 2. The practically achievable reflection phase, however, depends on the phase characteristics of that specific element, which are obtained through simulations or measurements. This phase error, which is somewhat analogous to phase quantization errors, can be mathematically given by

$$PE_{mn}(f) = \left|\varphi_{m,n,a}(f) - \varphi_{m,n,r}(f)\right|. \tag{7.1}$$

Here, mn is the element number on the aperture of the reflectarray, $\varphi_{m,n,r}(f)$ is the required or ideal element phase, and $\varphi_{m,n,a}(f)$ is the achievable element phase. While no optimization is necessary to minimize thits phase error due to the relatively simple nature of the problem, in general the goal is to minimize this error function. The cost function for this problem can be defined as

$$Cost\ Function = PE_{mn}(f_0). \tag{7.2}$$

This procedure is repeated one by one for all the elements, and thus an optimal phase characteristic of the reflectarray is achieved at the design frequency f_0.

In the second approach, referred to as the dual-frequency approach, one takes into account the gain performance at two properly separated frequencies f_1 and f_2. Different from the previous one, the goal is to minimize the sum of the phase errors at f_1 and f_2 of each element. The cost function for this problem is given by

$$Cost\ Function = PE_{mn}(f_1) + PE_{mn}(f_2). \tag{7.3}$$

Similarly, this procedure is repeated one by one for all the elements. Note that the phase error at any particular frequency may not be minimized. Similar to most wideband designs, this approach requires a complex element configuration and simultaneous tuning of multiple geometrical parameters to achieve good radiation performances at both design frequencies. If simple reflectarray elements such as square patches or rings are adopted, the minimization of phase errors becomes much less effective, and the reflectarray designed by this approach would suffer noticeable gain loss. In this section, we take a closer look at two broadband reflectarrays designed using the phase synthesis technique.

7.2.1 A Phase Synthesized Broadband Reflectarray

As discussed in Section 7.1, many wideband approaches require complex element geometries, resulting in a complicated and lengthy design process. A novel broadband phase synthesis approach, which is able to considerably increase the bandwidth of the reflectarray antenna and is independent of the element frequency behavior, was reported in [40]. Specifically, the reference phases at two design frequencies are optimized simultaneously to minimize the total phase error of the reflectarray.

A reflectarray with conventional square-ring element is considered for the study and is designed to operate at 12.5 GHz and 14.25 GHz. The substrate is Taconix TLX-8 with a thickness of 1.58 mm, dielectric constant of 2.55, and loss tangent of 0.0017. The element spacing is 11.2 mm with the ring width of 0.5 mm and the ring length varying from 1.5 mm to 10 mm. The phase behaviors at different frequencies are depicted in Figure 7.21.

The reflectarray aperture measures 400 × 400 mm with 1225 elements, and is illuminated by an offset feed horn and designed to radiate the beam at an angle of 25° from broadside direction. The dual-frequency approach is used to synthesize the array, and the cost function in (7.3) is calculated at 12.5 and 14.25 GHz. Once the elements are determined by the dual-frequency phase synthesis, the radiation pattern is computed using the array theory method and its gain variation is plotted, as indicated by the blue solid curve in Figure 7.22. Two peaks are observed as expected. However, noticeable gain losses exist due to the sub-optimal phase characteristics at each design frequency. Compared with the black dotted curve, which is obtained at each frequency using the single-frequency approach and represents the maximum achievable gain reconfigured for each frequency, the gain loss is about 2 ~ 3 dB at the two design frequencies. Moreover, throughout the frequency band in between, approximately 3 dB gain variation is clearly seen, which severely degrades the overall bandwidth performance of the reflectarray. This may be alleviated to some extent by reducing the separation between two design frequencies or by including a third design frequency in the phase synthesis process [36].

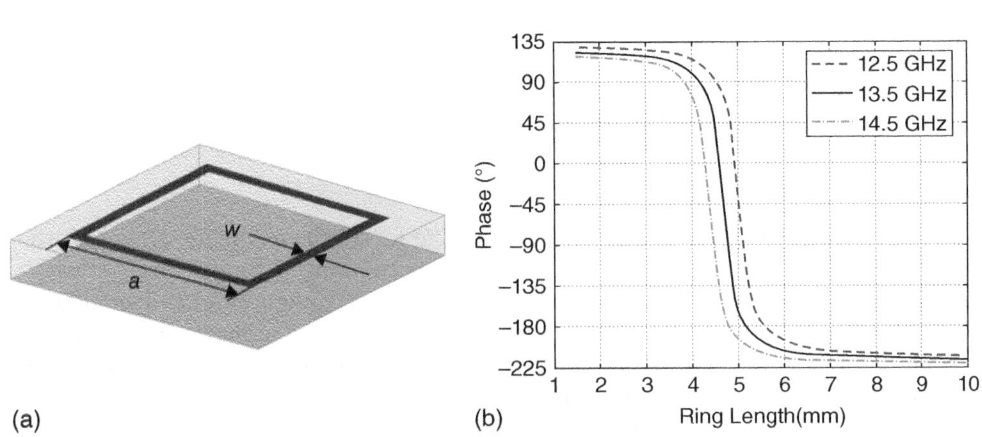

Figure 7.21 (a) Geometry of the reflectarray element. (b) Element reflection phase as a function of ring length. Reprinted with permission from [41].

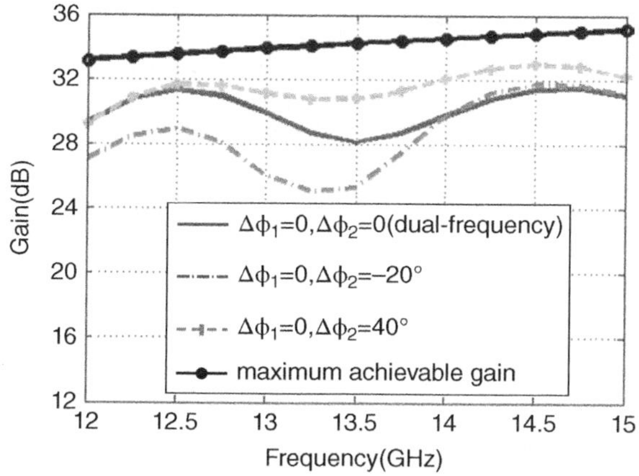

Figure 7.22 Computed gain using the dual-frequency phase synthesis approach. Reprinted with permission from [41].

It is well known that in reflectarray antennas phase is a relative term, that is, the relative phase difference between the elements on the aperture, not the absolute value, determines the radiation characteristics of the array. As such it is possible to add a constant reference phase to all the elements. Results are given in Figure 7.22 for two added phase constants. Here, $\Delta\varphi_1$ and $\Delta\varphi_2$ refer to the phase constants added to first and second operating frequencies, respectively. It can be seen that depending on these phase constants, the bandwidth performance of the system changes significantly.

Based on the observations in Figure 7.20, an improved approach for phase synthesis is to optimize the reference phases. To this end, a modified phase error term was defined in [41] as

$$PE_{mn}\left(f_1,\Delta\varphi_1\right)=\left|\varphi_{m,n,a}\left(f_1\right)-\varphi_{m,n,r}{}'\left(f_1,\Delta\varphi_1\right)\right|, \tag{7.4}$$

$$PE_{mn}\left(f_2,\Delta\varphi_2\right)=\left|\varphi_{m,n,a}\left(f_2\right)-\varphi_{m,n,r}{}'\left(f_2,\Delta\varphi_2\right)\right|. \tag{7.5}$$

Note that the reference phase is constant at a certain frequency for all the elements, therefore, this procedure cannot be done for each element independently. A cost function targeting the total phase error of the whole reflectarray was therefore proposed to minimize the errors as

$$\begin{aligned}&Cost\ Function\left(f_1,\Delta\varphi_1,f_2,\Delta\varphi_2\right)\\&=\sum_{m=1}^{M}\sum_{n=1}^{N}\left[w_{1-mn}\times PE_{mn}\left(f_1,\Delta\varphi_1\right)+w_{2-mn}\times PE_{mn}\left(f_2,\Delta\varphi_2\right)\right],\end{aligned} \tag{7.6}$$

where w_{1_mn} and w_{2_mn} are two weighting factors that depend on illumination intensity.

The phase synthesis process can be better graphically illustrated in Figure 7.23. These figures map the phase characteristics of each element at two design frequencies f_1 and f_2 onto a phase plane. The straight gray line is constructed by plotting all pairs of the required reflection phases of the reflectarray element. It passes the origin when $\Delta\varphi_1$ and $\Delta\varphi_2$ are configured to be zero as the initial values before optimization.

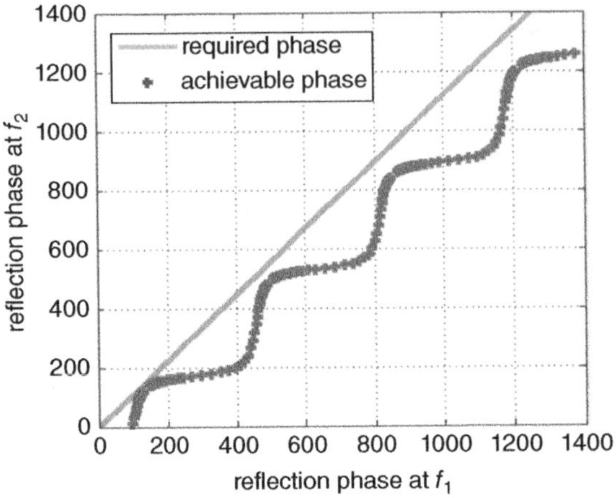

(a)

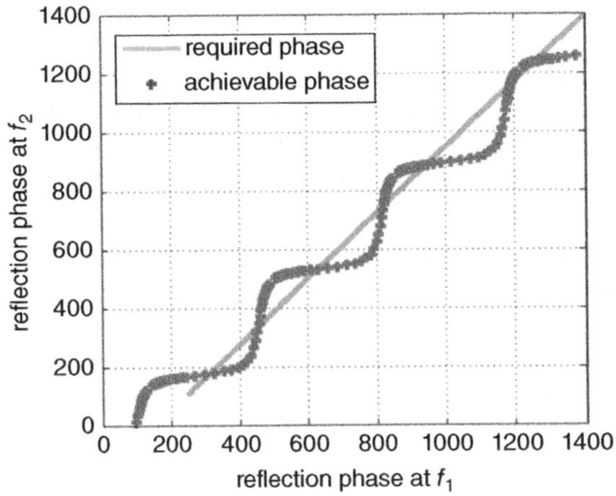

(b)

Figure 7.23 Graphical illustration of the proposed phase synthesis approach: (a) before and (b) after the optimization of the reference phases $\Delta\varphi_1$ and $\Delta\varphi_2$. Reprinted with permission from [41].

Each cross corresponds to a pair of achievable reflection phases of the reflectarray element obtained from simulations. The horizontal and vertical distances between the required phase curve and the achievable phase curve are the phase errors at the two frequencies. The improved phase synthesis process of minimizing phase errors is to shift the straight gray line in the phase plane to have a better matching between the two curves, as shown in Figure 7.23(b). To speed up the process, an iterative procedure based on a Particle Swarm Optimization (PSO) engine [42] was employed to minimize the cost function.

For verification and comparison purposes, three reflectarrays consisting of square-ring elements were designed using the three phase synthesis approaches, namely, the single-frequency approach, the dual-frequency approach, and the improved dual-frequency approach. Photos of the three prototypes are given in Figure 7.24.

The fabricated prototypes were measured in a compact-range anechoic chamber. Photos of the reflectarray system and the measured gain performances are given in Figure 7.25.

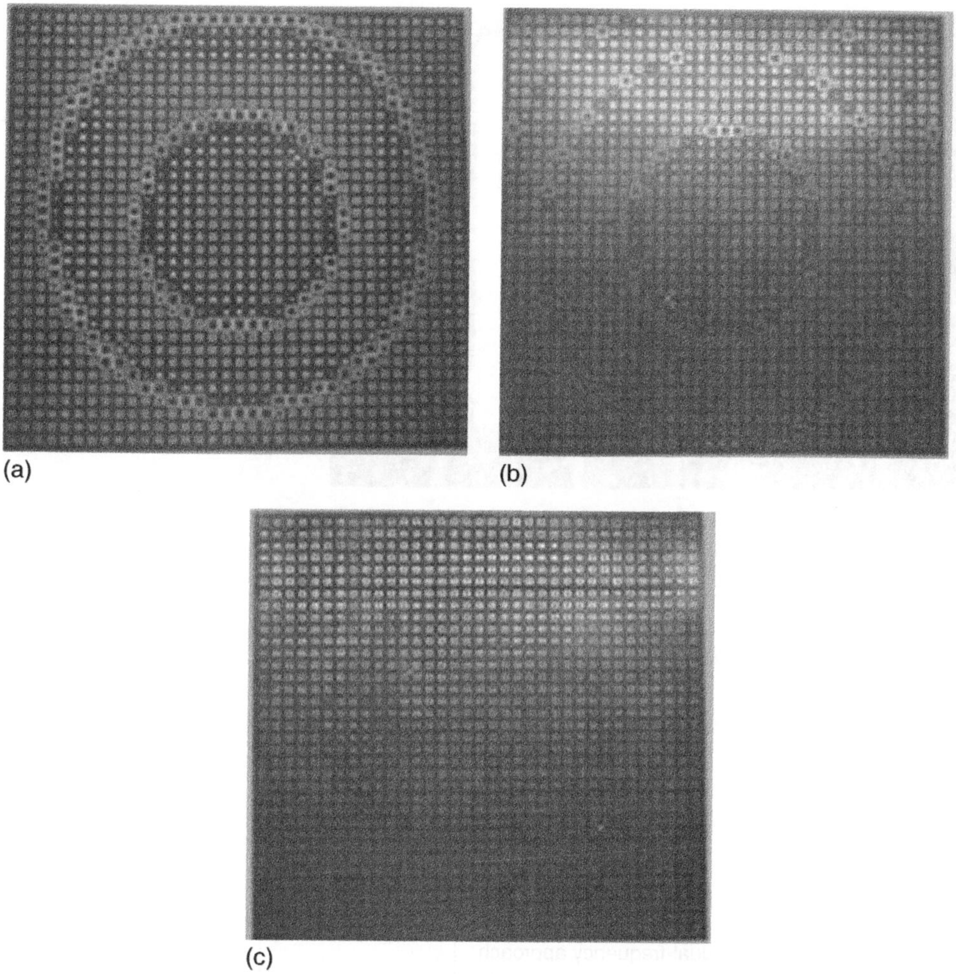

Figure 7.24 Photographs of the fabricated reflectarrays designed using (a) the single-frequency approach, (b) the dual-frequency approach, and (c) the improved dual-frequency approach in [40]. Reprinted with permission from [41].

Comparison of the results here clearly demonstrates the effectiveness of the improved dual-frequency approach for broadband reflectarray designs. A 16.7% measured bandwidth within 1.5-dB gain variation was achieved by a single-layer single-resonance design with a thin substrate. The proposed approach is applicable to general reflectarray elements, and the phase synthesis process is easy-to-implement and can be combined with other conventional wideband techniques for reflectarray designs.

7.2.2 A Dual-Frequency Broadband Reflectarray

To further improve the bandwidth, a dual-frequency broadband reflectarray antenna using triple-resonance elements was proposed in [43]. Essentially, this design combines the broadband element design approach with the improved dual-frequency phase

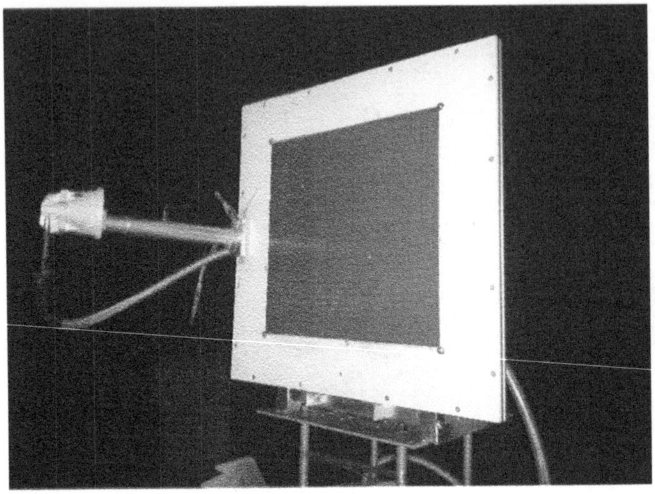

(a)

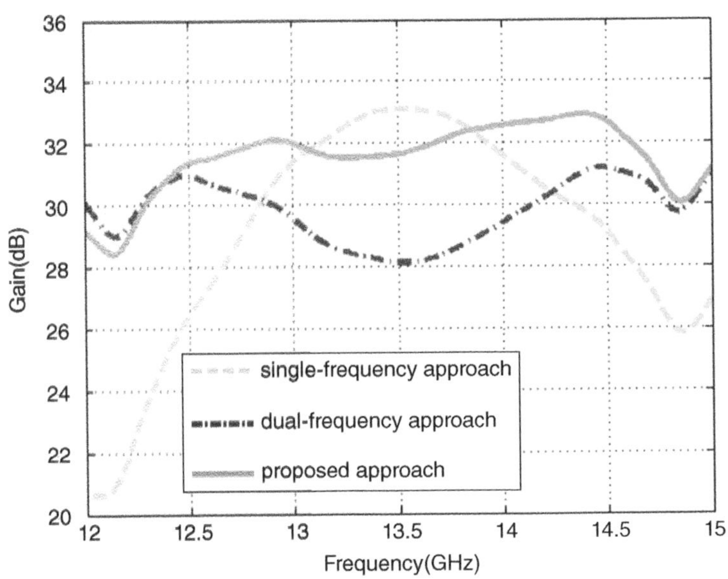

(b)

Figure 7.25 (a) Photograph of the measurement setup of the reflectarray, feeding horn, and the supporting structure. (b) Measured gain performances of the three reflectarrays designed using different phase synthesis approaches. Reprinted with permission from [41].

synthesis technique. The reflectarray was designed for operation at two frequencies bands for Ku-band satellite communications, namely, 12.25–12.75 GHz for receiving and 14.0–14.5 GHz for transmitting. The element chosen for the design is a triple-resonance element as depicted in Figure 7.26. It comprises two square loops and an inner square patch, which are etched on a Rogers 5880 substrate with thickness of 0.75 mm, dielectric constant of 2.2, and loss tangent of 0.0009. The element spacing is 11.0 mm (about 0.5λ at 13.5 GHz). In order to achieve a smoother phase curve and improve the

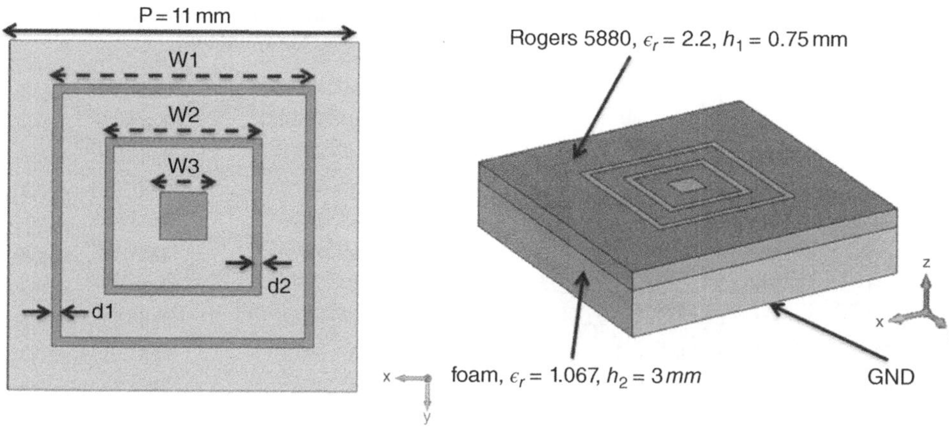

Figure 7.26 Geometry of the triple resonance phasing element. Reprinted with permission from [43].

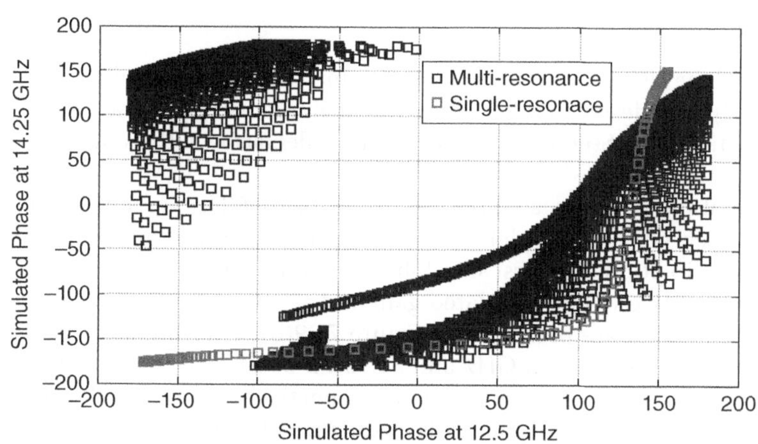

Figure 7.27 The coverage areas of two type of elements on the phase plane.

bandwidth performance, the Rogers substrate is backed by a grounded foam layer with thickness of 3.0 mm, dielectric constant of 1.067, and loss tangent of 0.0041.

There are five independent geometrical parameters in the element; however, in order to simplify the element design process, a special strategy was developed, so that only two parameters remain independent [43]. The achievable phase ranges for both frequencies (12.5 and 14.25 GHz) are larger than 700°. Note that compared to the single-resonance element, the multi-resonance element configuration provides a larger pool of possible element phase combinations for the phase synthesis procedure, as shown in Figure 7.27, which leads to a better optimized dual-frequency broadband reflectarray design.

The designed reflectarray aperture is square with side length of 350 mm, and is illuminated by a dual linearly polarized feed horn located at an offset angle of 25°. The desired main beam is also tilted 25° away from the boresight of the reflectarray aperture, which can fully take the advantage of the specular reflection. The improved dual-frequency phase synthesis method was then employed to optimize its gains at the two

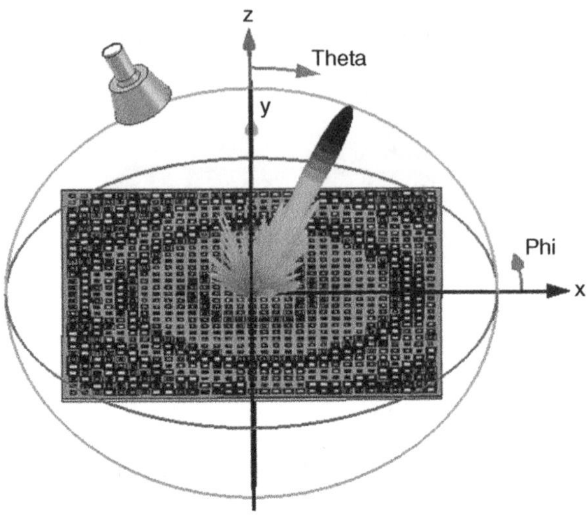

Figure 7.28 A 3D model of the reflectarray antenna.

center frequencies, and the element configurations with matching phase combinations are selected accordingly. The designed reflectarray was analyzed using full-wave simulation software, CST Microwave Studio. The model of the reflectarray in the software is given in Figure 7.28.

The simulated radiation patterns at 12.5 GHz and 14.25 GHz are plotted in Figure 7.29(a). It is observed that well-defined main beams at both frequencies are obtained in the desired direction. The first side-lobe levels are −20.0 dB and −22.8 dB at two frequencies, respectively. The simulated gain and aperture efficiency within the frequency band of interest are presented in Figure 7.29(b). The simulated gains of the reflectarray are 30.85 dB at 12.5 GHz and 31.95 dB at 14.25 GHz, which are equivalent to aperture efficiency of 50.3 and 49.8%, respectively. These results successfully demonstrate the capability of the triple-resonance element design and the phase synthesis method in dual-band reflectarray designs. It should be noted that, while this design is originally designed for dual-band operation, the combined use of broadband element and phase synthesis resulted in a fairly broad bandwidth for the reflectarray. The aperture efficiency is greater than 45% within the frequency band of 11.9–15.2 GHz, about 24.4%. In the transmitting band, the 1-dB gain bandwidth is 13.3–15.9 GHz, about 17.8%.

7.3 Multiband Reflectarray Designs

Although printed reflectarrays are limited by bandwidth performance, there are a variety of different techniques to provide coverage of multi-frequency bands. Intuitively, the characteristic of multiband operation can be implemented by the assembly of several sub-arrays, each operating at a single frequency. There are different approaches to combine multiple sub-arrays into a system. Most of the configurations are realized by exploitation of multiple layers, some are implemented by a single-layer structure, and a few configurations can be considered as hybrid types [44]–[58].

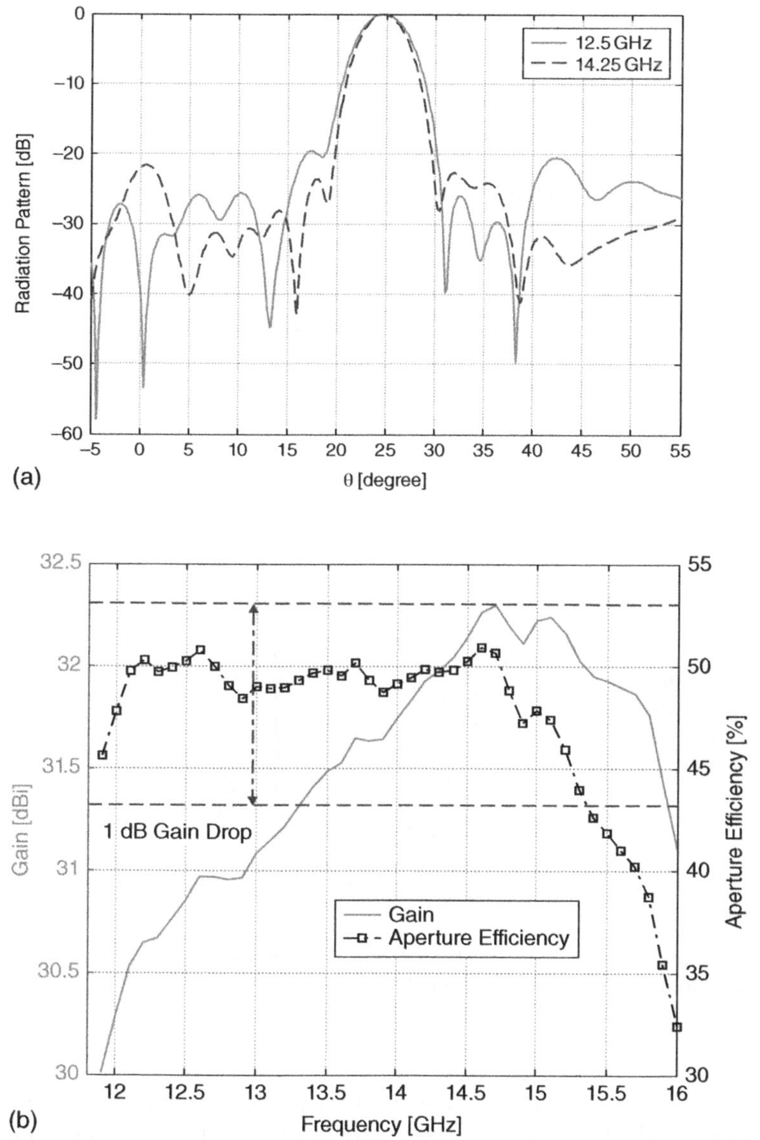

Figure 7.29 (a) Radiation patterns at the two design frequencies. (b) Gain and aperture efficiency as a function of frequency. Reprinted with permission from [43].

For the multilayer structures, the arrangement of the layers has to be carefully considered. In one case [44], the sub-array operating at the higher frequency is placed at the lower layer. In order to reduce the blockage of the incident energy, this layer arrangement is suitable for those elements having a relatively small metal area on the upper layer, for example, dipole or ring type elements. In another case, the order is reversed, where the sub-array elements at the lower frequency serve as the ground plane of the elements at

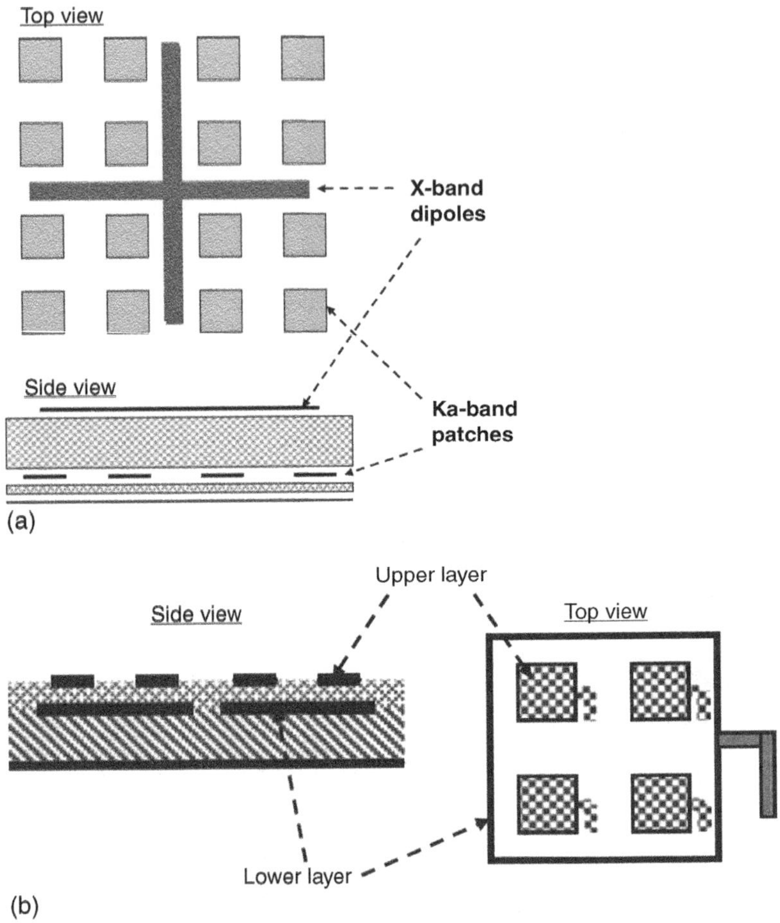

Figure 7.30 Dual-band reflectarrays: (a) low frequency layer above high frequency, (b) high frequency layer above low frequency. Reprinted with permission from [9].

the higher frequency [46]. This configuration is devised for patch elements since they have a bulk portion of metal, which operates as a solid ground plane for the elements on the upper layer. Examples of these two-layer configurations operating at two different bands are depicted in Figure 7.30.

Some multilayer reflectarray antennas make use of frequency selective surfaces (FSSs) as shown in Figure 7.31. For example, the idea introduced in [47] is to replace the solid ground planes by FSSs for all the sub-arrays. Each FSS is reflective at the operating frequency of the corresponding sub-array while transparent to other frequencies. This configuration yields the flexibility of arbitrarily arranging the sequence of the layers. FSS by itself may also be designed to possess the features of a reflectarray antenna, without ground planes. All these multilayer structures have apparent disadvantages such as the increased antenna volume and fabrication costs. More severely, if two operating frequencies close to each other are desired, for example, C band (7.1 GHz) and X band (8.4 GHz), the multilayer structure will encounter coupling problems.

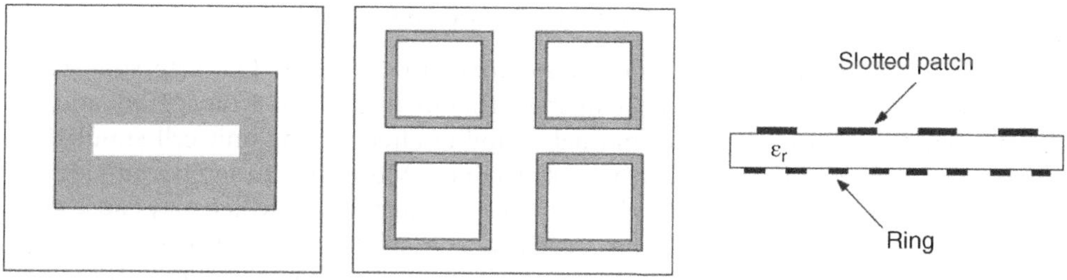

Figure 7.31 FSS-backed reflectarray antenna element. Reprinted with permission from [47].

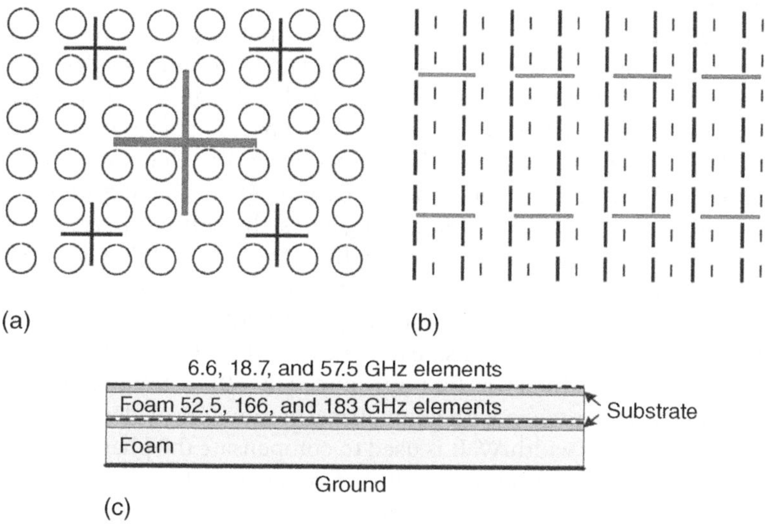

Figure 7.32 A six-band reflectarray element configuration: (a) top layer, (b) bottom layer, (c) side view. Reprinted with permission from [52].

To overcome these disadvantages, single-layer multiband reflectarray antennas have been investigated. The single-layer structure proposed in [48] operates at both 11.85 GHz and 13.88 GHz with orthogonal linear polarizations, which combines two sub-arrays with loop shaped elements. In [49], two reflectarray prototypes are demonstrated, both covering the frequencies of 20 GHz and 30 GHz. One of them is linearly polarized with patch elements whereas the other is circularly polarized with cross-dipole elements. Another example [50] covering 20.7 GHz and 44.5 GHz takes both patch elements and cross-dipole elements for its sub-array designs. A single-layer tri-band reflectarray antenna has been implemented with dipole elements to radiate linearly polarized waves at three close frequencies (12 GHz, 15 GHz, and 18 GHz) [51]. A six-band reflectarray concept is proposed in [52], where two single-layer tri-band reflectarray antennas are stacked to achieve the desired performance. Images of the six-band reflectarray element configuration are given in Figure 7.32.

In summary, various techniques exists to achieve multiband performance for a reflectarray, and the specific selection typically depends on the number of required frequency bands and their frequency ratios. In this section, we will briefly introduce two recent designs of multiband reflectarrays.

7.3.1 A Single-Layer Dual-Band Circularly Polarized Reflectarray

Although single-layer multiband reflectarrays are relatively easy to fabricate, some challenges still exist in these designs, such as the difficulty in element arrangement and the lack of complete consideration of mutual coupling effect in the unit-cell simulation, which will deteriorate the reflectarray performance. Moreover, an interlaced element grid in some cases will increase the element unit-cell periodicity, which may cause grating lobe at the higher frequency band. Recently, a complete investigation of a compact element configuration, namely, the concentric dual split ring, for designing a dual-band circularly polarized reflectarray [57]–[58], has demonstrated that the phasing element can give good radiation performance in terms of aperture efficiency, cross-polarization level, and side-lobe level.

In [59] a novel phasing element, composed of a concentric split ring and a modified Maltese Cross element, was proposed as an attractive candidate to design a single-layer dual-band circularly polarized reflectarray. The objective of this design was to achieve a left-handed circularly polarized (LHCP) pencil beam at 20 GHz, and a right-handed circularly polarized (RHCP) pencil beam at 30 GHz, respectively. The element geometry is depicted in Figure 7.33, which is printed on a 0.762 mm Arlon Di880 grounded substrate with $\epsilon_r = 2.2$ and $\tan\delta = 0.0009$. To avoid grating lobes, the unit-cell periodicity has been set to 5 mm, which is equal to half a wavelength in free space at 30 GHz and one-third free space wavelength at 20 GHz. It consists of two parts. The outer one is a concentric split ring, characterized by its outer radius r_o, gap width g_o, line width w_o, and rotation angle ϕ_o, which is used to compensate the phase shift at 20 GHz. The inner one is the modified Maltese Cross, which was first introduced in [60] to design a single-frequency broadband reflectarray antenna with geometrical parameters of element size L_1, the slot's length L_2, and the slot's width W. It is used to compensate the phase shift at 30 GHz. The reason for choosing the modified Maltese Cross element as the high frequency phasing element is that this configuration can achieve a wider phase range due

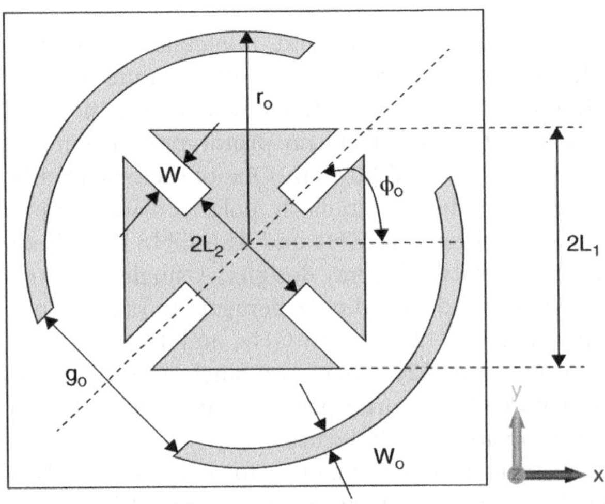

Figure 7.33 Geometry of the dual-band reflectarray element. Reprinted with permission from [59].

to its superior multi-resonance performance in comparison with that of square patch. If the patch is chosen as the inner element, because of space constraints, the maximum phase range at 30 GHz will only be 245°.

The angular rotation technique is used for the phase compensation at 20 GHz, hence gaps must be introduced in the concentric-ring element structure so that the incoming circular polarization (CP) signal is reflected into the same-sensed CP signal. As for the modified Maltese Cross element, it is a symmetric structure and reflects the incoming CP signal into the opposite-sense CP signal. Therefore, the variable element size technique is used to obtain the required reflection phase at 30 GHz. By using this element configuration, at each frequency band only one parameter, either ϕ_o or L_1, has to be considered for phase compensation, thus simplifying the analysis procedure.

The simulated reflection coefficients of the proposed element at 20 GHz and 30 GHz are presented in Figure 7.34. The phase range is 360° and 300° at 20 and 30 GHz, respectively. It can be clearly seen that the phase responses at two bands are well decoupled in the design, which are primarily determined by the rotation angle ϕ_o at 20 GHz and by the Maltese element size L_1 at 30 GHz, respectively. It should be pointed out that, however,

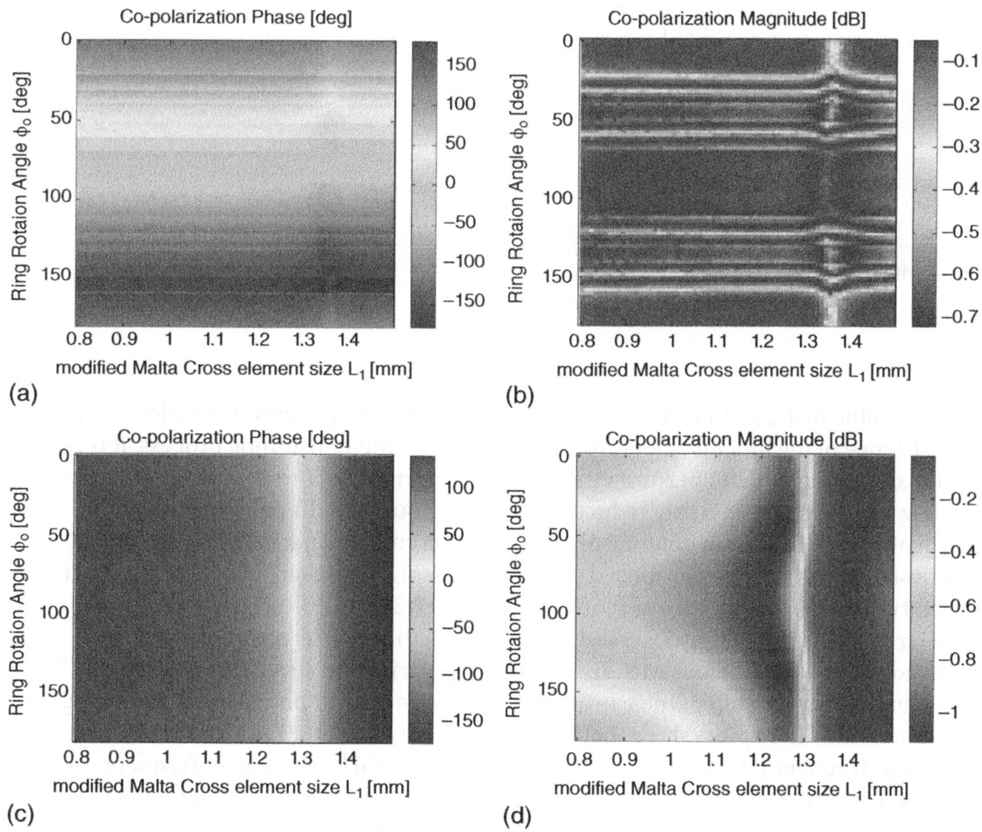

Figure 7.34 Co-polarization reflection coefficient at 20 GHz: (a) phase, (b) magnitude. Co-polarization reflection coefficient at 30 GHz: (c) phase, (d) magnitude. Reprinted with permission from [59].

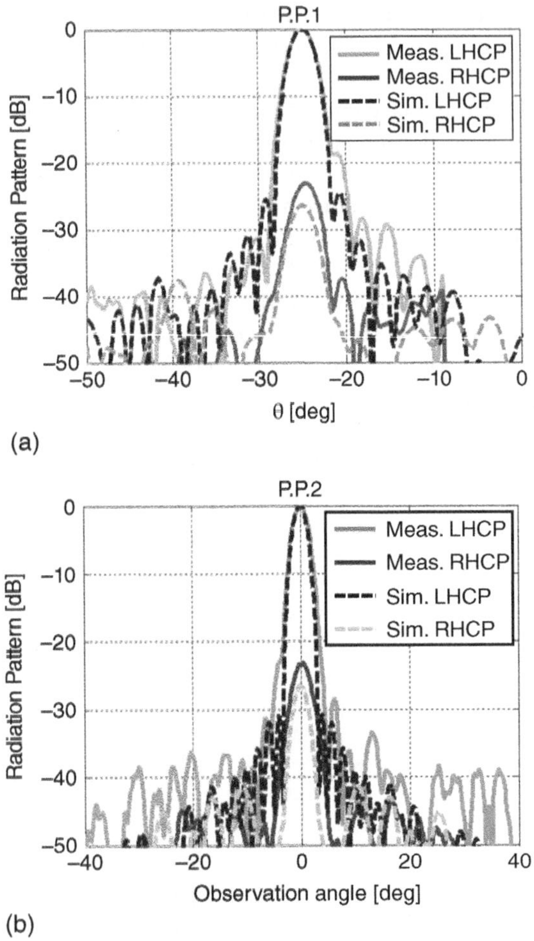

Figure 7.35 Radiation patterns at 20 GHz: (a) offset plane, (b) orthogonal plane. Reprinted with permission from [59].

for each value of ϕ_o, g_o must be optimized to reduce cross-polarization reflection due to the existence of the inner element. Although this is a time-consuming process, it is essential for designing a high aperture efficiency reflectarray. As a result, the element loss at 20 GHz has been considerably reduced, which is about −0.1 dB on average.

A dual-band reflectarray antenna prototype with a circular aperture diameter of 420 mm was designed to verify the proposed phasing element. Measured and simulated patterns in the two principal planes are plotted in Figure 7.35 for 20 GHz and Figure 7.36 for 30 GHz, respectively. It is observed that the radiation patterns are well-defined with peak side-lobe levels below −18 dB and cross-polarization levels below −20 dB. Plots showing the relationship of measured gain versus frequency bands of interest are given in Figure 7.37. These plots show that 1-dB gain bandwidth is 6.93% from 19.5~20.9 GHz for lower frequency band and 6.32% from 29.1~31 GHz for upper frequency band, which are comparable to those of a traditional single-frequency design. As for 3-dB axial ratio bandwidth, it is more than 10% for both frequency bands. The measured aperture efficiencies at the two designed frequencies are 66.5% and 50% at 20 GHz and 30 GHz, respectively.

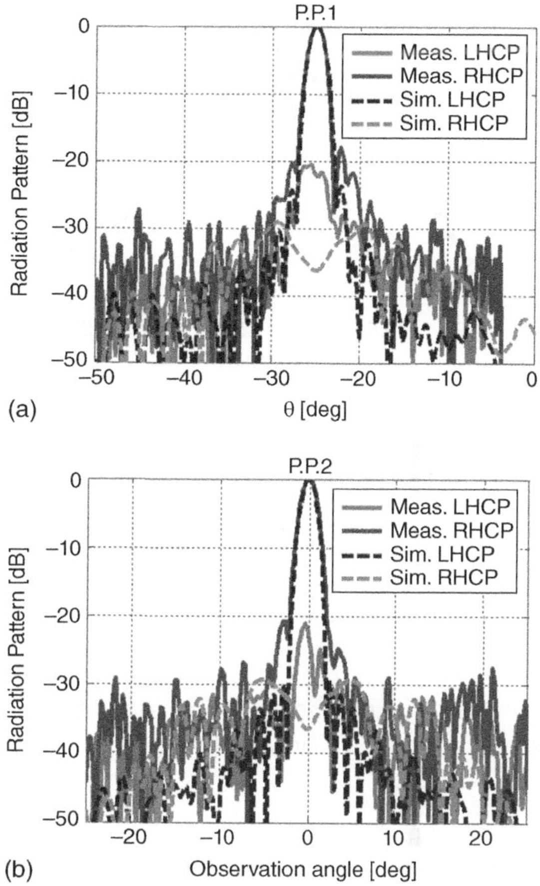

Figure 7.36 Radiation patterns at 30 GHz: (a) offset plane, (b) orthogonal plane. Reprinted with permission from [59].

7.3.2 A Single-Layer Tri-Band Reflectarray

A single-layer tri-band reflectarray was first proposed in [61] for NASA space applications. Circular ring, square loop, and cross dipole were designed to operate at 7.1 GHz, 8.4 GHz, and 32 GHz, respectively, and they are assembled on the same reflecting surface. The reflectarray design effectively utilizes the single reflecting surface area to realize tri-band (C/X/Ka) operation. This single-layer design is easy to fabricate, and it can be combined with a multilayer technique to further enhance the number of frequencies that a reflectarray can cover.

The geometry of the tri-band reflectarray design is shown in Figure 7.38. It consists of three different types of elements: split circular ring for Ka-band (32 GHz) operation, split square loop for C-band (7.1 GHz) operation, and cross-dipole element for the X band (8.4 GHz) operation. All three types of elements are mounted on a 62-mil thick RT/Duroid 5870 substrate. The C- and X-band elements share the same periodicity, which is around a half-wavelength, where the Ka-band element's periodicity is one-quarter of the previous one due to the shorter wavelength.

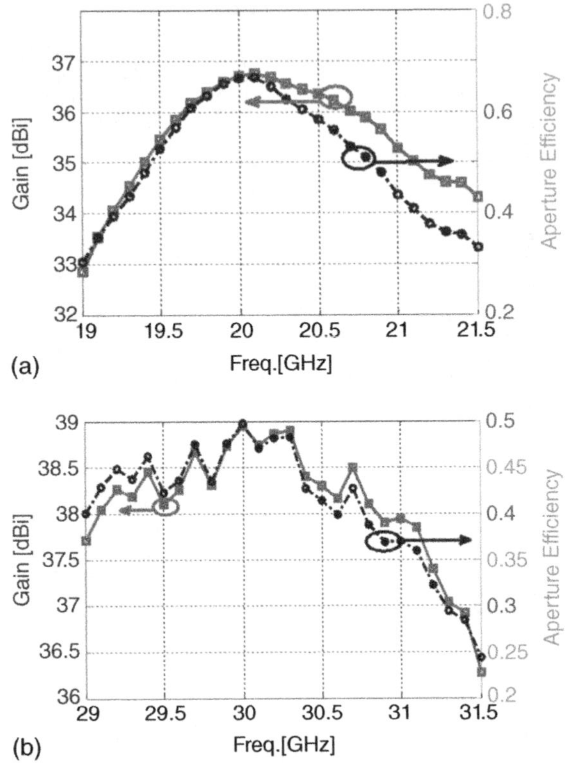

Figure 7.37 The measured gain and aperture efficiency versus frequency bands of interest: (a) 19~21.5 GHz, (b) 29~31.5 GHz. Reprinted with permission from [59].

Figure 7.38 Geometry of a single-layer tri-band reflectarray. Reprinted with permission from [61].

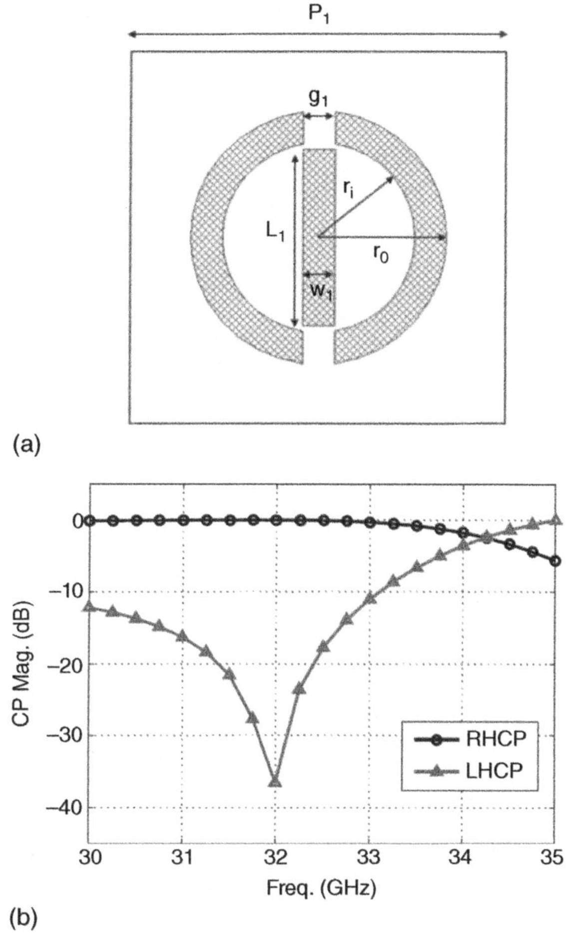

Figure 7.39 (a) Geometry of the Ka-band element. (b) Reflection phase as a function of frequency. Reprinted with permission from [61].

The Ka-band element is a split ring with a vertical bar [62] designed to reflect the RHCP wave with the same polarization state at 32 GHz. The angular rotation technique [63] was used to compensate the spatial phase delay. The geometry of the element and the reflection phase characteristics are given in Figure 7.39. Note that at 32 GHz, the cross-polarization (LHCP) is 36 dB lower than the co-polarization (RHCP).

The C-band element was a split square loop designed to reflect the RHCP wave with the same polarization state at 7.1 GHz. Note that the previous angular rotation technique could not be directly used here, because the rotation of square rings will collide with the split ring elements. To solve this problem, an alternative approach was proposed to move the position of slots along the perimeter of the square loop. The slot position was identified by an equivalent rotation angle Φ, shown in Figure 7.40. It was observed that when the slot position is moved along the perimeter of the square loop, the resonant frequency of the structure shifted. Thus, to maintain the same operating frequency the slot width was adjusted. Note that while this approach provided the desirable phase variation from −180° to 180° when the rotation angle changed from −90°

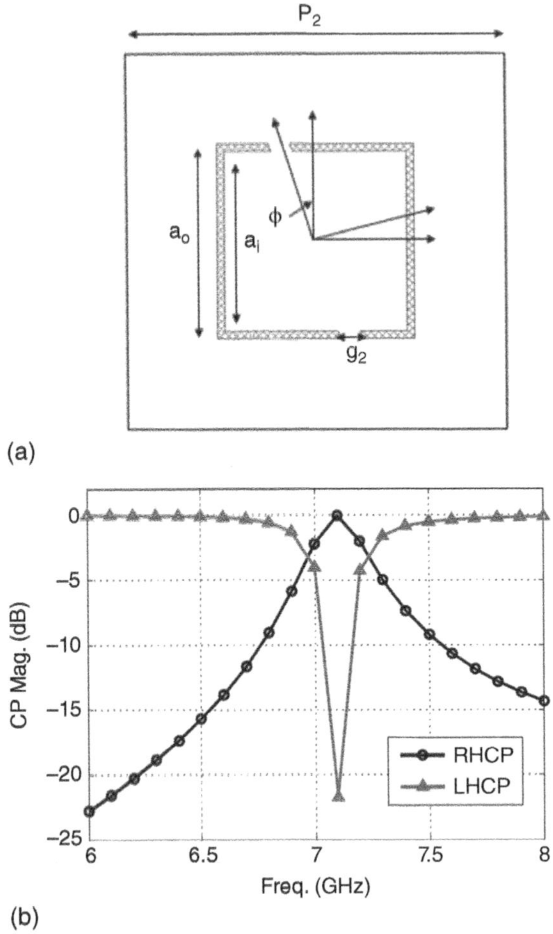

Figure 7.40 (a) Geometry of the C-band element. (b) Reflection phase as a function of frequency. Reprinted with permission from [61].

to 90°, the phase variation was no longer a linear function of the rotation angle as in the previous angular rotation technique for the circular ring element.

A cross-dipole element was designed to operate at 8.4 GHz, which has the same element spacing as the square loop but shifts half periodicity in space. To compensate for the spatial phase delay, the X-band element uses variable size method [64]. Note that in contrast to the previous two designs, this cross dipole reflects the incident CP wave with the opposite polarization state, namely, from LHCP to RHCP. This design will help to avoid the coupling effect between two close frequencies (7.1 GHz and 8.4 GHz). The geometry of the element and its reflection phase curve are given in Figure 7.41.

Next, a detailed investigation followed in order to optimize the design for tri-band performance, and in particular took into account the mutual coupling effects between the elements at all three bands [65]. A summary of the research on Ka- and X-band reflectarray antennas was reported in [66]–[69]. The final optimized geometries of the elements and the sub-arrays, as reported in [70], are shown in Figure 7.42. Note that the Ka-band element has been changed into a single split ring for the ease of fabrication.

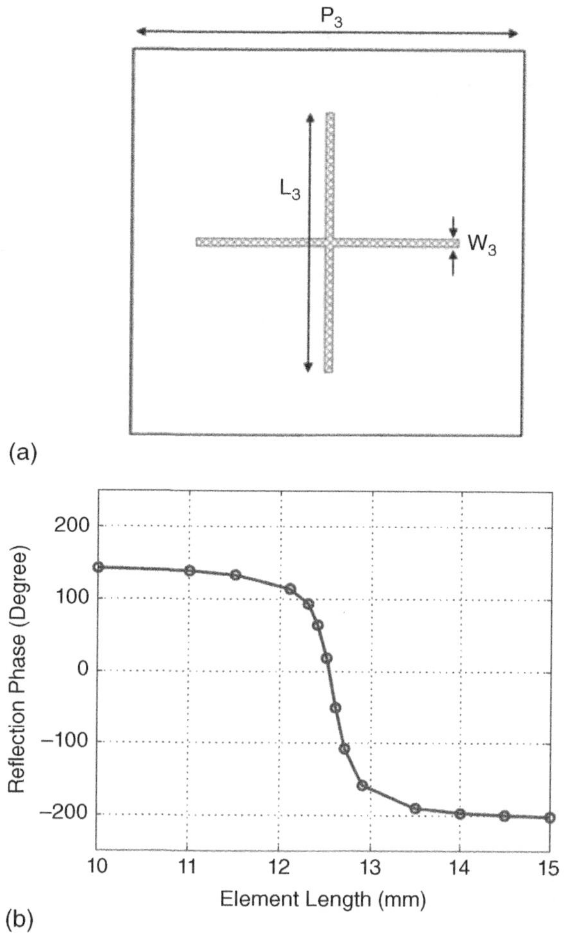

Figure 7.41 (a) Geometry of the C-band element. (b) Reflection phase as a function of frequency. Reprinted with permission from [61].

The C- and X-band elements share the same period of the lattice p = 18.75 mm, which is slightly smaller than half a wavelength at 7.1 GHz but larger than that at 8.4 GHz. The Ka-band lattice period is p/4, half a wavelength at 32 GHz. The reflection phases of the optimized elements as a function of frequency are given in Figure 7.43. The element coupling effects have been included in these simulations. The phase compensation schemes applied to all the three frequency ranges are shown in Figure 7.44. Note that the modified element rotation technique is applied for X-band [68], the conventional element rotation technique is used for the Ka band, and the variable element size method in applied for the C band.

Finally, a circular-shape reflectarray aperture was designed for the prototype with an aperture diameter of 566 mm. An offset feed was selected for the reflectarray system to reduce feed blockage and the aperture phase distribution was designed to focus the main beam in the specular reflection direction. The C band and X band shared the same feeding source, that is, a Narda 642 standard gain horn antenna. The Ka band is illuminated separately by a corrugated conical circularly polarized horn. The prototype was fabricated by dividing the array aperture into four quarters. The photo of the fabricated prototype is shown in Figure 7.45.

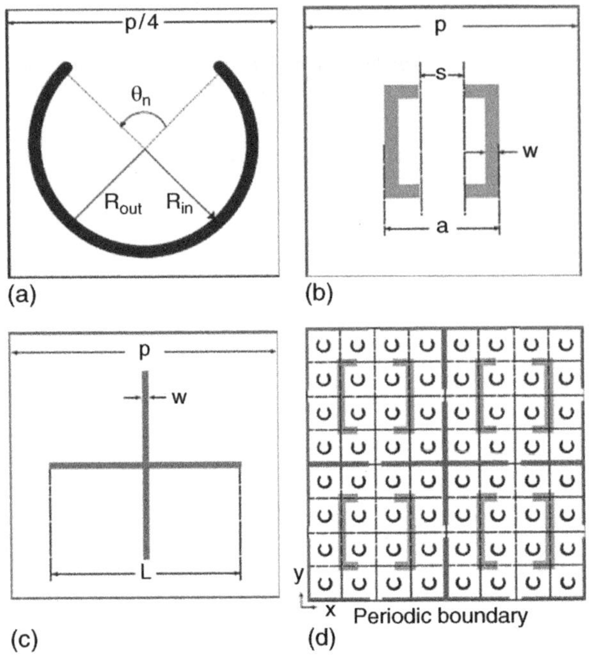

Figure 7.42 Element geometries of the (a) Ka band, (b) X band, (c) C band, with (d) the arrangement of the sub-arrays. Reprinted with permission from [70].

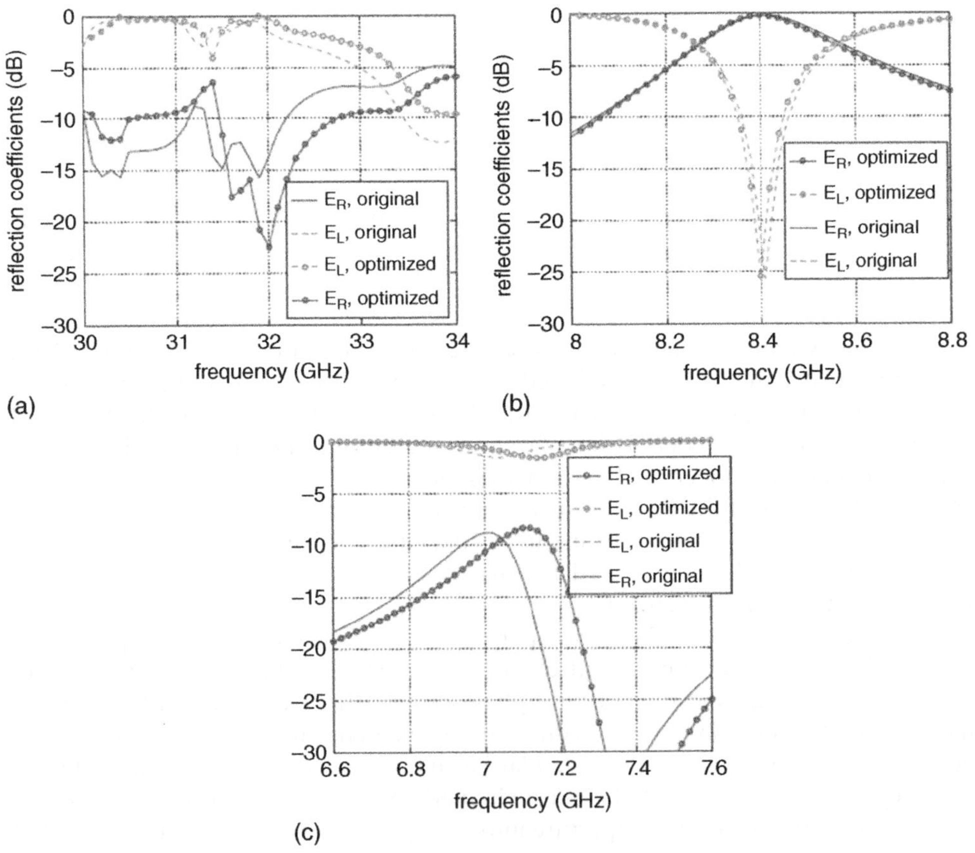

Figure 7.43 Element performance of (a) Ka band, (b) X band, and (c) C band. Reprinted with permission from [70].

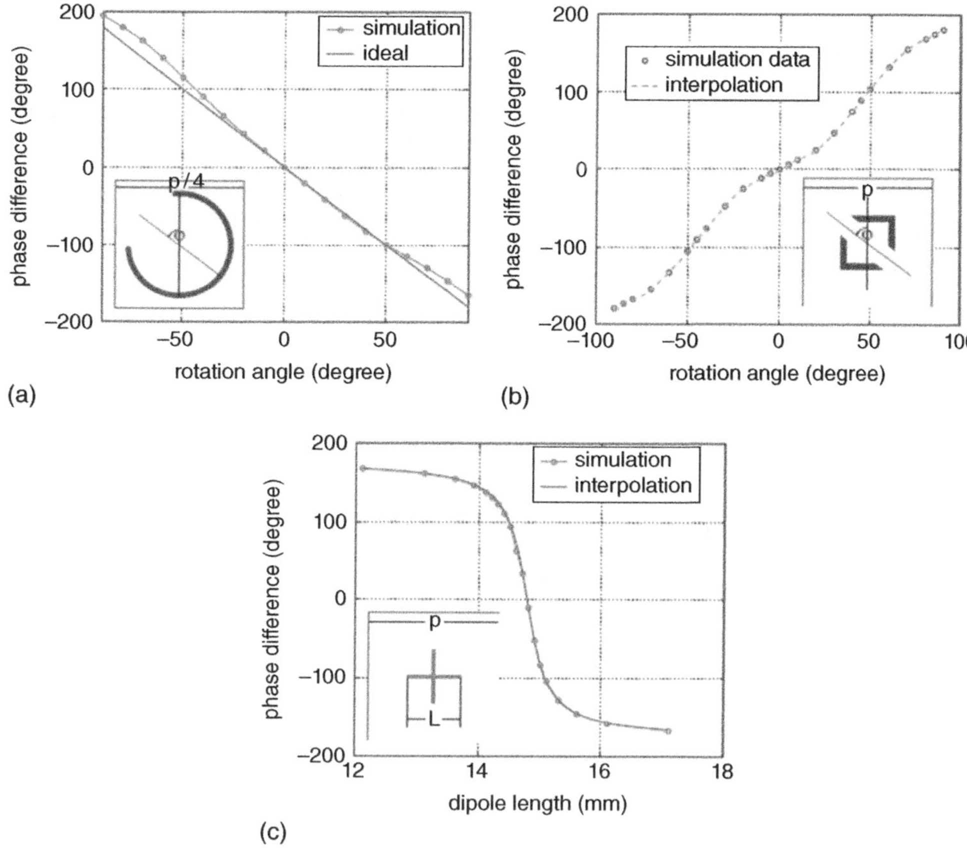

Figure 7.44 Element phase compensation schemes for the (a) Ka band, (b) X band, and (c) C band. Reprinted with permission from [70].

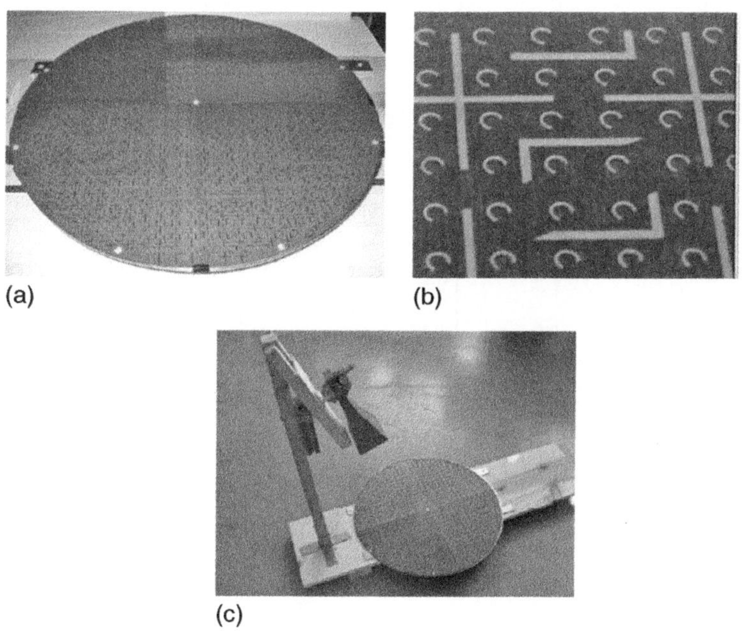

Figure 7.45 The fabricated prototype (a) reflectarray aperture, (b) local details, and (c) the reflectarray system with a C/X band feed horn. Reprinted with permission from [70].

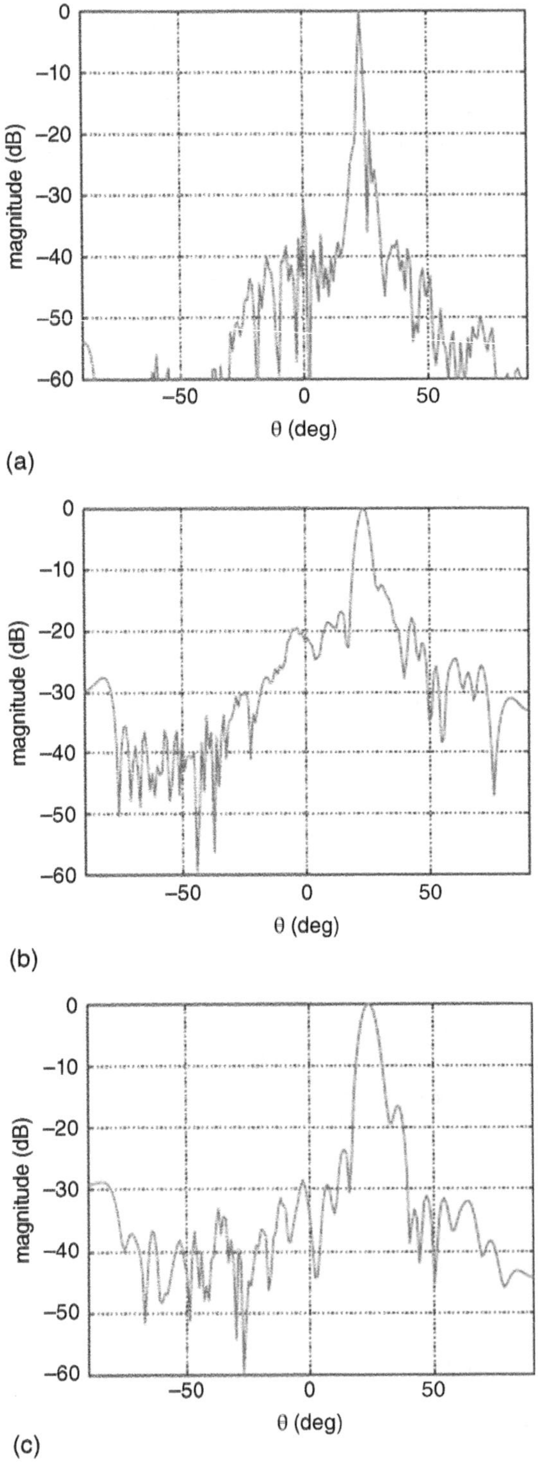

Figure 7.46 Measured far-field patterns of the reflectarray at (a) Ka band (32 GHZ, LHCP), (b) X band (8.4 GHZ, RHCP), and (c) C band (7.1 GHz, Eθ). Reprinted with permission from [70].

The antenna was tested with a planar near-field measurement system and the measured far-field patterns are given in Figure 7.46. The measured 3 dB main beam widths are 0.75° for the Ka band (32 GHz), 4.5° for the X band (8.4 GHz), and 5.9° for the C band (7.1 GHz), respectively. The peak gains are 28.4 dB for the C band, 29.1 dB for the X band, and 38.7 dB for the Ka band, corresponding to an aperture efficiency of 38.8, 33.6, and 20.6%, respectively. The relatively low efficiency observed in Ka-band is mainly due to the high cross-polarization level caused by the coupling from C-band and X-band elements. In order to address this problem, future research may focus on designing elements whose performance is robust by further minimizing the coupling effect.

References

1 J. Huang, "Bandwidth study of microstrip reflectarray and a novel phased reflectarray concept," presented at the *IEEE Antennas and Propagation Society International Symposium*, California, USA, June 1995.

2 D. M. Pozar, "Bandwidth of reflectarrays," *Electron. Lett.*, Vol. 39, No. 21, Oct. 2003.

3 D. M. Pozar and T. A. Metzler, "Analysis of a reflectarray antenna using microstrip patches of variable size," *Electron. Lett.*, Vol. 29, No. 8, pp. 657–658, Apr. 1993.

4 D. M. Pozar, S. D. Targonski, and H. D. Syrigos, "Design of millimeter wave microstrip reflectarrays," *Proc. IEEE Trans. Antennas Propagat.*, Vol. 45, pp. 287–295, Feb. 1997.

5 J. A. Encinar, "Design of two-layer printed reflectarrays using patches of variable size," *IEEE Trans. Antennas Propag.*, Vol. 49, No. 10, pp. 1403–1410, Oct. 2001.

6 N. Misran, R. Cahill, and V. Fusco, "Reflection phase response of microstrip stacked ring elements," *Electron. Lett.*, Vol. 38, No. 8, pp. 356–357, April 2002.

7 J. A. Encinar and J. A. Zornoza, "Broadband design of three-layer printed reflectarrays," *IEEE Trans. Antennas Propag.*, Vol. 51, No. 7, pp. 1662–1664, July 2003.

8 F.-C. E. Tsai and M. E. Bialkowski, "Designing a 161-element Ku-band microstrip reflectarray of variable size patches using an equivalent unit cell waveguide approach," *IEEE Trans. Antennas Propag.*, Vol. 51, No. 10, pp. 2953–2962, Oct. 2003.

9 J. Huang and J. A. Encinar, *Reflectarray Antennas*. New York, NY, USA: Wiley-IEEE, 2008.

10 M. E. Bialkowski and K. H. Sayidmarie, "Investigations into phase characteristics of a single-layer reflectarray employing patch or ring elements of variable size," *IEEE Trans. Antennas Propag.*, Vol. 56, No. 11, pp. 3366–3372, Nov. 2008.

11 J. Huang, "The development of inflatable array antennas," *IEEE Antennas Propag. Mag.*, Vol. 43, No. 4, pp. 44–50, Aug. 2001.

12 M. Bozzi, S. Germani, and L. Perregrini, "Performance comparison of different element shapes used in printed reflectarrays," in *Proc. IEEE Antennas Wireless Propag. Lett.*, 2003, Vol. 2, pp. 219–222.

13 N. Misran, R. Cahill, and V. Fusco, "Design optimization of ring elements for broadband reflectarray antennas," *Microwaves, Antennas and Propagation, IEE Proceedings*, Vol. 150, No. 6, pp. 440–444, Dec. 2003.

14 K. H. Sayidmarie and M. E. Bialkowski, "Investigations into unit cells offering an increased phasing range for single-layer printed reflectarrays," *Microwave and Optical Technology Letters*, Vol. 50, No. 4, pp. 1028–1032, Apr 2008.

15 P. Nayeri, M. E. Bialkowski, and K. H. Sayidmarie, "Design of unit cells of double circular rings for a single-layer microstrip reflectarray," *20th Asia Pacific Microwave Conference*, Hong Kong, China, Dec. 2008.

16 M. R. Chaharmir, J. Shaker, M. Cuhaci, and A. Ittipiboon, "A broadband reflectarray antenna with double square rings," *Microw. Opt. Technol. Lett.*, Vol. 48, No. 7, pp. 1317–1320, Jul. 2006.

17 F. Yang and Y. Rahmat-Samii, *Electromagnetic Band Gap Structures in Antenna Engineering*, ser. Cambridge RF Microw. Eng. Cambridge, UK: Cambridge University Press, 2008.

18 J. Shaker, M. R. Chaharmir, and J. Ethier, *Reflectarray Antennas, Analysis, Design, Fabrication, and Measurement*, Artech House, 2013.

19 D. M. Pozar, "Wideband reflectarrays using artificial impedance surfaces," *Electron. Lett.*, Vol. 43, No. 3, pp. 148–149, Feb. 2007.

20 P. Nayeri "Advanced design methodologies and novel applications of reflectarray antennas," Ph.D. dissertation, Department of Electrical Engineering, University of Mississippi, University, MS, 2012.

21 P. Nayeri, F. Yang, and A. Z. Elsherbeni, "A broadband microstrip reflectarray using sub-wavelength patch elements," presented at the *IEEE Antennas and Propagation Society International Symposium*, South Carolina, USA, June 2009.

22 P. Nayeri, F. Yang, and A. Z. Elsherbeni, "Bandwidth improvement of reflectarray antennas using closely spaced elements," *Progress in Electromagnetics Research C*, Vol. 18, pp. 19–29, 2011.

23 E. Almajali, et al., "Observations on the Performance of Reflectarrays with Reduced Inter-Element Spacings," *IEEE Intl. Symp. on Antennas and Prop.* (URSI/APS), Spokane, WA, July 3–8, 2011.

24 LPKF Website. Available at: www.lpkfusa.com/ (accessed July 2017).

25 P. Nayeri, F. Yang, and A. Z. Elsherbeni, "Broadband reflectarray antennas using double-layer subwavelength patch elements," *IEEE Antennas Propag. Lett.*, Col. 9, pp. 1139–1142, 2010.

26 J. Ethier, D. McNamara, M. R. Chaharmir, and J. Shaker, "Reflectarray design using similarity-shaped fragmented sub-wavelength elements," *Electron. Lett.*, Vol. 48, No. 15, July 2012, pp. 900–902.

27 G. Zhao, Y. C. Jiao, F. Zhang, and F. S. Zhang, "A subwavelength element for broadband circularly polarized reflectarrays," *IEEE Antennas Wireless Propag. Lett.*, Vol. 9, pp. 330–333, 2010.

28 J. Ethier, M. R. Chaharmir and J. Shaker, "Novel Approach for Low-Loss Reflectarray Designs," *IEEE International Symposium on Antennas and Propagation (URSI/APS)*, Spokane, WA, July 3–8, 2011.

29 J. Ethier, M. R. Chaharmir and J. Shaker, "Reflectarray design comprised of sub-wavelength coupled-resonant square loop elements," *Electron. Lett.*, Vol. 47, No. 22, Oct. 2011, pp. 1215–1217.

30 J. Ethier, M. R. Chaharmir, and J. Shaker, "Reflectarray thinning using sub-wavelength coupled-resonant elements," *Electron. Lett.*, Vol. 48, No. 8, March 2012, pp. 359–360.

31 E. Carrasco, M. Barba, and J. A. Encinar, "Reflectarray element based on aperture-coupled patches with slots and lines of variable length," *IEEE Trans. Antennas Propag.*, Vol. 55, No. 3, pp. 820–825, Mar. 2007.

32 E. Carrasco, J. A. Encinar, and M. Barba, "Bandwidth improvement in large reflectarrays by using true-time delay," *IEEE Trans. Antennas Propag.*, Vol. 56, No. 8, pp. 2496–2503, Aug. 2008.
33 A. Roederer, US6411255 "Reflector antenna comprising a plurality of panels," 2002-06-25.
34 R. Hodges and M. Zawadzki, "Wide swath ocean altimeter antenna electrical subsystem – preliminary design review," JPL internal document, Jan. 2003.
35 P. De Vita, A. Freni, G. L. Dassano, P. Pirinoli and R. E. Zich, "Broadband element for high-gain single-layer printed reflectarray antenna," *Electron. Lett.*, Vol. 43, No. 23, Nov. 2007.
36 M. R. Chaharmir, J. Shaker, and H. Legay, "Broadband design of a single layer large reflectarray using multi cross loop elements," *IEEE Trans. Antennas Propagat.*, Vol. 57, No. 10, pp. 3363–3366, Oct. 2009.
37 O. M. Bucci, G. D'Elia, G. Mazzarella, and G. Panariello, "Antenna pattern synthesis: A new general approach," *Proc. of the IEEE*, Vol. 82, No. 3, pp. 358–371, Mar. 1994.
38 A. Capozzoli, C. Curcio, G. D'Elia, A. Liseno, S. Sparice, "A design technique of reflectarrays with improved frequency performances," *Proc. 2nd Europ. Conf. Antennas Propagat.*, Edinburgh, UK, Nov. 11–16, 2007.
39 A. Capozzoli, C. Curcio, A. Liseno, and G. Toso, "Phase-only synthesis of flat aperiodic reflectarray," *Progr. Electromagn. Res.*, Vol. 133, pp. 53–89, 2013.
40 Y. Mao, C. Wang, F. Yang, and A. Z. Elsherbeni, "A single-layer broad-band reflectarray design using dual-frequency phase synthesis method," *Proc. Asia-Pacific Microw. Conf.* (APMC), Kaohsiung, Taiwan, Dec. 4–7, 2012.
41 Y. Mao, C. Wang, F. Yang, and A. Z. Elsherbeni, "A novel phase synthesis approach for wideband reflectarray design," *IEEE Trans. Antennas Propagat.*, Vol. 63, No. 9, pp. 4189–4193, Sep. 2015.
42 P. Nayeri, F. Yang, and A. Z. Elsherbeni, "Design of single-feed reflectarray antennas with asymmetric multiple beams using the particle swarm optimization method," *IEEE Trans. Antennas Propagat.*, Vol. 61, No. 9, pp. 4598–4605, Sep. 2013.
43 R. Deng, F. Yang, S. Xu, and M. Li, "Design of a dual-frequency broadband reflectarray using triple resonance elements," in *IEEE Antennas and Propagation Society Int. Symp.*, 2015, pp. 2169–2170.
44 H. Chulmin, J. Huang, and C. Kai, "A high efficiency offset-fed X/ka-dual-band reflectarray using thin membranes," *IEEE Trans. Antennas Propag.*, Vol. 53, No. 9, pp. 2792–2798, Sep. 2005.
45 H. Chulmin, C. Rodenbeck, J. Huang, et al., "A C/ka dual frequency dual Layer circularly polarized reflectarray antenna with microstrip ring elements," *IEEE Trans. Antennas Propag.*, Vol. 52, No. 11, pp. 2871–2876, Nov. 2004.
46 D. I. Wu, R. C. Hall, and J. Huang, "Dual-frequency microstrip reflectarray," in *IEEE Antennas and Propagation Society Int. Symp.*, pp. 2128–2131, July 1995.
47 J. Shaker, et al., "Investigation of FSS-backed reflectarray using different classes of cell elements," *IEEE Trans. Antennas Propag.*, Vol. 56, No. 12, pp. 3700–3706, Dec. 2008.
48 M. R. Chaharmir, et. al. "Broadband dual-band linear orthogonal polarization reflectarray," *IEE Electron. Lett.*, Vol. 45, No. 1, pp. 13–14, Jan. 2009.
49 J. Shaker, M. R. Chaharmir, M. Cuhaci, and A. Ittipiboon, "Reflectarray research at the communications centre Canada," *IEEE Antennas and Propagation Magazine*, Vol. 50, No. 4, Aug. 2008.

50 S. M. Duffy and S. D. Targonski, "Comparison of two flat reflector-type designs for dual-polarization, dual-band operation," in *IEEE Antennas and Propagation Society Int. Symp.*, pp. 288–291, July 2001.

51 K. Mayumi, et. al. "Wideband single-layer microstrip reflectarray based on multiple-resonance behavior," in *IEEE Antennas and Propagation Society Int. Symp.*, July 2008.

52 S. Oh, et. al. "A six-band reflectarray antenna," in *IEEE Antennas and Propagation Society Int. Symp.*, July 2008.

53 H. Shih-Hsun, H. Chulmin, J. Huang, et al., "An offset linear-array-fed Ku/Ka dual-band reflectarray for planet cloud/precipitation radar," *IEEE Trans. Antennas Propag.*, Vol. 55, No. 11, pp. 3114–3122, Nov. 2007.

54 M. R. Chaharmir, J. Shaker, N. Gagnon, et al., "Design of broadband, single layer dual-band large reflectarray using multi open loop elements," *IEEE Trans. Antennas Propag.*, Vol. 58, No. 9, pp. 2875–2883, Sep. 2010.

55 A. Yu, F. Yang, A. Z. Elsherbeni, et al., "Experimental demonstration of a single layer tri-band circularly polarized reflectarray," in *IEEE Antennas and Propagation Society Int. Symp.*, 2010, pp. 1–4.

56 C. Guclu, J. Perruisseau-Carrier, and O. Civi, "Proof of concept of a dual-band circularly-polarized RF MEMS beam-switching reflectarray," *IEEE Trans. Antennas Propag.*, Vol. 60, No. 11, pp. 5451–5455, Nov. 2012.

57 T. Smith, U. Gothelf, O. S. Kim, et al., "Design, manufacturing, and testing of a 20/30-GHz dual-band circularly polarized reflectarray antenna," *IEEE Antennas Wireless Propag. Lett.*, vol. 12, pp. 1480–1483, 2013.

58 T. Smith, U. Gothelf, O. S. Kim, et al., "An FSS-backed 20/30 GHz circularly polarized reflectarray for a shared aperture L- and Ka-band satellite communication antenna," *IEEE Trans. Antennas Propag.*, Vol. 62, No. 2, pp. 661–668, Feb. 2014.

59 R. Deng, Y. Mao, S. Xu, and F. Yang, "A single layer dual band circularly polarized reflectarray with high aperture efficiency," *IEEE Trans. Antennas Propagat.*, Vol. 63, No. 7, pp. 4189–4193, July. 2015.

60 P. De Vita, A. Freni, G. L. Dassano, et al., "Broadband element for high-gain single-layer printed reflectarray antenna," *Electron. Lett.*, Vol. 43, No. 23, 2007.

61 F. Yang, Y. Kim, J. Huang, and A. Z. Elsherbeni, "A single layer reflectarray antenna for C/X/Ka bands applications," in *IEEE Antennas and Propagation Society International Symposium Digest*, Honolulu, HI, pp. 5307–5310, Jun. 2007.

62 B. Strassner, C. Han, and K. Chang, "Circularly polarized reflectarray with microstrip ring element having variable rotation angles," *IEEE Trans. Antennas Propagat.*, Vol. 52, pp. 1122–1125, April 2004.

63 J. Huang and R. J. Pogorzelski, "A Ka-band microstrip reflectarray with elements having variable rotation angles," *IEEE Trans. Antennas Propagat.*, Vol. 46, pp. 650–656, May 1998.

64 D. M. Pozar and T. A. Metzler, "Analysis of a reflectarray antenna using microstrip patches of variable size," *Electron. Lett.*, pp. 657–658, April 1993.

65 A. Yu, "Microstrip reflectarray antennas: Modeling, design and measurement," Ph.D. dissertation, Dept. Elect. Eng., Univ. of Mississippi, Oxford, MS, 2010.

66 A. Yu, F. Yang, A. Z. Elsherbeni, and J. Huang, "A single layer broadband circularly polarized reflectarray based on the element rotation technique," in *IEEE Antennas and Propagation Society International Symposium Digest*, Charleston, SC, Jun. 2009.

67 A. Yu, F. Yang, A. Z. Elsherbeni, and J. Huang, "Design and measurement of a circularly polarized Ka-band reflectarray antenna," in *Proc. 3rd European Conference on Antennas and Propagation,* Berlin, Germany, Mar. 2009, pp. 2769–2773.

68 A. Yu, F. Yang, A. Z. Elsherbeni, and J. Huang, "An X-band circularly polarized reflectarray using split square ring elements and the modified element rotation technique," in *IEEE Antennas and Propagation Society International Symposium Digest,* San Diego, CA, Jul. 2008.

69 M. E. Cooley, J. F. Walker, D. G. Gonzalez, and G. E. Pollon, "Novel reflectarray element with variable phase characteristics," *IEE Proceedings Microwave and Antennas Propagations,* Vol. 144, No. 2, pp. 149–151, Apr. 1997.

70 A. Yu, F. Yang, A. Z. Elsherbeni, and J. Huang, "Experimental demonstration of a single layer tri-band circularly polarized reflectarray," in *IEEE Antennas and Propagation Society International Symposium Digest,* Toronto, Canada, Jul. 2010.

8

Terahertz, Infrared, and Optical Reflectarray Antennas

Electromagnetic waves cover a wide frequency spectrum, including microwaves and optics. The wave behavior at both microwave and optical frequencies follow Maxwell's equations; hence many microwave concepts have been extended to optics, and in turn a lot of optical designs have been translated to microwave frequencies. For example, the invention of optical fiber [1], a kind of dielectric waveguide, greatly propelled the development of optical communications and the Internet. In addition, resonant optical antennas [2], are recently being studied that could contribute to the improvement of light emitters in the future.

In this chapter, we present some of the latest progress in terahertz, infrared, and optical reflectarray antennas. Most reflectarray antenna research in the recent years has been in the microwave and sub-millimeter range [3], and while microwave concepts can generally be extended to higher frequencies, several factors come into play that complicate the antenna design. In this chapter, we will first study some basic material characteristics and design challenges, and then review several THz, infrared, and optical reflectarray antennas that have been developed over the years.

8.1 Above Microwave Frequencies

The region of frequency spectrum between microwave frequencies and visible light consists of terahertz (THz) and infrared bands, as shown in Figure 8.1. While certain applications such as far-infrared spectroscopy are well developed (since the 1950s), from an antenna engineering viewpoint, until recently these bands, and particularly the THz band, were not subject of much research. However, with the rapid advent of material science and fabrication technology, applications such as communications and imaging are now appearing, and the so-called terahertz gap has now emerged as a subject of great interest in antenna engineering. THz radiation has a great deal of amazing radiation properties, and some of them are listed here [4]:

1) Penetration: THz waves can get through ordinary clothing and packaging materials with a relatively small attenuation.
2) High-resolution imaging: Because of the shorter wavelengths of THz signals compared to that at microwaves, it can be used to construct imaging systems with a relatively high resolution.
3) Spectroscopy: Because many materials exhibit spectral characteristic features in the $0.5 \sim 3.0$ THz region, it enables different chemical substances to be detected.

Reflectarray Antennas: Theory, Designs, and Applications, First Edition.
Payam Nayeri, Fan Yang, and Atef Z. Elsherbeni.

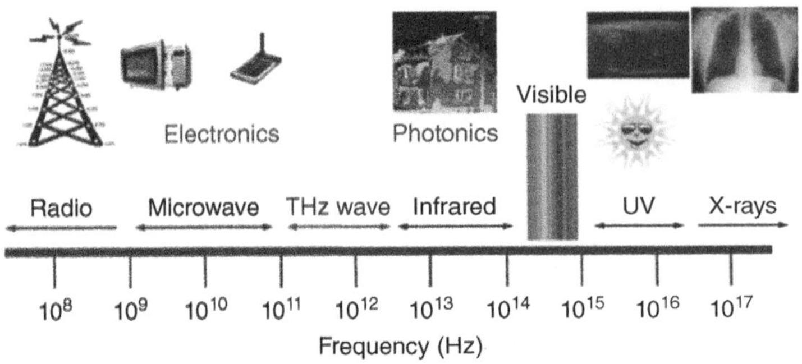

Figure 8.1 The electromagnetic spectrum [5].

These somewhat new characteristics, which do not exist in the microwave region, open a myriad of opportunities. Terahertz technology is rapidly emerging as a new frontier of electromagnetic research [6], while merging the gap between microwave and optical development communities. For antenna engineers, the special requirements of terahertz instruments, demand new antenna concepts and also new ways of implementing already established designs [7]. For example, in many applications of THz systems, such as radio astronomy, remote sensing, and radar, large reflector antennas with high surface accuracy and light weight are required. These designs, however, are typically high cost due to the high precision required for fabrication. Here is where reflectarray antennas can play a major role. While most reflectarray antenna research has been in the microwave range, during the past few years a significant interest has been observed in this relatively new and promising direction. At the short wavelength region, several factors come into play that complicate the antenna design, many of which arise from the material properties. In the next section, we will outline some of the fundamental characteristics of materials in this frequency range, and outline the basic design considerations regarding material selection for THz reflectarray design.

8.2 Material Characteristics at Terahertz and Infrared Frequencies

Reflectarray antenna design at terahertz and infrared frequencies is a rather new and challenging area for study [3], [4], and [8]. At microwave frequencies, since a high conductivity is obtained for conductors and a low-loss tangent can be achieved for substrates, the gain loss of a reflectarray due to the materials is relatively small. At optical frequencies, however, the losses in substrates and conductors can be severe, and their effects on reflectarray element designs become a critical issue. Thus, it is important to understand the loss mechanism for the elements and accurately model the losses in the system.

8.2.1 Optical Measurements and Electromagnetic Parameters

Ellipsometry is an optical technique for measuring the dielectric properties of materials. The analysis is based on the change of polarization of light that is reflected off a sample. From these measurements, the complex index of refraction of materials can be determined across the frequency band of interest. Electromagnetic simulations, however, require material property definitions as complex dielectric constants. Therefore, the measured data from the ellipsometer needs to be converted using

$$\tilde{\varepsilon}_r(\lambda) = \tilde{n}^2(\lambda), \tag{8.1}$$

where $\tilde{n} = n + jk$ is the complex index of refraction. For dielectric materials, the dielectric constant and loss tangent are calculated as

$$\varepsilon_c = \varepsilon_0(\varepsilon_r' + j\varepsilon_r'') = \varepsilon_0\left(n^2 - k^2 + j2nk\right), \tag{8.2}$$

$$\tan\delta_e = \frac{\varepsilon_r''}{\varepsilon_r'} = \frac{2nk}{n^2 - k^2}. \tag{8.3}$$

In these equations, the $e^{-j\omega t}$ time convention is assumed. For conductors, the electromagnetic property of interest is the conductivity, which is related to the index of refraction as

$$\sigma = j\omega\varepsilon_o\left(1 - \tilde{n}^2\right) = \omega\varepsilon_o\left[2nk + j\left(1 - n^2 + k^2\right)\right]. \tag{8.4}$$

8.2.2 Measured Properties of Conductors and Dielectric Materials

Measured material properties of two practical conductors for infrared operation, namely gold (Au) and aluminum (Al), are given in Figure 8.2. The optical properties of these materials were measured using a Woollam variable-angle spectroscopic ellipsometer (IR-VASE) [9], [10]. Note that in the short wavelength region, the conductivity of the materials is significantly lower than nominal values in the microwave region, which is a major issue for antenna design. To reduce metallic losses, gold is typically used as conductors at high frequencies; however, it can be seen that the conductivity of gold at infrared is still on the order of one tenth of its nominal value in the microwave region.

Measured properties of two dielectric materials, namely benzo-cyclo-butene (BCB) and silicon (Si) are given in Figure 8.3. In comparison between the dielectrics, although Si shows a superior low-loss performance, BCB tends to be a better candidate for the reflectarray substrate due to its lower dielectric constant. Note that the dielectric materials were also measured using IR-VASE, and the slight oscillations in the measured data for BCB is attributed to measurement and post processing errors. However, by detailed analysis of the standard deviation of the measured response and material model errors, the measured optical properties should be within ±5% of their actual values. It should be noted that for the dielectric materials, the loss tangent is not very high in this frequency range and they show less dispersive behavior in comparison with the conductors, which indicate that the major material concern at THz frequencies is the performance of the conductor.

8.2.3 Calculating Drude Model Parameters for Conductors

To study the propagation of electromagnetic waves in mediums, we consider the time-harmonic case. The applied electric field varies harmonically with the time factor $e^{j\omega t}$. The assumption is that the motions of the electrons in the medium have the same

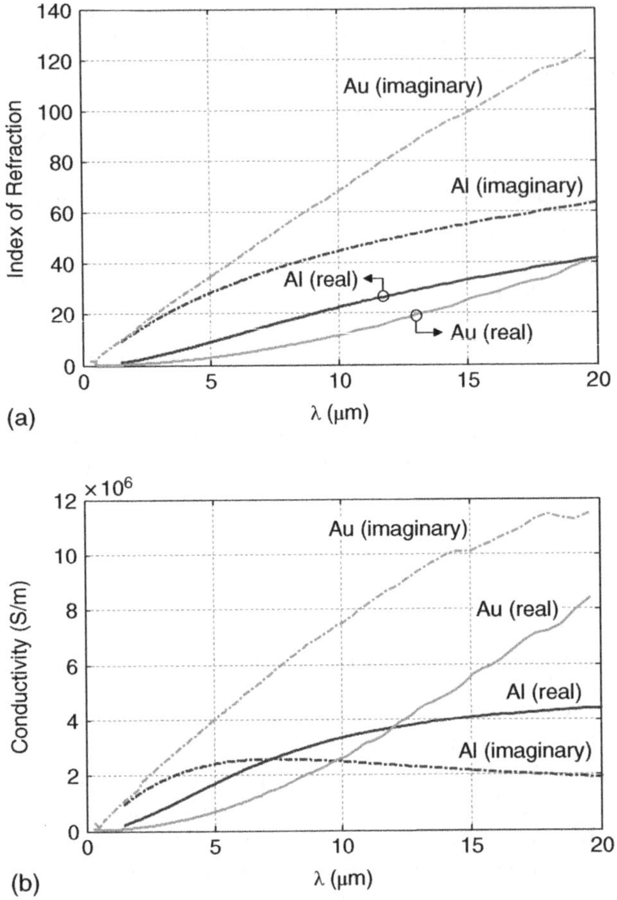

Figure 8.2 Measured properties of conductors at infrared: (a) index of refraction, (b) conductivity. *Source:* Yang 2012 [19]. Reproduced with permission of IEEE.

harmonic dependence. Usually for dielectrics, the effects of polarization on the propagation of light through a medium are studied. For conducting media, however, the effects of conduction on the propagation of light through a medium are studied. While different types of models such as Lorentz or Debye have been used to characterize dielectric media, in most cases the Drude model is more suitable for conductors. The Drude model was proposed to explain the transport properties of electrons in materials (especially metals). This classical model, which is an application of kinetic theory, assumes that the microscopic behaviors of electrons in a solid may be treated by considering the instantaneous collision between free electrons and ions (as solid spheres).

The significant results of the Drude model are: (1) equation of motion in the medium, and (2) a linear relationship between current density and electric field. The Drude model predicts the current as a response to a time-dependent electric field in which

$$\sigma = \frac{\sigma_0}{1 - j\omega\tau}. \tag{8.5}$$

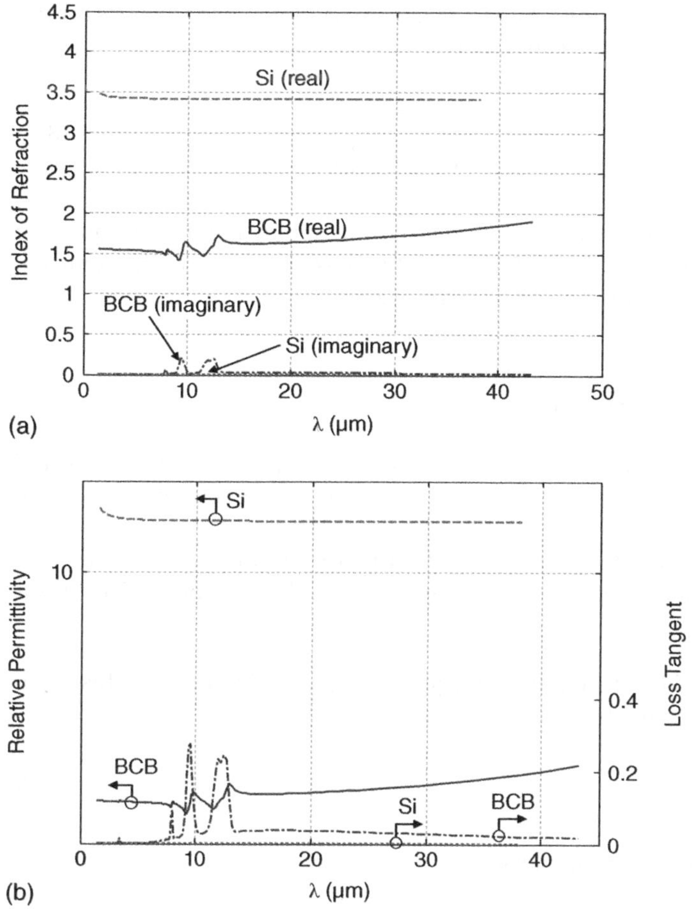

Figure 8.3 Measured properties of dielectrics at infrared: (a) index of refraction, (b) relative permittivity and loss tangent. *Source:* Yang 2012 [19]. Reproduced with permission of IEEE.

Here, $\sigma_0 = \dfrac{ne^2\tau}{m}$ is the DC conductivity, τ is the average time of electron collisions, n is the free electron density, e is the electron charge, and m is the effective mass of the electrons. The current can then be defined using

$$E(t) = \mathrm{Re}\left\{Ee^{-j\omega t}\right\},\ J(t) = \mathrm{Re}\left\{\sigma(\omega)Ee^{-j\omega t}\right\}. \tag{8.6}$$

The imaginary part indicates that the current lags behind the electric field, which happens because the electrons roughly need a time τ to accelerate in response to a change in the electric field.

The Drude model dielectric function is $\varepsilon(\omega) = \varepsilon_0 + i\dfrac{\sigma(\omega)}{\omega}$, where ε_0 is directly related to the DC conductivity $(\sigma_0 = \varepsilon_0\tau\omega_p{}^2)$ by the plasma frequency of metals. The plasma frequency is usually given in rad/s units. Separating the real and imaginary parts for the Drude model dielectric function, one can obtain

$$\varepsilon(\omega)=\varepsilon_0+i\frac{\sigma(\omega)}{\omega}=\varepsilon_0\left(1-\frac{\omega_p^2\tau^2}{1+\omega^2\tau^2}+i\frac{\omega_p^2\tau}{\omega\left(1+\omega^2\tau^2\right)}\right). \tag{8.7}$$

Typically, this Drude model dielectric function is used to obtain a fit for the data. The damping (collision) frequency for the material is defined as $\boldsymbol{\omega_\tau}=1/\tau$, with the unit s^{-1}. With this definition, the Drude model can be written as

$$\varepsilon_r(\omega)=\varepsilon_r'+i\varepsilon_r''=1-\frac{\omega_p^2}{\omega_\tau^2+\omega^2}+i\frac{\omega_p^2\omega_\tau}{\omega^3+\omega\omega_\tau^2}. \tag{8.8}$$

It should be noted that the plasma frequencies of metal are typically in the visible and near ultraviolet regions. The material is opaque (bulky) at frequencies below $\boldsymbol{\omega_p}$ and transparent at higher frequencies.

To determine the Drude model parameters for a material, one has to obtain the damping and plasma frequencies based on the measured results. The procedure [11] is as follows:

1) The equations are solved for the damping frequency ($\boldsymbol{\omega_\tau}$), eliminating $\boldsymbol{\omega_p}$. This equation is solved for ($\boldsymbol{\omega_\tau}$) at some frequency:

$$\omega_\tau=\frac{\omega\varepsilon_r''}{1-\varepsilon_r'} \tag{8.9}$$

2) The plasma frequency $\boldsymbol{\omega_p}$ is obtained using

$$\omega_p^2=\left(1-\varepsilon_r'\right)\left(\omega^2+\omega_\tau^2\right) \tag{8.10}$$

3) This is repeated for several values of $\boldsymbol{\omega}$ to obtain several pairs of $\boldsymbol{\omega_\tau}$ and $\boldsymbol{\omega_p}$, which produce the best eyeball fit to the data.

As an example, here we determine the Drude parameters for gold using the measured data in Figure 8.2. The plasma and damping frequency of gold are 1.3468×10^{16} rad/s and $6.501\times10^{13}\ s^{-1}$, respectively. These correspond to an average collision time of 1.5382×10^{-14} s and a DC conductivity of 2.4704×10^{7} s. Using these values, the material properties of gold are calculated from the Drude model and are given in Figure 8.4, where a close agreement between the measured data and Drude model is obtained.

Note that, depending on the electromagnetic simulation tool, modeling of conductors may require one to input the Drude model parameters, for example, CST Microwave Studio [12]. Alternatively, some software packages allow one to model the complex conductivity of the conductor using a dielectric model, for example, Ansys HFSS [13].

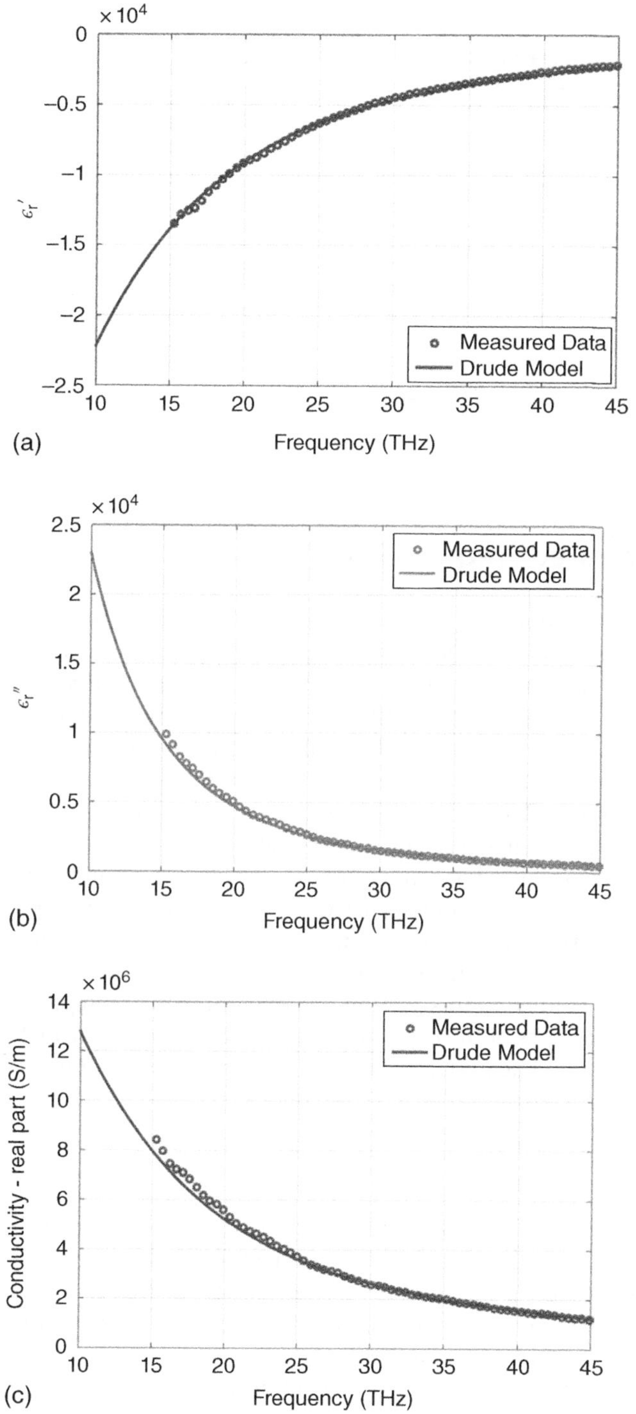

Figure 8.4 Comparison between the measured properties and the Drude model for gold.

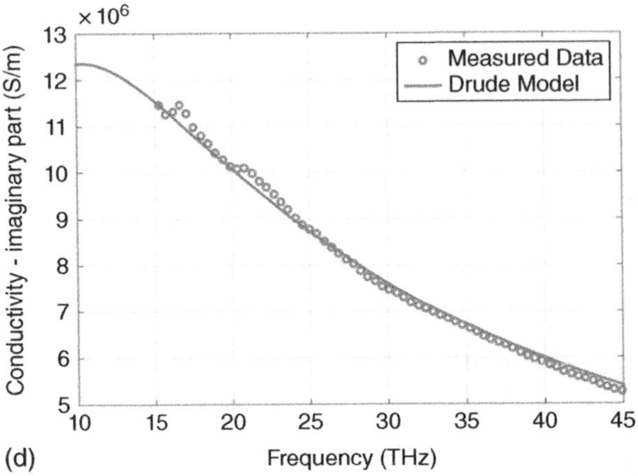

(d)

Figure 8.4 (Continued)

8.3 Element Losses at Infrared Frequencies

The aim of the study here is to understand the loss mechanism in reflectarray unit cells. To simplify the problem, the effects of the conductor loss in the metal patch and dielectric loss in the substrate are investigated separately. Furthermore, it is assumed that the ground plane is a perfect electric conductor (PEC). For the reflectarray elements studied here, the dielectric substrate ($\varepsilon_r = 2.2$) has a thickness of 450 nm, the unit-cell periodicity is 5.54 μm, and a 3 × 3 μm square patch is used as the phasing element. It should be noted that, in the analysis here, the dynamic reflection property of the element is studied with a fixed patch size when the frequency changes. For a practical design, however, the reflection property of the element is required at a fixed frequency when the patch size changes. While there is a strong relationship between these two methods, the main advantage of the dynamic analysis here is that from a circuit-model viewpoint, one value of C and L is sufficient to analyze the performance of the unit cells. In addition. for the resonance patch element (3 × 3 μm) studied in this section, the dynamic reflection properties essentially represent the highest loss of the element at the center frequency, which is quite advantageous for the loss study. The unit-cell simulations were carried out using commercial electromagnetic software Ansys HFSS and CST MWS.

8.3.1 Conductor Losses

To study the conductivity effect, we consider a lossless dielectric ($tan\ \delta_e = 0$) and investigate the reflection characteristics of the unit cells using different conductors for the patch. It is important to point out that when modeling conductors at THz and optical frequencies a zero-thickness approximation will be inaccurate for practical designs. Furthermore, the conductivity has to be modeled as a complex number, unlike microwave frequencies where typically only the real value of conductivity is used for simulations.

8.3.1.1 Effect of Conductor Thickness

As mentioned earlier, an important consideration in the simulations here is the impact of metal thickness on the performance of the elements. While thin films in the nanometer range can be fabricated, an important consideration is to ensure that the conductor is optically bulky (opaque). While this generally depends on both the plasma frequency of the metal and the thickness [14], in the frequency range of study here, a 100-nm thick patch is sufficient to ensure good reflection properties.

Both HFSS and CST simulations are studied to observe the reflection properties of elements with different conductivities and thicknesses. For both cases, only the real part of conductivity is used for the metal and it is assumed to be non-dispersive. A good agreement between the simulation results was observed with both software packages, hence only the results obtained by Ansys HFSS are reported for brevity.

As a reference, the reflection coefficients of reflectarray unit cells with zero thickness conductors are given in Figure 8.5. From these results, it can be seen that as the metal conductivity decreases, the magnitude of reflection coefficient decreases as well, indicating an increasing loss in the unit cell. When the conductivity (σ) is equal to 2.59×10^6 S/m, almost total loss is observed at the resonance frequency. The phase response of this case

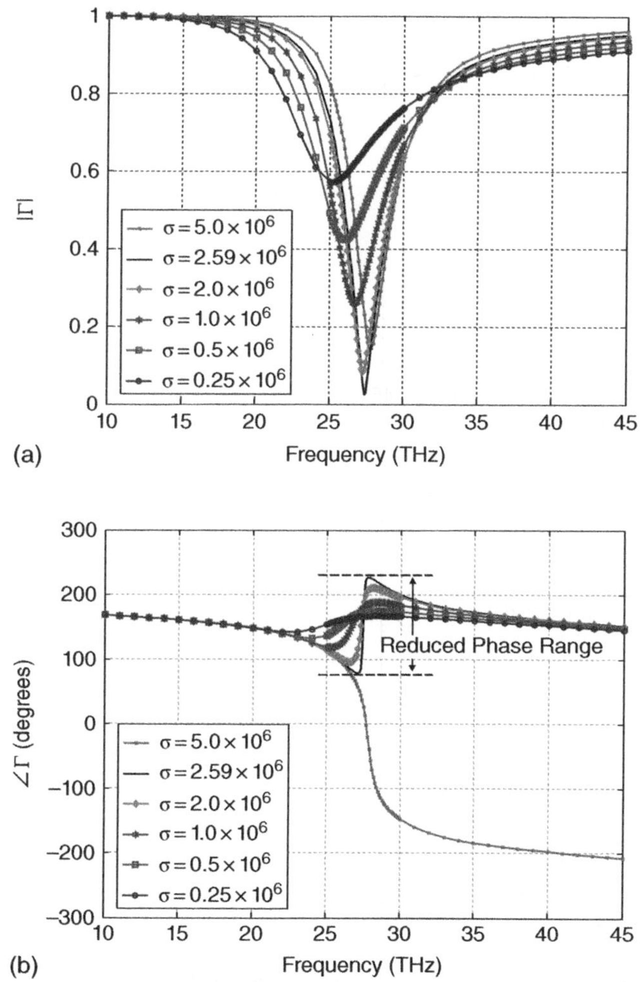

Figure 8.5 Reflection coefficients for infrared reflectarray elements with a zero-thickness patch: (a) magnitude, (b) phase.

also seems to be completely different from the typical S-curve. Instead of a 360° phase range, a reduced range of 150° is observed. When the real part of conductivity is further reduced to 1×10^6 S/m, it is interesting to observe that reflection magnitude increases now, indicating a decreasing energy loss. With this conductivity, the phase range becomes even smaller. It should be pointed out here that the values of conductivity used in the study here are not for any particular material. The aim here is to show the trend for reflection properties of elements for low-loss and high-loss cases. For the zero thickness patches studied here it can be seen that a proper phase curve is observed for high value of conductivity while an improper phase wrap is observed when the conductivity is below a certain value (referred to as critical conductivity value). For this case, the critical conductivity value is between 2.59×10^6 and 5.0×10^6.

The reflection coefficients of reflectarray unit cells with a patch thickness of 100 nm are given in Figure 8.6. Similar to Figure 8.5, it is observed that while for high values of

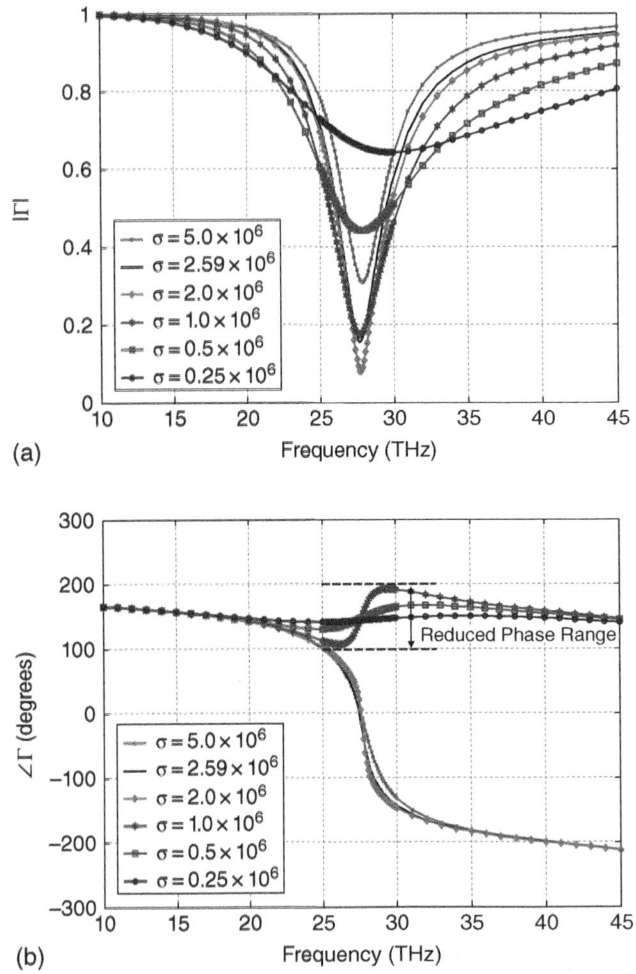

Figure 8.6 Reflection coefficients for infrared reflectarray elements with a 100-nm thick patch modeled with only the real part of conductivity (non-dispersive): (a) magnitude, (b) phase.

conductivity a proper phase curve is observed, as the conductivity decreases an improper phase wrap will be observed. However, the critical conductivity value for this 100-nm thick case is between 2.0×10^6 and 1.0×10^6, which is different than that of a zero-thickness patch. For a specific value of conductivity such as 2.59×10^6, the improper phase wrap occurs for zero thickness and does not appear for 100 nm thickness.

8.3.1.2 Effect of Complex Conductivity

As discussed earlier, for an accurate modeling of conductors in the optical range, the complex value of conductivity and the dispersive properties must be taken into account. To model complex and dispersive conductivity in the electromagnetic software, the Drude models obtained from measured data are used in this section. The reflection coefficients of reflectarray unit cells with a patch thickness of 100 nm are given in Figure 8.7 for various conductors.

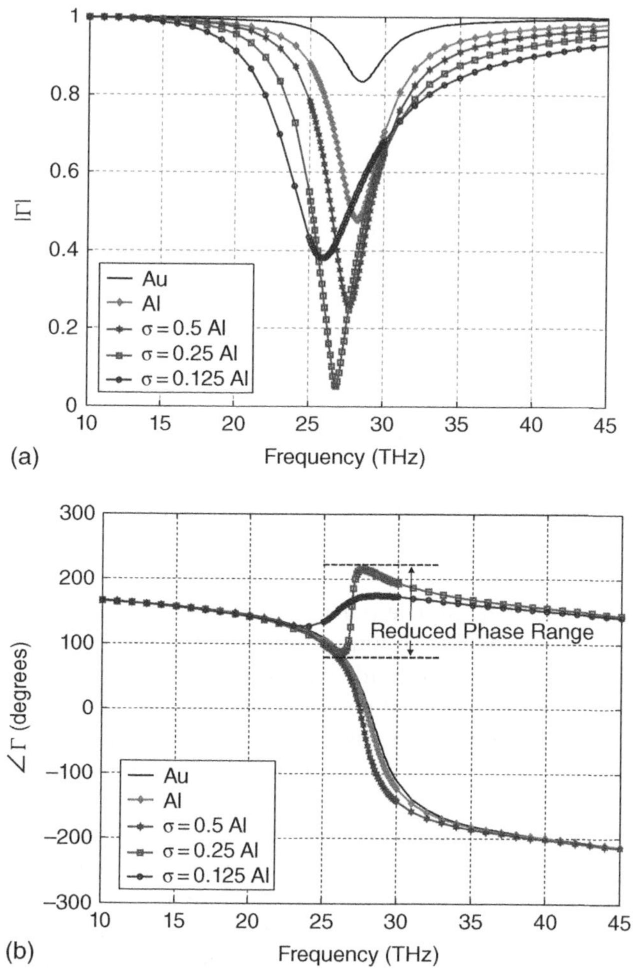

Figure 8.7 Reflection coefficients for infrared reflectarray elements with a 100-nm thick patch using the Drude model for conductors: (a) magnitude (HFSS), (b) phase (HFSS), (c) magnitude (CST), (d) phase (CST).

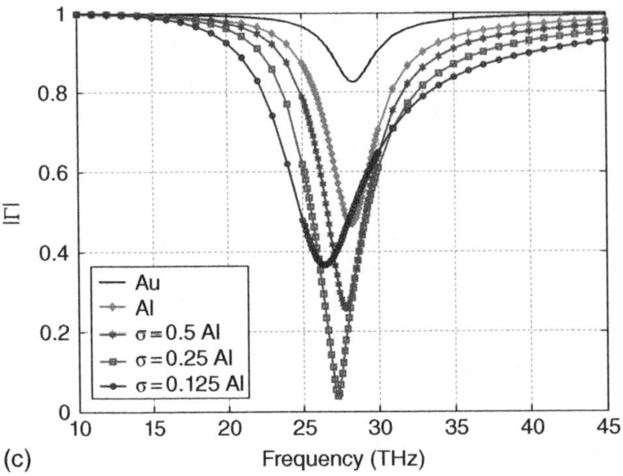

(c)

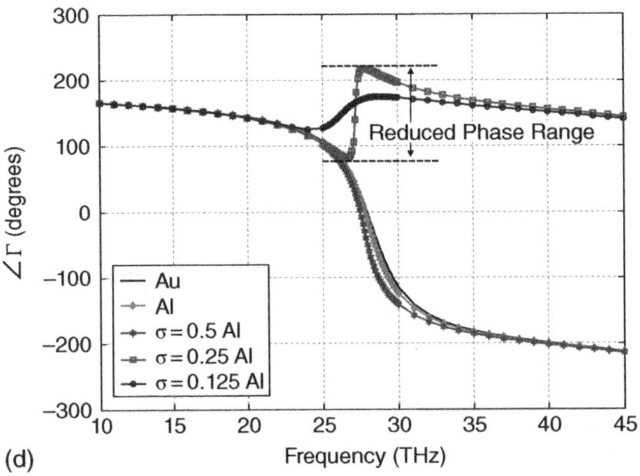

(d)

Figure 8.7 (Continued)

From these results, it can be seen that while the performance of a gold patch is quite acceptable in this frequency range, when an aluminum patch is used for the phasing element, less than half of the incident field will be reflected back from the elements. To study the effects of lower values of conductivity, we use the Drude models with values proportional to the aluminum case. It can be seen that as the metal conductivity decreases, the losses will increase. When the conductivity is reduced to one quarter of the value of conductivity for aluminum, almost a total loss of power will be observed. The phase response of this case is completely different from the typical S-curve. Instead of a 360° phase range, a reduced range of 150° is observed. Similar to the previous cases, it can be observed that while for high values of conductivity a proper phase curve is observed, as the conductivity decreases an improper phase wrap is observed for the elements. In this case, the critical value is between half and a quarter of the conductivity of aluminum, which is lower than the previous two cases.

A comparison between the results obtained using Ansys HFSS and CST MWS is also given in Figure 8.8 for a small-conductivity case ($\sigma = \frac{1}{4}\,\sigma_{aluminum}$) and a large-conductivity case ($\sigma = \frac{1}{2}\,\sigma_{aluminum}$). It can be seen that a close agreement is observed between the results, verifying the concept that when the conductivity is below a certain threshold level, an improper phase response will be observed.

From these studies, we can make following conclusions:

- When a large conductivity is used as the patch material, a proper phase curve that changes monotonically from 180° to −180° is observed. When the conductivity reduces below a threshold level, an improper phase wrap is observed and the phase range cannot cover the required 360° phase range.

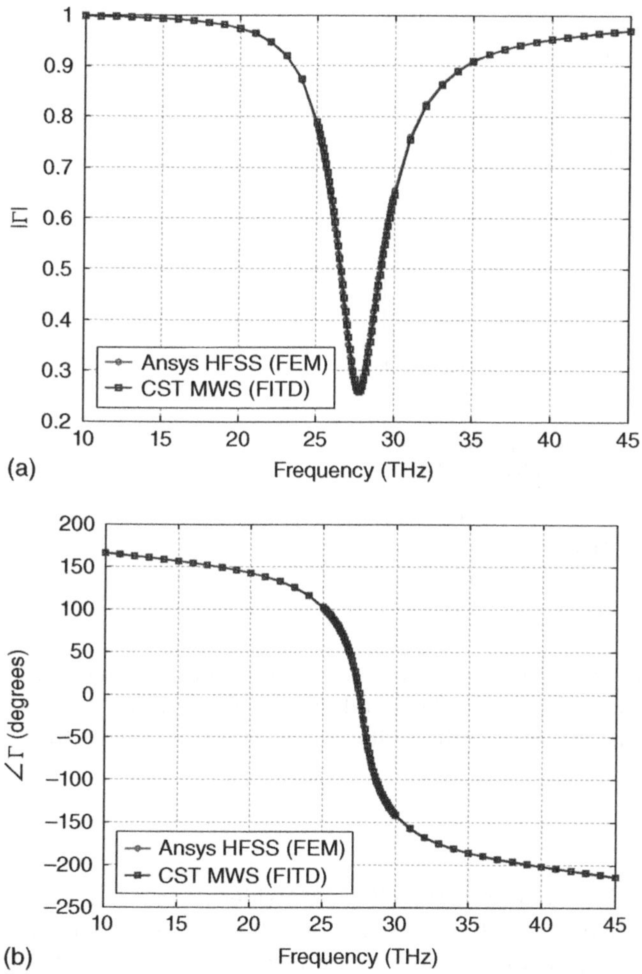

Figure 8.8 Comparison of reflection coefficients for infrared reflectarray elements with a 100-nm thick patch using the Drude model for conductors: (a) magnitude for the large-conductivity case, (b) phase for the large-loss case, (c) magnitude for the small-conductivity case, (d) phase for the small-conductivity case.

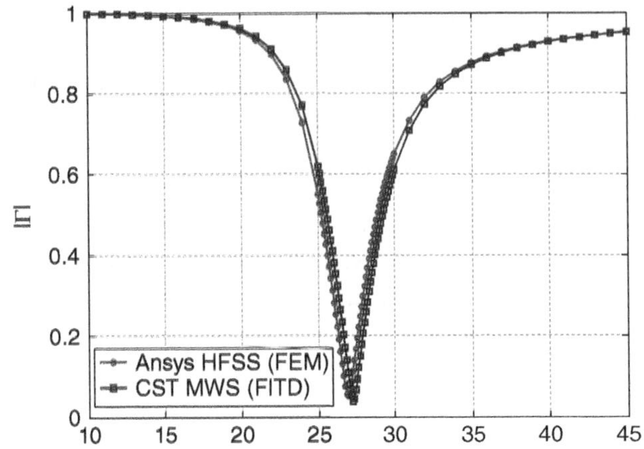

(c)

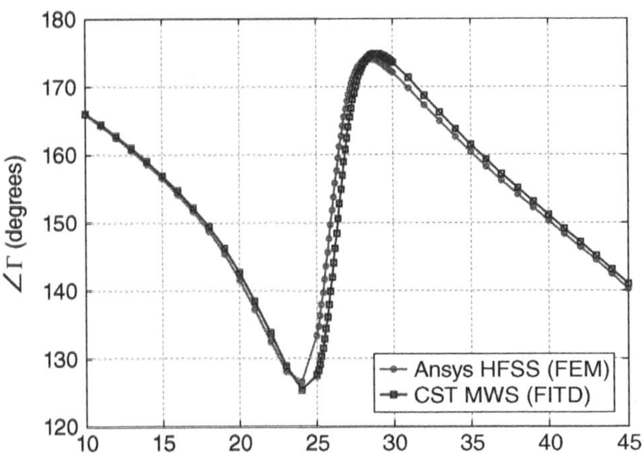

(d)

Figure 8.8 (Continued)

- When the metal patch is modeled using zero thickness, finite thickness with real conductivity, and finite thickness with complex conductivity, these behaviors can all be observed. However, the critical conductivity values are different in these three cases.

8.3.2 Dielectric Losses

To study the effect of dielectric losses in the unit cell, we consider the case of a PEC patch on different dielectric substrates with a thickness of 450 nm and various loss tangents. Similarly, we use a 3×3 μm square patch in a unit cell with periodicity of 5.54 μm. For this study, we have assumed that the dielectric properties are constant over the frequency band. It is implicit that in practical designs the dispersive nature of the parameters must be taken into account for accurate calculation of the reflection coefficients. However, for dielectrics this effect is not significant. The reflection

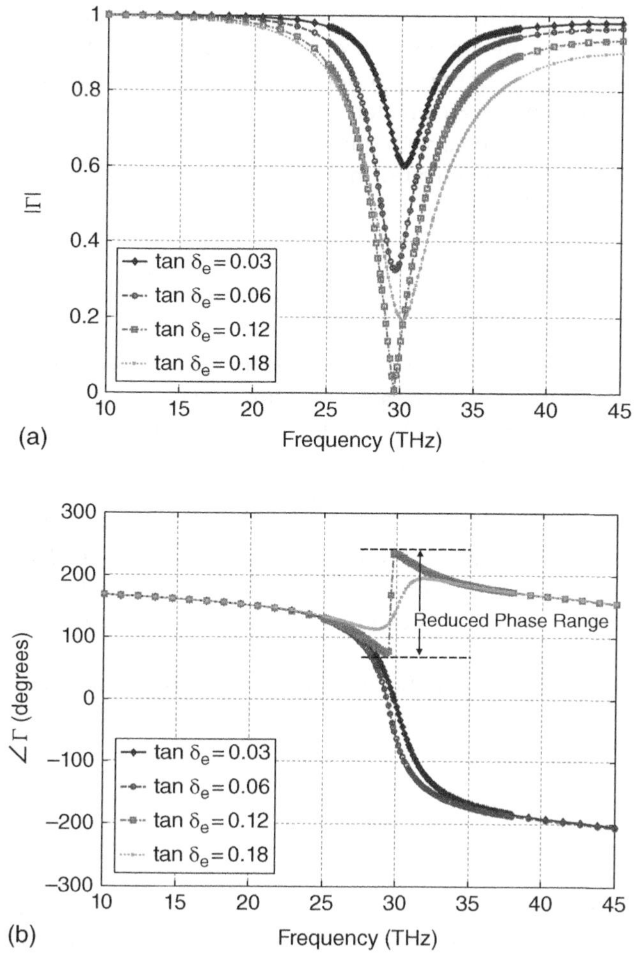

Figure 8.9 Reflection coefficients for infrared reflectarray elements with various values of dielectric loss tangent: (a) reflection magnitude and (b) reflection phase.

coefficients of reflectarray unit cells with different values of dielectric loss tangent are given in Figure 8.9.

It can be seen that as the loss tangent increases, the losses in the element increase, and for tan $\delta_e = 0.12$ almost total loss is observed at the resonance frequency. When the dielectric losses are further increased beyond this critical limit, the losses in the unit-cell decrease, however, the reflection phase does not return back to the typical S-curve.

8.3.3 Effect of Losses on Reflection Properties of Elements

In summary, two lossy situations exist in the reflectarray unit cells. When the metal conductivity is above a certain limit or the dielectric loss tangent is below a certain value, the phase of reflection coefficient varies with a 360° range. In this situation, when the conductor and dielectric losses increase but do not exceed the limit, the element loss will increase, but it will not affect the phase range.

On the other hand, when the metal conductivity is decreased or the dielectric loss tangent is increased beyond the limits, a new phase curve is observed with a limited angular range. In this situation when metal conductivity decreases or the dielectric loss tangent increases, the phase range becomes even smaller.

The traditional reflectarray design at microwave frequency has the loss situation in the first category. The reflectarray design at infrared may have both loss situations because of the high-loss material properties.

8.3.4 Circuit-Model Analysis

8.3.4.1 Circuit Theory and Loss Study

As discussed in Chapter 2, the unit-cell structure can be modeled as a transmission line circuit, where the square metallic patches act as parallel capacitors between adjacent cells and the short-circuited unit-cell ground can be modeled as a parallel inductor. This circuit model has previously been implemented for lossless reactive impedance substrates [15]. The loss in the dielectric substrate can be modeled by a parallel resistor, that is, the total unit cell is modeled as a parallel RLC circuit. To model the conductor loss in the patch, a resistor will be series connected to the capacitor as shown in [16]. In this section, we use the parallel RLC circuit model to understand the loss behavior associated with the dielectrics. Similar derivations can be obtained for conductor loss analysis.

Using this parallel RLC model, the reflection coefficient in the S-plane is calculated as

$$\Gamma(S) = -\frac{S^2 + S\left(-\frac{1}{Z_o C} + \frac{1}{RC}\right) + \frac{1}{LC}}{S^2 + S\left(\frac{1}{Z_o C} + \frac{1}{RC}\right) + \frac{1}{LC}}. \tag{8.11}$$

Using this equation, the reflection properties of a lossy unit cell can be calculated directly. Let's consider two cases, the low-loss ($\tan \delta_e = 0.0009$) and high-loss ($\tan \delta_e = 0.15$) dielectric substrates. The values of the lumped elements in the circuit model are obtained by initially designing an RLC circuit resonating at 28.3 THz and then tuning these values so that the circuit-model reflection curve fits the curve obtained by the full-wave simulations. The comparisons of full-wave simulation and circuit model are shown in Figure 8.10.

It can be seen that the circuit model shows good agreement with the full-wave simulations in both cases, where the values used for the capacitor and inductor are 0.134 *fF* and 0.215 *pH*, respectively. It should be pointed out that these lumped element values are for the case of normal incidence on the unit cell. However, a similar derivation can be obtained for the oblique incidence case. Comparing the resistor values in the circuit model, it can be seen that in the low-loss case the parallel resistor is quite large, so there is less power dissipated in the resistor. For the high-loss case, however, the resistance comes close to the free space impedance and as a result of the high-power transfer to the resistor, almost no power is reflected back from the system near the resonant frequency. As expected, the reflection phase of the circuit model also shows that this high loss will considerably change the reflection phase behavior of the unit cell.

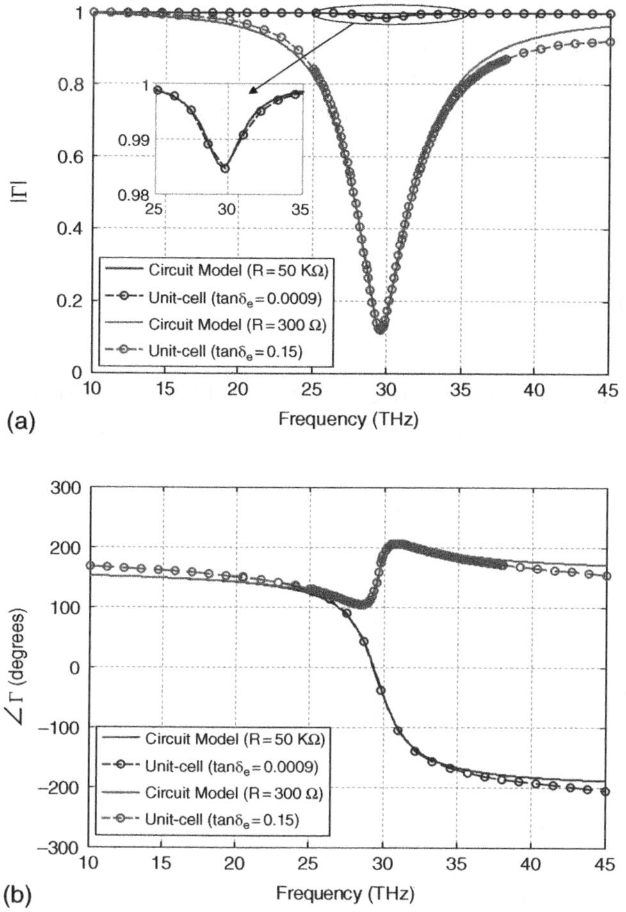

Figure 8.10 Comparison between circuit-model and full-wave simulations for low-loss and high-loss cases: (a) reflection magnitudes, (b) reflection phases. *Source:* Yang 2012 [19]. Reproduced with permission of IEEE.

8.3.4.2 Zero-Pole Analysis of Element Performance

In the previous sections, it was shown that when the losses in the unit-cell material increase beyond a certain limit, the typical S-curve of the reflection phase, which shows a phase range of about 360°, is no longer observed. The high losses change the traditional S-shape phase curve and create a new phase curve that has a narrow phase range. The unit-cell circuit model provides a helpful tool to obtain an insight into this unusual behavior. Three cases are considered for the value of the resistor, $R \to \infty$, $R = Z_0$ and $R < Z_0$.

In the first case, that is, the lossless model, the reflection coefficient can be simplified to

$$\Gamma(S) = -\frac{S^2 - S\dfrac{1}{Z_o C} + \dfrac{1}{LC}}{S^2 + S\dfrac{1}{Z_o C} + \dfrac{1}{LC}}. \tag{8.12}$$

The function has two poles located on the left side of the S-plane and two zeros on the right side of the S-plane. As the frequency goes from $\omega = 0$ to $\omega \to \infty$, the total phase changes from 180° to −180°, indicating a complete phase range and total reflection $|\Gamma(S)| = 1$. The phases of the numerator and denominator in (8.12) and the total phase are plotted in Figure 8.11(a).

When $R = Z_0$ the reflection coefficient will be

$$\Gamma(S) = -\frac{S^2 + \dfrac{1}{LC}}{S^2 + S\left(\dfrac{2}{Z_o C}\right) + \dfrac{1}{LC}}. \tag{8.13}$$

In this case the zeros of the function are on the imaginary axis in the S-plane. At the resonance frequency, the reflection coefficient will be zero ($|\Gamma(S)| = 0$), that is, a total loss and no reflected power.

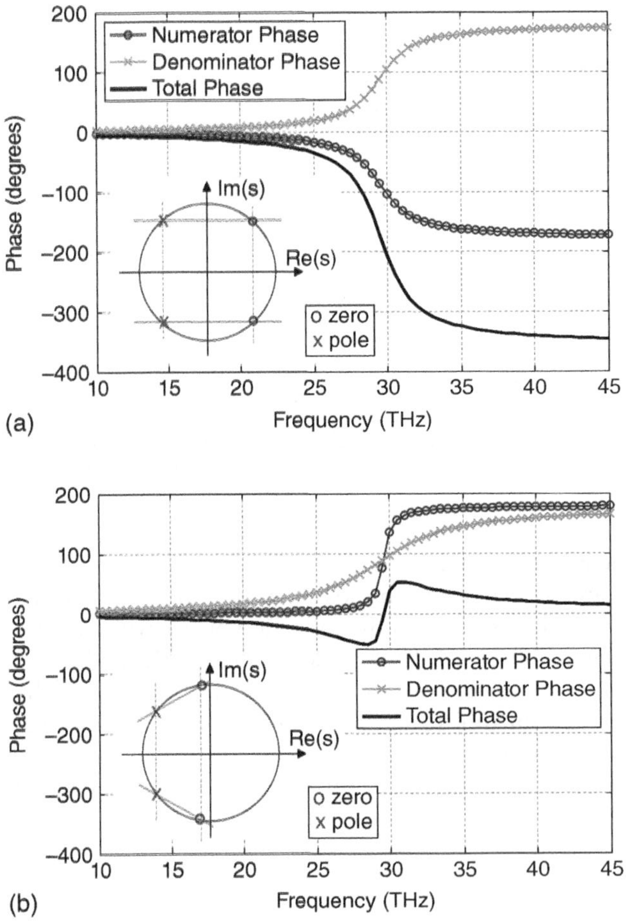

Figure 8.11 The positions of poles and zeros in the S-plane, and the corresponding phase versus frequency: (a) lossless case ($R \to \infty$), (b) high-loss case ($R < Z_o$). *Source:* Yang 2012 [19]. Reproduced with permission of IEEE.

When $R < Z_0$, the zeros will be located on the left side of the S-plane, as shown in Figure 8.11(b). The positions of the poles and zeros in this case indicate that although the magnitude of the reflection coefficient is non-zero, it will have a phase range below 180°. This means that when the losses in the unit cell go beyond the critical limit ($R = Z_0$), not only the reflected power but also the phase range will be insufficient for reflectarray unit cells.

8.4 Reflectarray Design Methodologies and Enabling Technologies

8.4.1 Reflectarrays with Patch Elements

In traditional reflectarray antennas, both conductors and dielectrics are used in element designs. While in general both metallic and dielectric losses can influence the performance of the reflectarray antenna, the dominant loss term depends on the materials (metal and dielectric) used in the design. At THz frequencies, the dielectric loss may be minimized by selecting proper materials such as BCB, but the conductor loss is a major concern. Using high quality conductors such as gold significantly reduces the losses, but this runs contrary to the goal of a low-cost system. On the other hand, low-cost conductors such as aluminum have a higher loss, which will decrease the efficiency of the system. In addition, while these low-cost conductors may achieve a complete phase range in the far-infrared range, at the short- and near-infrared range they will suffer from a limited phase range, which will further reduce the efficiency of the system.

The feasibility of a square-patch reflectarray elements operating at near-infrared frequencies was first demonstrated in [17]. Three different patch sizes were fabricated using electron-beam lithography, and the reflected phase as a function of patch size was characterized using an infrared interferometer at 28.3 THz. The materials were characterized using an infrared variable-angle spectroscopic ellipsometer, which allowed measurement of real and imaginary refractive index over the wavelength range of 3–30 m. Based on measured ellipsometric data, zirconium dioxide (ZrO_2) was chosen as the reflectarray standoff material because of its low values of loss tangent and relative permittivity at 10.6 μm. To reduce metallic losses, gold (Au) was used for both the reflectarray ground plane and patch elements because of its relatively high conductivity at infrared frequencies and its low chemical reactivity at room temperature. A patch thickness of 150 nm, approximately 15 times the skin depth of Au at 10.6 μm, was used to reduce unwanted skin-depth effects. Patch element spacing was chosen to be 5.54 m, approximately 60% the effective wavelength at 10.6 μm. For the standoff layer, a ZrO_2 thickness of 450 nm satisfied isolation conditions and limited the dielectric losses. The purpose of the design was to characterize the phase response as a function of element design for a given choice of materials, as a preliminary proof-of-concept demonstration. Thus, a simple 50-mm diameter substrate with three-stripe reflectarray prototype configuration was developed especially suited for interferometric characterization. The three reflectarray stripes are 6.3- by 28-mm rectangular arrays of square patches of uniform size, isolated from each other by 8-mm blank reference regions. The array comprising each stripe was made up of a single-sized patch element to demonstrate a unique phase shift upon reflection for comparison between the modeled and measured results.

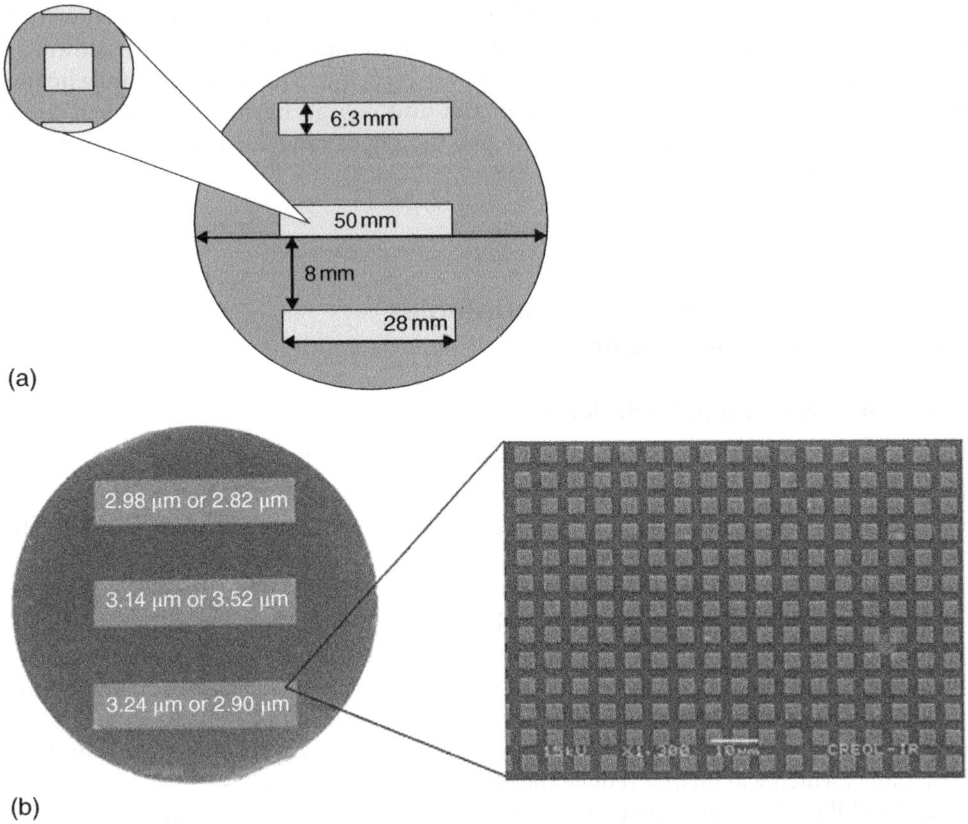

Figure 8.12 (a) Layout of the IR reflectarray. (b) Fabricated IR reflectarray prototype and a zoomed image of one patch strip. Reprinted with permission from [17].

To ensure that any phase shift on reflection was caused by the reflectarray and not the substrate, a 3-mm thick, fused-silica optical flat (peak-to-valley flatness variation less than 160 nm) was chosen as the mechanical support substrate for the reflectarrays. For investigation of reflectarray element behavior at IR, three multiple-stripe arrays were fabricated for testing at 10.6 μm. The first array (array #1) used the patch dimensions of 2.98, 3.14, and 3.24 μm and the second array (array #2) was fabricated with patches of size 2.82, 2.9, and 3.52 μm. Finally, a third array (array #3) was fabricated using a slightly modified layout to accommodate a larger number of stripes with patch sizes of 3.3, 3.4, 3.7, 3.9, and 4.1 μm. The layout of the reflectarray and an image of the fabricated arrays are given in Figure 8.12.

The phase shift on reflection of the reflectarray stripes, compared to that of the adjacent reference regions between the stripes was measured at 28.3 THz using a Wyko IR3 Twyman-Green interferometer. The measured interferogram results are given in Figure 8.13 for the substrate without a reflectarray, array #1, and array #2, respectively. It can be seen that the interferogram results for the bare substrate is essentially flat at 10.6 m, verifying that fringe deviations seen in Figure 8.12(b) and (c) are caused by the reflectarray patches. To determine the actual phase shift at each stripe, it is possible to spatially measure the fringe displacement at each reflectarray stripe relative to the nearest blank regions.

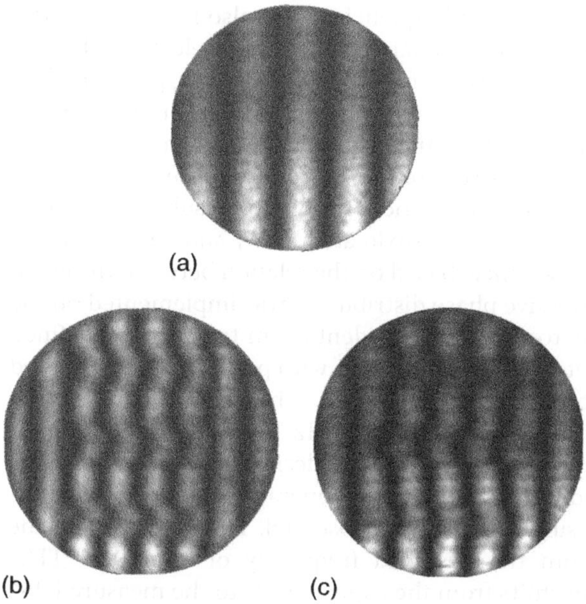

Figure 8.13 Interferograms of reference flat and both fabricated reflectarrays. (a) Bare substrate. (b) Array #1. (c) Array #2. Reprinted with permission from [17].

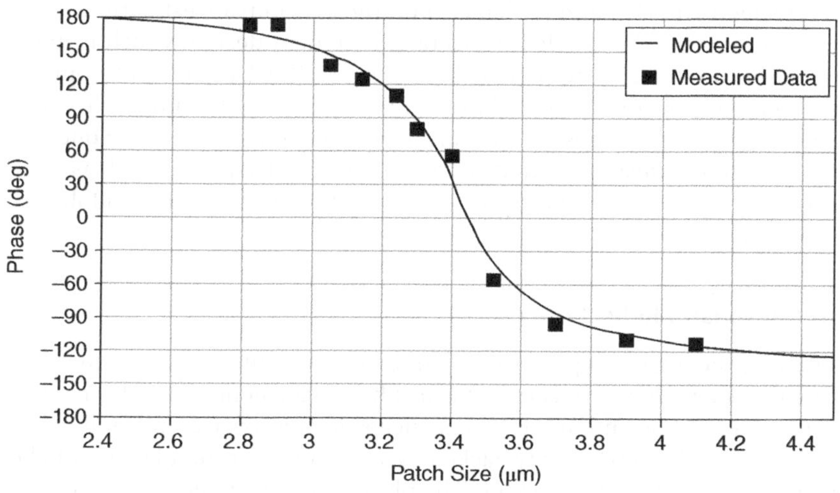

Figure 8.14 Comparison of measured and modeled reflection phase for the IR reflectarray elements. Reprinted with permission from [17].

In order to measure the relative phase response, a program was developed to follow the fringe patterns captured by the interferometer and to compute the relative phase response, while automatically accounting for the height difference between the patch rows and the ground plane. Reflectivity measurements were also taken of each reflectarray stripe. Measured and simulated results were compared and are given in Figure 8.14. The measured phase response of each of the stripes corresponds well with modeled results and a maximum phase shift of approximately 292° was measured.

Reflectarray antennas operating at terahertz frequencies were also reported in [18]. The reflectarrays were composed of microstrip square patches on a dielectric substrate operating at 1 THz. For the materials, gold was used for the patch elements, polydimethylsiloxane (PDMS) was used as the substrate due to its relatively low loss in the terahertz range, and platinum was used for the ground plane. The thickness of the PDMS substrate was 15 μm. Unit-cell periodicity was selected at 140 μm and the square-patch length varied from 10 to 136 μm. For this configuration, simulation results showed that the maximum range of the phase shift was approximately 330°, and the maximum reflection loss was about −1.2 dB at resonance. Based on the relation between the patch size and the reflection phase, a progressive phase distribution was implemented on the patch array to create a reflector able to deflect an incident beam toward a predefined angle off the specular direction. A set of reflectarrays each with periodically distributed 360 × 360 patch elements were fabricated and measured. The experimental results obtained through terahertz time-domain spectroscopy (THz-TDS) showed that up to nearly 80% of the incident amplitude is deflected into the desired direction. Despite a qualitatively satisfying demonstration of the reflectarray operations, some discrepancies between the simulated and measured results were observed. The frequency for the maximum deflection is shifted from the designed frequency of 1 to 0.93 THz. Meanwhile, the deflection angle also shifts from the expected 21° to the measured 25°. Possible causes were attributed to fabrication tolerance, normal incidence approximation in the design and subarray spacing.

While a few classic patch type reflectarray elements have been demonstrated at infrared and THz spectrum, [17] and [18], based on the previous investigations [19], [20], and [21], it is clear that the low-loss reflectarray element design, which is not a major concern at the microwave frequency, becomes a critical topic at the optical frequency. The element loss due to the finite value of the metal conductivity and high-loss tangent of the dielectric substrate becomes severe near the resonant frequency. Some conventional element design methods in the microwave frequency such as variable-size patches may suffer from significant losses, and new design methodologies need to be developed.

8.4.2 Dielectric Resonator Reflectarrays

A possible solution to the high-loss performance of patch elements at THz frequencies is to use resonant dielectric elements as reflectarray phasing elements. The idea is similar to the dielectric resonator antenna (DRA) concept in microwave and millimeter wave designs [22], [23]. A dielectric cavity backed by a conducting ground plane will be designed as the reflectarray phasing element. The phase control will be realized by the proper design of the shape and dimensions of the dielectric resonator, hence avoiding the use of a resonant conductor patch. This dielectric resonant element approach has some resemblance to the fiber optic design except that a cavity structure is designed instead of the transmission line structure.

In dielectric-type reflectarray elements, the resonant conductor patch is removed, and phase control is achieved by changing the geometrical parameters of the dielectric. In general, two different approaches are available depending on the availability of the material and the fabrication technique. In the first approach, the phasing elements of the reflectarray are dielectric resonators and phase tuning is typically achieved by changing the length of the dielectric cavity. These DRA elements exhibit high-gain performance; however, they require materials with high dielectric constant and very low loss, which makes them quite expensive.

In the second approach, one uses a dielectric slab for phase control of the reflectarray elements where phase tuning is achieved by varying the thickness (height) of the dielectric slab in each unit cell. These dielectric reflectarray elements put no constraint on the dielectric constant of the material, which would allow one to select low-cost materials for the design. Similar to optical mirrors, the ideal dielectric reflectarray would have a smooth profile; however, an important consideration for low-cost fabrication is the minimum size of the pixel (unit cell) that would yield satisfactory performance. This issue will be discussed later on in this section. Note that in the dielectric reflectarray design, the slab may cover the entire unit-cell surface, which is the case in this study, but this is generally not a necessary requirement for the elements.

It is also important to note that for both dielectric-type reflectarray elements, the reflective nature of the antenna still necessitates the use of a conductor ground plane; however, the losses on the conductor ground are quite small compared to those of resonant conductor elements.

Here, we design two types of reflectarray elements: (1) variable-size square metal patch elements on a low-loss dielectric substrate, and (2) variable-size low-loss dielectric resonator elements [24]. The aim of this study is to compare the loss performance of these two types of reflectarray phasing elements. The elements are designed for the operating frequency of 28.3 THz, and we use the same unit-cell size of 5.54×5.54 μm as the previous studies. While there is a strong relationship between the dynamic phase range study conducted in the previous sections and the variable-size element design here, the latter is necessary for a reflectarray design. For the patch design, we can use either gold or aluminum for the conductor, although as discussed earlier, aluminum patches have a higher loss. Also taking into account fabrication considerations, a thickness of 100 nm is used for the patch, and for accurate electromagnetic analysis, the complex conductivity model of the conductor is used in the simulations. For the substrate, we select a 450-nm BCB slab, which is sufficient to achieve a good phase range. For the dielectric resonator (DR) element design, BCB will not be practical, since it has a relatively low dielectric constant, thus Silicon (Si) is selected for the DR element. The height of the DR is selected at 2.5 μm, which is about five-sixths of the wavelength in the dielectric at the operating frequency. For both designs, the ground plane is assumed to be a perfect conductor. The geometry of both designs is shown in Figure 8.15.

While the fabrication precision depends on the technology used, for the numerical study here, we consider a precision of 0.25 μm for both designs. The reflection coefficients of these elements are obtained using HFSS software and are given in Figure 8.16.

These results clearly show that as expected, the variable-size DR elements outperform the patch elements. In addition to a significant improvement in the reflection loss with DR elements, they also show a much larger phase range, where the full phase cycle is achieved. It should be noted that the phase range obtained for the patch elements is about 320°, which is quite sufficient for a reflectarray design; however, the main disadvantage is the high loss of these elements. While the losses are substantially lower for gold, both patch configurations show a very high loss at the resonance dimension.

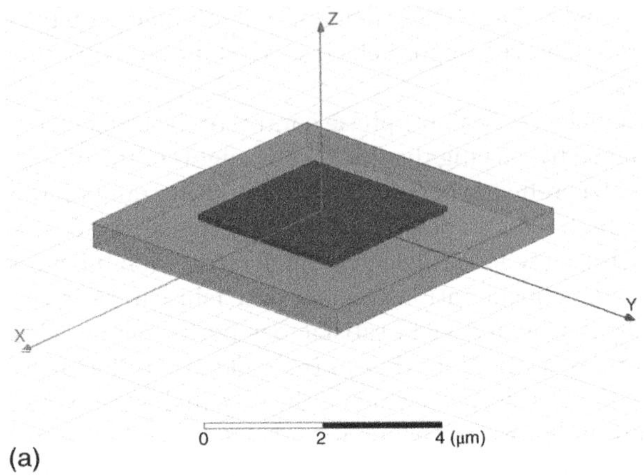

(a)

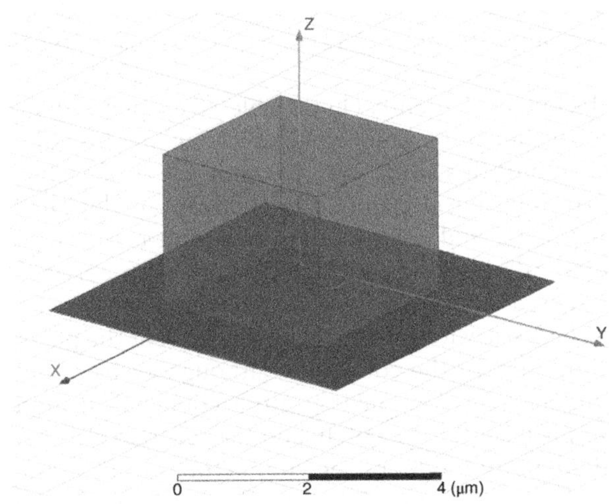

(b)

Figure 8.15 Variable-size elements for THz reflectarrays: (a) variable-size square patch, (b) variable-size square DR.

It is interesting to point out that similar to what was observed in the dynamic phase range study, the aluminum patch design here also shows a minimum reflection coefficient magnitude of about 0.5, that is, only half of the incident field will be reflected back from the element.

While DR reflectarrays are quite expensive, the study presented here showed that from an antenna performance viewpoint they are a practical solution for THz and higher frequencies. A nano-scale optical DRA operating in its fundamental mode was presented in [25]. The DRA operates via displacement current in a low-loss high-permittivity dielectric, resulting in reduced energy dissipation in the resonators. The non-uniform planar DRA array on a metallic plane was designed to impart a sequence of phase shifts across the wavefront to create beam deflection off the direction of specular reflection at 633 nm.

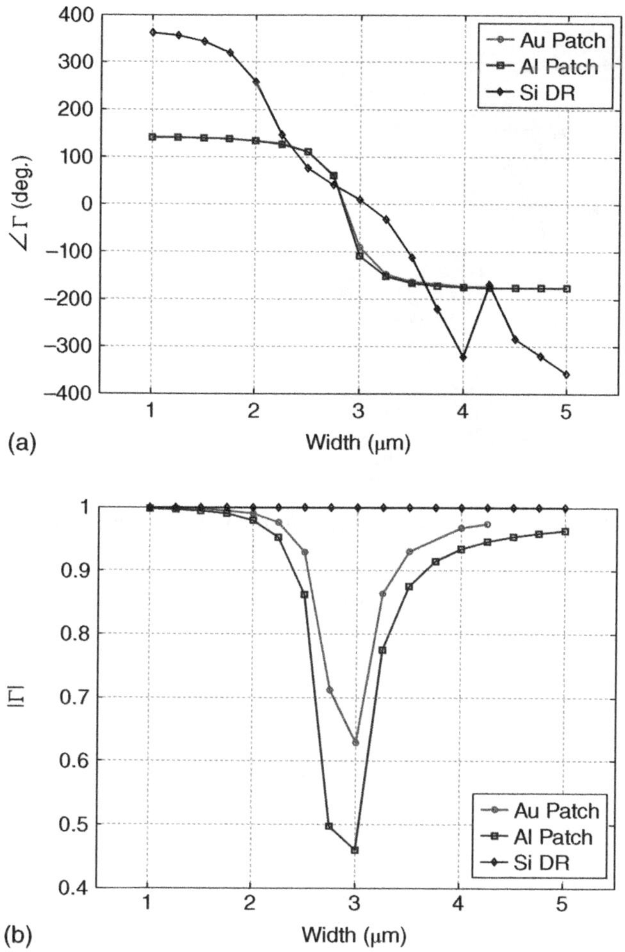

Figure 8.16 Reflection coefficients of the variable-size elements for THz reflectarrays: (a) phase, (b) magnitude.

8.4.3 Dielectric Reflectarrays

8.4.3.1 Dielectric Property and 3D Printing Technique

The dielectric reflectarray antenna was first reported in [26]. The first step in designing the dielectric reflectarray was characterization of the electromagnetic properties of the polymer material that was used for the design. This was done by measuring the transmission response of a uniform 3-mm thick slab using a THz time-domain spectrometer (THz-TDS). The system operates by propagating a picosecond-duration pulse through the material under test and extracting the frequency domain characteristics via Fourier transformation. The TDS spectral range is from 50 GHz to 1.2 THz, with a frequency resolution of 5 GHz. The THz-TDS system is shown in Figure 8.17. The measured dielectric properties for the sample are shown in Figure 8.18. The dielectric constant decreases slowly as the frequency increases, from 2.78 at 100 GHz to 2.70 at 600 GHz, while the loss tangent increases slowly from 0.02 to 0.05 across the measured spectrum.

Figure 8.17 The THz time-domain spectrometer system at University of Arizona. *Source:* Nayeri 2013 [26]. Reproduced with permission of IEEE.

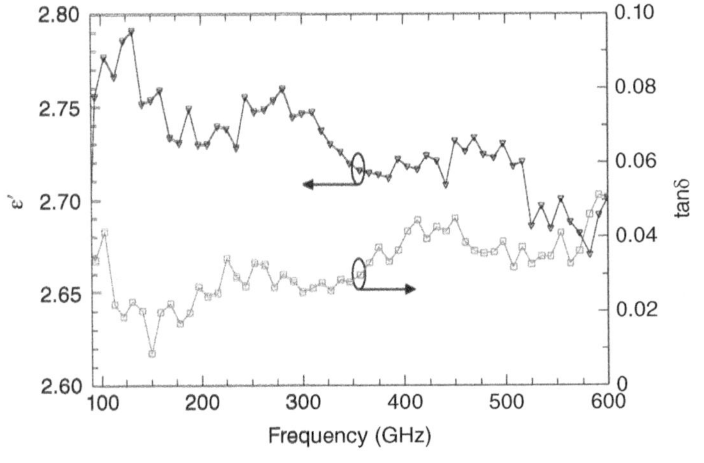

Figure 8.18 Measured dielectric properties of the polymer material. *Source:* Nayeri 2013 [26]. Reproduced with permission of IEEE.

Despite a slightly higher loss at the high end of the spectrum, the dielectric properties of this polymer material are stable, and therefore quite suitable for realizing dielectric reflectarrays.

In the THz spectrum (widely defined as 100 GHz to 10 THz), the corresponding wavelength is 3 mm to 30 μm, thus a fabrication technique that can provide feature sizes in the μm range is needed. The exact feature requirements, however, depend on the particular design and operating frequency, and will be discussed later on in this section.

A relatively new and promising fabrication technique that can offer resolutions in this range is 3D printing, also known as additive manufacturing. It is worthwhile pointing out that in comparison with traditional machining techniques, one of the distinct features of 3D printing is that it avoids the removal of materials by drilling or cutting methods. As such, 3D printing allows for rapid prototyping of arbitrary shapes with low

Figure 8.19 Photo of the polymer-jetting rapid prototyping machine. *Source:* Nayeri 2013 [26]. Reproduced with permission of IEEE.

cost for mass production. Polymer-jetting rapid prototyping machines are a class of 3D printers that can generate polymer objects of virtually any shape and arbitrary complexity from a digital model by laying down successive layers of polymer materials.

For our design, we used the Objet Eden 350 polymer-jetting rapid prototyping 3D printer [27], which has a fundamental resolution of $42 \times 84 \times 16$ μm in the *x*-, *y*-, and *z*-dimensions, respectively. To fabricate the object, the 3D model of the structure has to be generated and imported into a CAD program, which is then converted to a series of layered slices, each layer representing a 16-μm thick region of the model [28], [29]. Once the printer receives the data for each slice, a series of print heads, similar to the print head on an ink-jet printer, deposits a thin layer of ultraviolet-curable polymer on the construction stage. Then the ultraviolet lamps on the print head immediately cure the materials when they are being deposited. After one layer is completed, the construction stage is lowered by 16 μm, and the next slice is printed on top. Once the entire model is complete, the construction tray rises and the part can be removed. For traditional rapid prototyping, the final step is a high-pressure water spray to remove the water-soluble support material, leaving just the model material in the desired 3D shape. A photo of the prototyping machine Object Eden 350 is shown in Figure 8.19.

8.4.3.2 Dielectric Reflectarray Design

With the material properties and fabrication resolutions specified, the next task was to design the dielectric reflectarray element and system. Viewing this design as an array, we set the lattice size to be half-wavelength at the center design frequency. The height of the slab is then changed to provide the required phase shift on the reflectarray aperture. To demonstrate the feasibility of this design approach, the authors designed the dielectric reflectarrays for the operating frequency of 100 GHz. It is important to point out here that while this design methodology is readily scalable and 3D printers are also capable of fabricating the dielectric reflectarray for much higher operating frequencies, the availability of the feed antenna was the driving factor in selecting this design frequency in [26]. The measured electrical properties of the dielectric material at 100 GHz are $\varepsilon_r = 2.78$, and $\tan\delta_e = 0.039$. The unit-cell periodicity is selected to be 1.5×1.5 mm^2. For the element design, each unit cell is treated as an infinite slab of dielectric; therefore, either unit-cell analysis or analytical solutions for infinite slabs can be used to derive the reflection properties of the elements. One important consideration, however, is the position of reference plane for phase computation. Conventionally the reflectarray aperture has a flat surface, and the reflection phase is computed on that surface. In this design, elements have variable heights, so a reference plane that is placed at the top surface of the highest slab is selected, as depicted in Figure 8.20.

A full phase cycle (2π) is then achieved by changing the slab height from 0.3 to 2.32 mm with a resolution of 16 μm, which is shown in Figure 8.21. This reflection phase curve is for an oblique incident angle of $\theta = 25°$ under perpendicular polarization. Also note that while the phase range of this element is sufficient for designing a reflectarray antenna, the phase is limited to one phase cycle, so it would be necessary to zone the array. This is similar to zoned dielectric lenses where the overall profile and weight of the antenna is reduced, but would also result in a reduction of antenna bandwidth. It is also important to point out that this variable-height design has some resemblance to the grooved-dielectric, Fresnel zone plate antenna [30], with the added advantages of simpler (pixilated) design and increased subzone corrections.

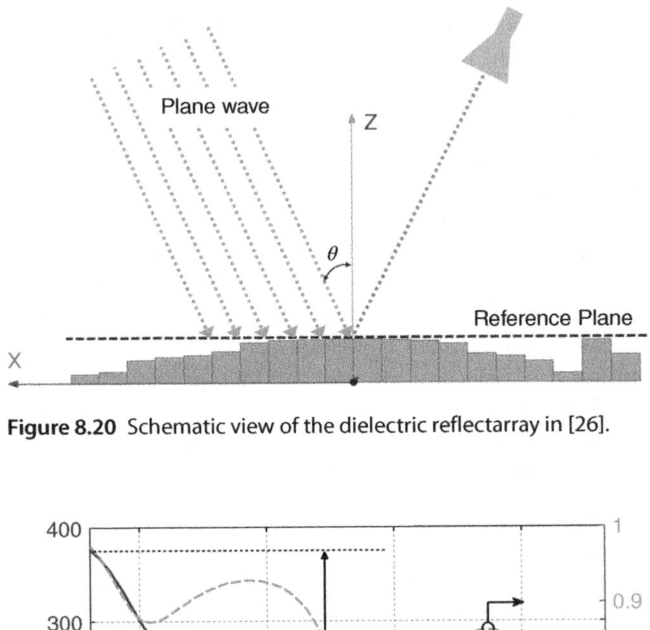

Figure 8.20 Schematic view of the dielectric reflectarray in [26].

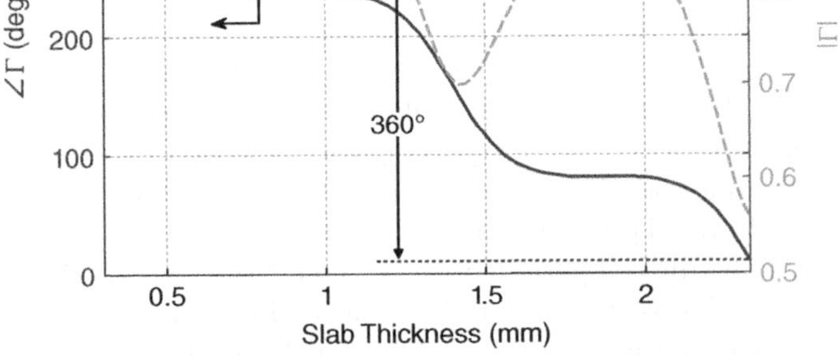

Figure 8.21 Reflection coefficients of the dielectric reflectarray elements at 100 GHz. *Source:* Nayeri 2013 [26]. Reproduced with permission of IEEE.

The dielectric reflectarray was designed to have a square aperture with a side length of 30 mm, and 20 × 20 variable-height dielectric elements are designed to provide the necessary phase shift on the aperture. An offset feed is selected to avoid blockage effects with a feed tilt angle of 25°. Based on the aperture efficiency analysis, the feed antenna (A-INFO LB-10–10) is placed at a distance of 22.5 mm from the aperture, and is pointing toward the geometrical center of the array. With this system setup, the edge taper at the upper edge center ($x = 15$ mm, $y = 0$), lower edge center ($x = -15$ mm, $y = 0$), and side edge centers ($x = 0$, $y = \pm 15$ mm), are −8, −11, and −11.5 dB, respectively. To minimize beam squint, the main beam direction is also set to 25° off broadside.

Different aperture phase distributions are studied for the dielectric reflectarrays. The initial phase distribution on the reflectarray aperture is then computed as described in [3]. At the center of the array, the required reflection phase is 103.7°. One concern for this offset dielectric reflectarray is the shadowing effect observed along the phase wraps on the aperture where the taller elements will intercept the incoming rays and shadow their adjacent elements. To minimize the shadowing effects, one can minimize the number of zones (phase wraps) by adding a phase constant to the aperture, which would also increase the bandwidth of the antenna. The solution is to have elements with maximum height at the center. Since the phase moves outward on the aperture, this ensures that all element sizes are used before the first phase wrap is observed. For this design (Design 1) this corresponds to a phase constant of −100°, and a reflection phase of 3.7° at the center of the array. This phase distribution is shown in Figure 8.22(a). It is worthwhile pointing out that for the extreme case (when the tallest

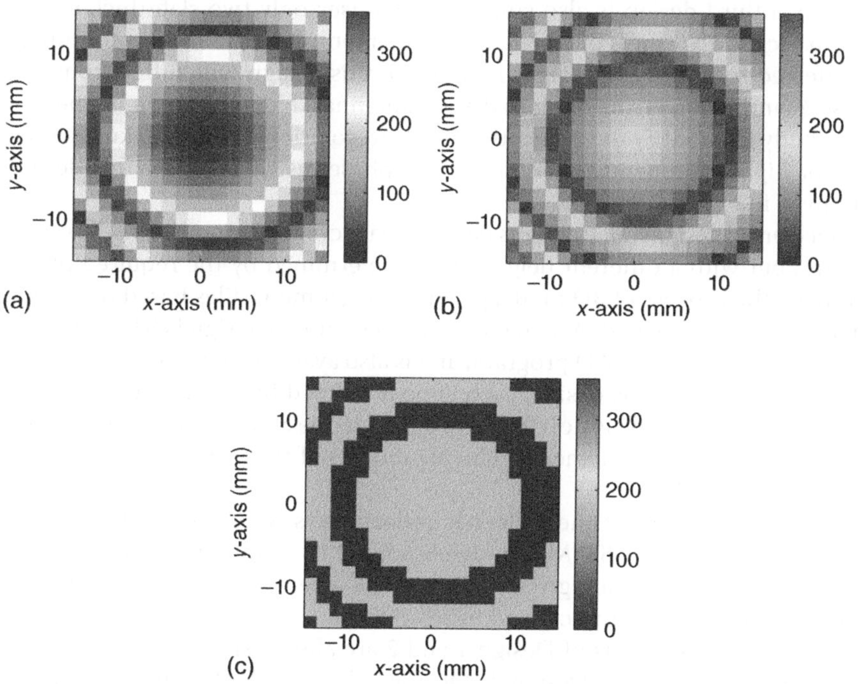

Figure 8.22 Aperture phase distributions for the dielectric reflectarrays: (a) design for minimum phase wraps (Design 1), (b) design for minimum element loss (Design 2), (c) 1-bit design (Design 3). *Source:* Nayeri 2013 [26]. Reproduced with permission of IEEE.

and shortest slabs are placed next to each other); the angular limit arising from this shadowing effect is about 20.5°. As such, in this design where the main beam is scanned to 25° off broadside, very few elements will observe this shadowing effect. Nonetheless this design approach (Design 1) will minimize the number of these shadowed elements. It is also anticipated that larger and spatially controllable dielectric constant range can be achieved in the near future so that all the unit cells will have the same height but different dielectric constants.

Another practical consideration is regarding the material losses of the elements. As shown in Figure 8.21, the element loss increases with the element thickness. In this case, the target is to minimize losses by using an element distribution that achieves the lowest antenna loss. Similarly, this is obtained by adding a phase constant. The function to be minimized is the sum of the weighted element reflection coefficients for the array. The weighted element loss, which takes into account both the element loss and the aperture illumination, is computed as

$$WEL = \frac{\sum_{M}\sum_{N}\left(\text{Illumination}(m,n)\cdot\left|\Gamma(m,n)\right|\right)^2}{\sum_{M}\sum_{N}\left(\text{Illumination}(m,n)\right)^2}. \tag{8.14}$$

For this design (Design 2), the minimum loss will be realized with a phase constant of +82° with the aperture phase distribution shown in Figure 8.22(b). This corresponds to a reflection phase of 185.7° at the center of the array.

As a reference, a third design is also studied which uses only two slab thicknesses. This design is basically a Fresnel zone plate reflector antenna, which in array terminology can be referred to as a 1-bit design. In general, this design will suffer from the classical phase quantization errors of phased array antennas. Nonetheless, it is a necessary reference study to quantify the fabrication limits (resolution in slice thicknesses for 3D printing) for higher frequency operations. The phase distribution for this design (Design 3) is given in Figure 8.22(c).

For these reflectarray designs, all elements have a square cross-section (unit-cell lattice) of $1.5 \times 1.5\ \text{mm}^2$, but with a different height that is determined by the required phase shift. To automate the process of 3D model generation, geometry files based on standard 3D formats need to be created. A suitable file format for this design is STL, which is used by the Objet Eden printer CAD program, and is also available in many commercial electromagnetic solvers. For this design, each cube is defined by specifying the vertex positions of 12 triangles. A code is developed to generate the STL files that contain the location and dimensions of the elements. The 3D models of these dielectric reflectarrays are shown in Figure 8.23.

The radiation performances of the dielectric reflectarrays were obtained using the full-wave simulation software CST Microwave Studio [12]. The radiation patterns of the three dielectric reflectarrays are given in Figure 8.24 where it can be seen that in all designs the main beam is correctly scanned to 25°.

Note that the radiation patterns of Design 1 and 2 are almost similar, with a slightly better side-lobe performance for Design 1. On the other hand, Design 2 achieves a lower element loss and higher gain and radiation efficiency as expected. Comparison of the performance of the 1-bit design (Design 3) with the other two dielectric

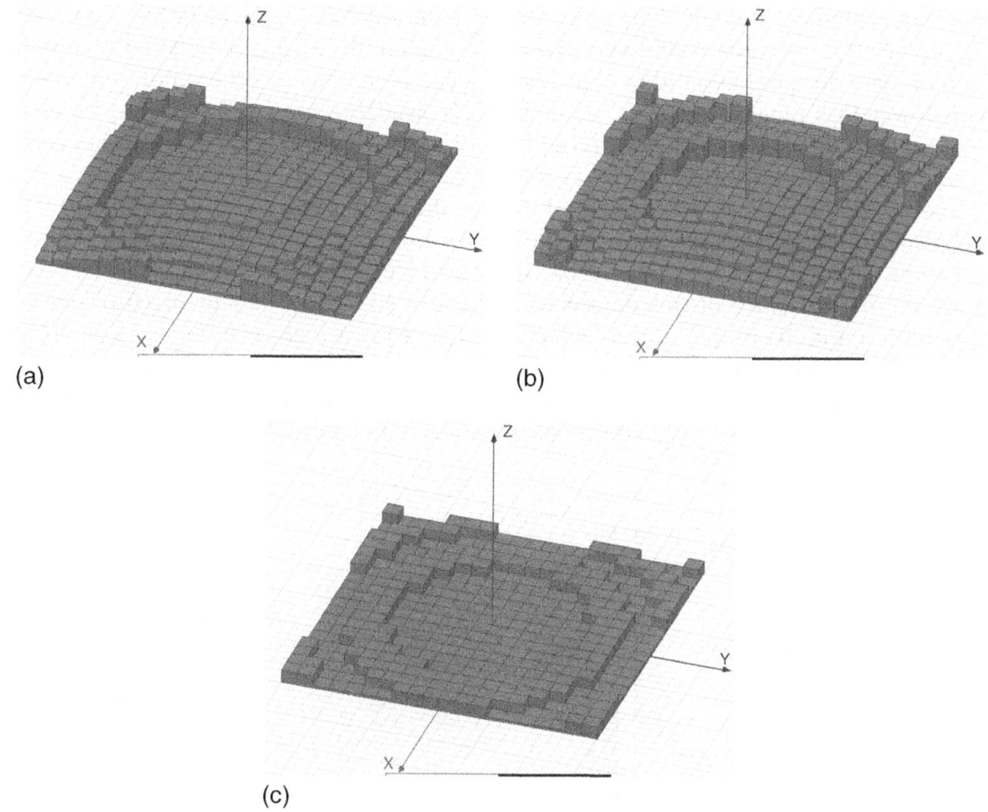

Figure 8.23 Three 3D models of dielectric reflectarrays in Ansys HFSS: (a) Design 1, (b) Design 2, (c) Design 3. *Source:* Nayeri 2013 [26]. Reproduced with permission of IEEE.

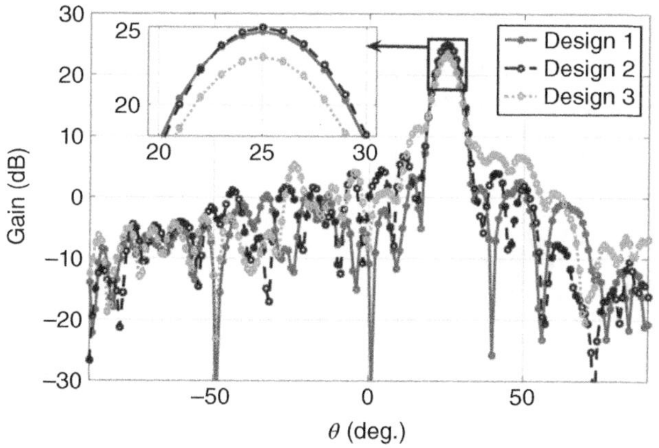

Figure 8.24 Simulated gain patterns of the dielectric reflectarrays at 100 GHz. *Source:* Nayeri 2013 [26]. Reproduced with permission of IEEE.

reflectarrays indicates a directivity loss of about 3 dB, which is due to the phase quantization errors. It is also important to point out here that in this work, the polymer material used in these designs has a rather high-loss tangent. Nonetheless, other photopolymers with lower loss tangent (<0.001) are potentially available (e.g., polytetrauorethylene (PTFE), high-density polyethylene (HDPE), and polypropylene (PP)), and can be used to achieve a higher radiation efficiency for these antennas.

To observe the effect of profile smoothness on the performance of the array, Design 1 was also studied with a finer lattice resolution. As discussed earlier, the dielectric reflectarrays were designed with a lattice size of λ/2 at the center design frequency. Here we study the performance of this design with a lattice size of λ/10. The aperture phase distribution and 3D model of the dielectric reflectarray are given in Figure 8.25. The simulated gain patterns for these two different lattice sizes are compared in Figure 8.26, where it can be seen that despite a slightly higher gain (about 0.3 dB), the radiation performance of these two designs is almost identical. This study reveals that while there is

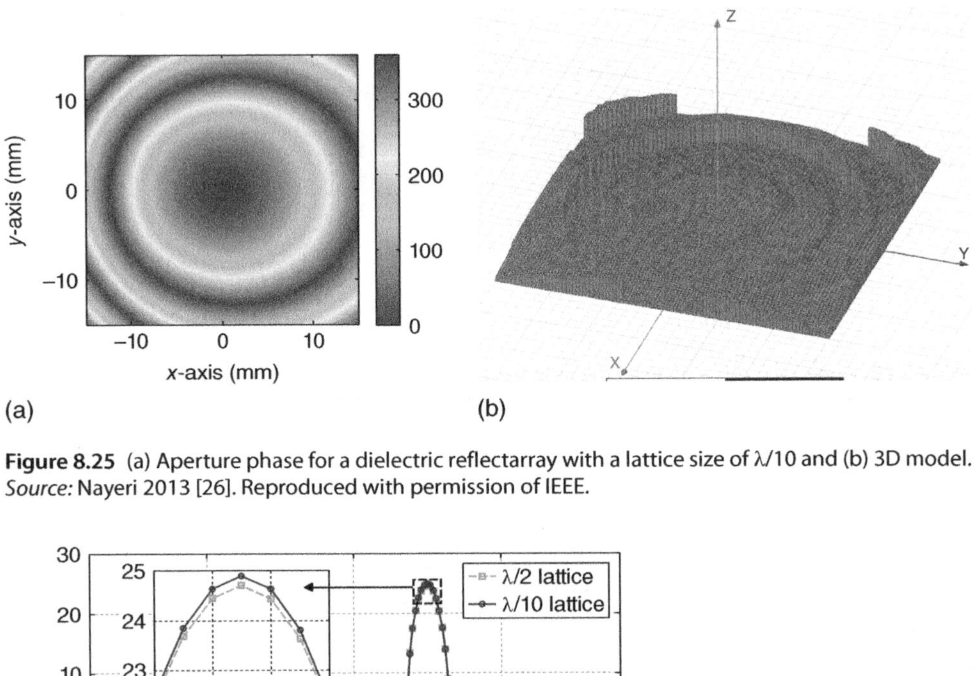

Figure 8.25 (a) Aperture phase for a dielectric reflectarray with a lattice size of λ/10 and (b) 3D model. *Source:* Nayeri 2013 [26]. Reproduced with permission of IEEE.

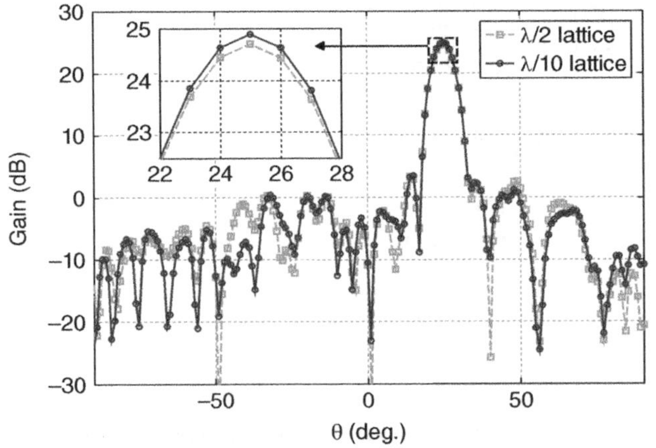

Figure 8.26 Effect of lattice size on the radiation performance of dielectric reflectarrays. *Source:* Nayeri 2013 [26]. Reproduced with permission of IEEE.

some advantage in increasing the accuracy of the model, a resolution of half-wavelength is quite sufficient to achieve a good performance with these designs. As such, this minimum lattice size can be used directly to determine the upper frequency limit based on the available fabrication capability.

8.4.3.3 Dielectric Reflectarray Prototypes and Measurements

The three dielectric reflectarrays are all fabricated using Objet Eden 350 polymer-jetting rapid prototyping 3D printer. After the three polymer structures are printed, a thin layer of gold with thickness about 100 nm is sputtered on the back surface of each polymer structure as a seed layer. Then, the polymer structure with the thin gold layer is inserted into a gold solution and a voltage is added between the gold layer on the surface of the polymer and the ground for plating. After 6 hours of plating, the thickness of the gold becomes about 6 μm and the reflectarray antenna is ready for test. Pictures of fabricated prototypes are shown in Figure 8.27.

The radiation patterns of the dielectric reflectarrays are measured using a vector network analyzer (Agilent E8361A) with W-band extension heads. A pyramidal horn antenna with a measured gain of 12 dB at 100 GHz is used to feed the dielectric reflectarray. The feeding horn is placed at a distance of 22.5 mm away from the aperture and pointing toward the center of the reflectarray with a tilted angle of 25° as described in the

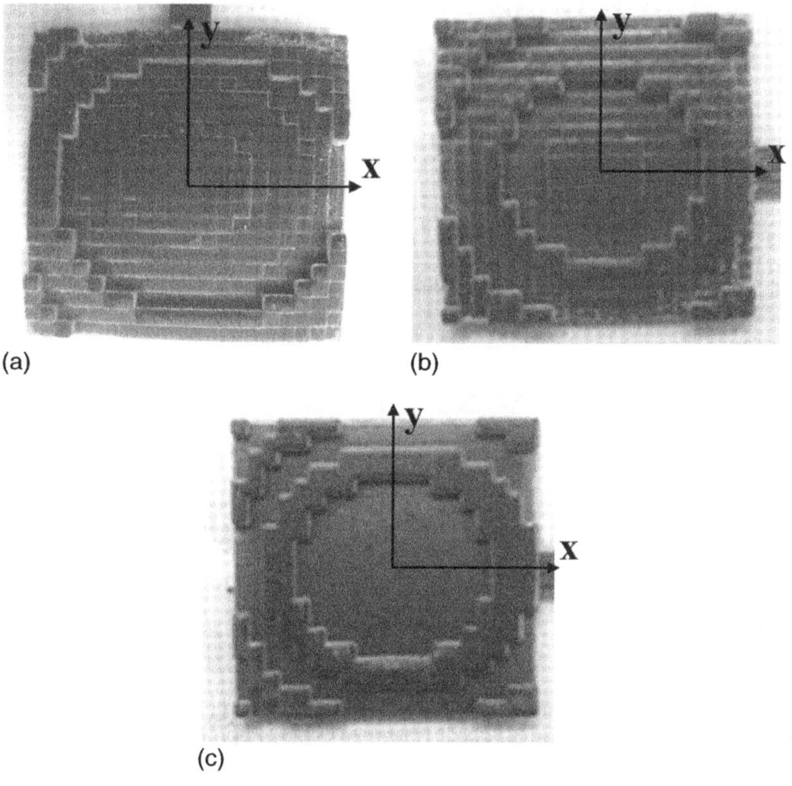

Figure 8.27 Top view of the fabricated dielectric reflectarray prototypes: (a) Design 1, (b) Design 2, (c) Design 3. The back side is gold plated. *Source:* Nayeri 2013 [26]. Reproduced with permission of IEEE.

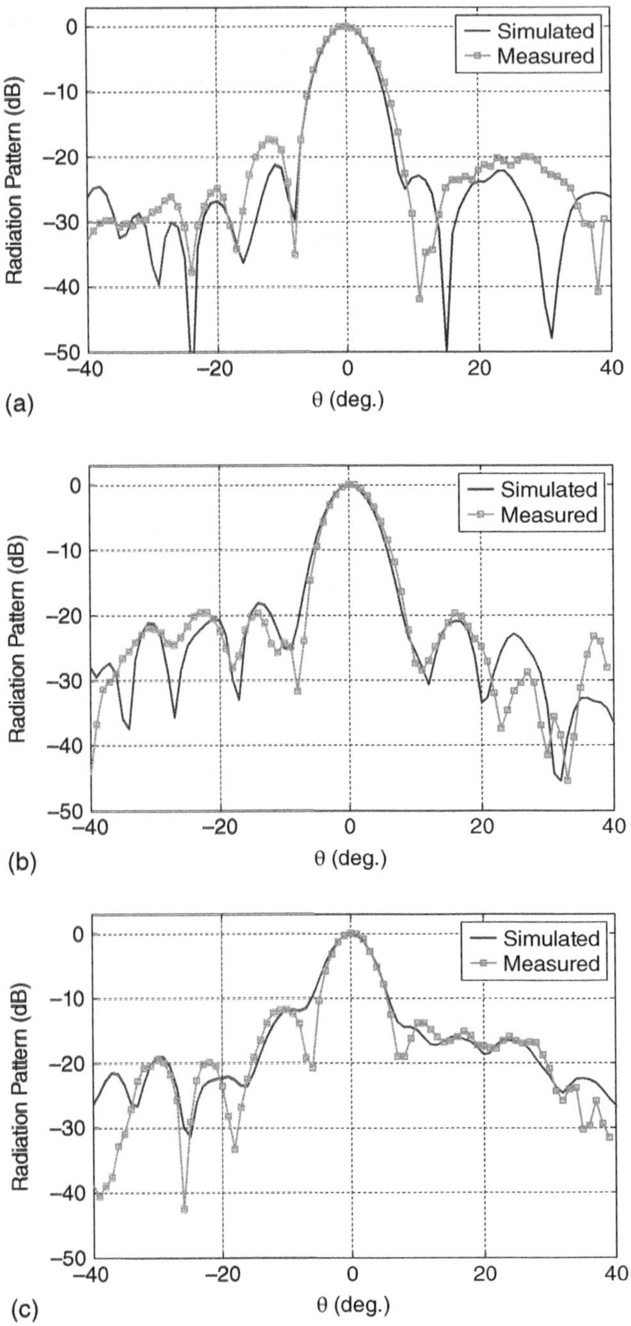

Figure 8.28 Comparison of measured and simulated radiation patterns of the dielectric reflectarrays at 100 GHz: (a) Design 1, (b) Design 2, (c) Design 3. *Source:* Nayeri 2013 [26]. Reproduced with permission of IEEE.

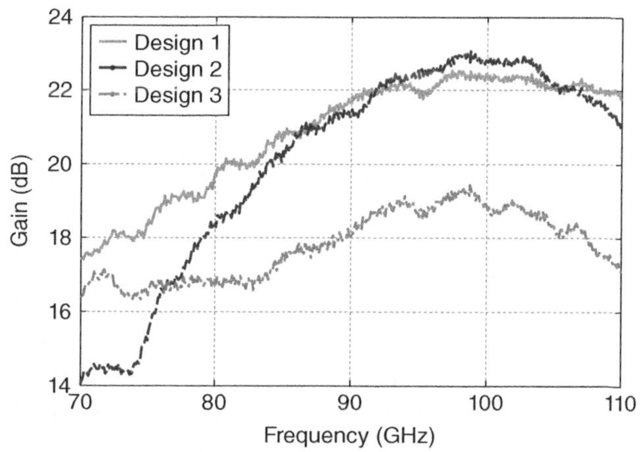

Figure 8.29 Measured gains versus frequency for the dielectric reflectarray prototypes. *Source:* Nayeri 2013 [26]. Reproduced with permission of IEEE.

previous section. Another W-band horn antenna is located at the far field distance to measure the radiation pattern. A standard gain horn is also used as the transmitter to calibrate the gain value of the dielectric reflectarray. Comparison between the simulated and measured radiation patterns in the xz-plane at 100 GHz are shown in Figure 8.28. Note that for the measurements the array is rotated 25° in the aperture plane, so the main beam is pointing to 0°. It can be seen that for all three designs, a close agreement between the measured and simulated radiation patterns is observed. The measured gains of the three prototypes at 100 GHz are 22.5, 22.9, and 18.9 dB, respectively. The measured half-power beam-widths are 6.95, 6.65, and 6.80°, respectively. The discrepancy between measured and simulated gain is attributed to material property uncertainty, alignment errors, and fabrication and measurement errors. Furthermore, Design 3 suffers more from the fabrication error, since the middle part of the aperture (0.3 mm) is too thin to maintain the flatness.

The gain of these dielectric reflectarray prototypes was also measured across the frequency range of 70–110 GHz. These results are given in Figure 8.29. As expected, Design 1 demonstrates the widest bandwidth that is attributed to the smaller number of phase wraps on the aperture. Design 2 on the other hand demonstrates the highest gain due to the lower loss of its elements. Despite a much lower gain in comparison with the previous two designs, Design 3 shows a slightly broader bandwidth than Design 2.

In summary, dielectric reflectarray antennas appear to be a low-cost solution for high-gain terahertz antennas. The methodology is readily scalable, and with the current material and fabrication technology, low-cost high-gain antennas up to 1.0 THz can be realized.

8.5 Future Trends

Reflectarray designs at THz and higher frequencies provide a variety of possibilities for new and exciting applications. For a practical system design, however, efficiency, bandwidth, and fabrication capabilities are of major concerns and must be investigated.

While several methods to improve the efficiency and bandwidth of reflectarrays are available in the microwave range [3], currently a major concern for optical designs is the fabrication limitations. There are, however, several promising technologies and materials that have more recently been investigated.

A very unique and promising material for THz reflectarrays is graphene. Graphene's unique electronic band structure leads to a complex surface conductivity at THz frequencies, which allows the propagation of very slow plasmonic modes. This leads to a drastic reduction of the electrical size of the array unit cell and thereby good array performance. A graphene reflectarray antenna was studied in [31]. The reflectarray was designed at 1.3 THz and comprises more than 25,000 elements of size about $\lambda/16$. Good performance is obtained in terms of bandwidth, cross-polarization, and grating lobes suppression, proving the feasibility of graphene-based reflectarrays and other similar spatially fed structures at terahertz frequencies. Another interesting design is the liquid crystal based THz reflectarray antenna reported in [32]. The element consists of three dipoles and a liquid crystal layer is placed under the radiating element. The reflection phase is reconfigured by varying the dielectric constant of the liquid crystal through a bias voltage. Images of the reflectarray element and reflection coefficients as a function of relative permittivity are given in Figure 8.30.

On the fabrication side, recent advances in nano-fabrication technique have made it possible to build artificial surfaces known as metasurfaces in the hopes of controlling scattered light [32]. Nanotechnology has emerged as a revolutionary technological breakthrough, which has led to the creation of many new materials and devices with a wide range of applications in medicine, electronics, and energy production. Thanks to the latest progress in nano-fabrication techniques, the fabrication accuracy has been significantly improved at an affordable cost. While several important factors need to be addressed for optical reflectarrays to be practical, with these rapid advancements, it seems likely that we will be able to build optical reflectarrays in the near future. Three-dimensional optical reflectarray nanoantennas with spherical elements were proposed in [33]. The concentric sphere element consists of a core made of ordinary dielectric and a shell made of plasmonic materials. The phase change is achieved by varying the dielectric constant of the core. A DR reflectarray nanoantenna was reported in [34] that consists of a silver ground and TiO_2 cylindrical DRs, where the phase chance is achieved by changing the diameter of the resonator. Another design for a 3D nanoantenna reflectarray involves using elements with perforated holes on the unit cell where phase change can be achieved by changing the radius of the hole [35], [36].

One interesting application could be concentrating solar power (CSP) systems. The practical barrier for a CSP system design is generally the fabrication capability at a competitive cost. CSP systems use mirrors or lenses to concentrate a large area of sunlight onto a small area. Compared to conventional solar panels, the CSP system has a significant cost advantage, since the solar collector is less expensive than an equivalent area of solar cells. In most CSP systems, mirrors are preferable to lenses, mainly because they have a lower cost. These mirrors are either parabolic or flat Fresnel reflectors, which focus the sunlight at the reflector's focal point. In comparison between these two types of reflectors, flat mirrors allow more reflective surface in the same amount of space as a parabolic reflector, thus capturing more of the available sunlight, and they are much cheaper than parabolic reflectors. The reflectarray has a flat plane and therefore a larger reflective surface compared to a parabolic reflector,

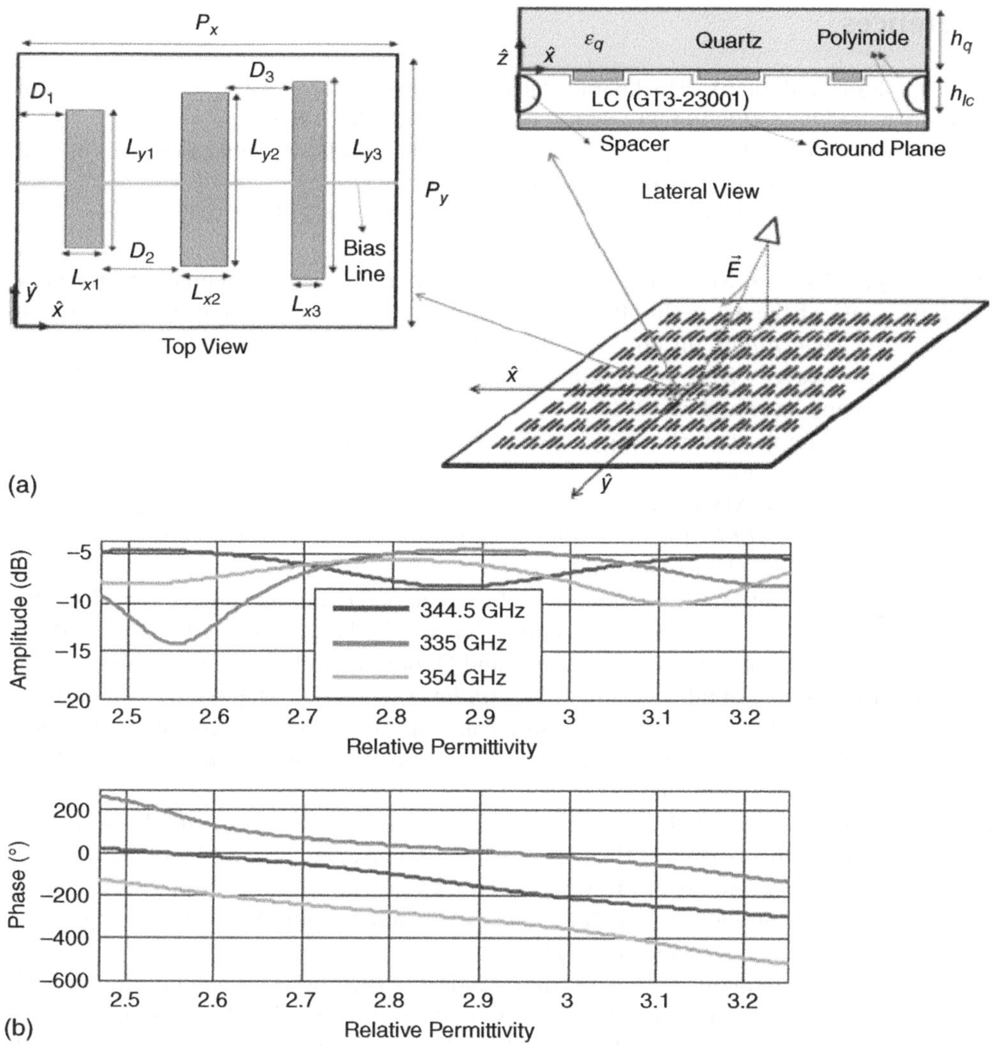

Figure 8.30 (a) Geometrical model of the liquid crystal reflectarray element. (b) Reflection coefficients as a function of relative permittivity. Reprinted with permission from [32].

and it has a much higher efficiency compared to a Fresnel reflector. In addition, low-profile and planar reflectarrays are easy to fabricate and integrate, and the individually controlled element phase increases the focusing flexibility and enhances the system capability. For example, reflectarrays can focus the solar energy onto a specified area with high concentration and uniform distribution instead of focusing all energy at a specific point. This feature is desirable in a CSP system, because it avoids over-illumination at a specific point and provides high efficiency over the entire solar cell surfaces. More recently infrared imaging systems using reflectarray were reported in [37], where wide-angle, broadband, efficient reflection holography is achieved by utilizing coupled dipole-patch nanoantenna cells.

References

1 K. C. Kao, and G. A. Hockham, "Dielectric-fibre surface waveguides for optical frequencies," *Proc. IEE*, Vol. 113, No. 7, pp. 1151–1158, Jul. 1966.

2 P. Mühlschlegel, H. J. Eisler, O. J. F. Martin, B. Hecht, and D. W. Pohl, "Resonant optical antennas," *Science*, Vol. 308, pp. 1607–1609, 2005.

3 J. Huang and J. A. Encinar, *Reflectarray Antennas*. New York: John Wiley & Sons, Inc., 2008.

4 R. Deng, F. Yang, S. Xu, and P. Pirinoli, "Terahertz reflectarray antennas: an overview of the state-of-the-art technology," *International Conference on Electromagnetics in Advanced Applications*, pp. 667–670, Aug. 2014.

5 Terahertz Wave, Website. Online, http://fir.u-fukui.ac.jp/thzlab/Images/EM%20 Spectrum.png (accessed July 2017).

6 P. H. Siegel, "Terahertz technology," *IEEE Trans. Microwave Theory Tech.*, Vol. 50, No. 3, pp. 910–928, Mar. 2002.

7 P. H. Siegel, P. de Maagt, and A. I. Zaghloul, "Antennas for terahertz applications," presented at the *IEEE Antennas and Propagation Society International Symposium*, Albuquerque, New Mexico, July 2006.

8 J. C. Ginn, B. A. Lail, and G. D. Boreman, "Phase characterization of reflectarray elements at infrared," *IEEE Trans. Antennas Propag.*, Vol. 55, No. 11, pp. 2989–2993, Nov. 2007.

9 J. M. López-Alonso, B. Monacelli, J. Alda, and G. Boreman, "Uncertainty analysis of the *spatia*l responsivity of infrared antennas," *Applied Optics*, Vol. 44, pp. 4557–4568, 2005.

10 T. Mandviwala, B. A. Lail, and G. D. Boreman, "Infrared-frequency coplanar striplines: design, fabrication and measurements," *Microwave and Optical Technology Letters*, Vol. 47, No. 1, pp. 17–20, Oct. 2005.

11 M. A. Ordal, L. L. Long, R. J. Bell, S. E. Bell, R. R. Bell, R. W. Alexander, Jr., and C. A. Ward, "Optical properties of the metals Al, Co, Cu, Au, Fe, Pb, Ni, Pd, Pt, Ag, Ti, and W in the infrared and far infrared," *Applied Optics*, Vol. 22, No. 7, April 1983.

12 CST Microwave Studio, CST Studio Suite 2014, Computer Simulation Technology AG.

13 ANSYS HFSS v15.0, ANSYS Inc., 2014.

14 G. R. Fowles, *Introduction to Modern Optics*, 2nd Edn, New York: Dover Publications, 1992.

15 H. Mosallaei and K. Sarabandi, "Antenna miniaturization and bandwidth enhancement using a reactive impedance substrate," *IEEE Trans. Antennas Propag.*, Vol. 52, No. 9, pp. 2403–2414, Sep. 2004.

16 M. Bozzi, S. Germani, and L. Peregrini, "Modeling of losses in printed reflectarray elements," *European Microwave Conference*, Oct 2004.

17 J. C. Ginn, B. A. Lail, and G. D. Boreman, "Phase characterization of reflectarray elements at infrared," *IEEE Trans. Antennas Propag.*, Vol. 55, No. 11, pp. 2989–2993, Nov. 2007.

18 T. Niu, W. Withayachumnankul, B. S.-Y. Ung, H. Menekse, M. Bhaskaran, S. Sriram, and C. Fumeaux, "Experimental demonstration of reflectarray antennas at terahertz frequencies," *Optics Express*, Vol. 21, No. 3, pp. 2875–2889, Feb. 2013.

19 F. Yang, P. Nayeri, A. Z. Elsherbeni, J. C. Ginn, D. Shelton, G. Boreman, and Y. Rahmat-Samii, "Reflectarray design at infrared frequency: effects and models of material loss," *IEEE Trans. Antennas Propag.*, Vol. 60, No. 9, pp. 4202–4209, Sep. 2012.

20 H. Rajagopalan, and Y. Rahmat-Samii, "Dielectric and conductor loss quantification for microstrip reflectarray: simulations and measurements," *IEEE Trans. Antennas Propag.*, Vol. 56, No. 4, pp. 1192–1196, Apr. 2008.

21 H. Rajagopalan and Y. Rahmat-Samii, "On the reflection characteristics of a reflectarray element with low-loss and high-loss substrates," *IEEE Antennas Propag. Magazine*, Vol. 52, No. 4, pp. 73–89, Aug. 2010.
22 D. Kajfez, *Dielectric Resonators*, Artech House Antennas and Propagation Library, 1990.
23 K. M. Luk and K. W. Leung, *Dielectric Resonator Antennas*, New York: Research Studies Press, 2002.
24 S. H. Zainud-Deen, S. M. Gaber, A. M. Abd-Elhady, K. H. Awadalla, and A. A. Kishk, "Perforated dielectric resonator antenna reflectarray," *ACES Journal*, Vol. 26, No. 10, 2011.
25 L. Zou, W. Withayachumnankul, C. M. Shah, A. Mitchell, M. Bhaskaran, S. Sriram, and C. Fumeaux, "Dielectric resonator nanoantennas at visible frequencies," *Optics Express*, Vol. 21, No. 1, pp. 1344–1352, Jan. 2013.
26 P. Nayeri, M. Liang, R. Sabory-García, M. Tuo, F. Yang, M. Gehm, H. Xin, and Atef Elsherbeni, "High gain dielectric reflectarray antennas for THz applications," presented at the IEEE Antennas and Propagation Society International Symposium, Orlando, Florida, July 2013.
27 The Stratasys Website. Available at: www.stratasys.com/3d-printers/design-series/objet-eden260vs (accessed July 2017).
28 Z. Wu, J. Kinast, M. Gehm, and H. Xin "Rapid and inexpensive fabrication of terahertz electromagnetic bandgap structures," *Optics Express*, Vol. 16, No. 21, pp. 16442–16451, Oct. 2008.
29 Z. Wu, A. Young, M. Gehm, and H. Xin "Investigation of several terahertz electromagnetic bandgap structures," *Microwave and Optical Tech. Lett.*, Vol. 52, No. 3, pp. 678–686, Dec. 2010.
30 D. R. Reid and G. S. Smith, "A full electromagnetic analysis of grooved-dielectric Fresnel zone plate antennas for microwave and millimeter-wave," *IEEE Trans. Antennas Propag.*, Vol. 55, No. 8, pp. 2138–2146, Aug. 2007.
31 E. Carrasco and J. Perruisseau-Carrier, "Reflectarray antenna at terahertz using graphene," *IEEE Antennas and Wireless Propagation Letters*, Vol. 12, pp. 253–256, Aug. 2013.
32 G. Perez-Palomino, J. A. Encinar, R. Dickie, and R. Cahill, "Preliminary design of a liquid crystal-based reflectarray antenna for beam-scanning in THz, *APSURSI*, 2013.
33 M. Kim, A. M. H. Wong, and G. V. Eleftheriades, "Optical Huygens' metasurfaces with independent control of the magnitude and phase of the local reflection coefficients," *Physical Review X*, Vol. 4, Dec. 2014.
34 A. Ahmadi, S. Ghadarghadr, and H. Mosallaei, "An optical reflectarray nanoantenna: The concept and design," *Optics Express*, Vol. 18, pp. 123–133, 2010.
35 L. Zou, W. Withayachumnankul, C. M. Shah, A. Mitchell, M. Bhaskaran, S. Sriram, et al., "Dielectric resonator nanoantennas at visible frequencies," *Optics Express*, Vol. 21, pp. 1344–1352, 2013.
36 S. H. Zainud-Deen, H. A. E. A. Malhat, S. M. Gaber, and K. H. Awadalla, "Perforated nanoantenna reflectarray," *Progress In Electromagnetics Research M*, Vol. 29, pp. 253–265, 2013.
37 Y. Yifat, M. Eitan, Z. Iluz, Y. Hanein, A. Boag, and J. Scheuer, "Highly efficient and broadband wide-angle holography using patch-dipole nanoantenna reflectarrays," *Nano Letters*, Vol. 14, pp. 2485–2490, May 2014.

9

Multi-Beam and Shaped-Beam Reflectarray Antennas

The conventional reflectarray antenna is designed to generate a pencil beam in a certain direction [1], [2]. While pencil beam patterns have numerous applications, for many other applications, such as fixed satellite services where the antenna is mounted on a spacecraft in a geostationary orbit, it is desirable to have a shaped beam (or multiple shaped beams) in order to efficiently illuminate selected geographical areas [3]. The classic choice for these applications has long been the shaped-beam reflector antenna structure. However, these antennas suffer from mechanical drawbacks such as bulkiness, the need for an expensive custom mold for each coverage specification, thermal problems, and others. As such, the numerous advantages that reflectarray antennas offer, make them a promising choice for these applications. It should be noted that while for wide-bandwidth applications, the standard parabola will always be the preferred design choice due to its true time-delay characteristics, for narrow-band applications (such as a direct broadcast satellite, which requires about 5% bandwidth) or for applications where electrical performance may be traded off for cost or mechanical advantages, a reflectarray antenna can be a good candidate.

The reflectarray allows for an individual control of the phase shift of each element in the array. As a result, the reflectarray can achieve shaped (contoured) beam performance without any additional cost. Similarly, multi-beam performance can also be realized by designing the phase shift of the elements appropriately. In this chapter, a systematic study on various design methods of multi-beam and shaped-beam reflectarray antennas is presented. Direct design methods for multi-beam reflectarrays, the geometrical method and the superposition method, are first described. It is demonstrated that although both methods can generate a multi-beam radiation pattern, neither approach provides satisfactory performance, mainly due to high side-lobe levels and gain loss in these designs. Synthesis design techniques are then introduced, and two different optimization techniques, the alternating projection method and the particle swarm optimization method, are then implemented to optimize the phase distribution on the reflectarray surface. Several examples of shaped and multi-beam reflectarray antennas are then presented.

Reflectarray Antennas: Theory, Designs, and Applications, First Edition.
Payam Nayeri, Fan Yang, and Atef Z. Elsherbeni.

9.1 Direct Design Approaches for Multi-Beam Reflectarrays

Multi-beam antennas have numerous applications, such as electronic countermeasures, satellite communications, and multiple-target radar systems [4]. These multi-beam antennas are typically based on reflectors with feed-horn clusters [5] or large phased arrays [6]. Horn array feeds for reflector antennas on communication satellites can provide multiple beams with tailored earth coverage patterns. For phased array antennas, multiple simultaneous beams can be generated by connecting the array to a beam-forming network with multiple ports. Considering the complexity of fabricating these antennas and deployment for space applications, these multiple beam designs are relatively high cost. The numerous advantages of reflectarrays, in particular the low-mass and low-cost features, makes the multiple beam reflectarray a suitable antenna candidate.

Reflectarrays can generate single or multiple beams with single or multiple feeds. Different design approaches have been introduced to achieve multi-beam performance. The main objective of this section is to provide a comprehensive and systematic comparison of multi-beam direct design approaches, their advantages, and shortcomings.

9.1.1 Geometrical Aperture Division

The basic idea behind this method is simply to divide the reflectarray surface into N sub-arrays where each sub-array can then radiate a beam in the required direction [7]. Although the array division and beam allocation can be arbitrary, it is feasible to define them based on the directions of the beams they are designed to generate. Since in most reflectarray designs the feed is pointing toward the geometrical center of the array, the sub-arrays can be defined by the zones corresponding to the beam direction they generate, that is

$$\varphi_n - \frac{\pi}{N} \le zone\, n \le \varphi_n + \frac{\pi}{N}. \tag{9.1}$$

Here, φ_n is the azimuthal direction of the nth beam, and it is assumed that the N beams are equally distributed in the azimuthal direction. With this definition, each sub-array will generate a beam in the direction of the zone center. It should be noted that with this approach each zone receives $1/N$ of the power from the feed horn while using $1/N$ of the aperture surface.

To demonstrate the multi-beam design capabilities of this approach, we will study two multi-beam designs: a double-beam and a quad-beam reflectarray antenna [8]. The antennas are designed for the operating frequency of 32 GHz and have a circular aperture with a diameter of 17λ at the design frequency. The element periodicity is λ/2 and we assume ideal phasing elements are used here. An axisymmetric design is considered here and the feed is positioned with an F/D ratio of 0.735. The feed horn is circularly polarized and the power q of the feed horn $cos^q(\theta)$ radiation pattern model is 6.5 at 32 GHz. The radiation patterns of the reflectarrays are calculated using the aperture field distribution method.

For the double-beam design, the element phase is adjusted to generate two beams in the directions of ($\theta_{1,2}=30°$, $\varphi_1=0°$, $\varphi_2=180°$). Since only two beams are required in this design the reflectarray surface can be divided into two half circles where each zone generates one beam. While different zone allocations can be considered for this design, here we study the case were the beams are generated in the direction of the zone center, that is, $-90° \le zone_1 \le 90°, 90° \le zone_2 \le 270°$. The radiation patterns of this design are given in Figure 9.1. From these results, it can be observed that although two beams are generated in the required directions, the side-lobe levels are quite high; that is, about 12 dB. In addition, the beams show broadening in one direction. The direction of beam broadening corresponds to the shorter length of the sub-arrays.

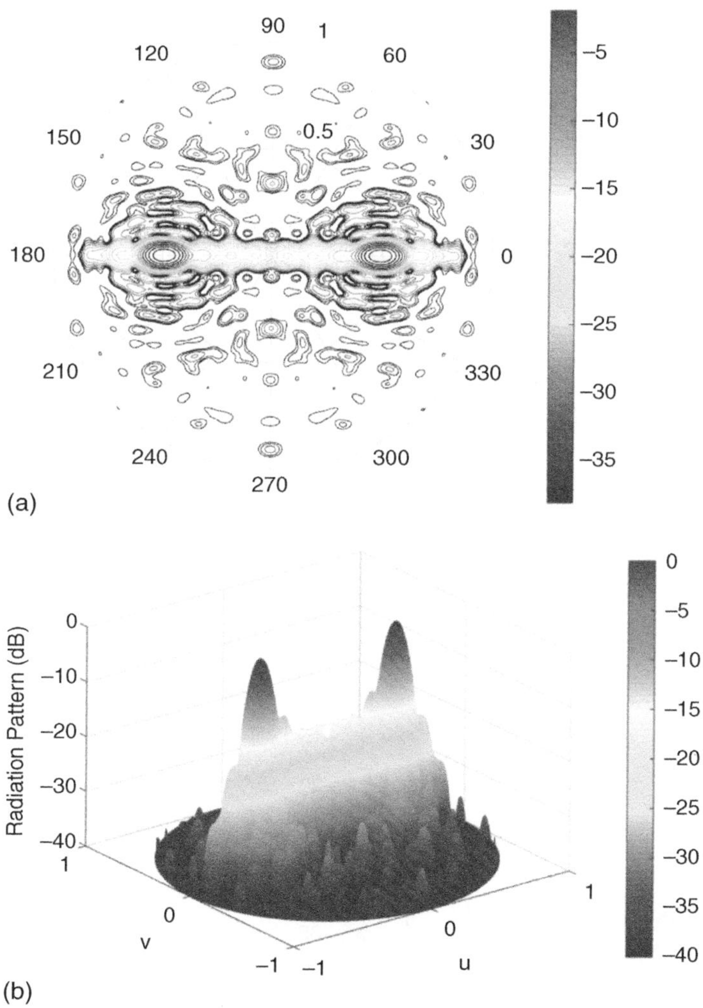

Figure 9.1 Radiation pattern of the double-beam reflectarray antenna designed with the geometrical approach: (a) 2D contour plot, (b) 3D pattern.

The quad-beam reflectarray is designed to generate four beams in the directions of ($\theta_{1,2,3,4}=30°$, $\varphi_1=0°$, $\varphi_2=90°$, $\varphi_3=180°$, $\varphi_4=270°$). Similarly, the beams are pointing in the zone center direction of the array. The radiation patterns for both these designs are given in Figure 9.2.

The results here show that four beams are generated in the required directions but with high side-lobe levels. While the examples here considered symmetric zone allocations with respect to the beam directions, it was observed that regardless of the choice of the zones, beam broadening will be observed in the radiation patterns, as shown in Figure 9.5 later. This is mainly due to the fact that a smaller array surface is being used, which would also reduce the antenna gain. Further discussions about the radiation performance of this approach will be given in Section 9.1.3.

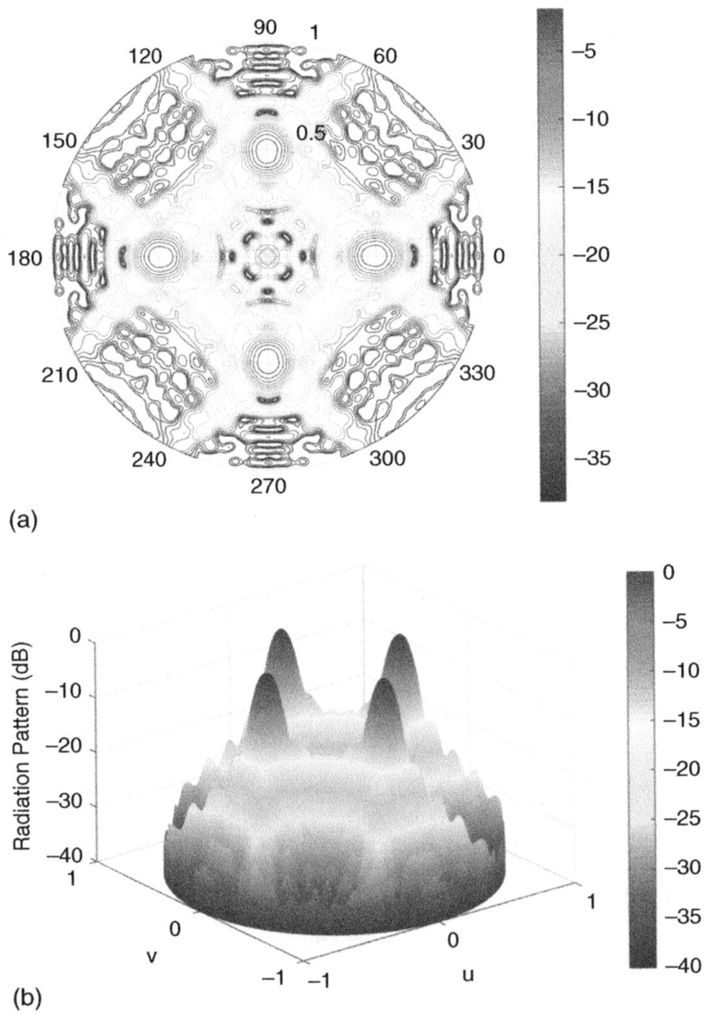

Figure 9.2 Radiation pattern of the quad-beam reflectarray antenna designed with the geometrical approach: (a) 2D contour plot, (b) 3D pattern.

9.1.2 Superposition of Aperture Fields

Another approach for multi-beam reflectarray designs is by using the superposition of the aperture fields associated with each beam on the reflectarray aperture [2]. To generate N beams with a single feed, the tangential field on the reflectarray surface can simply be written as

$$E_R(x_i,y_i)=\sum_{n=1}^{N}A_{n,i}(x_i,y_i)e^{j\phi_{n,i}(x_i,y_i)}. \tag{9.2}$$

Here, $A_{n,i}$ and $\Phi_{n,i}$ are the required amplitude and phase of the ith element that will radiate the nth beam. In reflectarrays the amplitude of each element is determined by the feed position and element location, which are independent of the beam direction, so

$$E_R(x_i,y_i)=A_i^{Feed}(x_i,y_i)\cdot\sum_{n=1}^{N}e^{j\phi_{n,i}(x_i,y_i)}. \tag{9.3}$$

The summation of the complex field distributions will give the overall required amplitude and phase distributions.

A basic problem exists here that is due to the fixed amplitude distribution imposed by the feed in reflectarray antennas. Although the required phase in (9.3) can be satisfied by proper element designs, the amplitude requirement cannot be satisfied in reflectarray antennas, that is,

$$\left|\sum_{n=1}^{N}e^{j\phi_{n(i)}(x_i,y_i)}\right|\neq 1. \tag{9.4}$$

As a result of this difference in the amplitude distribution on the aperture, reflectarrays designed using the superposition approach may show a degraded performance [9], [10]. This amplitude error may be minimized for some simple designs; however, complicated multi-beam reflectarrays will suffer from this error.

To demonstrate the multi-beam design capabilities of this approach, here we consider the quad-beam reflectarray system studied earlier. Three different normalized amplitude distributions on the aperture of the reflectarray antenna are given in Figure 9.3. Figure 9.3(a) shows the summation of the complex terms, that is, equation (9.4). Figure 9.3(b) shows the practical amplitude distribution of the reflectarray antenna, which is set by the feed-horn characteristics and its position. The ideal amplitude distribution is given in Figure 9.3(c), which is the practical amplitude distribution multiplied by the summation of the complex terms. It is implicit that such a taper cannot be realized on the reflectarray antenna aperture and this ideal amplitude distribution is not practical. This typically results in a degradation of antenna performance, mainly by an increase in the side-lobe levels.

For the double-beam antenna, the periodic variations in the x-direction coincide with the element periodicity of $\lambda/2$. As a result, the amplitude variation is actually not seen by the phasing elements and is avoided on the x-axis. On the other hand, the quad-beam design shows amplitude variations that are not avoided and thus significantly alter the illumination taper of the aperture. The radiation patterns of both these multi-beam antennas are given in Figure 9.4.

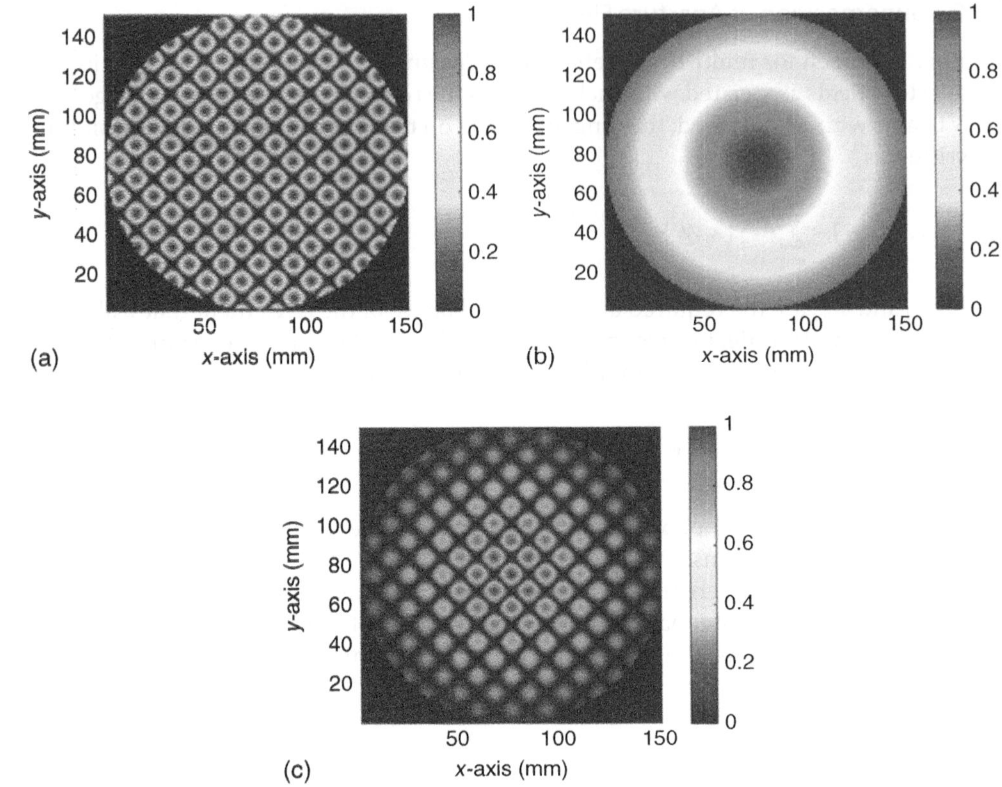

Figure 9.3 (a) Amplitude distribution from equation (9.4). (b) Practical amplitude distribution. (c) Ideal amplitude distribution.

From these results, it can be seen that, as expected, the double-beam design shows an ideal performance with side-lobe levels below $-25\,\text{dB}$ and beams correctly scanned to the designed directions. The quad-beam design, however, shows a poor performance. Although four beams are generated, the side-lobe levels are quite high and the beams are not exactly scanned to the designed directions. More discussions on the pattern characteristics of the antennas are given in the next section.

9.1.3 Comparison of Direct Design Approaches

In order to compare the performances of these two direct design approaches for multi-beam reflectarray designs, let's study the quad-beam reflectarray antenna introduced in the previous sections. For reference purpose, we also design a single-beam reflectarray antenna whose beam is in the direction of ($\theta = 30°$, $\varphi = 0°$). In Figure 9.5, the normalized radiation pattern of this single-beam design is compared with the multi-beam reflectarrays. Since the reference beam is in the $\varphi = 0°$ plane, only this plane is given here, and a similar pattern was observed in the orthogonal plane. Comparison of the radiation patterns shows small beam deviations in the multi-beam reflectarray designs. The radiation patterns also show that the side-lobe level is below $-11\,\text{dB}$ for the geometrical design and

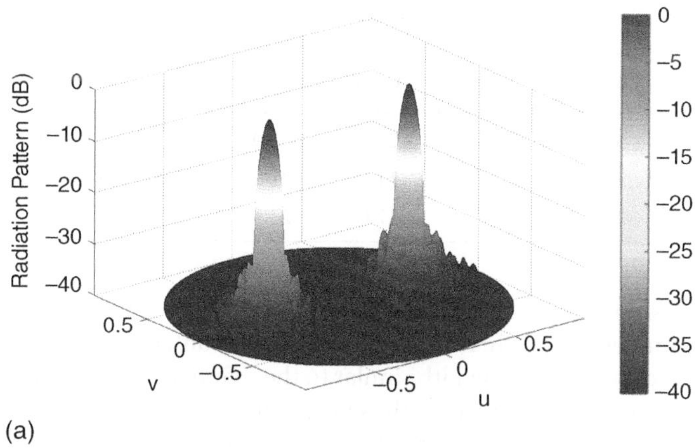

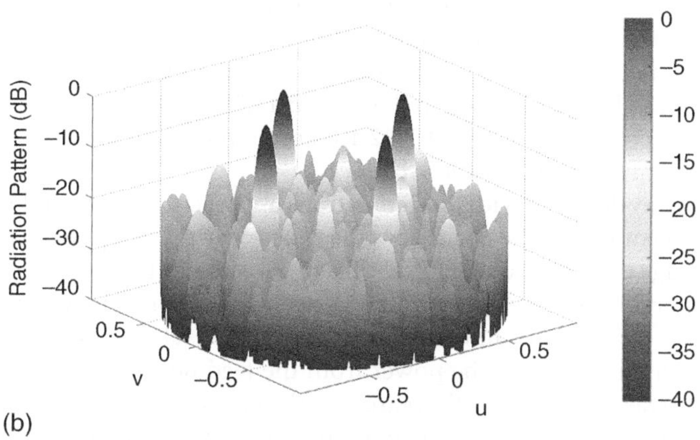

Figure 9.4 A 3D radiation pattern of the multi-beam reflectarray antennas designed with the superposition approach: (a) double-beam, (b) quad-beam.

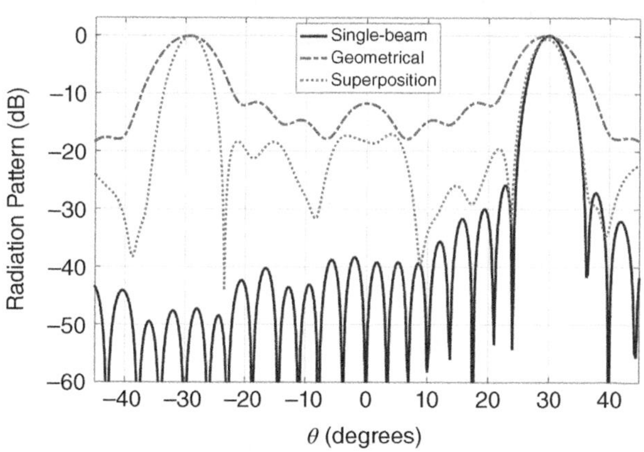

Figure 9.5 Normalized radiation patterns of the single- and quad-beam designs.
Source: Nayeri 2012 [10]. Reproduced with permission of IEEE.

Table 9.1 Calculated radiation characteristics of the single- and quad-beam reflectarrays.

Reflectarray	Design Method	Beam Direction (°)	SLL (dB)	Directivity (dB)	HPBW (°)
Single-beam		30	−25.9	32.94	5.0
Quad-beam	Geometrical	29.44	−11.05	21.21	8.4
	Superposition	29.57	−17.05	25.92	4.9

below −17 dB for the superposition design. The side-lobe levels (SLL) are much higher than the reference single-beam design. Furthermore, the quad-beam antenna designed using the superposition method shows a beamwidth similar to the reference single-beam design, while in the geometrical design the beamwidth is much wider.

The antenna directivity is an important measure to compare the radiation performance of these multi-beam design methods. For multi-beam reflectors using a single feed, theoretically, the power of each beam will be reduced by $1/N$. Therefore, ideally it is expected that generating four beams will reduce the antenna directivity by 6 dB. However, the geometrical design has a directivity reduction of 11.73 dB and the superposition design exhibits a directivity reduction of 7.02 dB. Some important results of these three reflectarrays are summarized in Table 9.1.

For the multi-beam reflectarray designed with the geometrical approach, the amplitude distribution in each zone is maximum at the corner of that zone (near the array center) and minimum at the outer edge, which results in a significant increase in the side-lobe level. The wider beamwidth and lower directivity in the geometrical designs, however, requires further attention. The reduction in antenna directivity is the result of using one fourth of the array surface and one fourth of the power from the feed horn to generate each beam. This reduction of array surface is also the reason for the increase in beamwidth.

For the multi-beam reflectarray designed using the superposition method, the high side-lobes are due to the amplitude errors, which alter the required illumination taper on the aperture. Also, in comparison of the calculated antenna directivity with the reference, this design approach shows a directivity loss about 1 dB higher than the ideal directivity reduction (6 dB), which is mainly due to the high SLL of this design.

In summary, the shortcomings of both these direct design approaches are tabulated here:

Geometrical design method:

1) High side-lobe due to illumination taper.
2) Gain loss and beam broadening due to dividing the array surface into sub-arrays.
3) Small beam deviation.

Superposition design method:

1) High side-lobe due to amplitude error.
2) Gain loss due to the increase in side-lobe level.
3) Small beam deviation.

As a result of these problems associated with the direct methods for multi-beam reflectarray designs, it is necessary to implement some form of optimization routine to achieve desirable performance [11].

9.2 Synthesis Design Approaches for Shaped- and Multi-Beam Reflectarrays

9.2.1 Basics of Synthesis Techniques

Antenna synthesis in its broadest sense is the procedure of using a systematic method to arrive at an antenna configuration that produces a specified radiation performance [12]. In the case of antenna arrays, the synthesis problem is that of determining the excitation of the elements. While there is no single synthesis method that can yield the optimum array with the most efficient computation time and resources for any desired radiation pattern, several synthesis methods have been developed over the years. In this section, we will study the basics of the common synthesis methods for reflectarrays that are currently in use.

In reflectarray antennas, the element magnitude is fixed by the properties of the feed and the element location. Therefore, to realize a shaped or multi-beam radiation pattern, the only parameter that can be controlled is the reflection phase of the elements. The phase shift on a reflectarray aperture is obtained by controlling the geometrical parameters of its phasing elements. Owing to this feature any phase distribution can be attained on the aperture, which provides the means for a reflectarray to achieve simultaneous multiple beams or shaped-beam patterns. The challenge is how one determines the required phase shift for such performance.

Several different synthesis techniques have been developed for antenna arrays over the years, however, it has been shown that the most practical approach is to implement some form of a search algorithm also known as optimization [13]. In optimization design approaches, an optimum phase distribution on the reflectarray aperture that can achieve the desirable performance is obtained using a phase-only optimization technique. Optimization techniques are typically categorized into two groups; local and global-search algorithms. While both techniques are model-based optimization routines, the primary difference is that in a local search, the algorithm only searches for the solution in a limited space, as determined by the initial values. On the other hand, a global algorithm, searches the entire solution hyperspace. The advantages and disadvantages are therefore quite explicit. A local search can only find the optimal solution if it's discoverable from the starting points. Otherwise, the optimization would get trapped in local minima and will not converge. In contrast, a global-search has the capability of finding the optimum solution from any starting point; however, the drawback is that the computation time could be very high.

The pattern requirements for the synthesis problem are usually defined by a mask, that is, two sets of bound values, between which the pattern must lie. The general form of the radiation patterns that satisfy the mask requirements is

$$\text{set } M \equiv \left\{F(u,v) : M_L(u,v) \leq \left|F(u,v)\right| \leq M_U(u,v)\right\}, \tag{9.5}$$

where F is the far-field radiation pattern of the array and (u,v) are the angular coordinates. M_U and M_L set the upper and lower bound values of the desired pattern in the entire angular range.

In all optimization routines, it is necessary to define a cost or fitness function that should be minimized and can also control the number of iterations required for the convergence of the solution. In its simplest form, the fitness function may only evaluate the performance of the upper bound given by (9.5). In this case, the mathematical formula for this function can be given by

$$\begin{aligned} &\text{If} \quad \left|F(u,v)\right| > M_U(u,v) \\ &Cost = \sum_{u^2+v^2\leq 1} \sum \left(\left|F(u,v)\right| - M_U(u,v) \right)^2. \end{aligned} \tag{9.6}$$

where the cost is evaluated over every point in the (u,v) space that does not belong to the main beams.

9.2.2 Local-Search Techniques

Several local-search phase-only techniques have been developed for antenna arrays, and the same techniques can be applied to reflectarrays. A practical consideration, however, is the very large number of elements in a reflectarray antenna configuration. Very few efficient techniques are able to deal with thousands of variables. The alternating projection method (APM), also known as the intersection approach, has been shown to be very efficient for synthesis of large antenna arrays [14]–[17]. Compared to other phase synthesis methods developed for arrays, the main advantage of APM is the significantly reduced computational time for convergence of the solution, which makes it suitable for large reflectarray antennas. Here, we will briefly describe the basics of this technique for reflectarray synthesis.

In order to implement the APM to obtain the desired radiation pattern, in the first step, one needs to define the lower and upper bounds (set M) as given by (9.5). The APM is basically an iterative process that searches for the intersection between two sets, that is, the set of possible radiation patterns that can be obtained with the reflectarray antenna (set R) and the set of radiation patterns that satisfy the mask requirements (set M). The synthesis problem is then finding an element in set R that also belongs to set M, or has the minimum distance from set M. This is done by alternative projection from the two sets, which is schematically shown in Figure 9.6.

In the APM optimization, in most cases, the requirements in the main beam will be satisfied by the projection operators with the bounds set in (9.5), therefore the cost function need only to take into account the side-lobe performance of the array [18].

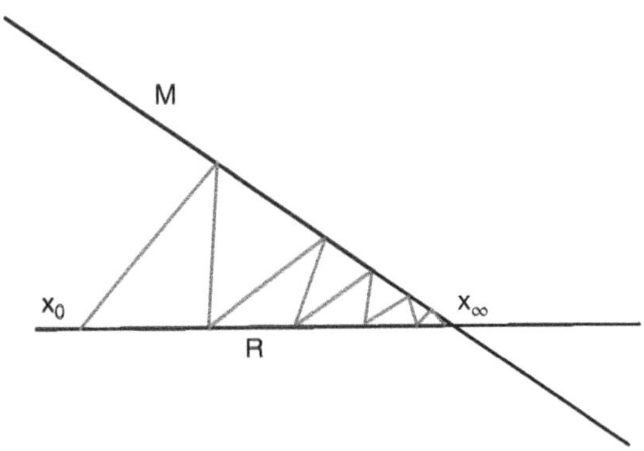

Figure 9.6 Schematic model of alternating projections on two intersecting sets M and R.

It should be noted that this technique is typically only applicable to convex sets. In the case where the two sets intersect (Figure 9.6), this method will find a point in the intersection of these sets. For two convex sets that don't intersect, this method yields a pair of points in M and R that have minimum distance.

Implementing the APM requires definition of two projection operators: the mask projector (P_M) and the inverse projector (P_I). These operators allow one to impose the constraints in the aperture amplitude that is imposed by the feed antenna, and also improve the efficiency of the algorithm.

For reflectarray synthesis, the iterative procedure is then implement as

$$x_{n+1} = P_I P_M x_n \tag{9.7}$$

where x_n is the radiation pattern of the array.

The mask projector (P_M) uses the upper and lower bounds of the mask to correct the radiation pattern. Mathematically this is given by

$$P_M F(u,v) = \begin{cases} M_U(u,v)\dfrac{F(u,v)}{|F(u,v)|} & |F(u,v)| > M_U(u,v) \\ F(u,v) & M_L(u,v) \le |F(u,v)| \le M_U(u,v) \\ M_L(u,v)\dfrac{F(u,v)}{|F(u,v)|} & |F(u,v)| < M_L(u,v) \end{cases} \tag{9.8}$$

As discussed earlier, in reflectarrays the synthesis of radiation patterns is restricted by the fact that the amplitude of each reflectarray element is fixed by the feed properties and element location. Therefore, a function needs to be defined in order to calculate new phase values for the reflectarray elements while the elements amplitude remain unchanged. The inverse projection (P_I) consists of a series of functions, mathematically given by

$$P_I = F_S F_C F_I F_S^{-1}. \tag{9.9}$$

Here, F_S and F_S^{-1} are the Fourier and inverse Fourier transforms, respectively. F_I multiplies the set by the characteristic function. For reflectarrays this resets the extended grid used for the far-field pattern computation (see Chapter 4) and also corrects the shape of the array (for non-square apertures). F_C projects an excitation sequence onto the set with the excitation constraints, which is used to reset the amplitude to the value imposed by the feed antenna. A flowchart of the optimization procedure is given in Figure 9.7.

As discussed earlier, in synthesis of shaped or multi-beam reflectarrays, the required masks are typically given based on the coverage requirements. In most cases, these masks are given in the satellite coordinates that must first be transformed to the reflectarray coordinate systems [2]. Typical masks for different contour beams can be found in the literature [15] and we refer the reader to those publications.

For multi-beam designs the mask definition is different. The required masks for multi-beam radiation patterns are typically circular contours defined in the direction of each beam. Since typically one doesn't want to change the beamwidth, which is directly related to the aperture size and illumination, the mask upper and lower bounds in the

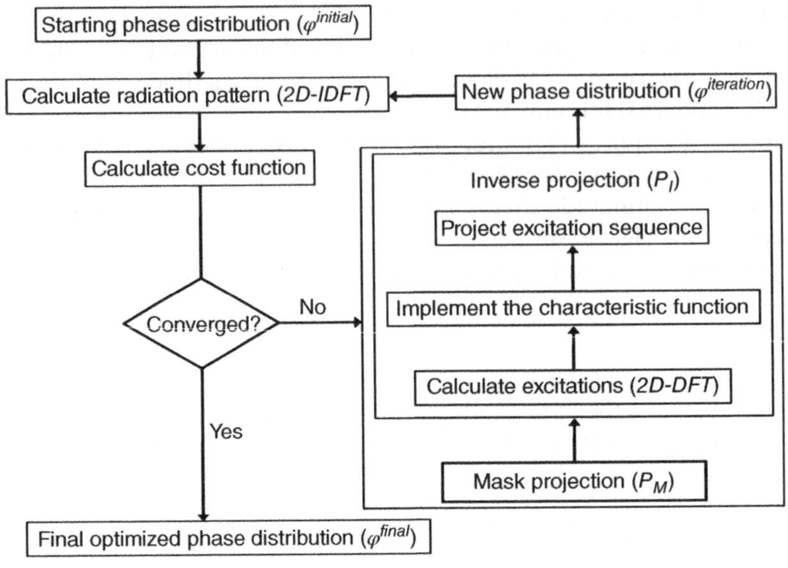

Figure 9.7 Flowchart of the APM optimization.

beam area can be defined according to the reference single-beam design. These upper and lower bounds are defined as

$$\begin{aligned} &\text{If } (u,v) \in \text{main beam}: && M_U(u,v) = 0\,\text{dB}, \\ &\text{If } (u,v) \in -3\,\text{dB beamwidth}: && M_L(u,v) = -3\,\text{dB}. \end{aligned} \tag{9.10}$$

In the case of multi-beam designs, an important objective of the optimization is to minimize the side-lobe level. While it is possible to control the side-lobe level by defining an upper bound (M_U) at certain values, it was shown that to minimize the side-lobe level, both upper and lower bounds in the side-lobe area can be set to zero ($-\infty$ dB). Thus, a 2D figure of this multi-beam mask model is given in Figure 9.8.

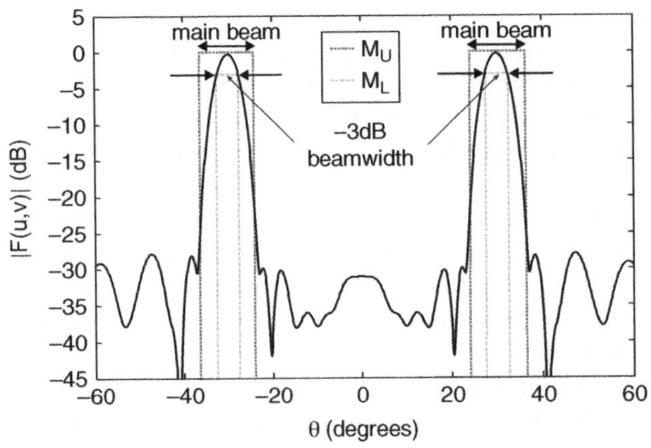

Figure 9.8 A 2D view of the mask model for the quad-beam reflectarray in the *xz*-plane. *Source:* Nayeri 2012 [10]. Reproduced with permission of IEEE.

With the mask and cost function defined, the optimized phase distribution of the reflectarray elements can be obtained with an iterative procedure. It should be noted here that the optimization is considered to be converged when the cost function becomes stable. In most cases the optimization converges with only a few iterations; however, a suitable starting point can reduce the number of iterations. Typically, the phase distribution obtained by the superposition method yields a good starting point for local optimizations. In the case of shaped beams, multiple beams directed in the desired coverage area can be used for the initial phase values.

It is important to note that in many cases, APM converges into a local minimum due to the non-convexity of the optimization problem and does not satisfy the mask requirements. Typically, the number of local minima rapidly increases with the number of variables in the optimization (number of reflectarray elements) or the complexity of the pattern mask. One approach that has shown to be effective for reflectarrays with large number of elements is a process where the illumination on the reflectarray edge is gradually (multiple stages) increased until it matches the practical taper. In the beginning stage, a very high taper illuminates the elements at the center of the reflectarray aperture with high amplitude, while the outer elements are excited with a negligible amplitude, which consequently reduces the number of local minima in the optimization. In the next stage, the excited area is gradually increased and the phase distribution obtained in the previous stage is used as the starting point. The procedure is repeated until the practical taper is realized on the reflectarray aperture [2].

9.2.3 Global-Search Techniques

While several multi-beam and contour beam reflectarray antennas have been demonstrated using local-search techniques [10], [11], [19]; in many cases a local search is incapable of finding the solution, due to the complexity (non-convex) of the optimization problem and the resulting optimization traps or local minima. As such, the application of a global and more powerful optimization routine for reflectarray synthesis may be required [20].

Global optimization algorithms search for the solution in the entire hyperspace, therefore it is necessary to have an efficient and high-performance algorithm for the optimization. Global-searches can be categorized into three groups: deterministic, stochastic, and heuristic search methods [20]. Of these approaches, heuristic methods such as genetic algorithm (GA) and particle swarm optimization (PSO) are efficient and widely used techniques in array optimizations [18]. GA relies on Darwin's theory of natural selection and the competition between individual chromosomes [21]. On the other hand, in PSO, the swarm behavior in nature is modeled by fundamental Newtonian mechanics for optimization purposes [22]–[25]. Both techniques search the entire solution hyperspace and typically achieve similar performances; however, in comparison between these two optimizations, PSO is generally considered to be simpler to program and also requires less computational bookkeeping [18], [26], [27]. Here we will briefly describe the basics of PSO search technique.

The optimization problem in PSO is a simultaneous optimization of N variables. A collection or swarm of particles are defined, where each particle is assigned a random position in the N-dimensional problem space. Each particle's position corresponds to a candidate solution to the optimization problem. Note that in the PSO terminology each individual in the swarm is referred to as a particle or agent. Each of these particle positions is scored to obtain a scalar cost based on how well it solves the problem. The particles then fly through the N-dimensional solution hyperspace subject to both deterministic and stochastic update rules to new positions, which are subsequently scored. As the particles traverse the problem hyperspace, each particle remembers its own personal best position that it has ever found, called its local best. Each particle also knows the best position found by any particle in the swarm, called the global best. On successive iterations, particles move toward these prior best solutions with linear attraction forces.

To implement PSO, first an initial set of random positions and velocities are defined for the particles in the swarm. For an optimization problem with M particles and N variables, if the velocities and positions at the nth iteration of all M particles are represented by $M \times N$ matrices v_n and x_n, the swarm behavior is modeled as an iterative process by applying Newton's second law

$$\begin{aligned} v_n &= \omega * v_n + c_1 \mathrm{rand}()*\left(p_{best,n} - x_n\right) + c_2 \mathrm{rand}()*\left(g_{best} - x_n\right), \\ x_n &= x_n + \Delta t * v_n, \end{aligned} \tag{9.11}$$

where c_1 and c_2 are the self- and group-knowledge constants, that determine the relative pull, and ω is the inertial weight. As particles explore the problem hyperspace, each particle remembers its own personal best position (p_{best}) and also knows the best position found by any particle in the swarm (g_{best}). The swarm of particles explores the problem hyperspace, and eventually settles down near a good solution.

A flowchart illustrating the steps and update equations of PSO is shown in Figure 9.9. Similar to the local-search techniques, one must define a fitness function for the problem, which in the case of reflectarray synthesis also requires the definition of a desired pattern mask. It is also important to point out that while PSO and GA have been extensively implemented in array optimizations, in most cases the number of array elements was around 100. A major challenge in practically implementing a global-search for a reflectarray antenna is the very large solution hyperspace due to the several hundreds of elements included in a reflectarray, which requires one to develop a very efficient far-field radiation pattern computation code. Please refer to Chapter 4 for detailed discussions on this topic.

9.2.4 Full-Wave Optimization Design Approaches

The local and global optimization techniques described in the previous two sections, view the optimization as a classic synthesis problem. As such the optimal phase distribution on the reflectarray aperture is determined for each of its elements. The second step in the design is to select practical elements that can achieve this desired phase shift. Typically, this is done using classical unit-cell analysis and a local periodicity approximation. Due to the approximation in such an approach, however, discrepancies between simulations and measurements have been observed in some cases [28].

A direct optimization technique in which the geometrical parameters of the reflectarray elements are simultaneously optimized can potentially produce more optimal designs. In [29], this method was implemented, under the local periodicity approximation, thus maintaining a direct relation between optimization goals and optimization variables, that is, element dimensions. The authors showed that the designs obtained using this direct optimization technique are much better in performance when compared to the

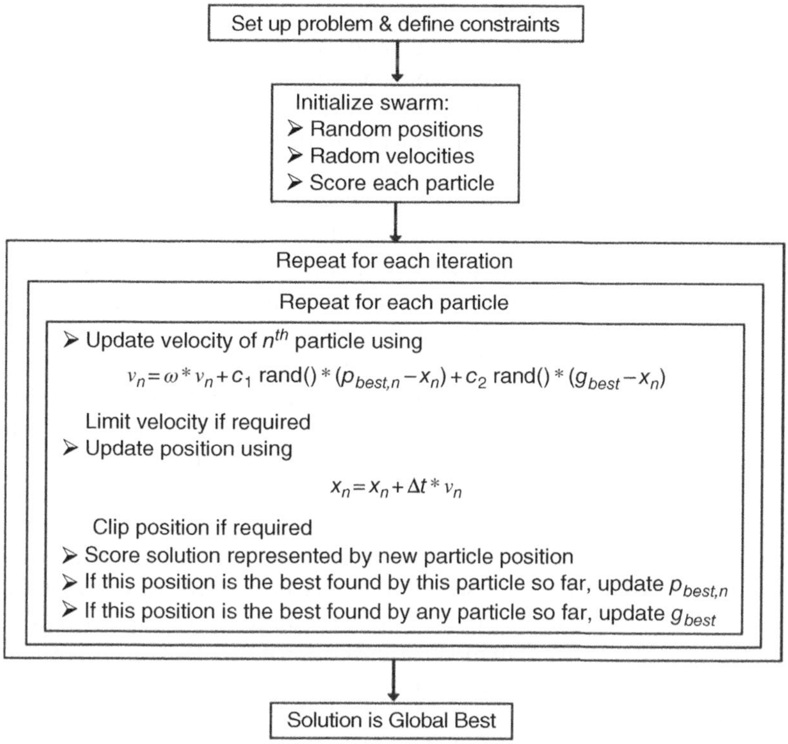

Figure 9.9 A flowchart of the PSO algorithm.

phase-only technique, both for multi-frequency and dual-polarized designs. An optimization where the geometrical parameters of all the elements are directly optimized in the full-wave software is the ideal solution, however, implementing a full-wave optimization for a reflectarray proved to be a very time-consuming synthesis [30], [31].

9.3 Practical Reflectarray Designs

Reflectarrays can generate single or multiple pencil- or shaped beams with single or multiple feeds, and several designs have been presented over the years. A contour beam reflectarray was first demonstrated in [32]. A two-beam reflectarray prototype using a single feed was demonstrated in [11] while [7], [10] present a single-feed reflectarray generating four simultaneous beams. Multi-feed multi-beam reflectarrays with shaped beams were also studied in [19]. In addition, multi-feed single-beam reflectarray antennas were investigated in [33]. In this section, we will take a closer look at some of these advanced reflectarray designs.

9.3.1 Single-Feed Reflectarray with Multiple Symmetric Beams

A single-feed reflectarray antenna generating four simultaneous pencil beams was demonstrated in [10]. The antenna is designed for the operating frequency of 32 GHz and has a circular aperture with a diameter of 17λ at the design frequency. A centered prime focus left hand circularly polarized (LHCP) feed is used for this design and positioned with an F/D ratio of 0.735 that gives an edge taper of −12.4 dB. The reflectarray elements are square patches with variable sizes, and since by nature these elements are dual-linear polarized, they reflect the LHCP incident wave to a right hand circularly polarized (RHCP) wave. This quad-beam reflectarray is designed to generate four beams in the directions of ($\theta_{1,2,3,4} = 30°$, $\varphi_1 = 0°$, $\varphi_2 = 90°$, $\varphi_3 = 180°$, $\varphi_4 = 270°$).

In this design, the phase distribution obtained by the superposition method was used as the starting point for the optimization. The APM was then implemented to improve the reflectarray performance by optimizing the phase distribution on the reflectarray aperture. The mask definition and cost function used for this optimization were as given in Section 9.2.2. A far-field pattern of 400 × 400 points evenly spaced in the angular coordinates was computed for each candidate reflectarray at each cost evaluation. As discussed earlier, the APM optimization might converge to local minima for some problems, however, for this symmetric quad-beam design the optimization did not get trapped and the solution converged smoothly. Similar radiation pattern results were observed when the phase distribution obtained by the geometrical approach was used for the starting point. Furthermore, the solution convergence was studied for different side-lobe level requirements in the design. The convergence curves for different mask levels are given in Figure 9.10 where it can be seen that in all cases, APM converges with a very few iterations. Although the number of iterations required for the optimization generally depend on the problem at hand, in most cases the APM will converge with just a few iterations. The radiation pattern of the optimized design with ideal elements and the −60 dB mask SLL is given in Figure 9.11.

A quad-beam performance with side-lobes below −26 dB is obtained with the optimized reflectarray. It should be noted that, while the optimization has converged, and the cost is minimized, it does not approach zero and a −60-dB side-lobe is not achieved. Compared to the initial design (superposition approach), the side-lobe level has been reduced by about 9 dB and all four beams are exactly scanned to 30° off broadside. Also, the calculated directivity for this antenna is 26.95 dB, which is about 1 dB higher than the initial superposition design (Table 9.1). These results clearly demonstrate the effectiveness of the phase optimization process. The computational time for the APM

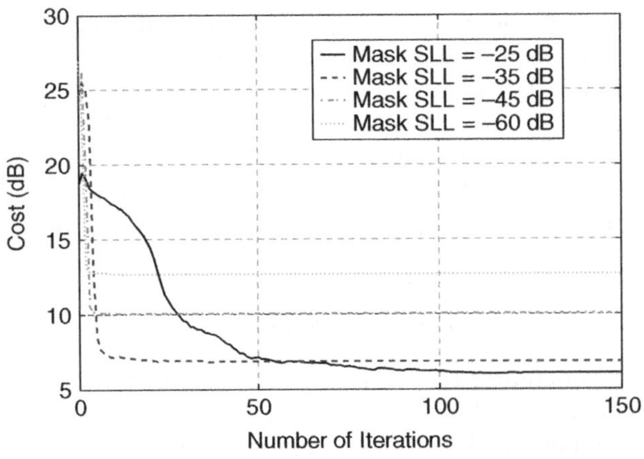

Figure 9.10 Convergence curves of the APM optimization for different SLL requirements.

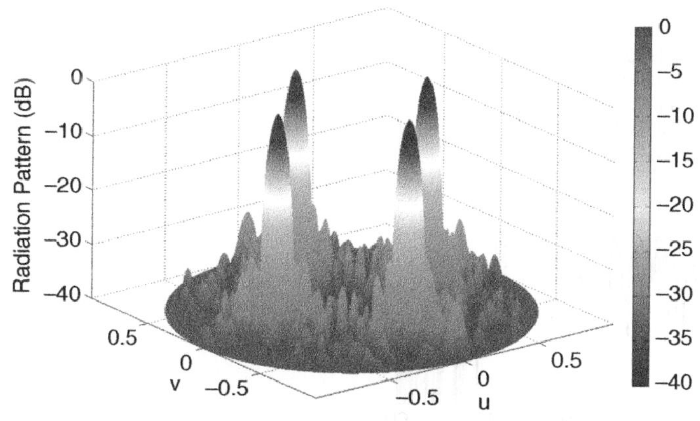

Figure 9.11 Radiation pattern of the optimized quad-beam reflectarray at 32 GHz. *Source:* Nayeri 2012 [10]. Reproduced with permission of IEEE.

optimization with 30 iterations was 456 s on a 2.2 GHz Intel core Duo CPU with 4GB RAM. This is to be compared with 16.5 s for both direct design methods.

The optimized quad-beam prototype was fabricated on a 20 mil Rogers 5880 substrate. The optimized phase distribution for the reflectarray elements and the photograph of the fabricated array with 848 square patch elements are shown in Figure 9.12.

For pattern measurements, in order to avoid blockage from the supporting strut of the feed horn, the array was rotated 45° in the reflectarray plane so the main beams are in the directions of ($\theta_{1,2,3,4} = 30°$, $\varphi_1 = 45°$, $\varphi_2 = 135°$, $\varphi_3 = 225°$, $\varphi_4 = 315°$). Since dual-linear square patch elements are used in this design, the reflected co-polarized radiation of the reflectarray system is RHCP. The measured co-polarized radiation patterns are shown in Figure 9.13. The simulated radiation pattern of the single-beam reference design is also plotted for comparison. For the quad-beam design, the simulated radiation

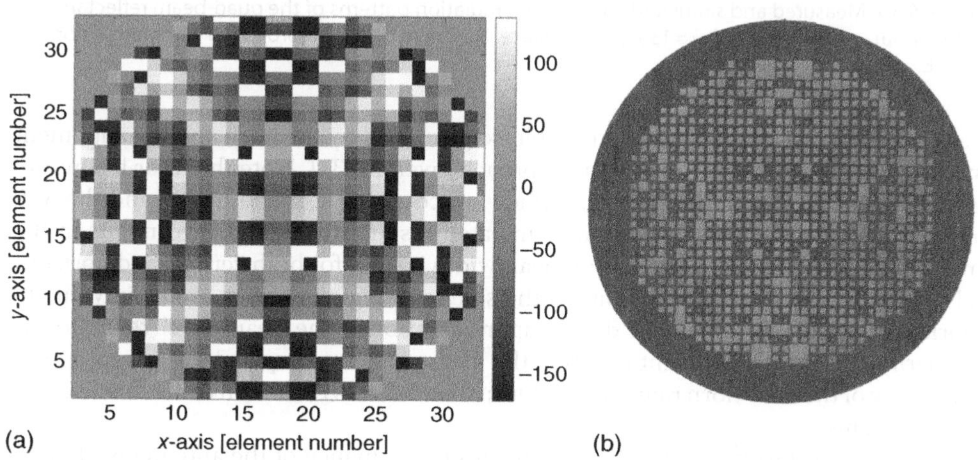

Figure 9.12 (a) Optimized phase distribution of the reflectarray elements. (b) Fabricated quad-beam reflectarray. *Source:* Nayeri 2012 [10]. Reproduced with permission of IEEE.

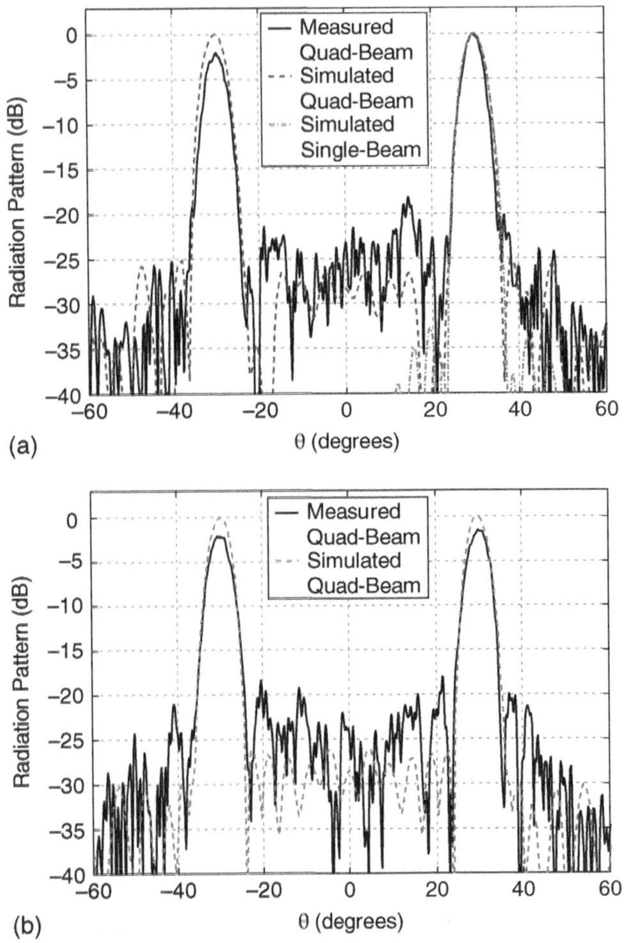

Figure 9.13 Measured and simulated co-polarized radiation patterns of the quad-beam reflectarray antenna: (a) $\varphi = 45°$ plane, (b) $\varphi = 135°$ plane. *Source:* Nayeri 2012 [10]. Reproduced with permission of IEEE.

patterns here also include the aperture blockage caused by the horn and the alignment system, which is calculated using the approach given in [34]. Note that four beams are generated in the required directions, which are correctly scanned to 30°, and the SLL are below −18 dB. The measured and simulated results show good agreements in the main lobes where the measured −3 dB beamwidth is 4.35° for both vertical and horizontal planes. Some discrepancies exist in the side-lobe regions, which are mainly due to fabrication errors and element design approximations. The beam level reduction is primarily due to the alignment errors of the measurement setup and the azimuth non-symmetry of the feed-horn radiation pattern; however, this reduction is less than 2.15 dB for any beams.

As in all reflectarray designs, the bandwidth performance of the antenna is of paramount importance. In a multi-beam design the average performance of all the beams needs to be considered for gain, efficiency, and bandwidth analysis. In [10], the authors

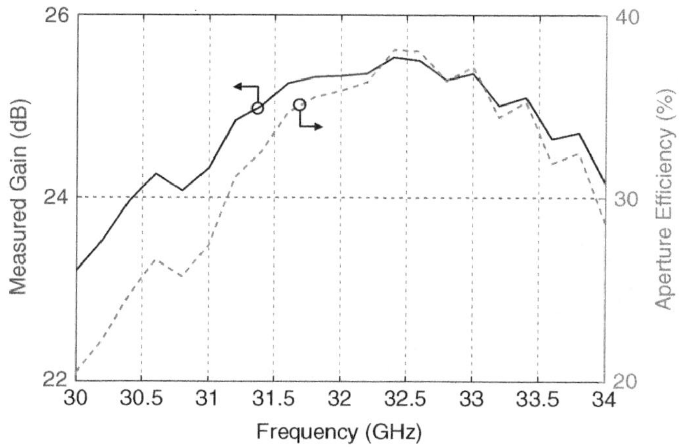

Figure 9.14 Measured gain and efficiency of the quad-beam reflectarray antenna. *Source:* Nayeri 2012 [10]. Reproduced with permission of IEEE.

proposed a modified definition of aperture efficiency for multi-beam designs that takes into account the performance of all beams. This is given by

$$\eta_a = \sum_{i=1}^{N} \frac{\lambda^2 G_i}{4\pi A}, \tag{9.12}$$

where N represents the number of beams and A is the aperture area. This definition takes into account the measured gains of all four beams, and the aperture efficiency for this quad-beam antenna was calculated to be 35.26%. The measured gain and aperture efficiency versus frequency are given in Figure 9.14. The authors mentioned that all beams showed an almost similar gain bandwidth performance, thus the results for the beam in $\theta = 30°$, $\varphi = 45°$ was shown as a reference. At 32 GHz, the measured gain is 25.3 dB, and the 1 dB gain bandwidth is 8.6%.

Another interesting concern in bandwidth analysis of multi-beam antennas is the beam performance over the band. The bandwidth of a reflectarray antenna is usually defined by the 1-dB gain bandwidth [35], [36]. For multi-beam reflectarrays, however, the practical bandwidth of the antenna is also limited by the fact that the beams shift with frequency. This is due to the fact that the main beam direction depends on the total progressive phase on the aperture (including all time-delay effects from feed radiation and total reflection phase of the elements). It was shown in [37] that beam squint can be minimized in reflectarray antennas by enforcing the condition $\theta_o = \theta_i$, where θ_i is defined as the angle from the phase center of the feed to the center of the array and θ_o is the main beam direction. It is clear that this condition cannot be satisfied for multi-beam designs and beam squint would become a limiting factor. The measured radiation patterns at the center and extreme frequencies are given in Figure 9.15 for the $\varphi = 45°$ plane. Similar results were observed in the orthogonal plane. The radiation patterns show quad-beam patterns across the entire band with a slight increase in side-lobe level at the extreme frequencies. The beam squint across the 1 dB gain bandwidth of the antenna is about 2.5°. Considering that the beamwidth of the antenna is more than 4°, the effect of this beam squint is acceptable for this quad-beam prototype.

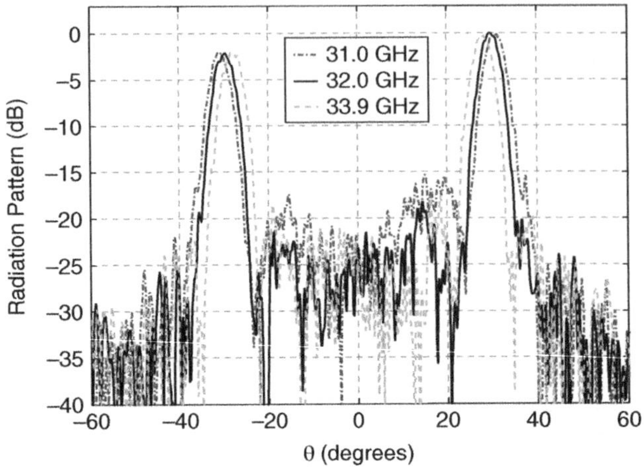

Figure 9.15 Measured radiation patterns of the reflectarray antenna across the 1 dB gain band in the $\varphi = 45°$ plane. *Source:* Nayeri 2012 [10]. Reproduced with permission of IEEE.

While the quad-beam prototype here did not show a very large beam squint across the band, it should be pointed out that in general multi-beam reflectarrays, beam squint could limit the operating band of the antenna. Thus, for general multi-beam designs where a minimum beam squint requirement is specified, additional constraints over the frequency bandwidth needs to be imposed when optimizing the phase distribution of the elements.

9.3.2 Feed Reflectarrays with Multiple Asymmetric Beams

Multi-beam reflectarrays with asymmetric beams, that is, beams in arbitrary non-symmetric directions and/or with different beam levels, present a very complex design problem. The non-convexity of such optimization problems typically requires one to use global-search methods for synthesis. In [8], [38], and [39], several designs of non-symmetric beam directions were studied where the PSO method was used to synthesize the phase distributions on the aperture.

In [38], three designs were presented. A symmetric quad-beam design, similar to the one studied in the previous section was designed using the PSO method, which showed a similar performance as the APM design. A second reflectarray was designed to have four simultaneous beams in one plane. This type of multiple beam antenna is primarily designed for radar applications. While this design is also symmetric, the optimization problem is quite complex, and the APM technique was not able to achieve the required performance. A multi-beam reflectarray with asymmetric beam directions and different gain levels was also demonstrated in [38], where again PSO showed to be a very effective synthesis approach. The results presented in [38] clearly illustrated the need for implementing global-search approaches for complex multi-beam reflectarray designs. While a local-search technique such as APM can be used for simpler problems, for complex designs the problem with local minima is more challenging, and some approaches that can circumvent the local minima problem may be required. In this

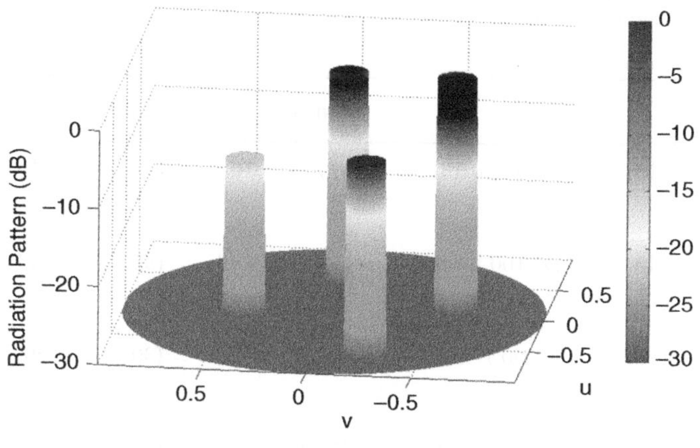

Figure 9.16 Upper bound mask (M_U) for an asymmetric multi-beam reflectarray antenna. *Source:* Nayeri 2013 [39]. Reproduced with permission of IEEE.

section, we will systematically study the design of asymmetric multi-beam reflectarrays and present a practical example of a single-feed quad-beam asymmetric design using the powerful PSO method.

As discussed earlier, in order to implement the optimization, the first step is to define the far-field radiation pattern mask based on the design requirements, where the mask definition follow the formulations presented earlier. A 2D figure of a mask model for a multi-beam design with different beam levels is given in Figure 9.16. In [39], the authors used a modified fitness function for the optimization, which is given by

$$Cost = \sum_{\substack{(u,v)\in \text{mainbeam}\\ \text{and}\,|F(u,v)|<M_L(u,v)}} \sum \left(|F(u,v)| - M_L(u,v) \right)^2 + \sum_{\substack{(u,v)\notin \text{mainbeam}\\ \text{and}\,|F(u,v)|>M_U(u,v)}} \sum \left(|F(u,v)| - M_U(u,v) \right)^2. \tag{9.13}$$

This function takes into account the performance of every point in the visible space, that is, $u^2 + v^2 \leq 1$. The first term in the fitness function calculates the absolute difference between the radiation pattern and the lower bound; in other words, the beam performance. The second term evaluates the side-lobe performance of the array for every point that does not belong to the main beam area.

Four different designs were studied in [39]. In all designs the reflectarray has a circular aperture with a diameter of 17λ at the design frequency, corresponding to 848 phasing elements with an element spacing of λ/2. The feed is prime focus and placed with an F/D = 0.725. The power q of the feed-horn $\cos^q \theta$ radiation pattern model is 6.5, which will give an edge taper below −12 dB. The beam requirements for these multi-beam designs are summarized in Table 9.2. It is important to note that, while in general, the beamwidth and gain level for multi-beam antennas are determined based on the design requirements, such as coverage area for satellite applications, in the designs studied here all beams are set to have beamwidths identical to a reference single-beam design.

Table 9.2 Design requirements of quad-beam reflectarrays (beam directions and normalized gain levels).

Design	Beam 1	Beam 2	Beam 3	Beam 4
Case I: Symmetric beam directions and equal beam levels	(30°, 0°) 0 dB	(30°, 90°) 0 dB	(30°, 180°) 0 dB	(30°, 270°) 0 dB
Case II: Asymmetric beam directions and equal beam levels	(30°, 0°) 0 dB	(25°, 100°) 0 dB	(45°, 200°) 0 dB	(35°, 280°) 0 dB
Case III: Symmetric beam directions but different beam levels	(30°, 0°) −3 dB	(30°, 90°) −10 dB	(30°, 180°) −5 dB	(30°, 270°) 0 dB
Case IV: Asymmetric beam directions and different beam levels	(30°, 0°) −3 dB	(25°, 100°) −10 dB	(45°, 200°) −5 dB	(35°, 280°) 0 dB

In addition, for all designs the mask side-lobe level was set to −30 dB. A 3D figure of the upper bound mask model for the multi-beam design with asymmetric beam directions and different beam levels (Case IV) is given in Figure 9.16.

With the design requirements specified, the task was then to initialize the PSO parameters for the optimization problem. The inertial weight was varied linearly from 0.9 to 0.4 over the course of the run and the self- and group-knowledge constants were both set equal to 2 throughout the optimization. A swarm population of 400 particles, roughly half the size of the 848-dimensional solution hyperspace, was selected for the optimization and reflecting wall boundary conditions were used [40]. While other boundary conditions such as invisible or damping walls may improve the performance of the optimization, the reflecting walls used in these designs yielded satisfactory results. For each design 100,000 iterations were performed, corresponding to 40 million fitness evaluations. The computation time for pattern calculation depends on the far-field resolution. For arrays with about 1000 elements, nominally 30 dB gain, a far-field pattern with 400 × 400 points evenly spaced in the angular coordinates is usually sufficient and this value was used for all designs studied in [39]. The lower and upper bounds of the variable; that is, element phases, are kept within $[0, 2\pi]$. The authors mentioned that with their reflectarray radiation analysis code, the computation time for each pattern calculation was about 4 ms on a 2.33 GHz Intel I7 computer, corresponding to a total computation time of about 44 h on a single core. It is also important to note that depending on the design requirements it is possible to utilize the symmetry of the problem to reduce the dimension of the optimization problem; however, the focus of the study in [39] was on asymmetric multiple beam designs, therefore no symmetry boundary was implemented in the optimizations.

The convergence curves for the optimizations are given in Figure 9.17. It can be seen that in all cases, although the solution doesn't convergence initially, eventually it escapes the optimization traps. It is worthwhile to point out that it is possible to improve the solution convergence by using a larger swarm population; however, this will come at the cost of more fitness evaluations and computation time.

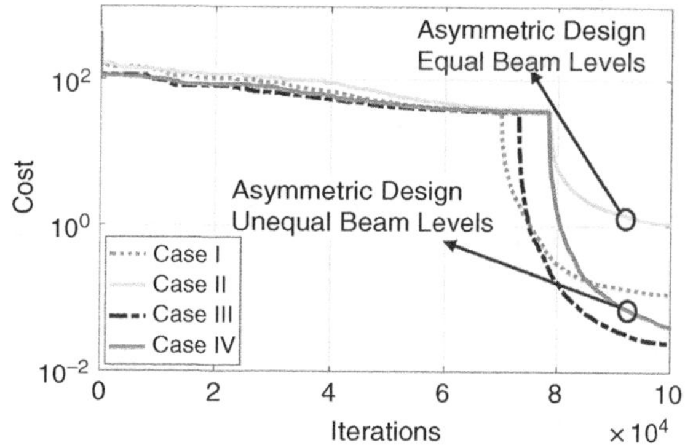

Figure 9.17 Convergence curves of the PSO for multi-beam reflectarray antenna design. *Source:* Nayeri 2013 [39]. Reproduced with permission of IEEE.

The first design (Case I) is a symmetric quad-beam reflectarray, similar to the one reported in [10]. The second design (Case II) is an asymmetric quad-beam radiation performance. With the fixed number of iterations used in the optimizations, the fitness improvements for these two designs were significant. Different beam levels for multi-beam reflectarrays were also studied in the last two designs. The monotonous decrease in fitness improvement for the optimizations is a good indicator of the complexity of the asymmetric design problems.

The optimized phase distributions and radiation patterns for the four multi-beam designs are given in Figure 9.18 and Figure 9.19, respectively. It can be seen that all multi-beam reflectarrays achieved the design requirements, which demonstrates the effectiveness of the PSO for multi-beam designs.

In [39], the authors also showed the shortcomings of a local search by implementing the APM method for synthesis of an asymmetric design (Case IV). The aperture phase distribution and radiation pattern obtained using the APM optimization are given in Figure 9.20. It can be seen that for this case, the radiation performance of the optimized design is not satisfactory. Beam direction, gain levels, and SLL cannot satisfy the design requirements, and in particular, some undesired lobes are generated that have a high beam level. In general, while the APM technique is quite advantageous for simple designs such as Case I, the optimization problem is quite challenging for asymmetric cases, and a good multi-beam performance cannot be achieved with this approach. Moreover, while in comparison with the APM method, PSO requires a much higher computation time, the main strength of this global-search approach is that it has the capability to escape the traps in the optimization process, which makes it a generalized approach for general multi-beam designs.

The quad-beam reflectarray antenna with asymmetric beam directions and unequal gain levels (Case IV) was fabricated and tested. The prototype is fabricated on a 20-mil Rogers 5880 substrate and the reflectarray has a circular aperture with a diameter of 16 cm. The phasing elements are variable size square patches, identical to the ones used in [10], with a unit-cell periodicity of 4.7 mm. The patch dimensions were selected assuming a normal incidence excitation for the elements, and the phase on the reflectarray aperture with these patch elements is shown in Figure 9.21(a). Comparing this with the ideal phase distribution in Figure 9.18(d), it can be seen that despite some minor differences, these patch elements are providing the necessary phase shift on the reflectarray aperture. The mean value of the aperture phase error is 7.6°. Photographs

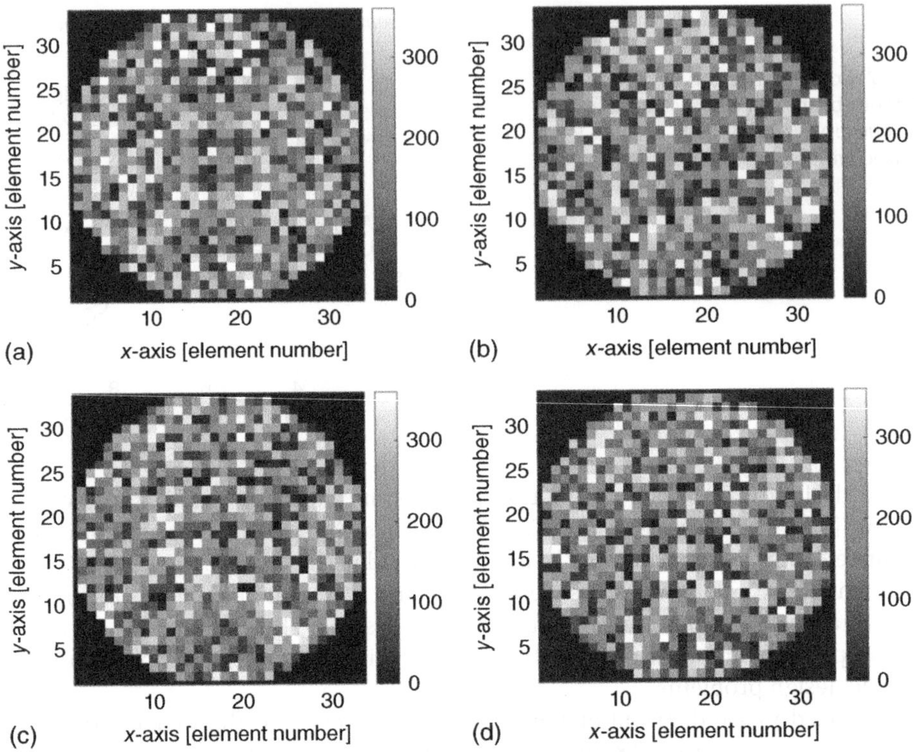

Figure 9.18 Optimized phase distributions for quad-beam reflectarrays: (a) Case I, (b) Case II, (c) Case III, and (d) Case IV. *Source:* Nayeri 2013 [39]. Reproduced with permission of IEEE.

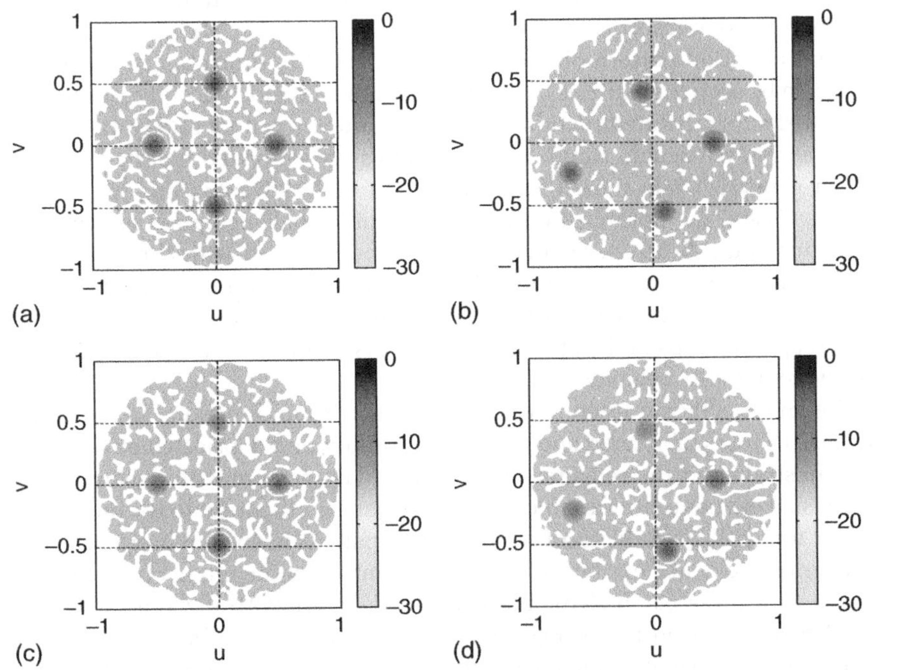

Figure 9.19 Radiation patterns of optimized multi-beam reflectarrays: (a) Case I, (b) Case II, (c) Case III, and (d) Case IV. *Source:* Nayeri 2013 [39]. Reproduced with permission of IEEE.

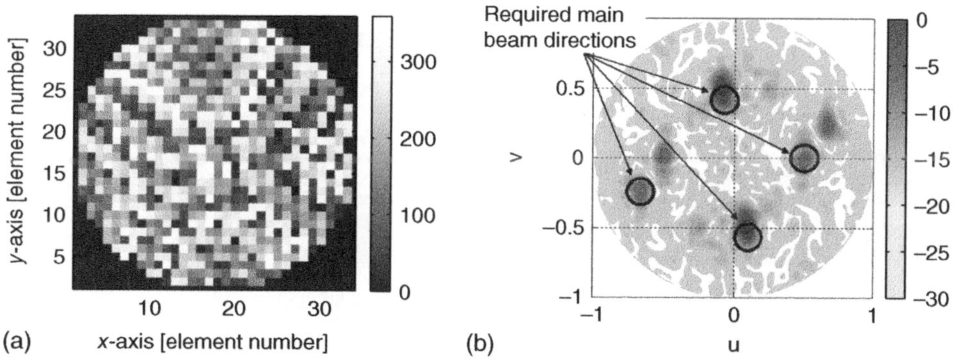

Figure 9.20 An asymmetric quad-beam reflectarray antenna designed using the APM method, which does not satisfy the design requirements: (a) phase distribution, (b) radiation pattern. *Source:* Nayeri 2013 [39]. Reproduced with permission of IEEE.

of the fabricated reflectarray with 848 square patch elements and the reflector system are shown in Figure 9.22.

The simulated and measured radiation patterns of this prototype at 32 GHz are compared in Figure 9.23, where a close agreement between these results can be observed. The measured side-lobe level around each beam is about −22 dB; however, some areas show slightly higher side-lobes. A similar performance was observed for the symmetric quad-beam prototype reported in [10], which was primarily attributed to the blockage effects of the feed and alignment system.

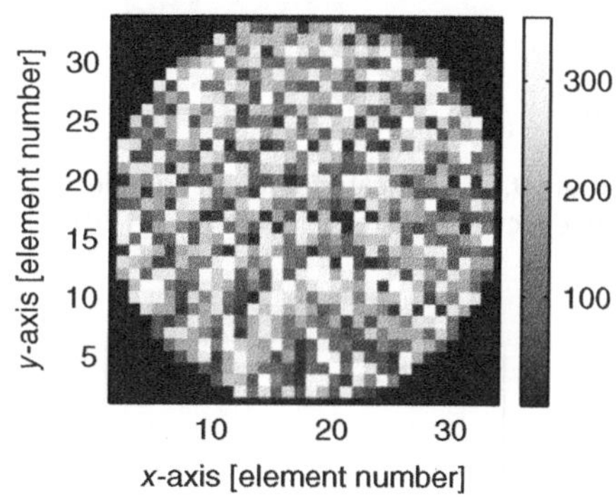

Figure 9.21 Phase distribution on the reflectarray aperture using variable size square patch elements. *Source:* Nayeri 2013 [39]. Reproduced with permission of IEEE.

As with the symmetric case, bandwidth analysis of multi-beam reflectarrays is a bit more complex. In addition, while for symmetric multiple beams, the gain performance of all the beams are similar, this is not the case for asymmetric designs and the frequency behavior of each beam must be studied. The measured gain performance of the asymmetric multi-beam prototype in [39] is given in Table 9.3, where it can be seen that the widest bandwidth is observed for the beam with highest gain, that is, beam 4. Comparison of these results also shows that the narrowest bandwidth is observed for the beam scanned furthest from broadside, beam 3. It is worthwhile to point out that the beam with lowest gain also shows a narrow bandwidth. As a summary, the results in Table 9.3 indicate that for the single-feed asymmetric multi-beam design, beams scanned further from broadside or with lower gain level will show a narrow bandwidth. The measured gain for beam 4 and the aperture efficiency (computed using 9.12) as a function of frequency are given in Figure 9.24. The peak aperture efficiency is about 36%, which is slightly lower than the efficiency reported for the symmetric design in [10].

As discussed earlier, to accurately determine the bandwidth, it is necessary to also study the beam directions as a function of frequency. The measured beam squint performance of the prototype is summarized in Table 9.4 where f_L and f_H represent the

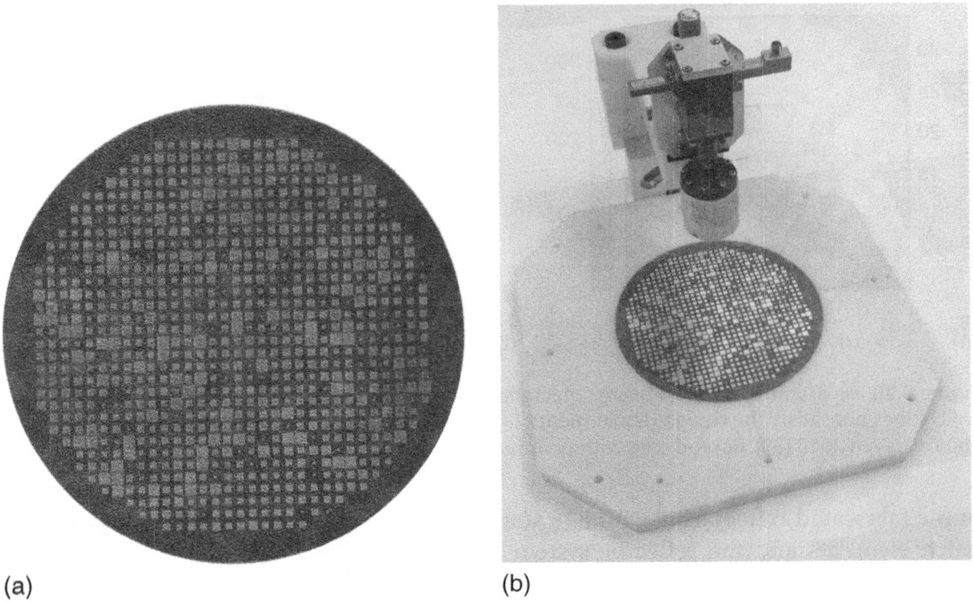

Figure 9.22 Fabricated single-feed reflectarray antenna with asymmetric multi-beam: (a) array, (b) reflectarray system. *Source:* Nayeri 2013 [39]. Reproduced with permission of IEEE.

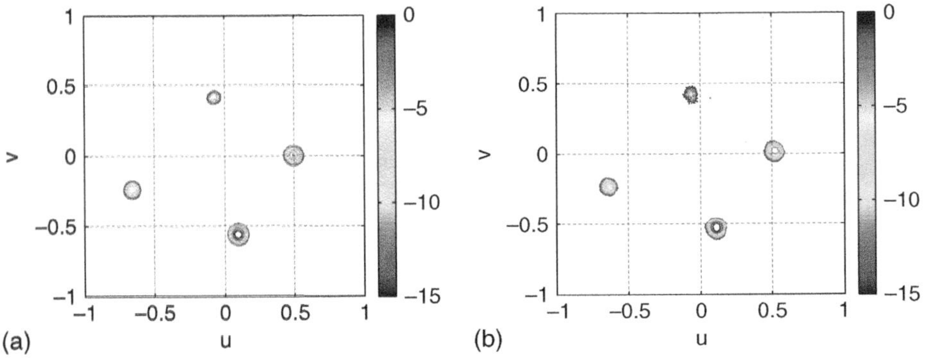

Figure 9.23 Radiation pattern of the asymmetric quad-beam reflectarray prototype at 32 GHz: (a) simulated result, (b) measured result. *Source:* Nayeri 2013 [39]. Reproduced with permission of IEEE.

Table 9.3 Measured gain of the asymmetric multi-beam reflectarray prototype.

Beam	Gain @ 32 GHz (dB)	Max Gain	1 dB Gain Bandwidth
1	24.18	24.41 dB @ 32.4 GHz	31.3–33.2 GHz (5.89%)
2	17.89	17.89 dB @ 32.0 GHz	31.1–32.7 GHz (5.02%)
3	22.04	22.29 dB @ 32.4 GHz	31.5–33.1 GHz (4.95%)
4	26.91	27.27 dB @ 32.4 GHz	31.3–33.4 GHz (6.49%)

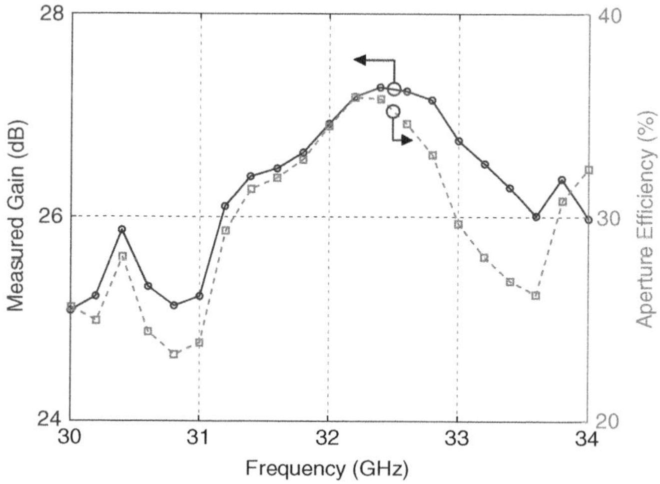

Figure 9.24 Measured gain and efficiency of the asymmetric quad-beam reflectarray antenna prototype. *Source:* Nayeri 2013 [39]. Reproduced with permission of IEEE.

Table 9.4 Measured beam performance across the band for the prototype.

	Beam Direction ($\theta°,\varphi°$)			Beam Squint	
Beam	f_L	**32.0 GHz**	f_H	$\Delta\theta$	$\Delta\varphi$
1	(32.0, 1.4)	(31.5, 0.7)	(30.2, 1.3)	1.8°	0.7°
2	(26.8, 97.6)	(24.9, 98.5)	(24.6, 95.0)	2.2°	3.5°
3	(44.0, 199.8)	(42.9, 199.2)	(41.7, 199.3)	2.3°	0.6°
4	(33.2, 281.1)	(32.4, 280.7)	(30.6, 281.7)	2.6°	1.0°

lower and upper frequencies across the 1 dB gain bandwidth of each beam. These results show that the beam squint observed along the elevation direction θ is almost similar in all cases; however, a noticeable difference is observed in the beam squint along φ direction for beam 2. Nonetheless, the maximum beam squint observed across the gain bandwidth of each beam is less than 3.5° in either direction. Considering that the beamwidth of the antenna is more than 4°, the effect of this beam squint is quite acceptable for this quad-beam prototype and beam squint is not limiting the operational bandwidth of the design. The measured radiation patterns of this prototype at two off-center frequencies are given in Figure 9.25.

The bandwidth studies presented in [39] showed that for this single-feed asymmetric multi-beam reflectarray, the beam scanned furthest from broadside direction observed the narrowest gain bandwidth and a large beam squint. In addition, beam level difference also influences the bandwidth performance, and the beam with lowest gain level showed a narrow gain bandwidth and the largest beam squint. Based on this study, it is expected that these two factors are the primary causes of bandwidth reduction in single-feed

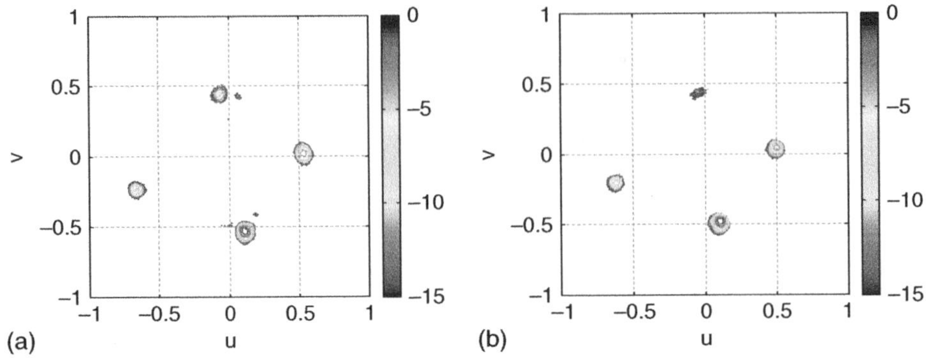

Figure 9.25 Measured radiation patterns of the asymmetric quad-beam reflectarray prototype at off-center frequencies: (a) 31 GHz, (b) 33 GHz. *Source:* Nayeri 2013 [39]. Reproduced with permission of IEEE.

asymmetric multi-beam reflectarray antennas. As such, if it is possible, it would be advantageous to have a higher gain for the beams that are scanned further from broadside. It is also worthwhile to point out that if further improvement of bandwidth is required for the design, one may use the procedure outlined in this chapter (local or global-search techniques) to obtain the ideal aperture phase profile across the band of interest, and use the approach given in [41] to select the appropriate elements.

9.3.3 Shaped-Beam Reflectarrays

The shaped reflector antenna is a mature technology, both in terms of manufacturing and simulation tools, and is therefore used in many space missions to fulfill the coverage, cross-polarization, and isolation specifications [29]. However, as discussed earlier, these reflectors suffer from large volume and mass, as well as high cost of the manufacturing, in particular the mold, which depends on the antenna requirements and can therefore not be reused for other missions.

Printed reflectarrays have a flat surface, they are light, and for a specific coverage only the array elements are varied, thus many of the recurring costs associated with shaped reflector antennas can be eliminated. The concept and demonstration of the shaped-beam microstrip reflectarray was originally reported in 1994 [32], but publication was postponed in order to protect the proprietary nature of this work. A microstrip reflectarray with a shaped-beam pattern was experimentally demonstrated in [3]. The aim of this study was to compare the performance of this reflectarray with that of a conventional shaped reflector antenna designed to produce the same shaped beam.

The shaped-reflector antenna was designed at Spar Aerospace, and coordinates defining the contour of the shaped reflector were available for use in the design of the microstrip reflectarray. This shaped reflector was designed to illuminate continental Europe with a shaped coverage pattern at two 5% frequency bands at 14 and 18 GHz. The original design employed two superimposed gridded and shaped reflectors, with each gridded shell handling a separate polarization. The reflectors had elliptical projected apertures with major and minor axes of 110 and 90 cm, respectively, with focal lengths of 76 cm. The commercial reflector design software (Ticra's POS program) was used to design the shaped-reflector antenna.

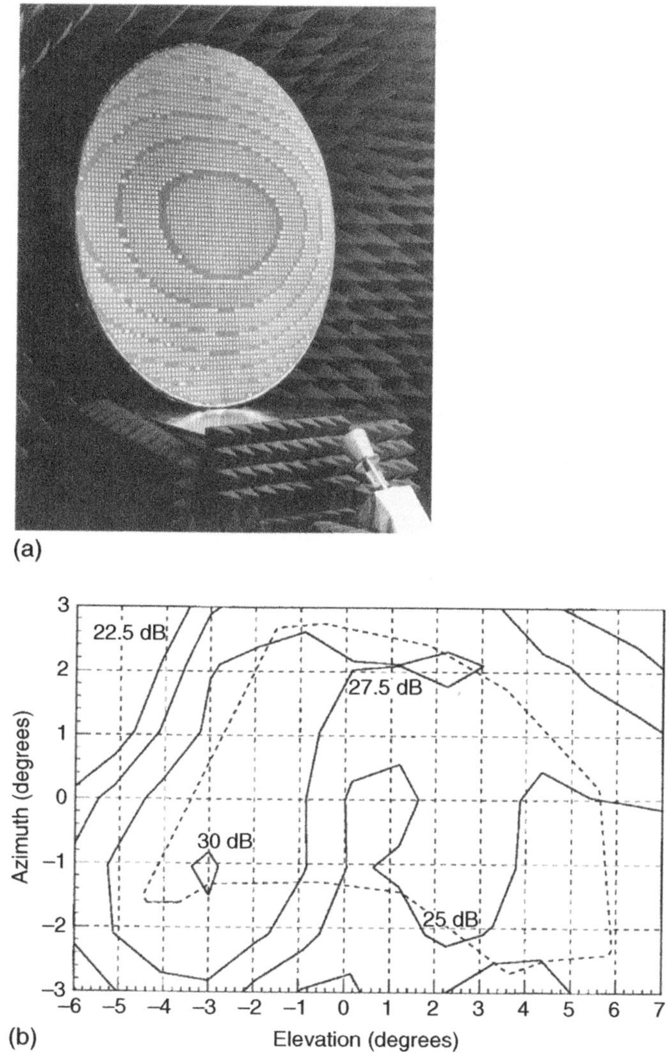

Figure 9.26 Shaped-beam reflectarray: (a) fabricated prototype, (b) measured radiation patterns. *Source:* Pozar 1999 [3]. Reproduced with permission of IEEE.

The Ku-band reflectarray was designed with the same aperture size and coverage specification as a traditional shaped reflector. Ray tracing was used to determine the necessary reflection phases of the reflectarray elements based on the shaped-reflector antenna, and the phase at each reflectarray element was obtained from the distance between the shaped surface and the planar reflectarray. Then, an elliptical reflectarray with axes 110 and 90 cm was designed, manufactured and measured. Variable size patches were used for the elements. Photograph of the prototype and measured patterns of the antenna at 14.15 GHz are shown in Figure 9.26. While this shaped-beam reflectarray demonstrated the feasibility of the concept, it was suggested that a direct array synthesis method to obtain the phase distribution on the reflectarray without the previous design of a shaped reflector will improve the performance.

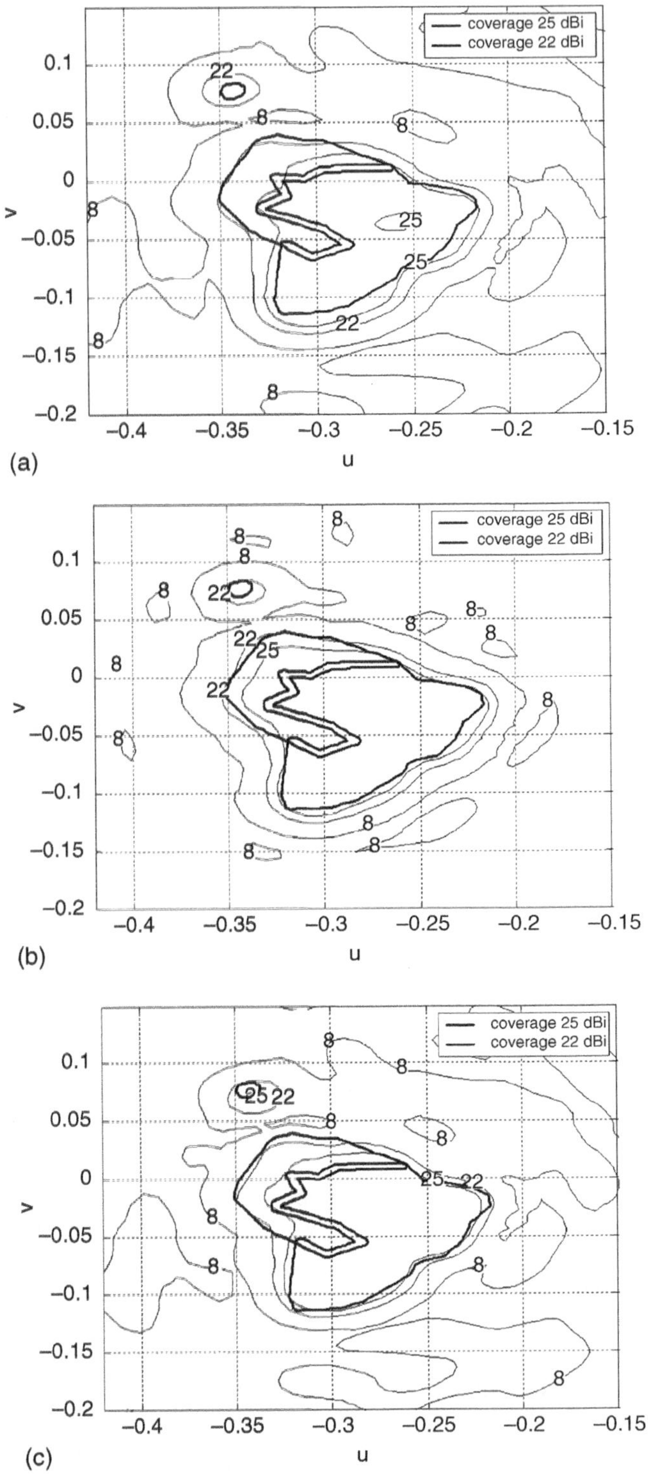

Figure 9.27 Radiation patterns at central and extreme frequencies for the reflectarray optimized in the 12.8–14.2 GHz band. (a) 12.8 GHz. (b) 13.5 GHz. (c) 14.2 GHz. *Source:* Encinar 2004 [11]. Reproduced with permission of IEEE.

Applications of contoured beam reflectarrays are especially interesting since the manufacturing cost and process is the same as for a pencil beam, and several other designs of printed contoured beam reflectarrays have been presented over the years. Most shaped-beam reflectarrays are designed using phase-only synthesis methods [11], although a number of designs using direct approaches have also been demonstrated [29].

As with most reflectarray designs, a major concern is the narrow band of the reflectarray elements, which limits the antenna performance. This limitation can be overcome by using stacked patches of variable sizes [41]. In [11], a three-layer contoured beam reflectarray was designed for DBS application to radiate a beam for dual-polarization coverage of South America and Florida with a 10% bandwidth. First, the patch dimensions were adjusted to produce the objective phase distribution at the central frequency, and then an optimization technique based on the Fletcher–Powell algorithm was implemented that adjusts all the dimensions of the stacked patches simultaneously to match both objective phase and phase delay difference for the two linear polarizations. Radiation patterns at central and extreme frequencies are shown in Figure 9.27.

Multi-frequency single-polarized and single-frequency dual-polarized shaped reflectarrays were demonstrated in [29] using a direct optimization approach. For the latter design, a high-gain beam for a European coverage with cross-polar suppression within the same coverage, and side-lobe suppression within the southern African contour was considered, and the reflectarray was optimized for two orthogonal linear polarizations, H- and V-polarization at 10 GHz. A photo of the antenna in the test setup, and simulated and measured radiation patterns at 10 GHz are shown in Figure 9.28 and Figure 9.29, respectively. The agreement between simulations and measurements is extremely good for both polarizations, and the high-gain contours curves practically coincide.

9.3.4 Multi-Feed Multi-Beam Reflectarrays

In addition to the multiple beam designs introduced earlier that use a single-feed antenna, reflectarrays can also achieve a multi-beam performance by using multiple feed antennas. In the case of one feed per beam, the configuration is essentially similar

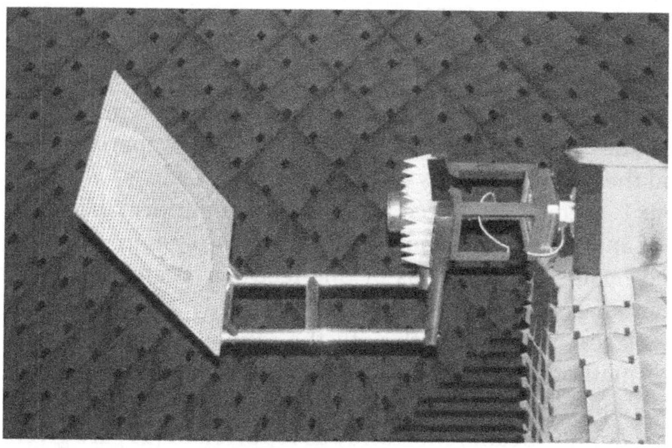

Figure 9.28 Shaped-beam reflectarray antenna in the DTU-ESA spherical near-field antenna test facility. *Source:* Zhou 2013 [29]. Reproduced with permission of IEEE.

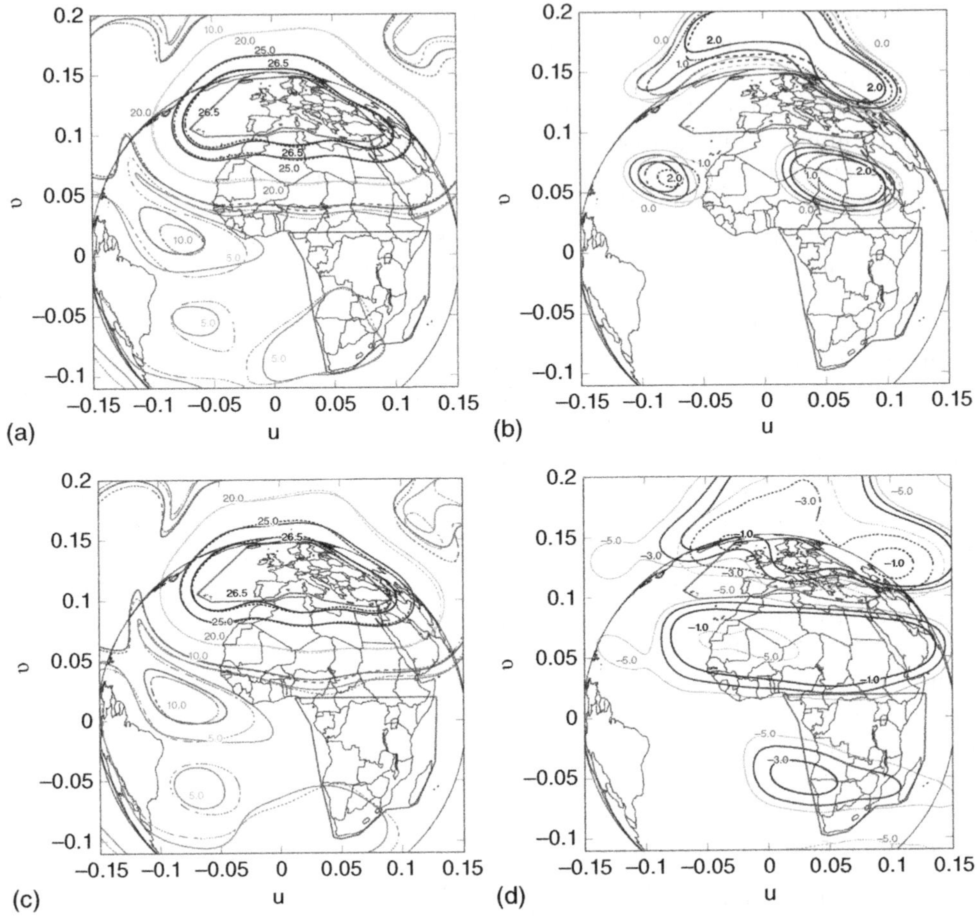

Figure 9.29 Simulated (solid lines) and measured (dotted lines) radiation patterns of the manufactured breadboard for both H- and V-polarizations at 10 GHz. (a) H-polarization, co-polar; (b) H-polarization, cross-polar; (c) V-polarization, co-polar; and (d) V-polarization, cross-polar. *Source:* Zhou 2013 [29]. Reproduced with permission of IEEE.

to multi-fed reflectors. In [2], a four-beam X-band design was demonstrated for SAR applications in microsatellites. The phase distribution on the reflectarray aperture was designed for a single beam, and the rest of the feeds were positioned off-focus. The four beams covered a range of about 7° with a rather acceptable radiation performance, primarily due to the small range of angular coverage and large F/D of the system.

Several simultaneous shaped beams, on the basis of one feed per beam were also demonstrated in [19]. The antenna was designed to generate three independent shaped beams that covered adjacent 30° sectors in azimuth with the same squared cosecant pattern in elevation. The APM method was used for synthesis, and patch dimensions were adjusted to produce the required phase for both polarizations. Measured and simulated patterns for this shaped multi-beam reflectarray as reported in [19] are given in Figure 9.30. This antenna again demonstrates the multiple shaped-beam capability of reflectarray antennas.

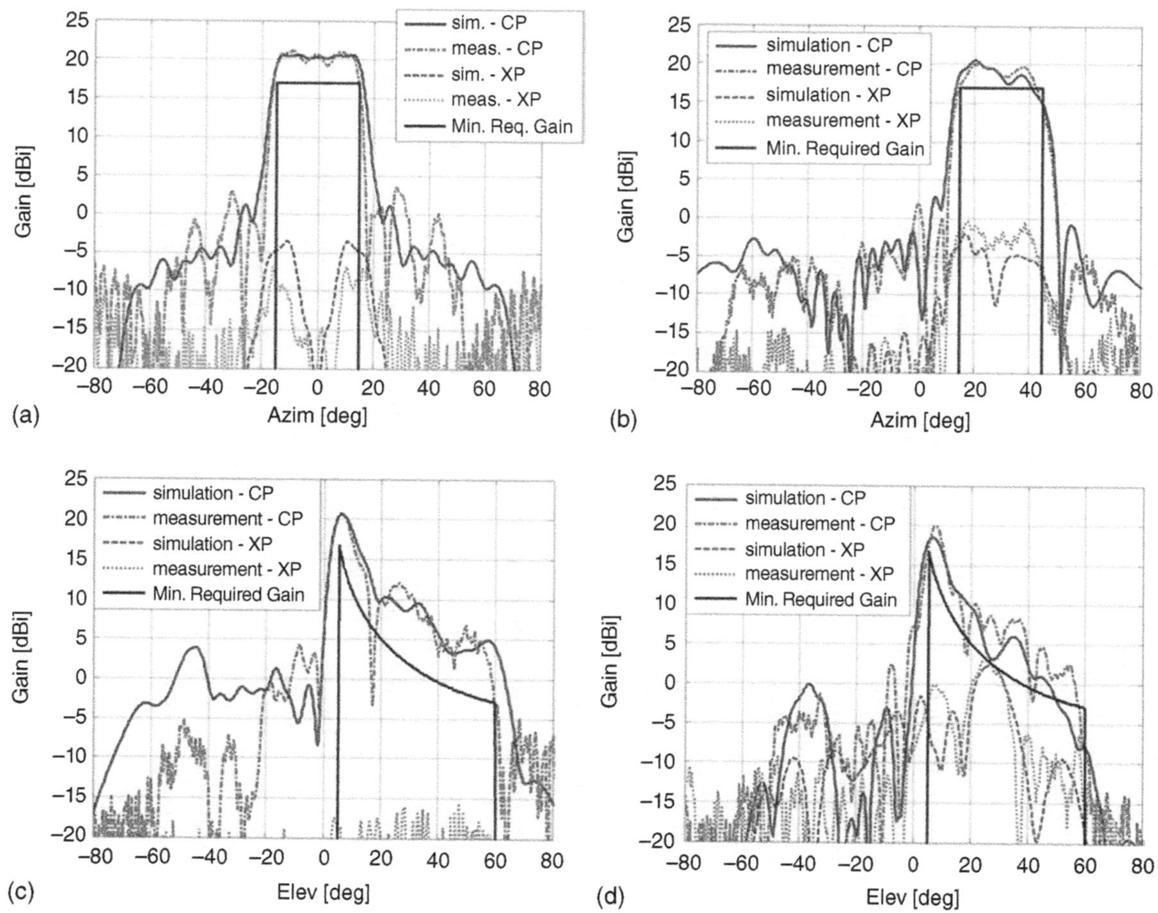

Figure 9.30 Simulations (co-polar and cross-polar in solid and dashed lines, respectively) and measurements (co-polar and cross-polar in dash-dot and dotted lines, respectively) at the central frequency (25.5 GHz). Central beam in H-polarization (a) and (c) and lateral beam in V-polarization (b) and (d). Main cuts in azimuth (a) and (b) and elevation (c) and (d). Minimum required gain in solid line. *Source:* Yang 2012 [19]. Reproduced with permission of IEEE.

References

1 D. M. Pozar, S. D. Targonski, and H. D. Syrigos, "Design of millimeter wave microstrip reflectarrays," *IEEE Trans. Antennas Propag.*, Vol. 45, No. 2, pp. 287–296, Feb. 1997.

2 J. Huang and J. A. Encinar, *Reflectarray Antennas*, by Institute of Electrical and Electronics Engineers, John Wiley & Sons, Ltd, 2008.

3 D. M. Pozar, S. D. Targonski, and R. Pokuls, "A shaped-beam microstrip patch reflectarray," *IEEE Trans. Antennas Propag.*, Vol. 47, pp. 1167–1173, July 1999.

4 R. C. Hansen, *Phased Array Antennas*, Wiley Series in Microwave and Optical Engineering, Chichester, UK: John Wiley & Sons, Ltd, 1998.

5 P. Balling, K. Tjonneland, L. Yi, and A. Lindley, "Multiple contoured beam reflector antenna systems," *IEEE Antennas and Propagation Society International Symposium*, Michigan, USA, Jun. 28–Jul. 2, 1993.

6 L. Schulwitz and A. Mortazawi, "A compact dual-polarized multibeam phased-array architecture for millimeter-wave radar," *IEEE Trans. Microw. Theory Tech.*, Vol. 53, No. 11, pp. 3588–3594, Nov. 2005.

7 J. Lanteri, C. Migliaccio, J. Ala-Laurinaho, M. Vaaja, J. Mallat and A. V. Raisanen, "Four-beam reflectarray antenna for mm-waves: design and tests in far-field and near-field ranges," *EuCAP 2009*, Berlin, Germany, pp. 2532–2535, March 2009.

8 P. Nayeri, *Advanced Design Methodologies and Novel Applications of Reflectarray Antennas*, Ph.D. dissertation, Dept. Eng. Sci., Univ. of Mississippi, Jul. 2012.

9 P. Nayeri, F. Yang, and A. Z. Elsherbeni, "Single-feed multi-beam reflectarray antennas," *IEEE Antennas and Propagation Society International Symposium*, Toronto, Canada, July 2010.

10 P. Nayeri, F. Yang, and A. Z. Elsherbeni, "Design of a single-feed quad-beam reflectarray antenna," *IEEE Trans. Antennas Propag.*, Vol. 60, No. 2, pp. 1166–1171, Feb. 2012.

11 J. A. Encinar and J. A. Zornoza, "Three-layer printed reflectarrays for contoured beam space applications," *IEEE Trans. Antennas Propag.*, Vol. 52, No. 5, pp. 1138–1148, May 2004.

12 W. L. Stutzman and G. A. Thiele, *Antenna Theory and Design*, 3rd Edn, Hoboken, NJ: John Wiley & Sons, Inc., 2012.

13 R. Fletcher, *Practical Methods of Optimization*, 2nd Edn, Hoboken, NJ: John Wiley & Sons, Inc., 1987.

14 O. M. Bucci, G. Franceschetti, G. Mazzarella, and G. Panariello, "Intersection approach to array pattern synthesis," *IEE Proceedings*, Vol. 137, No. 6, pp. 349–357, Dec. 1990.

15 O. M. Bucci, G. Mazzarella, and G. Panariello, "Reconfigurable arrays by phase-only control," *IEEE Trans. Antennas Propag.*, Vol. 39, No. 7, pp. 919–925, July 1991.

16 O. M. Bucci, G. D'Elia, G. Mazzarella, and G. Panariello, "Antenna pattern synthesis: A new general approach," *Proc. IEEE*, Vol. 82, No. 3, pp. 358–371, Mar. 1994.

17 R. Vescovo, "Reconfigurability and beam scanning with phase-only control for antenna arrays," *IEEE Trans. Antennas Propag.*, Vol. 56, No. 6, pp. 1555–1565, June 2008.

18 D. W. Boeringer and D. H. Werner, "Particle swarm optimization versus genetic algorithms for phased array synthesis," *IEEE Trans. Antennas Propag.*, Vol. 52, No. 3, pp. 771–779, Mar. 2004.

19 M. Arrebola, J. A. Encinar, and M. Barba, "Multifed printed reflectarray with three simultaneous shaped beams for LMDS central station antenna," *IEEE Trans. Antennas Propag.*, Vol. 56, No. 6, pp. 1518–1527, June 2008.

20 R. Horst, P. M. Pardalos and N.V. Thoai, Introduction to Global Optimization, 2nd Edn, Kluwer Academic Publishers, 2000.

21 R. L. Haupt and S. E. Haupt, *Practical Genetic Algorithm*, 2nd Edn, Hoboken, NJ: John Wiley & Sons, Inc., 2004.

22 R. Eberhart and J. Kennedy, "A new optimizer using particle swarm theory," in *Proc. 6th Int. Symp. Micro Machine and Human Science* (MHS '95), pp. 39–43, 1995.

23 J. Kennedy and R. Eberhart, "Particle swarm optimization," *in Proc. IEEE Int. Conf. Neural Networks*, Vol. 4, pp. 1942–1948, 1995.

24 J. Kennedy and R. C. Eberhart, *Swarm Intelligence*. San Francisco, CA: Morgan Kaufmann, 2001.

25 M. Clerc and J. Kennedy, "The particle swarm – Explosion, stability, and convergence in a multidimensional complex space," *IEEE Trans. Evol. Comput.*, Vol. 6, pp. 58–73, Feb. 2002.

26 J. Robinson, and Y. Rahmat-Samii, "Particle swarm optimization in electromagnetics," *IEEE Trans. Antennas Propag.*, Vol. 52, No. 2, pp. 397–407, Feb. 2004.

27 N. Jin, and Y. Rahmat-Samii, "Advances in particle swarm optimization for antenna designs: real-number, binary, single-objective and multiobjective implementations," *IEEE Trans. Antennas Propag.*, Vol. 55, No. 3, pp. 556–567, Mar. 2007.

28 O. M. Bucci, A. Capozzoli, G. D'Elia, P. Maietta, and S. Russo, "An advanced reflectarray design technique," *in Proc. 28th ESA Antenna Workshop*, Noordwijk, The Netherlands, 2005.

29 M. Zhou, S. B. Sørensen, O. S. Kim, E. Jørgensen, P. Meincke, and O. Breinbjerg, "Direct optimization of printed reflectarrays for contoured beam satellite antenna applications," *IEEE Trans. Antennas Propag.*, Vol. 61, No. 4, pp. 1995–2004, April 2013.

30 A. Capozzoli, C. Curcio, A. Liseno, M. Migliorelli, and G. Toso, "Aperiodic conformal reflectarrays," *in Proc. IEEE AP-S Int. Symp.*, Spokane, WA, USA, 2011.

31 A. Capozzoli, C. Curcio, A. Liseno, M. Migliorelli, and G. Toso, "Power pattern synthesis of advanced flat aperiodic reflectarrays," *in Proc. 33th ESA Antenna Workshop*, Noordwijk, The Netherlands, 2011.

32 D. M. Pozar and S. D. Targonski, "Design of a shaped beam microstrip reflectarray," Tech. Report, Antenna Design Associates, Inc., Amherst, MA, Nov. 1994.

33 F. Arpin, J. Shaker, and D.A. McNamara, "Multi-feed single-beam power-combining reflectarray antenna," *Electronics Letters*, Vol. 40, No. 17, pp. 1035–1037, Aug. 2004.

34 L. Diaz and T. Milligan, *Antenna Engineering Using Physical Optics*, Artech House, 1996.

35 J. Huang, "Bandwidth study of microstrip reflectarray and a novel phased reflectarray concept," *IEEE Antennas and Propagation Society International Symposium*, California, USA, June 1995.

36 D. M. Pozar, "Bandwidth of reflectarrays," *Electronics Letters*, Vol. 39, No. 21, Oct. 2003.

37 S. D. Targonski and D. M. Pozar, "Minimization of beam squint in microstrip reflectarrays using an offset feed," *IEEE Antennas and Propagation Society International Symposium*, MD, USA, July 1996.

38 P. Nayeri, F. Yang, and A. Z. Elsherbeni, "Design of single-feed reflectarrays with asymmetric multi-beams," presented at the *IEEE Antennas and Propagation Society International Symposium*, Illinois, U.S., July 2012.

39 P. Nayeri, F. Yang, and A. Z. Elsherbeni, "Design of single-feed reflectarray antennas with asymmetric multiple beams using the particle swarm optimization method," *IEEE Trans. Antennas Propag.*, Vol. 61, No. 9, pp. 4598–4605, Sep. 2013.

40 S. Xu, and Y. Rahmat-Samii, "Boundary conditions in particle swarm optimization revisited," *IEEE Trans. Antennas Propag.*, Vol. 55, No. 3, pp. 760–765, Mar. 2007.

41 J. A. Encinar and J. A. Zornoza, "Broadband design of three-layer printed reflectarrays," *IEEE Trans. Antennas Propag.*, Vol. 51, No. 7, pp. 1662–1664, Jul. 2003.

10

Beam-Scanning Reflectarray Antennas

In addition to the numerous advantages of reflectarrays over reflectors and arrays for fixed beam applications, the reflectarray antenna also becomes a new paradigm for beam-scanning applications. High-gain beam-scanning architectures impose stringent requirements on the antenna radiation capabilities [1], [2]; as such, very few types of antennas are capable of achieving the required performance. In most high-gain beam-scanning applications, the conventional choice is between a reflector (or lens) and a phased array, which is driven by factors relating to scan rate, scan volume, and cost [3]. Reflector antennas are typically more suitable when a very high-gain aperture is required and a phased array is cost prohibitive [4]. For reflectors, beam-steering can be accomplished through mechanical scanning, or by using sophisticated feed cluster architectures, where the latter is typically high cost [5]. On the other hand, electronic scanned phased arrays rely on phase shifter technology. For high-gain beam-scanning applications where the antenna aperture is large, active phased arrays using transmit/receive (T/R) modules are employed [6]. While the phase shifter and T/R module technology have greatly matured over the years and the cost of active phased arrays have dropped dramatically, they still represent a considerable portion of the overall antenna cost [7]. Typically, the primary disadvantages of phased arrays are the large hardware footprint and high cost.

Owing to their hybrid nature, the reflectarray antenna provides advantages over these two types of antennas, and is well suited for applications requiring high-gain beam-scanning antennas. The beam of a reflectarray antenna can be scanned by means of the reflector nature or the array nature of the antenna [8]. In addition, it is possible to utilize both approaches in a single design to improve scan performance or reduce system cost. It should be noted that similar to reflectors it is also possible to scan the beam of a reflectarray by mechanical movement of the aperture [9]; however, the focus of this chapter is to realize dynamic beam-scanning.

In this chapter, we will review the available design methodologies for beam-scanning reflectarray antennas and present a comparative study on the performance of these approaches, with the aim of providing guidelines for antenna engineers to select a suitable design for their systems.

Reflectarray Antennas: Theory, Designs, and Applications, First Edition.
Payam Nayeri, Fan Yang, and Atef Z. Elsherbeni.

10.1 Beam-Scanning Approaches for Reflectarray Antennas

10.1.1 Design Methodologies

To realize a dynamic radiation pattern, the phase distribution on the reflectarray aperture has to be tuned corresponding to the direction of the scanned beam. Let us consider the reflectarray system shown in Figure 2.1, which consists of a single reflector and a single feed. The phase distribution for each element on the reflectarray aperture consists of two components, that is

$$\phi(x_i, y_i) = -k_0 R_i + \phi_R(x_i, y_i). \tag{10.1}$$

In this equation, the first term corresponds to the spatial delay, which is the electrical distance between the phase center of the feed and the element position on the array, that is $k_0 R_i$. Here, k_0 is the free space wave number. The second term is the reflection phase of the ith element on the aperture. Since in a reflectarray antenna one can control these two components almost independently, different approaches are available for beam-scanning reflectarray antennas. These can be categorized as:

- Feed-tuning techniques: where one changes $-k_0 R_i$;
- Aperture phase-tuning techniques: where one changes $\phi_R(x_i, y_i)$.

An illustrative diagram, showing the various design methodologies for beam-scanning reflectarrays is given in Figure 10.1.

In the feed-tuning technique, one takes advantage of the space feed system, and changes the phase on the reflectarray aperture by tuning the spatial delay. To address the system design requirements such as scan speed, coverage, and resolution, proper selection of the reflector and feed system is of paramount importance. In general, scan coverage typically depends on the choice of reflector, while scan speed and resolution depend on the feed system. For beam-scanning applications with these systems, the phase center of the feed antenna should be displaced. The displacement path (or focal arc) depends on the system design and is typically determined using ray optics analysis. For feed-tuning, movable feeds provide continuous scan coverage, however, they generally have a slow scan speed. Multiple feed arrays can provide fast beam-switching, essentially on the basis of one feed per beam, but not a continuous scan. Phased array feeds (PAFs) are typically considered the best choice, and although they are relatively expensive, they can provide continuous high-speed scanning. Various designs of planar array and cluster feeds have been developed for reflector antennas [10]–[13], and in general all these designs are applicable for beam-scanning reflectarray antenna systems. However, very little has been done to design feed arrays specifically for beam-forming reflectarrays, and the focus in this area has primarily been on the scan capabilities of different reflectarray systems. Recent developments in beam-scanning reflectarrays that utilize the feed-tuning technique will be reviewed in Section 10.2.

In the second approach, the elements on the reflectarray aperture are equipped with a phase-tuning mechanism [14], essentially utilizing the array nature of a reflectarray antenna. This is basically similar to a phased array antenna design [1], [2]; with the added advantage of replacing the feed network, which is a major challenge in phased array

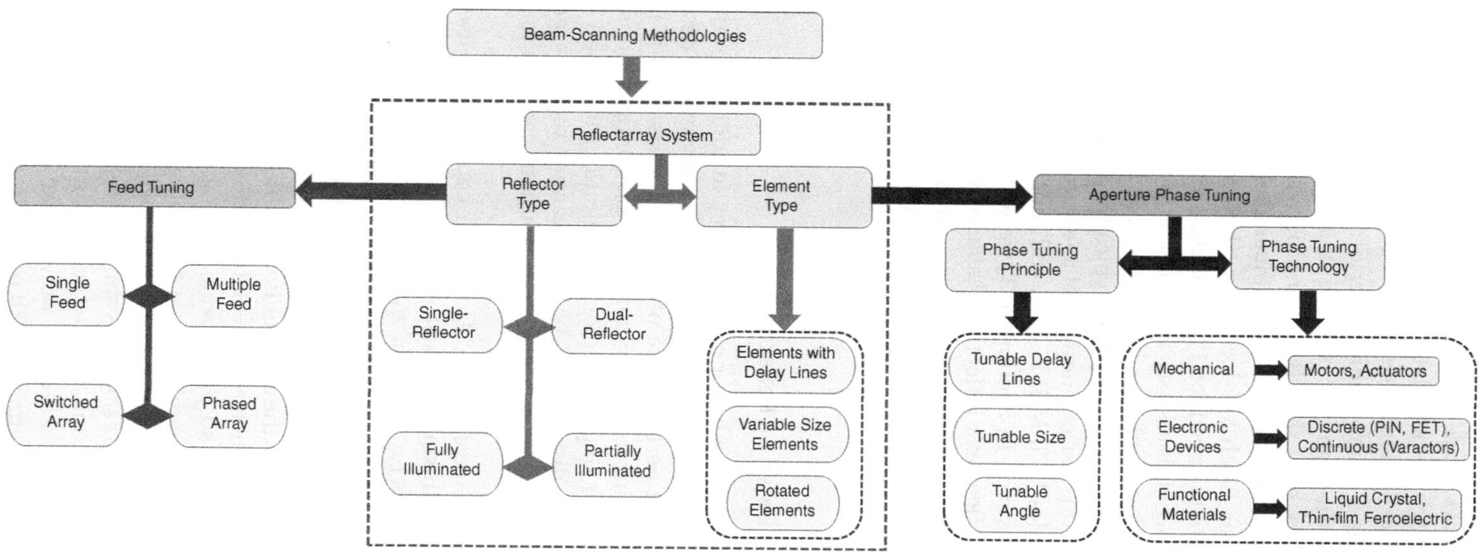

Figure 10.1 Design methodologies for beam-scanning reflectarray antennas. *Source:* Nayeri 2015 [40]. Reproduced with permission of IEEE.

design, by a space feed system. This feature is particularly very advantageous for high-gain beam-scanning arrays, since while small phased arrays can be designed by only using phase shifters for each element, large phased arrays require the use of T/R modules to compensate for the distribution losses in the feed network. These active phased arrays [6] can achieve a high-gain beam-scanning performance; however, the main disadvantage of these designs is the high cost of the system. With the spatial feed system of a reflectarray antenna, not only the challenging part of designing a feed network is removed, the space feed also reduces the cumbersome distribution losses associated with large array antennas. With low-loss phase shifters, an aperture phase controlled reflectarray can eliminate the need for T/R modules. It is worthwhile pointing out that, while there are some distribution losses in a reflectarray antenna due to spillover losses [15], in most cases, a spillover efficiency above 90% can be achieved for large arrays.

In the phase-tuning technique, the system typically relies on the enabling technology, as illustrated in Figure 10.1. Various designs of aperture phase tuned reflectarrays with micro-motors [16], diodes [17], MEMS [18], and functional materials [19] have been demonstrated over the years. In Section 10.3, we will review the recent advances in these dynamically reconfigurable reflectarray antennas. It should be noted that generally for these beam-scanning reflectarays, apart from the spatial feed network, a control board has to be placed behind the array to supply control voltages to each element, and a microcontroller is used to interface the system with an external computer, which synthesizes the array.

10.1.2 Classifications Based on Reflector Type

The choice of single- or dual-reflector configuration, in addition to the choice of aperture phase distribution, or in other words the reflectarray system design plays a major role in the beam-scanning performance of these antennas. This is particularly more important for the feed-tuning technique, since in these configurations, the array has no phase-tuning capability. For single-reflector configurations that use a single-feed architecture, a parabolic-phase distribution typically cannot realize a good scan performance. In these cases, non-parabolic-phase distributions can help to improve the scan capability of the antenna.

Dual-reflector antennas have evolved from their counterparts used in optical telescopes. In comparison with single-reflectors, one of the most advantageous features of dual-reflector systems is provision of an additional degree of freedom to optimize a particular characteristic of these reflectors; for example, to reduce the cross-polarization [20], [21]. They also have some mechanical advantages since they allow the feed to be placed near the main reflector. In addition, they can provide a higher accuracy for beam pointing, thus they are quite suitable for high-gain beam-scanning applications. Configurations such as bifocal dual-reflectors [22], Gregorian corrected spherical reflectors [23], and so on, provide wide-angle scan coverage with higher aperture efficiency. As far as the reflectarray technology goes, they can be designed as the sub-reflector, main reflector, or both in these dual-reflector configurations [24].

In addition, oversized apertures, such as spherical reflectors that are partially illuminated for any given beam direction, are well suited for wide-angle scan coverage with a passive reflectarray. In the next section, we will take a closer look at some of these beam-scanning reflectarray antennas, which utilize the capabilities of different reflector classes in the feed-tuning context.

10.2 Feed-Tuning Techniques

As discussed in the previous sections, in the feed-tuning technique, scan coverage typically depends on the choice of reflector. Here we will review some of the different reflector configurations that have been developed for beam-scanning reflectarray antennas.

10.2.1 Fully Illuminated Single-Reflector Configurations

10.2.1.1 Parabolic-Phase Apertures

Conventionally, the phase on a reflectarray antenna aperture is designed based on the phase compensation of a comparable parabolic reflector antenna with the same subtended angle [25]. As discussed earlier, it is possible to scan the beam of a reflectarray antenna by changing the phase center of the feed antenna. To systematically study the effect of feed displacement on the radiation performance of reflectarray antennas, we follow the procedure developed for reflector antennas [21], and study axial and lateral displacements of the feed independently. An image illustrating these two feed displacement paths is given in Figure 10.2.

A circular reflectarray aperture with a diameter of 20λ is considered for the study here, and in all studies the F/D and radiation pattern model (q) are designed to achieve an edge taper equal to −10 dB when the feed is at focal point. For the axial displacement study, we study the gain loss versus displacement (δ/λ) for various F/D ratios. The feed is displaced along the axial direction from −3λ to +3λ, where the negative sign indicates

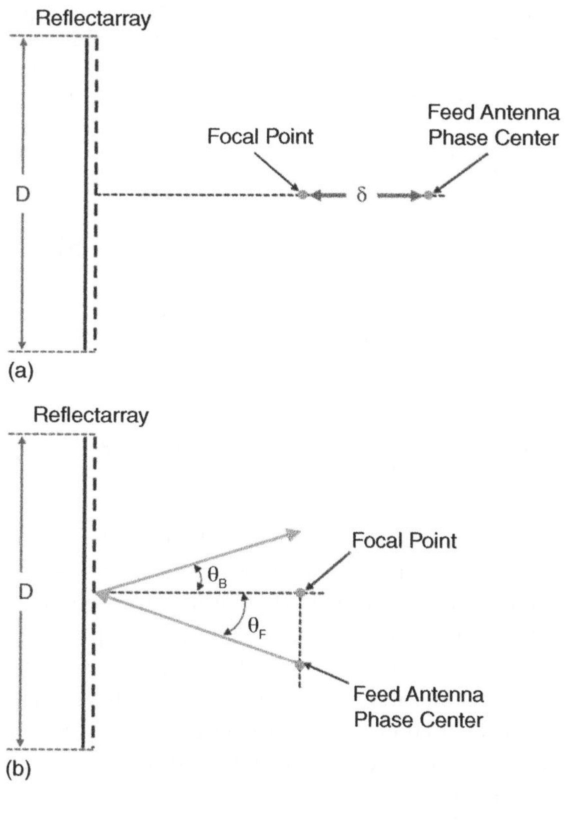

Figure 10.2 Feed displacement paths for reflectarray antennas: (a) axial displacement, (b) lateral displacement.

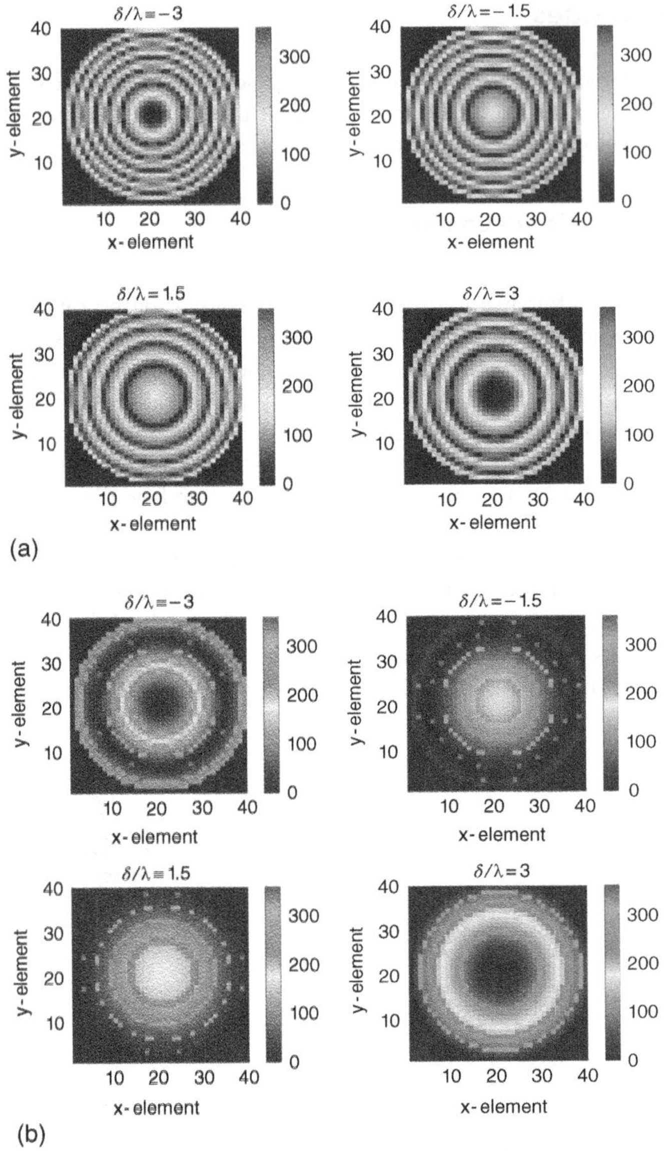

Figure 10.3 Effect of axial displacement of the feed on the phase distribution on the aperture: (a) Ideal phase shift, (b) corresponding phase error.

that the feed is moving toward the reflector. The ideal phase shift and the corresponding phase error on the aperture is given in Figure 10.3 for F/D = 0.4.

These studies showed that similar to reflector antennas, the dominant effect of axial defocusing is the generation of quadratic phase error across the aperture [26], [27]. The quadratic phase error (order 2) does not scan the beam but results in a reduction of antenna gain and a change in the side-lobe levels and the depth of the nulls between them. To further observe the effect of axial feed displacement on the radiation performance of the antenna, the gain patterns of reflectarray antennas with various values of F/D are also given in Figure 10.4.

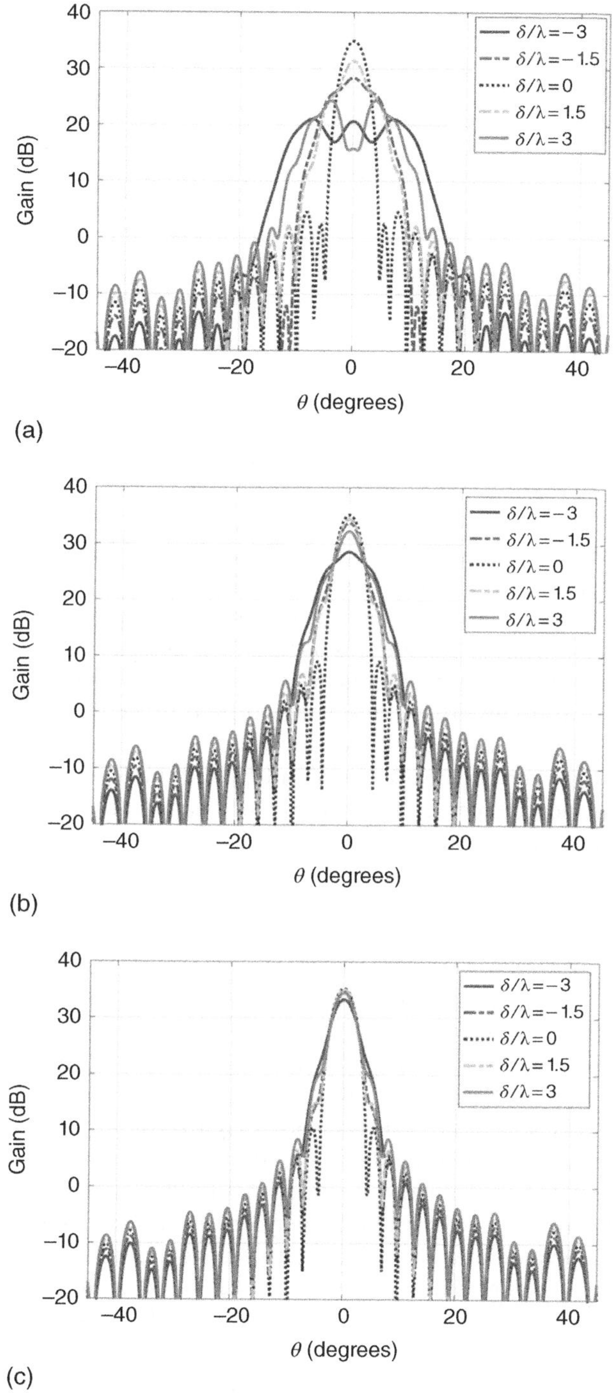

Figure 10.4 Effect of axial displacement of the feed on the radiation performance of reflectarray antennas: (a) F/D = 0.4 (q = 1.447), (b) F/D = 0.7 (q = 4.586), (c) F/D = 1.0 (q = 9.318), (d) normalized gain versus axial displacement.

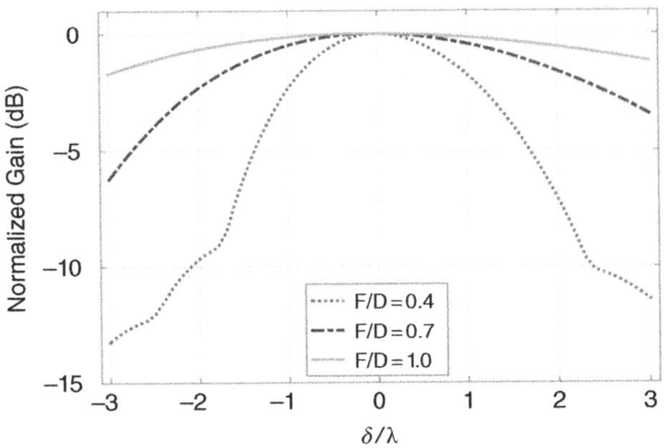

(d)

Figure 10.4 (Continued)

These numerical studies showed that although axial feed displacement cannot be used to scan the beam, increasing the F/D ratio for the design will improve the performance of the design. In particular, from a fabrication and measurement viewpoint, the alignment sensitivity will be reduced if a larger F/D ratio is used for the design.

Next let's study the effect of lateral feed displacements. With reference to reflector antennas, the parameter of interest is the gain loss versus beamwidths scanned (bmws). An important parameter defined in this study is the beam deviation factor. The beam deviation factor (BDF) is the ratio between the beam direction (θ_B) and the feed angle (θ_F). Using this definition of BDF and calculating the half-power beamwidth of the antenna, the beamwidths scanned can be calculated using

$$\text{bmws} = \frac{\theta_F.\text{BDF}}{\text{HPBW}}. \tag{10.2}$$

The procedure to calculate the radiation pattern in terms of bmws then requires one to: (1) obtain the BDF for various feed angles (θ_F), and (2) use the values of BDF and HPBW for the design and setup a simulation to scan the beam to the required directions. Using this procedure, the effect of lateral feed displacement on the radiation performance of the antenna with various values of F/D is given in Figure 10.5.

The numerical studies presented here showed that lateral feed displacement can be effectively utilized to scan the beam of a reflectarray antenna. In addition, increasing the F/D ratio for the system significantly improves the scan performance. While in the numerical studies presented here the value of q was adjusted to realize a certain edge taper for the design, in practice, for an available horn antenna one has to determine the maximum F/D possible based on other design criteria such as aperture efficiency and edge taper.

An additional consideration in feed displacement technique is the path that the feed antenna is displaced along. While moving the feed antenna along the lateral direction can be used to scan the beam of the reflectarray antenna, depending on the system design, better performances may be achieved by using a different path. One practical

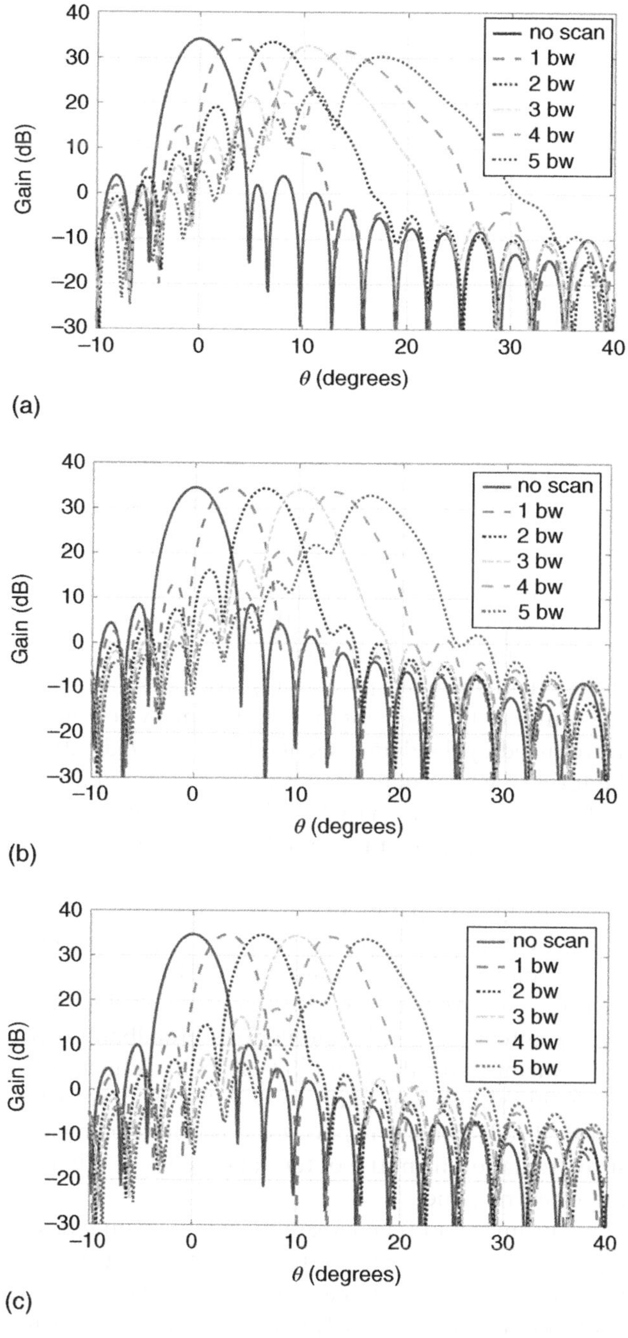

Figure 10.5 Effect of lateral displacement of the feed on the radiation performance of reflectarray antennas: (a) F/D = 0.4 (q = 1.447), (b) F/D = 0.7 (q = 4.586), (c) F/D = 1.0 (q = 9.318), and (d) normalized gain loss versus bmws.

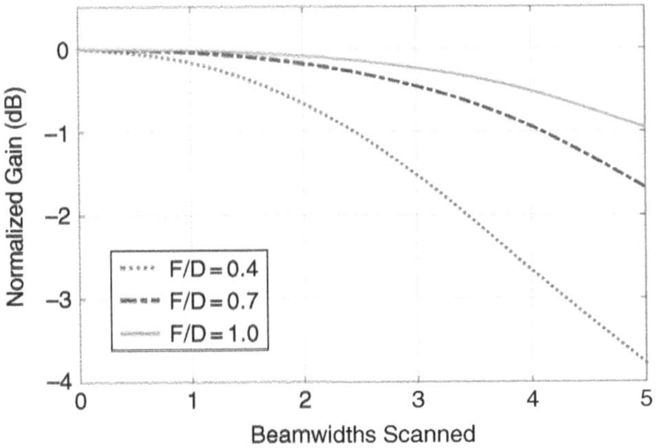

(d)

Figure 10.5 (Continued)

option is a circular arc shown in Figure 10.6. A notable advantage of the later design is that the reflectarray focal depth remains almost constant, and the spillover efficiency is also improved, which results in a better scanning performance.

It is important to note that, while the main beam of a parabolic-phase reflectarray can be scanned by displacing the phase center of the feed antenna, the primary concern is that due to the phase aberration introduced by the defocused feed, the far-field pattern can be substantially degraded, the level of which depends on the F/D ratio and scan angle in terms of number of bmws. In [26] it was shown that the beam-scanning performance of a reflectarray by means of feed displacement is limited to a few beam widths, and is slightly inferior to a parabolic reflector.

In another study on the scanning capability of parabolic-phase reflectarrays, it was shown that the scan performance can be improved by using sub-wavelength elements [28]. Moreover, scanning properties of piecewise planar parabolic (PPP) reflectarrays were studied in [29]. PPP reflectarrays were proposed in [30] to overcome the bandwidth limitations of large reflectarray antennas [31]. While various broadband techniques have been developed for reflectarrays over the years [32–38], PPP reflectarrays are still desirable for very high-gain applications. The study in [29], however, revealed that the scan performance of tilted PPP reflectarrays is poor even for one beamwidth scan, which is due to the fact that the radiation from the outer panels peak at different directions compared to that from the central panel.

Despite the fact that this technique is limited in scan capability, it still is quite desirable for applications that don't require wide scan coverage, that is limited field-of-view systems. A practical implementation of this approach is a multi-fed reflectarray with four beams for X-band synthetic aperture radar (SAR). A 1.6 m offset reflectarray with four feeds, and F/D around 1 was designed to generate four beams SS1, SS2, SS3, and SS4, in the directions of 20.91°, 18.41°, 15.91°, and 13.41°, respectively. The reflectarray is designed for one beam (SS3), and the scanned beams (with about 7.5° of angular coverage) are achieved by feed displacement [39].

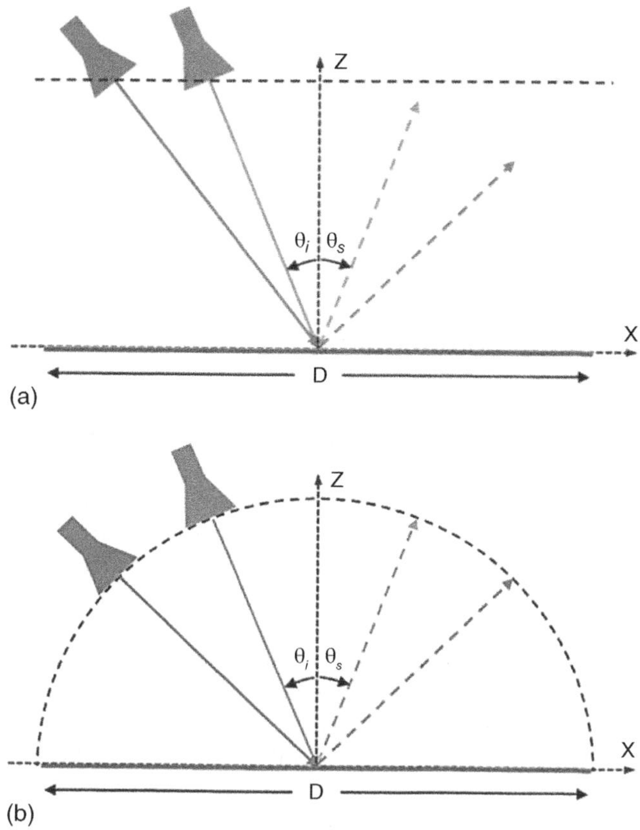

Figure 10.6 Different paths for feed displacement: (a) lateral movement, (b) circular arc movement.

10.2.1.2 Non-Parabolic-Phase Apertures

One prominent feature of the reflectarray antenna is that it allows for individual phase control of each element in the array. Owing to this feature, any phase distribution can be attained on the aperture and one is not bound to use a parabolic-phase distribution. In other words, one can design an aperture that provides a better phase distribution for the scan range required.

Based on the concept of single-reflector multi-focal apertures, different aperture phase distributions were proposed in [40]. The basic idea in these designs is that a reflectarray antenna designed to compensate for multiple focal points will have a better scan performance than a uni-focal design. It is well known that a bifocal lens has a wider scan capability than a lens with a single focus, since it focuses quite well nearby and between the two focal points [41], [42]. Similarly, it was shown that a bifocal dual-reflector antenna has a better scan capability than a classical Cassegrain reflector [22]. A bifocal folded dual-reflectarray antenna using a seven-element feed array was presented in [43], which showed a scan range of about 27° (±13.5°). The scan range was further improved to 49° (±24.5°) by designing a system with restricted apertures for each beam. More recently, a dual reflectarray with dual-offset feeds was also demonstrated [44].

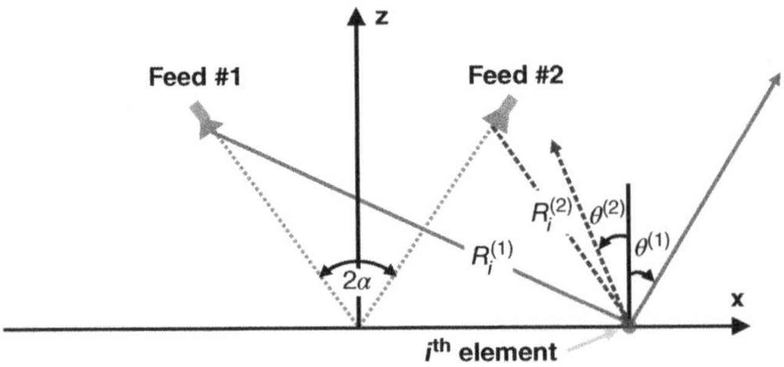

Figure 10.7 Coordinate system for the bifocal reflectarray antenna. *Source:* Nayeri 2013 [8]. Reproduced with permission of IEEE.

In all these designs, the ideal bifocal reflector or lens requires two surfaces for phase correction. While the beam-scanning performance that can be achieved with these designs is a significant improvement over the conventional single-reflector parabolic-phase configurations, the main disadvantage is the design complexity and fabrication difficulty of dual-reflector systems. The designs presented in [40] are single-reflectarray systems based on the concept of having two focal points, which increase the scan range of the beam-scanning reflectarray antenna.

To formulate the problem let us consider the system shown in Figure 10.7, where the reflectarray aperture lies on the xy-plane and two feed antennas are placed in the xz-plane, each with an offset angle of α degrees. In this design, the feeds can be turned on concurrently to realize a multi-beam performance, or separately to achieve a beam-switching performance. For this two-feed system, the required phase distribution on the aperture corresponding to each feed and beam direction can be given by

$$\phi_i^{(f)} = k_o\left(R_i^{(f)} - \sin\theta^{(f)}\left(x_i\cos\varphi^{(f)} + y_i\sin\varphi^{(f)}\right)\right). \tag{10.3}$$

In this equation, the superscript (f) denotes the corresponding feed, and θ and φ represent the direction of the scanned beam. In comparison with a single-feed system, the difference here is that the aperture phase distribution has to compensate for two different feed positions and beam directions. To simplify the problem, we consider a symmetric system, that is, when the first feed is turned on, the scanned beam is in the direction of $(\theta = \theta^{(1)}, \varphi = 0°)$. Similarly with the second feed turned on, a scanned beam will be generated in the direction of $(\theta = \theta^{(2)}, \varphi = 180°)$. With such a design, the phase requirements can then be written as

$$\begin{aligned}\phi_i^{(1)} &= k_o\left(R_i^{(1)} - x_i\sin\theta^{(1)}\right),\\ \phi_i^{(2)} &= k_o\left(R_i^{(2)} + x_i\sin\theta^{(2)}\right).\end{aligned} \tag{10.4}$$

Note that for this symmetric system we also have $\theta = \theta^{(1)} = \theta^{(2)}$. It is clear that in general, the ith element cannot realize two different phase requirements with the same polarization in a passive design. Therefore, an average phase is set for the ith element,

$$\varphi_i = \frac{\varphi_i^{(1)} + \varphi_i^{(2)}}{2}. \tag{10.5}$$

Due to the symmetry of the system, the terms $x_i \sin \theta^{(f)}$ are cancelled out in (10.5), and the direction of the main beam of the reflectarray antenna will now depend on the feed offset angle, α. As discussed earlier, an important parameter in dealing with scanned beams is the beam deviation factor, [21], [26], which is defined as BDF $= \theta/\alpha$, where θ and α are the main beam direction and the feed offset angle, respectively. With displaced feeds, typically θ is smaller than α, which results in a BDF less than 1. Nonetheless, it is possible to achieve the required scan range by compensating for this factor. It is also worthwhile pointing out here that the phase requirement in (10.4) can be realized by using different polarizations for each feed, as demonstrated in [45]; however, this would not be a suitable approach for a beam-scanning system.

A Ka-band reflectarray antenna with main beams pointing 30° off-broadside was designed to investigate the feasibility of this approach. By using the feed displacement technique this reflectarray system can be used for a 60° scan coverage (−30° to +30°). A circular aperture with a diameter of 17 wavelengths at the center frequency of 32 GHz was selected for the reflectarray. This corresponds to 848 phasing elements with half-wavelength spacing on the reflectarray aperture. The feed displacement path is along a circular arc with a radius of F, which shows a better performance than a lateral feed displacement path. To further improve the beam-scanning performance of the reflector, the maximum attainable F/D was selected for the design. The maximum F/D of 0.725 was determined based on the available feed horn antenna, which has a $\cos^q\theta$ radiation pattern with $q = 6.5$, to ensure that the maximum edge taper is below −8 dB when the feed is tilted α degrees. The performance of this design was also compared with a reflectarray antenna designed using the conventional parabolic equation. The aperture phase distribution for both designs is given in Figure 10.8.

While in both designs, the main beam is scanned by displacing the feed antenna, the required offset feed angle is different, and has to be obtained experimentally to compensate for the BDF. For a 30° scan, the BDF for the parabolic and bifocal designs is 0.93 and 0.90, respectively. The scanned gain patterns of both the parabolic-type and bifocal-type reflectarray antennas are given in Figure 10.9. It can be seen that for this scanned beam, the bifocal design shows a much better performance. A gain variation less than 1 dB is observed across the entire scan range for the bifocal design.

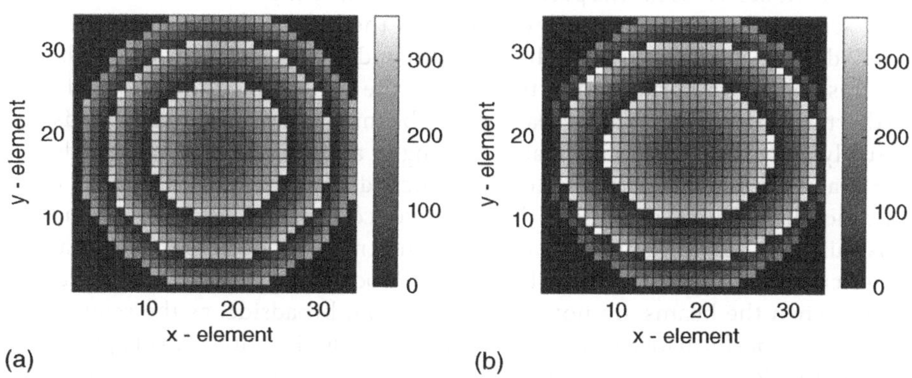

Figure 10.8 Phase distribution on the reflectarray antenna aperture: (a) conventional parabolic design, (b) bifocal design. *Source:* Nayeri 2013 [8]. Reproduced with permission of IEEE.

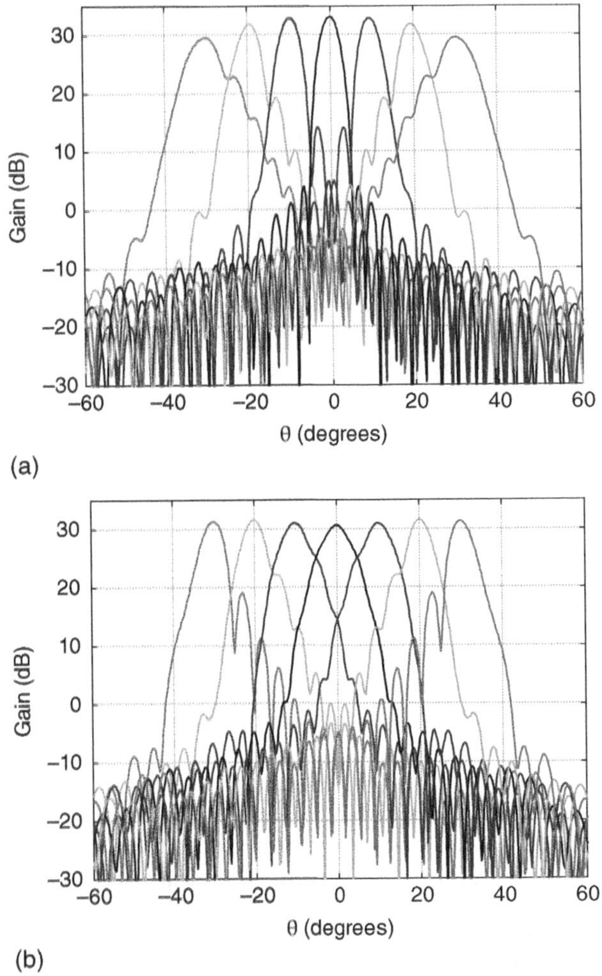

Figure 10.9 Scanned gain patterns of reflectarray antennas: (a) parabolic-type, (b) bifocal-type. *Source:* Nayeri 2013 [8]. Reproduced with permission of IEEE.

In comparison, the scan loss for the parabolic design is more than 3.5 dB. Furthermore, for the 30° scan angle, the side-lobe level (SLL) of the bifocal design is reduced by more than 7 dB, and a much narrower beam width and higher gain are observed.

Despite this good scan performance, the SLL for the bifocal design is still relatively high. Furthermore, the side-lobes increase as the beam is scanned toward broadside, and eventually merge with the main beam resulting in a very wide beam at broadside. While the beam-scanning performance of this bifocal design could be acceptable for a communication link where scan loss is the major concern, radar systems also impose strict constraints on SLL [3], thus such a performance may not be acceptable. In addition, numerical studies showed that although the bifocal designs can achieve a good performance when the beams are not scanned far from broadside, as the scan range increases, the SLL performance of the system degrades significantly. This is primarily due to the fact that for larger scan angles, the dissimilarity in the spatial delay associated with the two feeds increases, thus averaging these values results in a higher error in the aperture phase distribution. Similarly, this approach will be less effective as the aperture size increases.

Nonetheless, for limited field-of-view systems, a bifocal design shows a significant improvement over the conventional parabolic design. It should be noted that in comparison with an ideal dual-reflectarray bifocal design, the single-reflector design here shows a gain reduction of about 1 dB.

Further improvement of the performances of single-reflectarray bifocal designs can be achieved by means of optimization. The phase distribution obtained for the bifocal design was based on averaging the spatial phase delay associated with two feed positions. As such, it is implicit that this simple average can not provide the optimum phase distribution. On the other hand, one prominent feature of the reflectarray antenna is that the elements have the ability to achieve any value of reflection phase. As such any phase shift distribution can be realized on the aperture, and unlike reflector antennas [21], one is not bound to any physically realizable geometry. By utilizing this flexibility in the design, optimization of the phase distribution on the reflectarray aperture could prove to improve the performance of the bifocal design. In addition, it is possible to further improve the scan performance of these systems by optimizing the antenna for several beams along the scan range.

Two general approaches are available for aperture phase optimization of reflectarray antennas. In the first approach, one minimizes the phase error by comparing it with an ideal phase distribution [46]. In the second approach, one optimizes the radiation pattern by defining a far-field pattern mask based on the design requirements, and determines the corresponding phase of the elements indirectly [47]. The advantage of the first approach is that it doesn't require calculating the radiation pattern in the optimization process, which is computationally advantageous. On the other hand, the latter approach allows for a direct control over the radiation pattern shape, which is quite advantageous, particularly if one aims at controlling the radiation characteristics such as the SLL.

In [40] the authors used the second approach and optimized the phase of the reflectarray elements in order to improve both scan loss and SLL. The particle swarm optimization method, [48], [49], was selected for phase optimization of the reflectarray elements. A far-field radiation pattern mask was defined for the problem which are defined by two sets of bound values, namely M_U and M_L, which set the upper and lower bounds of the desired pattern in the entire angular space. The cost function defined for the problem was defined as

$$\begin{aligned} Cost = & \sum_{\substack{(u,v)\in \text{mainbeam} \\ \text{and} |F(u,v)|<M_L(u,v)}} \sum \left(|F(u,v)| - M_L(u,v) \right)^2 + \\ & \sum_{\substack{(u,v)\notin \text{mainbeam} \\ \text{and} |F(u,v)|>M_U(u,v)}} \sum \left(|F(u,v)| - M_U(u,v) \right)^2 + \\ & P \times \sum_{\substack{(u,v)\in \text{first sidelobe area} \\ \text{and} |F(u,v)|<M_L(u,v)}} \sum \left(|F(u,v)| - M_U(u,v) \right)^2 . \end{aligned} \tag{10.6}$$

Here, F is the far-field radiation pattern of the array and the term P in the cost function is the penalty term added to account for the performance of the first side-lobe. Although defining an optimum mask and fitness function for the problem proves to be invaluable, the main challenge in implementing PSO is the very large number of elements that need to be optimized for a reflectarray antenna. For this problem however, one takes advantage of the symmetry of the system to reduce the dimension of the solution hyperspace. The bifocal reflectarray system shown in Figure 10.7 is symmetric along both x- and y-directions, thus the number of elements that have to be optimized are one quarter of the total array.

The same reflectarray antenna system designed for 60° scan coverage was used and two optimized designs were presented. For the PSO parameters, the inertial weight was varied linearly from 0.9 to 0.4 over the course of the run, and the self- and group-knowledge constants were both set equal to 2 throughout the optimization. A swarm population of 100 particles was selected for the optimization, and 100,000 iterations were performed in the design. In the first design, referred to as the PSO design, the phase of the reflectarray elements are optimized for two scanned beams, ±30°, and their corresponding feed positions, as shown in Figure 10.7. While basically two objectives are addressed, taking advantage of the symmetry reduces this problem to a single-objective optimization search. For the second design, referred to as the MOPSO design, the phase of the reflectarray elements is optimized for four scanned beams, ±10°and ±30° and their subsequent feeds. A multi-objective optimization with two objective functions is implemented for this design. This design is basically a quadrufocal system, and is expected to provide a better beam-scanning performance.

Several different masks were also studied for the optimizations to determine a suitable setup for the problem and to ensure the array can achieve maximum directivity while maintaining a SLL of −20 dB. The initial mask was set to have a constant SLL in the entire (u, v) space; however, it was observed that with this type of mask, the array cannot achieve a good directivity. To improve the directivity, a shaped mask was defined such that the side-lobe envelope decreases linearly outside the main beam area. The optimum mask envelopes and the penalty coefficient P were then determined experimentally for the optimization. In the design here, the mask decreases linearly outside the main beam area from −25 to −45 dB. A cross-sectional view of the shaped mask used for the optimization here is given in Figure 10.10. The optimum value of P was

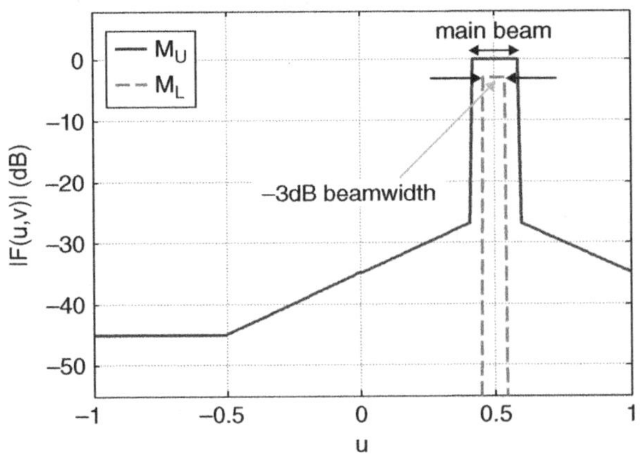

Figure 10.10 Shaped far-field mask used for the optimization. *Source:* Nayeri 2013 [8]. Reproduced with permission of IEEE.

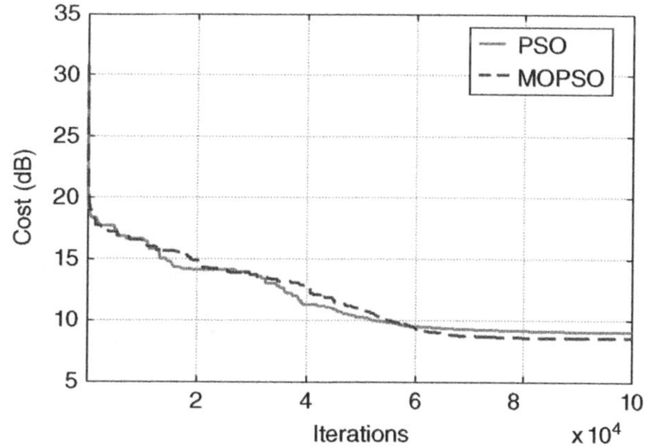

Figure 10.11 Convergence curves for the single-objective (PSO) and multi-objective (MOPSO) optimizations. *Source:* Nayeri 2013 [8]. Reproduced with permission of IEEE.

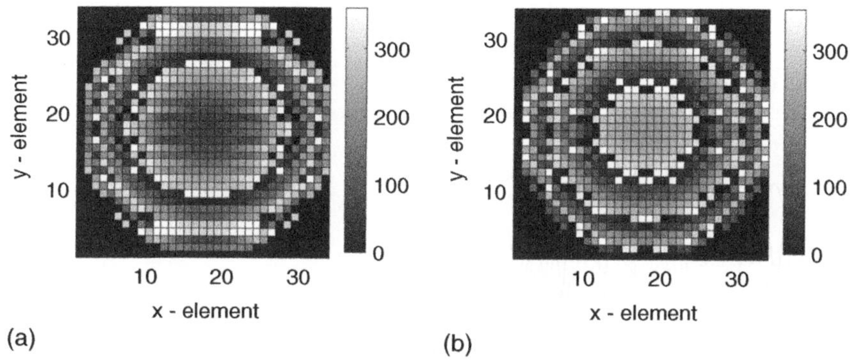

Figure 10.12 Phase distribution on the aperture of the optimized reflectarray antennas: (a) PSO design, (b) MOPSO design. *Source:* Nayeri 2013 [8]. Reproduced with permission of IEEE.

determined to be 6. The convergence curves for both optimizations are shown in Figure 10.11. The optimized aperture distributions for these two designs are given in Figure 10.12.

The scanned gain patterns of the PSO design are given in Figure 10.13(a), where a gain variation less than 0.35 dB is observed across the scan range. Furthermore, these results show that not only the SLL has been reduced by more than 8 dB for the ±30° beams; an almost symmetric beam is observed along these scan directions. Despite the notable improvements the PSO design achieved over the bifocal design, the beam broadening effect observed around the broadside is still not desirable for a beam-scanning array, and such a design is more suitable for a two-beam beam-switching system.

Meanwhile, the MOPSO design is targeting four optimized beams across the scan range, and as such it can be a good candidate for a beam-scanning application. The scanned gain patterns of this design are given in Figure 10.13(b), where a much better side-lobe performance is observed across the scan range. While in general some compromise has to be made in multi-objective optimizations, the SLL achieved across the entire scan range is below −17.5 dB. For this design, the gain variation across the scan coverage is about 1.35 dB, which is quite acceptable for a beam-scanning antenna.

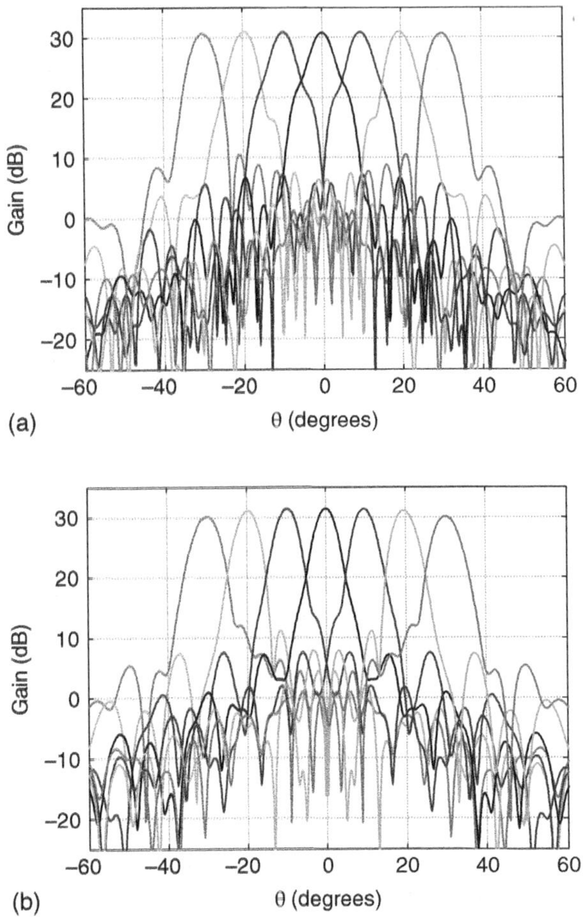

Figure 10.13 Scanned gain patterns of the optimized beam-scanning reflectarray antennas: (a) PSO, (b) MOPSO. *Source:* Nayeri 2013 [8]. Reproduced with permission of IEEE.

The studies show that the scan limitation of conventional reflectarray antennas is primarily attributed to the parabolic-phase distribution on the aperture. With a single-reflector bifocal or multi-focal aperture, one can improve the scan performance beyond those of the parabolic-phase distribution. Furthermore, in comparison with an ideal dual-reflectarray bifocal antenna, the PSO design (pattern optimized bifocal single reflector) can achieve an almost similar SLL at the expense of some compromise in the gain and a broader beam. The MOPSO design on the other hand demonstrates a better scanning performance, but some further degradation in gain and SLL at the maximum scan angle. Meanwhile, once the new aperture phase profile is determined, it can be further improved by using sub-wavelength elements and increasing the aperture phase accuracy [28].

An important consideration in feed-displaced beam-scanning reflectarrays is the sensitivity of the elements reflection coefficient with respect to the excitation angle. In general, with a fixed feed phase center, each element is excited with a specific incident angle that can be accurately taken into account by using appropriate element design tables [39]. However, for a feed displacement beam-scanning system that requires a movable feed phase center, the incident angle for each element will change when the feed is displaced. Consequently, this would result in some phase error due to element sensitivity to excitation angle. As such, for a practical design it would be necessary to have elements with low sensitivity toward excitation angles. To this end, a simple yet effective approach is to use single-layer patch elements with very thin substrates for the design.

In [40], the authors used a 10-mil Rogers 5880 substrate for the Ka-band reflectarrays to minimize the sensitivity of the element reflection phase toward oblique excitation. Variable size square patches were used for the reflectarray phasing elements, and larger patch sizes were avoided in the element design, since higher order mode excitation was observed for some oblique angles. With normal incidence excitation, these elements provide 330° of phase range. To accurately take into account the performance of the elements in the analysis, the reflection coefficients in the array configuration were calculated with the exact angle of excitation as the feed moves along the displacement arc.

The antenna gain versus scan angles is given in Figure 10.14 for all four designs. It can be seen that while the parabolic-type design achieves a higher gain in the broadside direction, the scan loss of this design is quite high (in excess of 3 dB), and cannot be used as a beam-scanning antenna for a 60° coverage. On the other hand, despite having a lower peak gain, all three other designs achieved an acceptable level of scan loss for a beam-scanning application. In particular, the PSO design achieves a scan loss less than 0.5 dB across the entire scan range. It should be noted that while the patch elements used in the designs here have a limited phase range and generate some phase quantization errors, the radiation pattern of the reflectarray antennas were not significantly affected and similar patterns as given in Figure 10.9 and Figure 10.13 were observed with the patch elements. It is, however, important to note that a monotonic increase in both element and phase loss was observed as the beam was scanned off-broadside. While a slight difference was observed in the four designs, the peak element and phase loss was in the order of 0.3 and 0.4 dB, respectively.

Photographs of the four fabricated prototypes each with 848 square patch elements are shown in Figure 10.15. A corrugated conical left-hand circularly polarized horn is used for the reflectarray feed, and a mechanical beam-scanning performance was demonstrated. The feed horn was mounted on a mechanical alignment system, and to

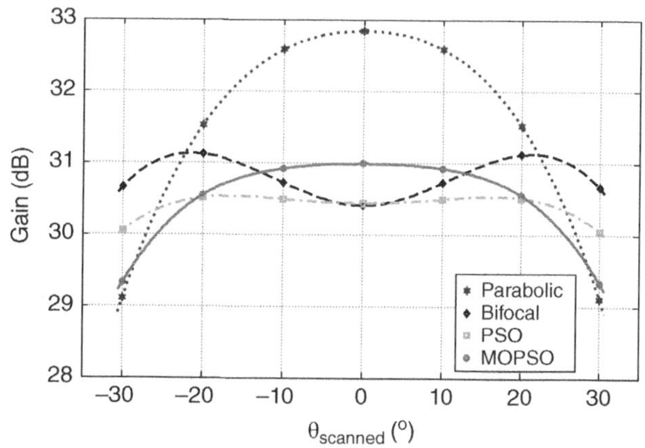

Figure 10.14 Simulated gain versus scan angle for reflectarray antennas. *Source:* Nayeri 2013 [8]. Reproduced with permission of IEEE.

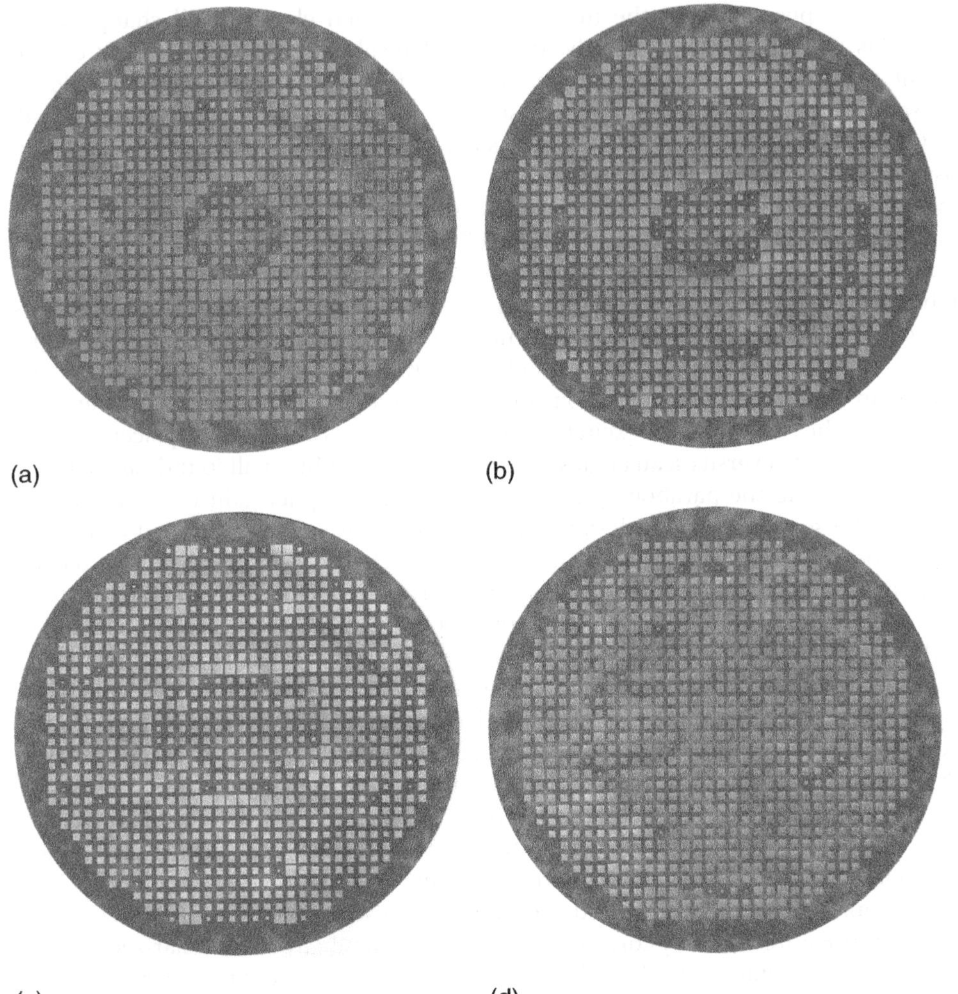

Figure 10.15 Fabricated reflectarray prototypes: (a) parabolic design, (b) bifocal design, (c) PSO design, and (d) MOPSO design. *Source:* Nayeri 2013 [8]. Reproduced with permission of IEEE.

accurately move the feed along the circular arc, a frame was fabricated that allows for a mechanical movement of the feed. The fabricated beam-scanning reflectarray system is shown in Figure 10.16.

The antennas were measured for four scanned beams across the 60° scan range with 20° scan resolution, and the results are given in Figure 10.17. The measured scanned beam patterns for the bifocal design showed that, as expected, a rather stable gain pattern is observed across the scan range. In addition, in comparison with the parabolic design that shows very high coma lobes for the scanned beams, the measured SLL has been reduced by more than 7 dB with this bifocal design. A further improvement is observed with the PSO design, where a measured SLL below −18 dB is obtained for the antenna, when the beams are scanned to ±30°. This design shows a very desirable bifocal performance, however, as the beams are scanned toward broadside, side-lobes

Figure 10.16 Fabricated mechanical beam-scanning reflectarray prototype. *Source:* Nayeri 2013 [8]. Reproduced with permission of IEEE.

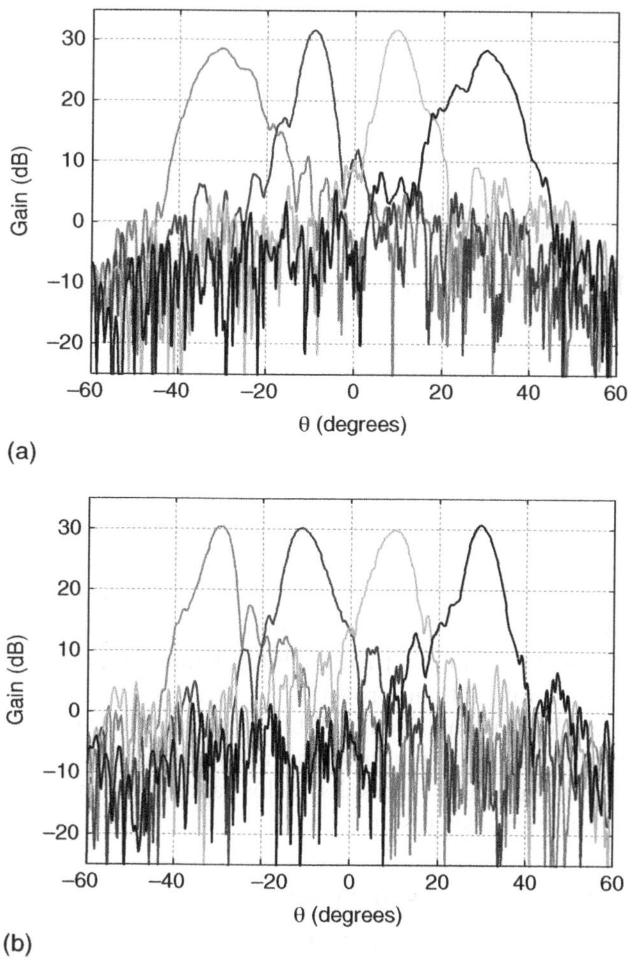

Figure 10.17 Scanned gain pattern of the fabricated beam-scanning reflectarray antennas: (a) parabolic-type design, (b) bifocal design, (c) PSO design, and (d) MOPSO design. *Source:* Nayeri 2013 [8]. Reproduced with permission of IEEE.

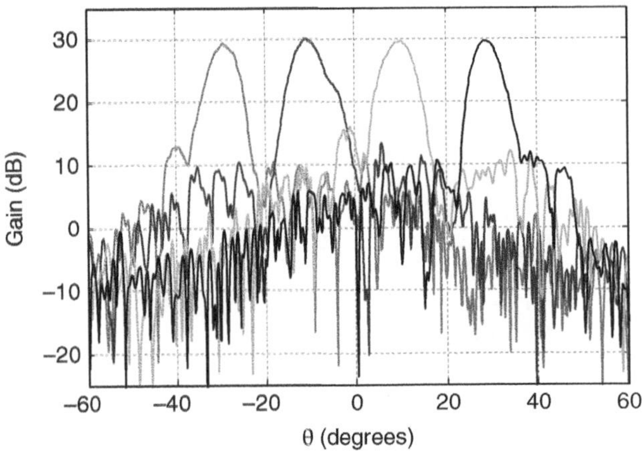

(c)

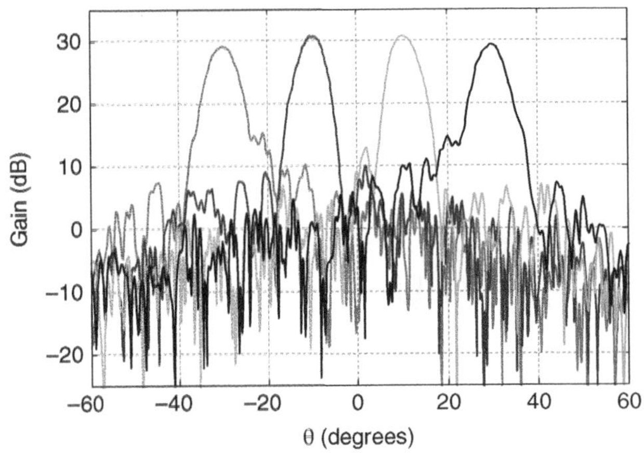

(d)

Figure 10.17 (Continued)

increase, and some beam broadening effects are observed, as expected from simulations. The MOPSO design on the other hand shows a very suitable beam-scanning performance, where the measured SLL is below −15 dB across the entire 60° scan range. While this design exhibits a slightly higher scan loss than the bifocal and PSO designs, the beam broadening effect has been significantly improved.

The measured gain versus scan angle is summarized in Table 10.1 for all four prototypes. Here, ΔG is the gain variation along the measured scan directions. Despite a slightly lower gain in comparison with the analytical results, a close agreement in scan-loss performance is observed with the measured results.

10.2.2 Partially Illuminated Single-Reflector Configurations

10.2.2.1 Parabolic Cylindrical-Phase Reflectarray Antennas (PCPRA)

It has been well known that a parabolic cylindrical reflector fed by a linear source along its focal line can achieve wide-angle beam-scanning capability [3]. A Ku-band parabolic cylindrical reflector was introduced in [50] that was made of a polymer

Table 10.1 Measured gain of the beam-scanning reflectarray prototypes.

Design	Scan −30° (dB)	Scan −10° (dB)	Scan +10° (dB)	Scan +30° (dB)	ΔG (dB)
Parabolic	28.57	31.67	31.76	28.50	3.26
Bifocal	30.32	30.15	29.93	30.68	0.75
PSO	29.44	30.05	29.74	29.73	0.61
MOPSO	29.04	30.71	30.70	29.30	1.67

membrane and fed by a microstrip array. The experimental results showed that the antenna achieved a ±30° scan range with 3-dB gain loss. To achieve a low, planar profile, a reflectarray was used to replace the curved parabolic cylindrical reflector [51], [52]. With two array feeds to imitate two states of scan beam directions, respectively, it demonstrated that the scan angle reached ±20° with 1.8 dB directivity loss. In summary, the wide-angle beam-scanning capability of the parabolic cylindrical reflector/reflectarray has been verified; however, the scan performance is limited by the radiation performance degradation such as gain loss and broadened main beam with the increasing scan angle. The useful scan range is around ±30° with 3dB gain loss, which may not be sufficient for certain advanced applications. In order to enhance the beam-scanning performance and increase the scan range, a novel reflectarray design, which evolves from the conventional parabolic cylindrical design, was proposed in [51] to compensate for the spatial phase differences at different scan angles by phase optimization. The design and experimental results for this beam-scanning reflectarray is summarized in this section.

The conventional parabolic cylindrical reflectarray outlined here, emulates its reflector counterpart with a cylinder axis in the x-direction. An array feed, polarized in the y-direction in this case, is placed along its focal line, and the rotating axis of the array is perpendicular to the H-plane. When the array rotates, different portions of the aperture is illuminated, and consequently the main beam direction scans in the H-plane, namely, the scanning plane. To imitate a parabolic cylindrical reflector, the element compensation phase varies in the y-direction to achieve a parabolic-phase distribution, and the main beam of the reflectarray in the E-plane is collimated by the reflectarray aperture. In the x-direction the element compensation phase remains the same and acts as a mirror. Hence, the main beam in the H-plane is primarily determined by the array. In the design procedure, the compensation phases of the elements in the center column are first computed using the ray tracing method; those phase values are then simply re-used to construct the elements in other columns. Illustrative images of a parabolic cylindrical reflector and reflectarray are given in Figure 10.18.

A parabolic cylindrical reflectarray with a 600 × 400 mm dimension fed by a Ka-band (32.5 GHz) 12 × 2 array was designed using the conventional approach. The compensation phase distribution using the conventional design method is shown in Figure 10.19(a). When the array feed rotates, the incident E-fields on the reflectarray aperture changes accordingly. The main beam scans as the array rotates, however, at large scan angles, 30° in this case, the radiation pattern deteriorates considerably. The beamwidth in the E-plane substantially increases with the increasing scan angle, resulting in considerable directivity loss and pattern deformation. This is due to the fact that the phase errors in illumination area greatly increase with the scan angle and may even be more than 180°

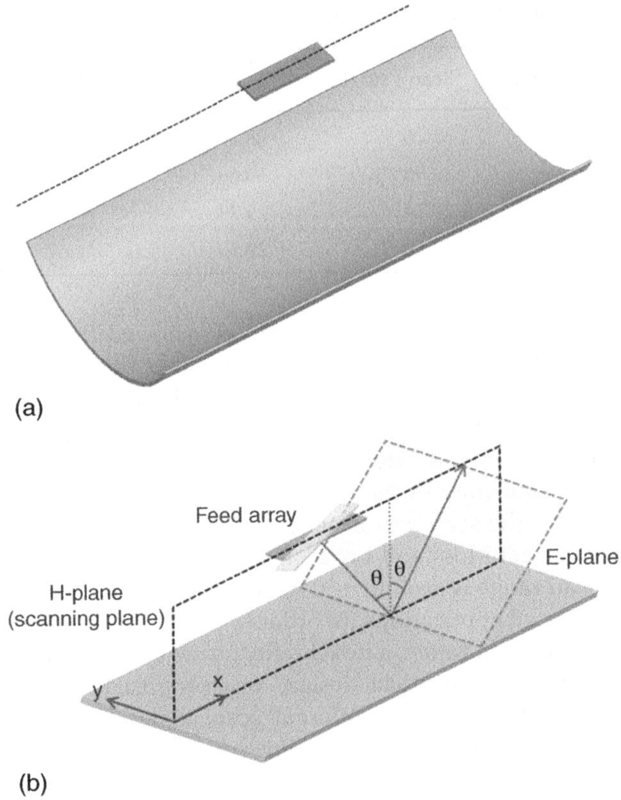

Figure 10.18 Illustration of a (a) parabolic cylindrical reflector, and (b) parabolic cylindrical reflectarray.

when the array rotated 30°. The main reason is that the spatial phase delays to the elements in different columns do not follow the assumption of cylindrical distribution. In order to calibrate the phase errors, the elements in each column are designed individually. Each column has a specific incident angle, and then the element phases in that column are designed to collimate the main beam in the corresponding specular reflection direction. After calculating the element phases in each column, the optimized compensation phase distribution can be obtained as given in Figure 10.19(b). Note that the elements in each row in the x-direction are no longer identical as in the conventional design.

Two reflectarrays are designed using the conventional and optimized design methods with simple square patches of variable sizes. Images of the fabricated prototypes are given in Figure 10.20(a) and (b). A waveguide slot array antenna operating at the center frequency of 32.5 GHz, as shown in Figure 10.20(c), is used as the array feed to illuminate the reflectarray aperture.

The measured radiation patterns in the H- and E-planes for the scan beams at 0°, 15°, and 30° scan angles are given in Figure 10.21 and Figure 10.22. Both reflectarrays successfully collimate the main beam at desired directions and demonstrate good beam-scanning capabilities. The beamwidth and SLLs in the H-plane are similar to those of

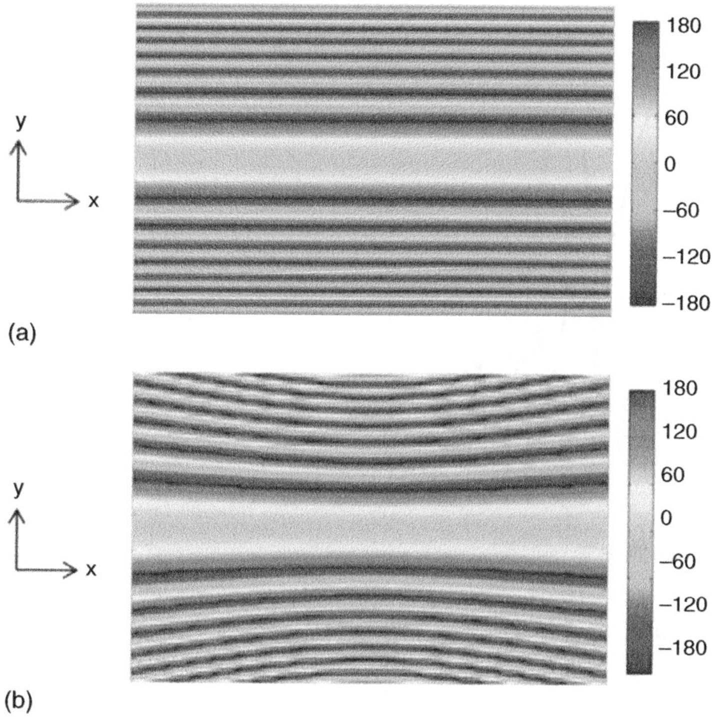

Figure 10.19 Compensation phase distribution of the parabolic cylindrical reflectarrays: (a) conventional, (b) optimized.

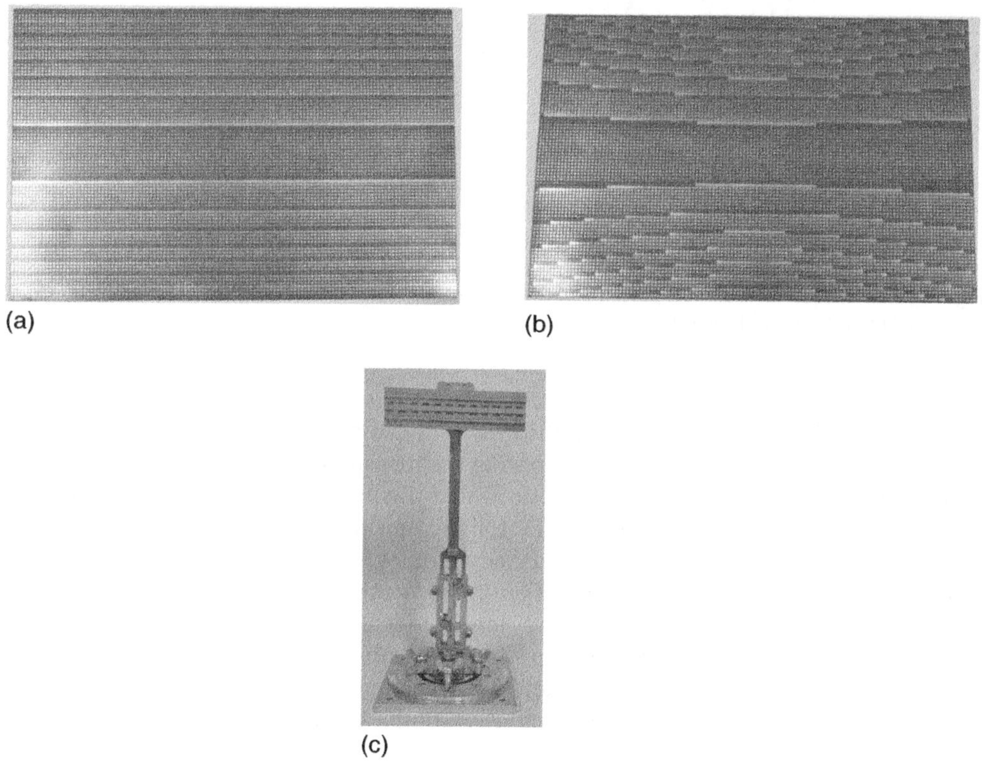

Figure 10.20 Photographs of the fabricated parabolic cylindrical reflectarrays: (a) conventional, (b) optimized, and (c) photograph of the waveguide slot array antenna.

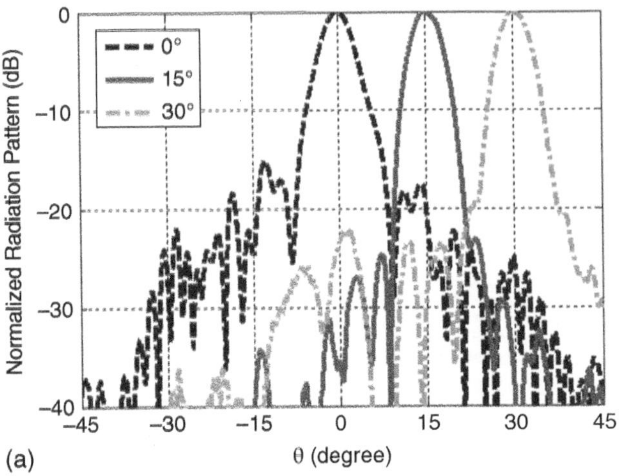

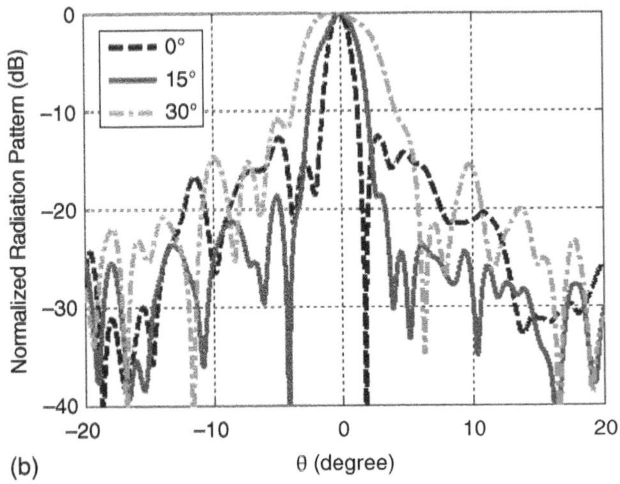

Figure 10.21 Measured radiation patterns of the conventional design at 0°, 15°, and 30° scan angles in the (a) H-plane and (b) E-plane.

the feed array, showing that the reflectarray functions as a mirror in the x-direction. The beamwidth in the E-plane is significantly reduced due to the focusing property of the aperture in the y-direction. By comparing the measured results of the reflectarrays using different design methods, it is noticed that the beamwidth in the E-plane maintains a stable value of approximately 1.5° for the optimized design, while it is significantly broadened for the conventional design from 1.6° at 0° scan angle to 5.3° at 30° scan angle. Accordingly, the gain for the optimized design remains stable within a large scan range by phase optimization, while a significant gain loss of 6.0 dB at 30° scan angle is observed for the conventional design due to its inherent increased phase errors at large scan angles. The gains at 0°, 15°, and 30° scan angles for the optimized design are 31.1 dBi, 31.9 dBi, and 31.2 dBi, respectively. Compared to those of the conventional

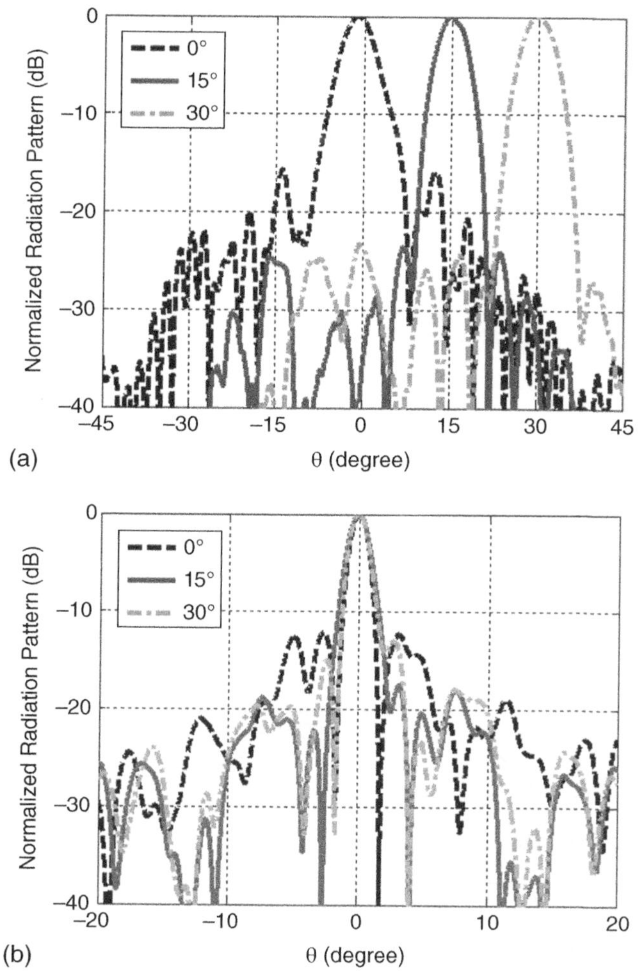

Figure 10.22 Measured radiation patterns of the optimized design at 0°, 15°, and 30° scan angles in the (a) H-plane and (b) E-plane.

design, an improvement of 0.5 dB, 0.6 dB, and 5.9 dB is achieved, respectively. Note that the relatively low gain at 0° scan angle is due to the feed blockage, which can easily be avoided by using an offset configuration.

10.2.2.2 Parabolic Torus-Phase Reflectarray Antennas (PTPRA)

Another reflector antenna system that is used for wide-angle scanning is the parabolic torus. The parabolic torus is generated by rotating a parabola about an axis that is perpendicular to the axis of the parabola [53], [54]. A 3D image of a parabolic torus reflector is given in Figure 10.23(a). The beam of this hybrid reflector can scan many beamwidths along a circular arc in one angular plane, making it a suitable choice for wide-angle 1D scanning. Actually, the parabolic cylindrical reflector discussed in the

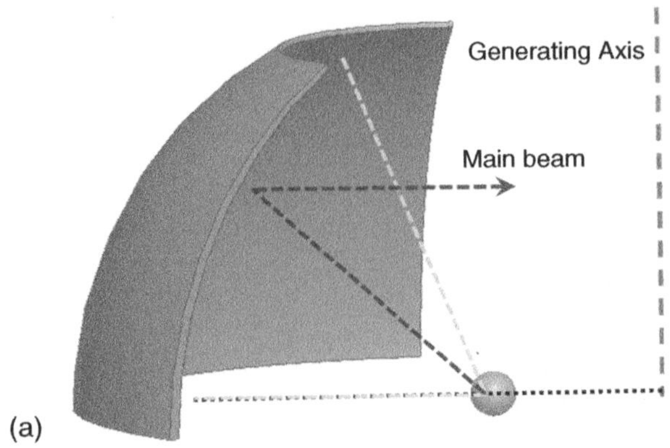

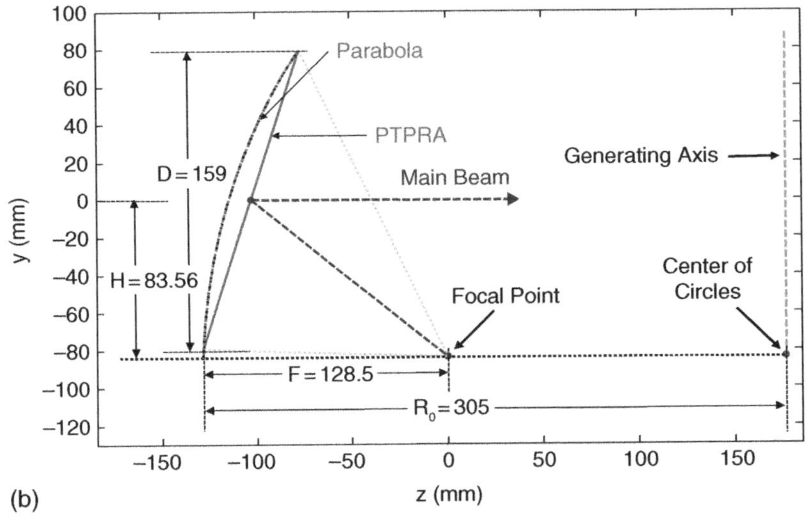

Figure 10.23 (a) Image of a parabolic torus reflector antenna. (b) Geometrical setup of the PTPRA in the parabolic (*yz*) plane. Reprinted with permission from [55].

previous section is a special case of the parabolic torus whose radius goes to infinity. Taking advantage of these unique features of the parabolic torus, one can design a planar printed reflector such that it mimics the parabolic torus, and incorporate the many favorable features of the reflectarray antennas into the beam-scanning system.

To design a reflectarray antenna with parabolic torus phase distribution, the elements on the reflectarray aperture have to compensate for the phase of the parabolic torus surface. The parabolic torus geometry is parabolic in one direction and circular in the orthogonal direction. The geometrical setup of the PTPRA in the parabolic plane (yz) as described in [55] is given in Figure 10.23. The phase term for every element of the PTPRA that lies along this plane can be computed using the conventional parabolic equations. To account for the rotation along the generating axis, a phase term in the orthogonal direction (x) needs to be added, which is similar to the case of the spherical-phase reflectarray. The phase distribution on the aperture of a Ka-band PTPRA designed

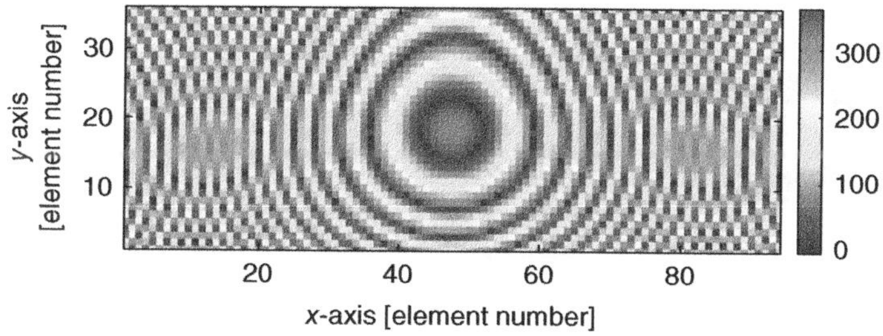

Figure 10.24 The aperture phase distribution on the PTPRA. Reprinted with permission from [55].

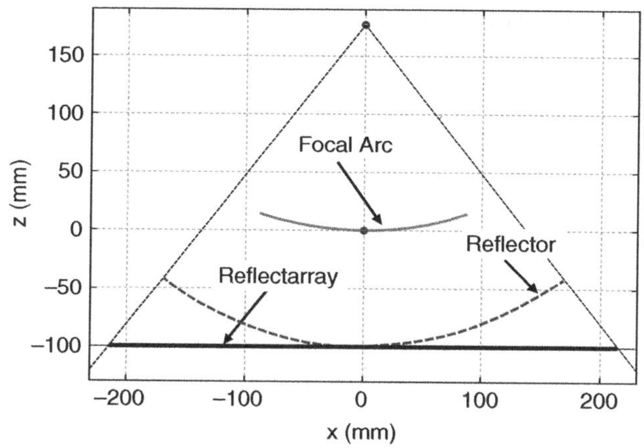

Figure 10.25 Feed displacement path for the PTPRA. *Source:* Nayeri 2014 [55]. Reproduced with permission of IEEE.

with the dimensions shown in Figure 10.23, is depicted in Figure 10.24. The reflectarray has 36 elements along the parabolic direction and 90 elements along the circular direction with a unit-cell periodicity of 4.7 mm. The feed offset angle is 20°.

To scan the beam of the offset PTPRA, the feed antenna should be moved along the focal arc of the reflector system as depicted Figure 10.25. The PTPRA system studied in [55] is designed for 40° scan coverage, that is −20° to +20°. The feed horn used for this reflectarray is a linearly polarized corrugated conical horn with a q value of 6.5 at 32 GHz. The scanned beam patterns from 0° to 20° are shown in Figure 10.26, where it can be seen that a rather stable pattern is obtained with this design over the entire scan range.

10.2.2.3 Spherical-Phase Reflectarray Antennas (SPRA)

Conventionally, the elements on the reflectarray antenna aperture are designed with a parabolic-phase distribution, essentially creating a planar surface that mimics a parabola. However, similar to parabolic reflectors, the scan range of these parabolic-phase planar reflectarrays by means of feed displacement is limited to a few beamwidths [27], [40]. In order to increase the scan range, a circular arc is used to replace the parabolic arc. For one dimensional beam scan, the parabolic torus is discussed in the previous

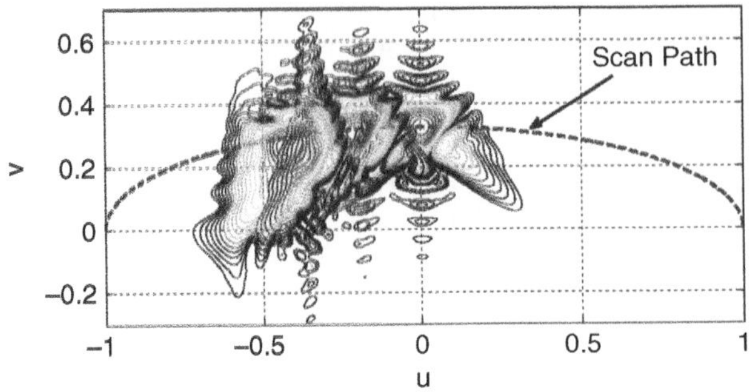

Figure 10.26 Scanned patterns of the PTPRA when the feed is moved along the focal arc of the reflector. *Source:* Nayeri 2014 [55]. Reproduced with permission of IEEE.

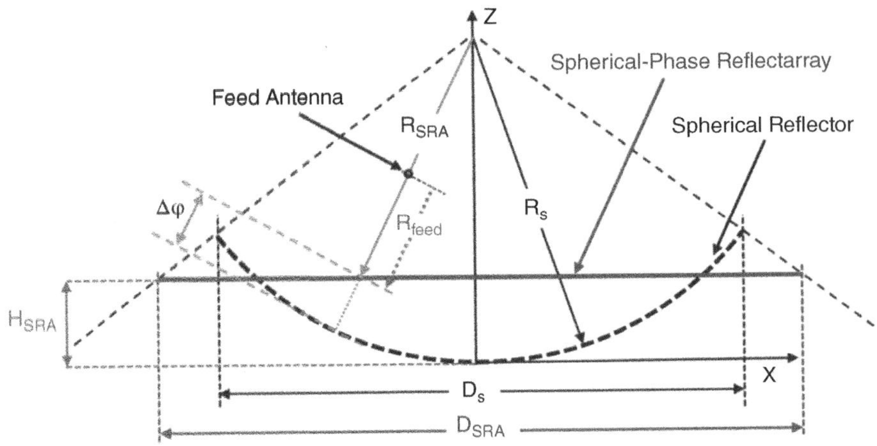

Figure 10.27 The cross-sectional geometry of a spherical-phase reflectarray antenna. *Source:* Nayeri 2014 [57]. Reproduced with permission of IEEE.

section. Spherical reflector antennas have been recognized for years as a suitable design for wide-angle high-gain beam-scanning applications [56]. As such, it could be advantageous if one designs the phase distribution on the reflectarray aperture based on the phase compensation of a comparable spherical reflector.

To design a reflectarray antenna with spherical-phase distribution, the elements on the reflectarray aperture have to compensate for the phase of the sphere ($\Delta\varphi$). To achieve a similar performance as the spherical reflector, the reflectarray is designed with the same subtended angle. A geometrical model of the spherical-phase reflectarray antenna system is given in Figure 10.27, where R_S is the radius of the sphere and D_S is the diameter of the spherical reflector [57]. R_{SRA} and R_{feed} are the vectors from the center of the sphere and the feed antenna to any point on the aperture of the reflectarray, respectively. D_{SRA} is the diameter of the spherical-phase reflectarray.

The required element phase shift for the elements on the spherical-phase reflectarray aperture is then calculated using $\psi = 2k_0\left(R_{SRA} - R_S\right)$. It is worthwhile pointing out that for the spherical-phase reflectarray, once the radius of the corresponding sphere is selected; the elements phase is independent of the feed position. Thus, the next task is to determine the optimum position of the feed.

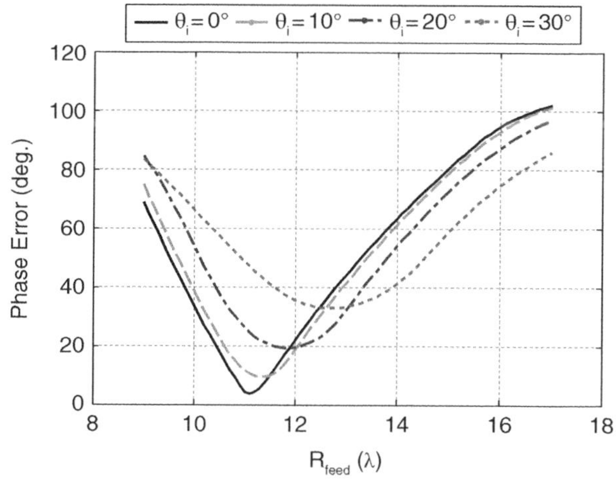

Figure 10.28 Phase error over the aperture of a spherical-phase reflectarray antenna. *Source:* Nayeri 2014 [57]. Reproduced with permission of IEEE.

Similar to spherical reflectors, spherical-phase reflectarrays will also exhibit poor collimating properties due to spherical aberrations. To minimize the adverse effects of spherical aberration, two general approaches have been developed over the years. The first approach uses a restricted aperture and a reflector with a sufficiently large radius. The second approach uses techniques for compensation of spherical aberration, which allows one to utilize a larger aperture. However, the later approach is generally more complex and it also reduces the useful angle of scan [50]. In [57] the authors implemented the first approach, and only illuminated a small portion of the aperture (D_{ill}) that has the minimum phase error.

The optimal focal length for the spherical reflector is determined by the minimum of the phase error curve for any given aperture, which is typically close to the paraxial focus [50]. A similar approach was used to determine the optimal focal of the spherical-phase reflectarray. Numerical results are given in Figure 10.28 for a spherical-phase reflectarray with a diameter of 48λ, which is designed for an illuminated aperture of 15λ and 30° elevation coverage. The corresponding radius of the sphere is $R_S = 23.125\lambda$. It can be seen that for the broadside beam ($\theta_i = 0°$), the minimum phase error is achieved for a feed that is placed at $Z = 11.1\lambda$. Unlike spherical reflectors however, the focal length for a spherical-phase reflectarray is dependent on the beam direction and both the optimal focal and the minimum phase error increase monotonously as the scan angle is increased. This is due to the fact that for scanned beams, the circular aperture with minimum phase error is transformed into a conic section with a somewhat larger aperture. Moreover, unlike a sphere, the beam direction is not normal to the local surface. As a result of these effects, some degradation in pattern shape, beam deviation, and scan loss will be observed with spherical-phase reflectarrays; however, the optimal focal point will ensure that these effects are minimized.

The scanned patterns of this spherical-phase reflectarray with the feed placed at focal optima (the curve minimums from Figure 10.28) is given in Figure 10.29, where it can be seen that the SLL increase is less than 8 dB over the entire scan range. An almost similar pattern is observed in all azimuth planes. The gain variation across the scan

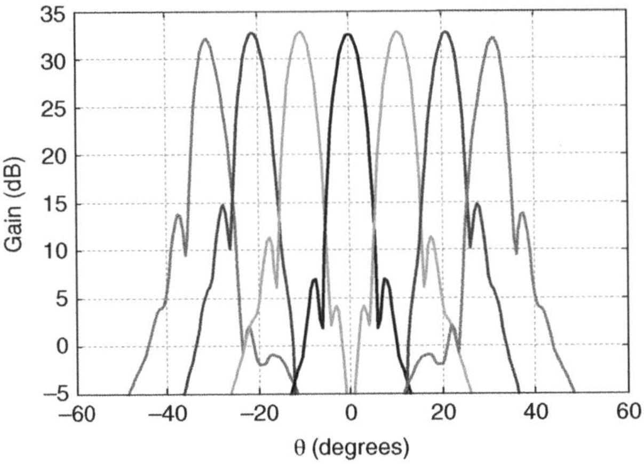

Figure 10.29 Scanned patterns for the spherical-phase reflectarray antenna. *Source:* Nayeri 2014 [57]. Reproduced with permission of IEEE.

coverage is also less than 0.7 dB, which is a notable improvement over the conventional parabolic-type reflectarrays that typically exhibit a scan loss in excess of 3 dB for such coverage.

While the spherical reflector is an ideal candidate for wide-angle scanning, the primary drawback is its slow scan speed. Typically for spherical reflectors, the feed is moved along a spiral path, where all points on the spiral path have an equal distance from the center of the sphere. Consequently, this requisite mechanical movement limits the scan speed of these reflectors. Nonetheless, for many applications such as radio astronomy that don't require a high scan speed, a spherical-phase reflectarray can be a very suitable choice.

10.2.3 Dual-Reflector Configurations

10.2.3.1 Parabolic Reflector/Reflectarray Antennas

In these configurations, beam-scanning is typically achieved by tuning the sub-reflector or the feed antenna. This can be done mechanically, which will be discussed here, or using reconfigurable sub-reflectors that will be reviewed in the next section. A 3-m dual-band Cassegrain reflectarray (X/Ka band) with a reflectarray main reflector was demonstrated in [58]. At the X band (8.4 GHz) and K band (32 GHz), 19,000 and 275,000 annular ring elements provided the phase shift on the reflectarray aperture, respectively. The feed antenna was a microstrip array, which was moved mechanically to scan the beam. A 10-cm movement of the feed is translated to more than 1.5° scan for the main beam at both bands. A photograph of the dual Cassegrain reflectarray antenna is given in Figure 10.30(a). A Ka-band Gregorian reflectarray with a parabolic sub-reflector was also studied in [59], where a similar beam-scanning performance was observed.

A 94 GHz parabolic reflector with a reflectarray sub-reflector was demonstrated in [60]. The initial design used mechanically rotated flat metal sub-reflectors to scan the beam. A 28 × 28 element patch reflectarray fabricated on a metal backed quartz wafer was designed to replace the metal sub-reflectors, and the reflectarray phase control

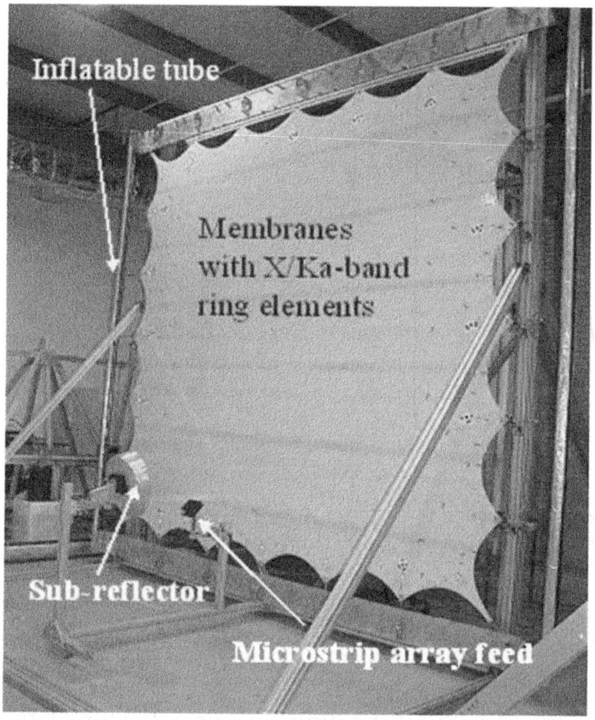

(a)

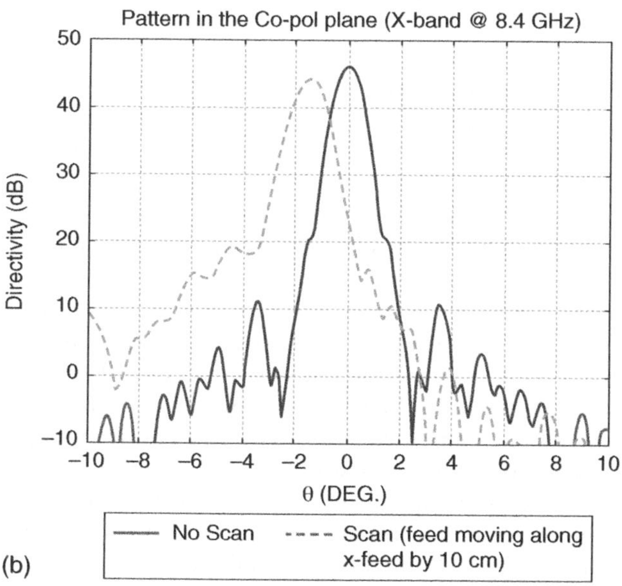

(b)

Figure 10.30 Dual-band beam-scanning dual-reflectarray antenna with the reflectarray as the main reflector: (a) image of the dual-reflectarray antenna, (b) scanned beam in the X band, (c) scanned beam in the Ka band. Reprinted with permission from [58].

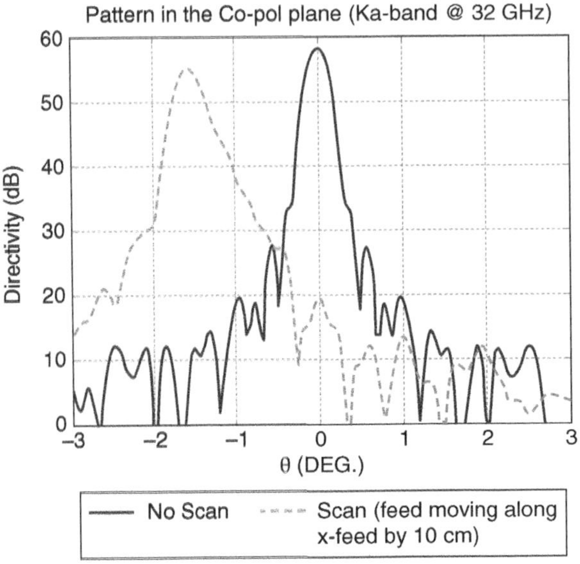

(c)

Figure 10.30 (Continued)

capability was used to scan the beam 5° without rotating the sub-reflector plate. A photo of the prototype and scanning performance of this antenna is given in Figure 10.31.

10.2.3.2 Non-Parabolic Reflector/Reflectarray Antennas

For passive reflectarray designs, a notable improvement in scan coverage can be realized with non-parabolic configurations. One of the best configurations is that of the bifocal dual-reflector antenna [22]. In contrast to the single-reflector bifocal design discussed in the previous section that cannot achieve two perfect foci, the dual-reflector configuration allows for two degrees of phase control, thus enabling a perfect bifocal performance.

A bifocal folded dual-reflectarray antenna was presented in [61], and a seven-element feed array is placed on a focal arc between the two focal points. In this design, the sub-reflector also acts as a polarizer, thus no blockage effects are observed. The prototype was designed for a multi-beam performance at 76.5 GHz; however, it can be used to scan the beam by switching from one feed to the next. The scan range for the design was about 27° (±13.5°); however, this was further improved to 49° (±24.5°) by designing a system with restricted apertures. A photo of the fabricated antenna and a diagram showing the measured scanned beam patterns are shown in Figure 10.32. Recently, a dual reflectarray with dual-offset feeds was also demonstrated [62].

Dual-reflectors with a spherical main reflector are also a very practical choice for wide-angle scanning. The sub-reflector provides a degree of freedom to minimize the effects of spherical aberrations, although this generally reduces the scan coverage.

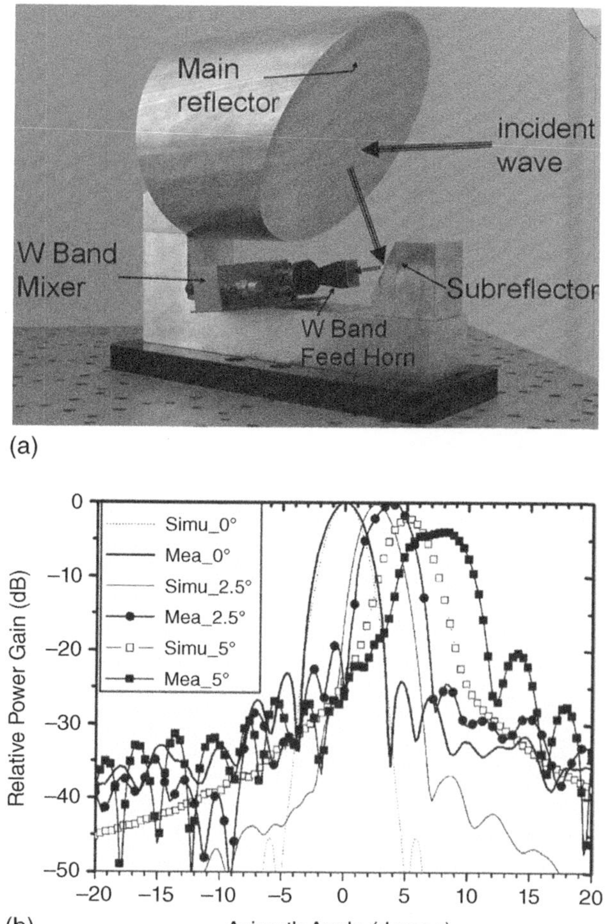

Figure 10.31 A 94 GHz beam-scanning dual-reflectarray antennas with the reflectarray as sub-reflector. (a) Image of the dual-reflectarray antenna. (b) Measured and simulated radiation patterns of dual-reflector antenna with three flat metal sub-reflectors in the azimuth plane at 94 GHz. Reprinted with permission from [60].

A 35.5-m Ka-band spherical reflector antenna with a 0.3-m sub-reflectarray was analyzed in [63]. The main beam is scanned 4° (about 200 beam widths) by rotating the feed and the sub-reflectarray together. The scanned beam patterns for this antenna are given in Figure 10.33, where it can be seen that a very stable radiation pattern is obtained across this wide scan area.

10.2.4 Summary of Feed-Tuning Techniques

In summary, a multitude of reflector configurations, including the single reflector or dual reflector, fully illuminated aperture or partially illuminated aperture, parabolic-phase or non-parabolic-phase, are available for beam-scanning reflectarrays that can

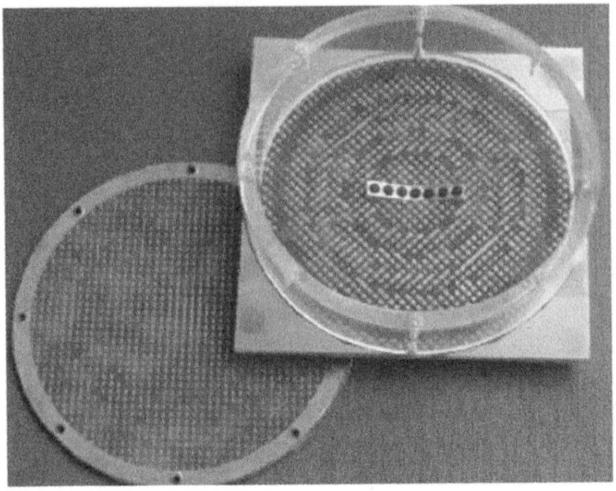

(a)

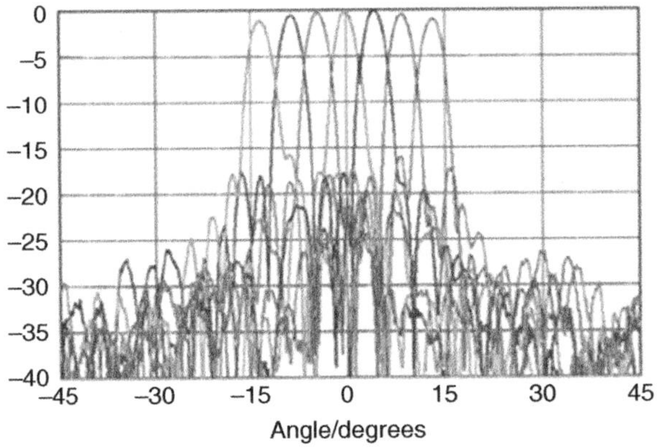

(b)

Figure 10.32 Bifocal dual-reflectarray antenna for 27° scan coverage: (a) prototype, (b) measured scanned beam patterns. Reprinted with permission from [61].

effectively utilize the feed-tuning technique. The primary advantage of the feed-tuning technique is the ease of fabrication for passive arrays. This feature also allows for very high-gain and high-power operation at low cost. Some of the key characteristics of this beam-scanning technique are tabulated here.

Advantages:

- Ease of fabrication for passive arrays
- Very high-gain operation
- High RF power handling

Limitations:

- Depending on the design, it may have slow scan speed
- Feed array may be complicated and expensive

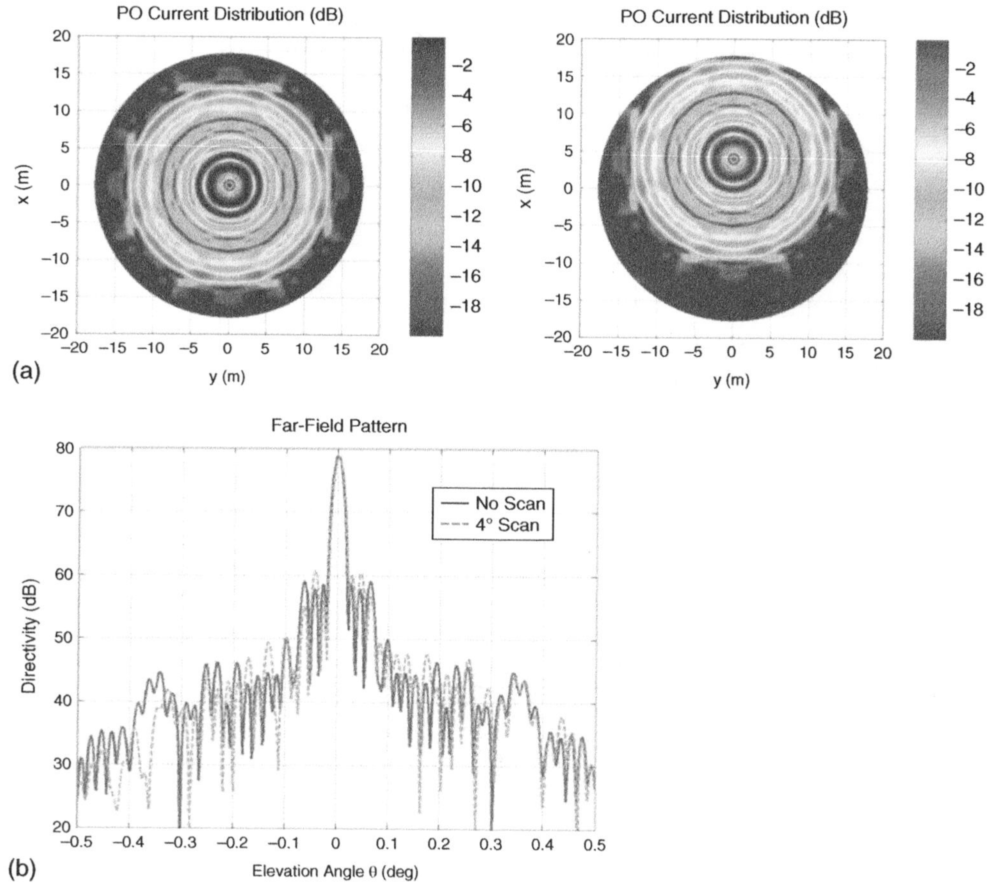

Figure 10.33 Spherical reflector antenna with sub-reflectarray: (a) PO current distribution on the main reflector, (b) scanned beam patterns. Reprinted with permission from [63].

10.3 Aperture Phase-Tuning Techniques

10.3.1 Basics of Aperture Phase Tuning

As discussed in the earlier sections, the aperture of a reflectarray antenna consists of phasing elements that are designed to realize a certain reflection phase response. Different approaches are available for tuning the phase of reflectarray elements, which in general can be categorized into three groups:

1) Elements with phase/time delay lines,
2) Variable size elements,
3) Rotated elements.

In the first group, the incident wave is first received by the element, then phase shifted using a delay-line microwave network, and finally re-radiated. This methodology is known as the guided-wave technique, and is essentially the concept Berry [64] used

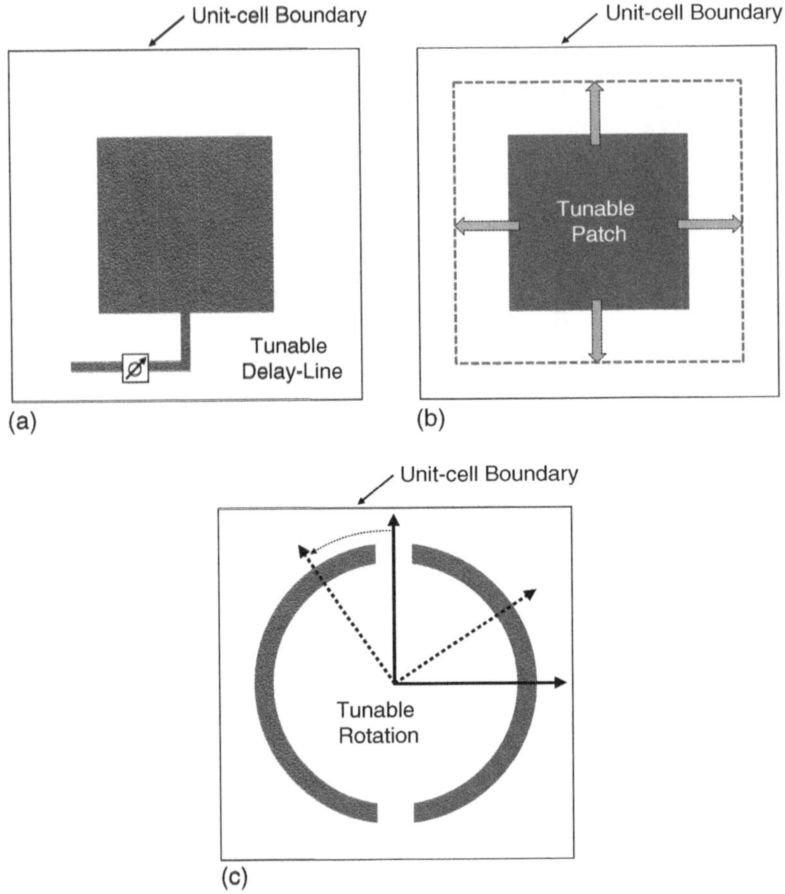

Figure 10.34 Phase-tuning approaches for reconfigurable reflectarray elements: (a) tunable size elements, (b) tunable delay-line elements, and (c) tunable rotated elements.

when introducing the first reflectarray antenna in 1963 that utilized short-circuited waveguide elements. With the second group, variable size elements, the phase of the reflected field is controlled by changing the dimensions of the element, for example the patch length. The last group, which is applicable only to circularly polarized (CP) designs, takes advantage of the unique property of an incoming CP wave, that is the 90° phase difference between the two orthogonal electric field components. It is shown that by rotating the element on the aperture plane by ψ, the phase of the reflected field will be rotated by 2ψ; hence this approach is known as element rotation. While there is a fundamental difference in the principal of operation between the second and third group, in essence both approaches directly modify the radiating structure and the phase shifting mechanism is directly integrated with the element.

In passive (or fixed) reflectarray configurations, such as the ones reviewed in the previous section, the reflection phase for each element is a fixed quantity. For dynamic phase apertures, however, a tunable phase shifting mechanism must be incorporated into the elements. The three general approaches discussed in this section for reflectarray elements can be realized in a reconfigurable form, which are schematically illustrated in Figure 10.34.

In the tunable delay-line approach, the incoming wave is received by the element (receiver patch), and then guided into a transmission line of length l, which is equipped with a microwave phase shifter ($\Delta\phi$), and open circuit at the end, as shown in Figure 10.34(a). The guided wave then reflects back through the line, going through the phase shifter again, and finally re-radiating from the patch, which now acts as a transmitter. The change in the re-radiated phase is then equal to $2(\beta l + \Delta\phi)$. If all delay lines are of equal length, which is typically the case, the relative phase difference between the elements will be controlled by the tunable phase shifter, that is $2\Delta\phi$. One notable advantage of this tunable guided-wave phase shifting technique is that it simply allows for integration with microwave components [65]. A drawback, however, is that typically for a good performance; this approach requires using multilayer configurations.

In the second approach, the effective electrical dimension of the patch (or resonator) is changed, which is illustrated in Figure 10.34(b), for example by using varactor diodes. The concept is similar to frequency-agile antennas where a tunable device changes the resonance frequency of the antenna. In the case of reflectarray elements, however, this concept is utilized to provide phase agility at the operating frequency. Typically, a fixed physical dimension that provides the maximum tunable range is selected for all the array elements, however, a major challenge with this approach is achieving a complete phase range for the elements. Nonetheless, different loading techniques and element configurations have been proposed to solve this problem, which will be discussed further in this section.

Tunable element rotation, depicted in Figure 10.34(c), can be realized by means of both physical and electrical rotation. In the former, micromotors are used to physically rotate the element, thus controlling the phase of the reflected CP wave. Electrical rotation on the other hand essentially rotates the currents on the element, hence avoiding a physical movement. This can be realized by selectively activating appropriate sections of a rotationally symmetric element. It is worthwhile pointing out here that similar to tunable resonators, a good advantage of this approach is that that it can be realized with a single-layer configuration. The drawback of both these integrated phase-shifting techniques is that it is very complicated to integrate them with microwave components.

10.3.2 Enabling Technologies

Similar to phased array antennas, and regardless of the choice of element phase-tuning approach, a primary challenge is how the element is integrated with the phase-tuning mechanism. Here we will take a close look at various enabling technologies that have been used in designing phase-tuned reflectarray antennas. Tuning techniques based on mechanical actuation are presented in Section 10.3.2.1. In Section 10.3.2.2, various electronic devices that can be integrated into the reflectarray element to achieve either discrete or continuous tuning are presented. Functional materials are then described in Section 10.3.2.3, where several promising designs are reviewed. In each given class, the tuning mechanisms share common attributes, for example, low scan speed with mechanically actuated designs, or limited RF power handling with electronic devices. A summary of the advantages and limitations of various tuning methods is presented at the end of each part.

10.3.2.1 Mechanical Actuators/Motors

In this approach, phase agility is achieved by means of a mechanical actuation, which ultimately changes the physical dimension or orientation of the elements. For reflectarray elements both linear (translation) and rotary (rotation) motions can be utilized to achieve aperture phase tuning. For high frequency designs, micro-actuators and motors, typically fabricated using micromachining techniques, with maximum dimensions in the order of a few millimeters are available. While various types of actuators such as electric, magnetic, electromechanical, thermal, hydraulic, pneumatic, and chemical are available in the market [66], only a handful of designs using electrostatic and electromechanical type actuators have been experimentally demonstrated for reflectarray antennas. The following paragraphs take a closer look at some of the mechanically actuated beam-scanning reflectarray designs that have been developed over the years.

10.3.2.1.1 Mechanical Rotation

Utilizing the element rotation technique, mechanical phase tuning can be achieved by placing micromotors beneath each element as proposed in [14], [67], [68]. This element design approach is applicable for CP elements and is capable of achieving a full phase range. Experimental results were demonstrated in [16] for a small prototype, achieving 10° of beam scan. A schematic model of the design concept and a fabricated prototype are shown in Figure 10.35.

More recently, a 15 × 15 element reflectarray operating at 8.4 GHz, using dual concentric split ring elements and equipped with micromotors was demonstrated in [69]. An image of the reflectarray prototype is given in Figure 10.36. Each element is connected to a micromotor through a hole on ground plane. A supporting pedestal made of Kapton is specifically designed to suspend the substrate above the ground plane, which also forms an air layer to improve the element performances. This design achieved reconfigurability with low cross-polarization and low element loss. The simulated maximum gain of this reflectarray was 26.43 dB.

10.3.2.1.2 Mechanical Translation

Micro electrostatic actuators were proposed in [70] to adjust the height of each patch element in the reflectarray over the ground plane. The electrostatic fields are generated through a DC voltage bias applied between the patch and the ground plane, which moves the patch vertically and changes the fringing fields, thus allowing for phase control of the reflected fields by controlling the DC bias. In general, however, a full phase cycle is hard to be achieved with this design. A schematic model of this reflectarray element is given in Figure 10.37(a).

Another linear motion mechanical actuation was demonstrated in [71] with a multilayer configuration, where the entire middle layer, consisting of two sets of slots, was displaced with a digital caliper to change the relative position of slots and collimate the beam at −30°, 0°, +30°. It should be noted that with this design the elements don't have independent phase control, and this approach is more suitable for beam-switching applications. A photo of the fabricated prototype is shown in Figure 10.37(b).

In general, the primary drawbacks of mechanical phase-tuning techniques arise from the requisite physical displacement. As such, these configurations have received little attention and only a handful of designs have been demonstrated. Some of the key characteristics of these types of beam-scanning reflectarrays are tabulated here.

Advantages:

- High RF power handling
- Low insertion loss and continuous phase tuning

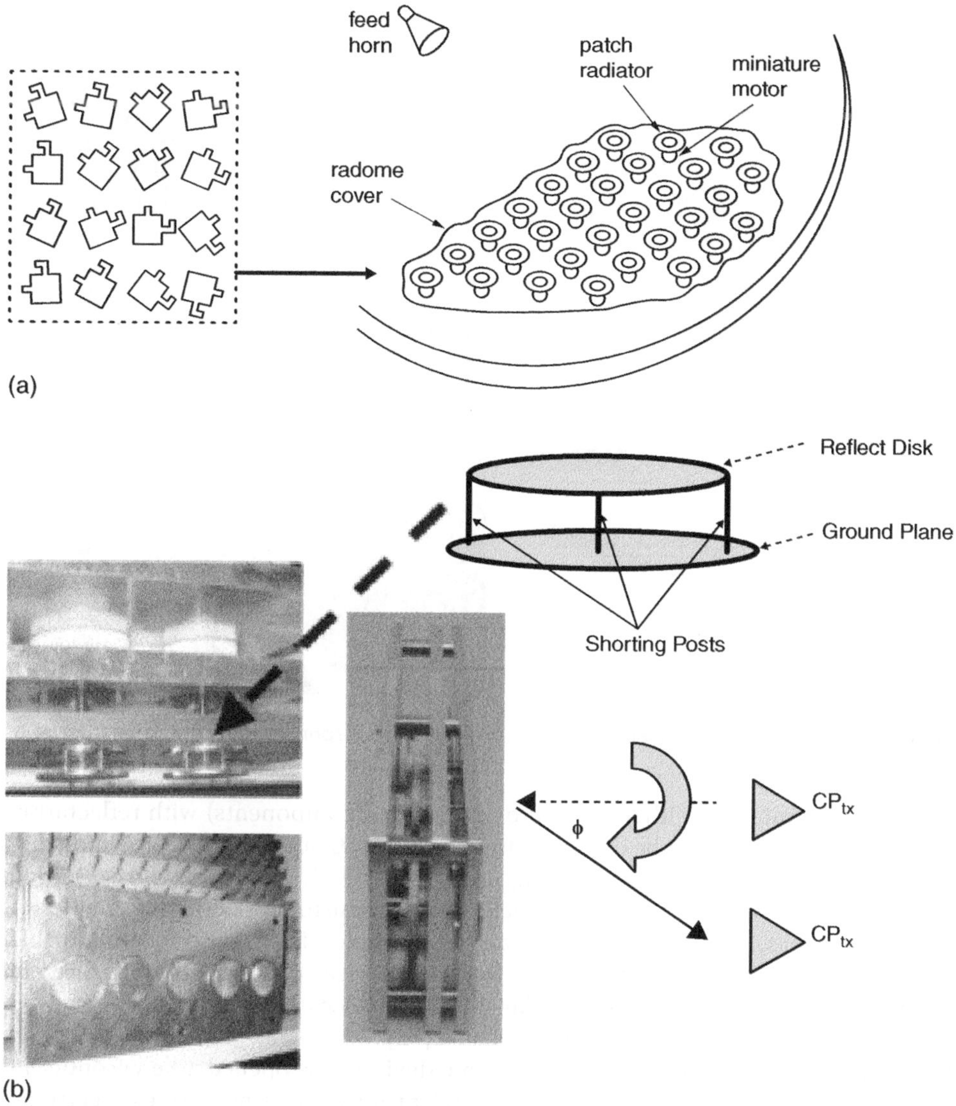

Figure 10.35 (a) The concept of mechanically rotated elements. (b) A mechanically beam-scanning reflectarray prototype demonstrating a scan range of 10°. *Source:* Fusco 2005 [16]. Reproduced with permission of IEEE.

Limitations:

- Slow scan speed (milliseconds)
- Typically require high bias levels for MEMS components

10.3.2.2 Electronic Devices

Electronic phase tuning provides the highest scan speed, and is therefore a subject of great interest for beam-scanning applications. Thanks to the recent advances in packaging and miniaturization, a variety of electronic devices are commercially available

Figure 10.36 A CP beam-scanning reflectarray using micromotors. Reprinted with permission from [69].

that can be integrated (as either lumped or distributed components) with reflectarray elements. These devices can be divided into two classes, that is switches providing discrete phase states (digital phase control), and devices with continuous phase tuning (analog phase control). PIN diodes and field-effect transistors (FETs) are examples of electronic switches, and varactor diodes are an example of an electronically tunable device. Micro-electromechanical systems (MEMS) devices can provide both discrete (MEMS switch) and continuous (tunable MEMS devices) phase control; however, digital MEMS devices are typically more reliable [72].

The promising capabilities of these electronic devices, has opened a new frontier in reconfigurable reflectarray designs, and several prototypes using various electronic devices have been demonstrated. In this section, we will review some of the electronic beam-scanning reflectarray antennas that have been developed over the years.

10.3.2.2.1 PIN Diode Switches

The PIN diode switch is a semiconductor device that can be made to behave as either a short (On state) or open circuit (Off state) by controlling the biasing dc voltage across its terminals. Phase shifting reflectarray antennas that use a series of half-wavelength dipoles connected to the periphery of a circular metal layer by means of diodes were proposed in [73]. This characteristic makes it a very good candidate for delay-line phase shifters.

A 60 GHz electronically reconfigurable reflectarray with 160×160 elements using PIN diode switches, shown in Figure 10.38, was experimentally demonstrated [74]. The phasing elements are rectangular microstrip patches connected to a short-circuited stub through a PIN diode. This 1-bit design provides two states for each element, which

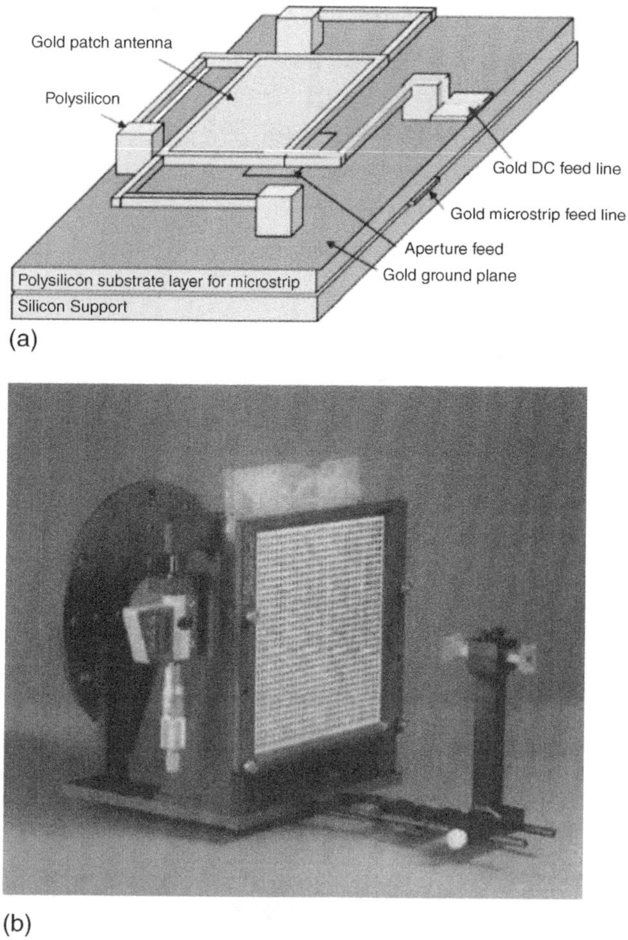

Figure 10.37 Mechanically actuated beam-scanning/switching reflectarrays: (a) Electrostatic actuators for reflectarray patch elements. Reprinted with permission from [70]. (b) Surface translation. Reprinted with permission from [71].

could ideally realize a 180° step between successive phases, although a slight difference between the phase steps was observed in the experimental results.

A number of other designs using a similar 1-bit phase control mechanism have also been demonstrated. A wideband reconfigurable reflectarray element consisting of a circular patch, aperture coupled to a phase shifting circuit was demonstrated in [75]. In this design, phase control has been implemented at sub-array level, using aperture-coupled patches gathered by pairs to a common delay line. Dual polarization operation and a flat phase shift response over a 50% relative bandwidth were obtained with this element. The element configuration is given in Figure 10.39(a). A 244-element beam-switching X-band reflectarray antenna using PIN diodes was also demonstrated in [76]. The antenna was designed to switch the beam between ±5°, in a tilted scanning plane. The scanned beams are obtained by forward biasing half the diodes, while the other half remained in reverse bias. Broadside beam was obtained when all the diodes are in reverse bias. The schematic model of the element is given in Figure 10.39(b).

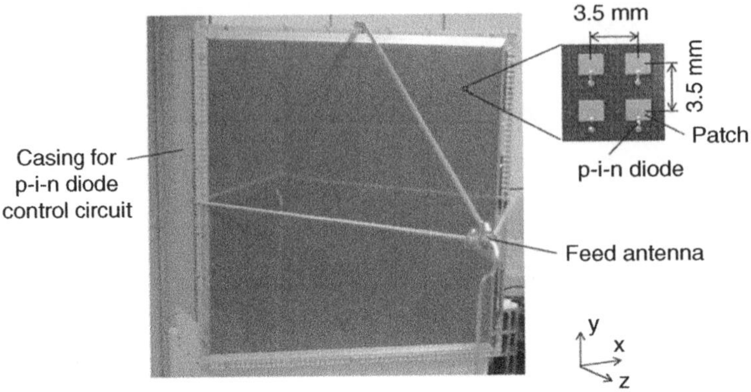

Figure 10.38 A 25,600 element electronic beam-scanning reflectarray antenna using PIN diodes demonstrating 50° scan coverage in both elevation and azimuth planes. Reprinted with permission from [74].

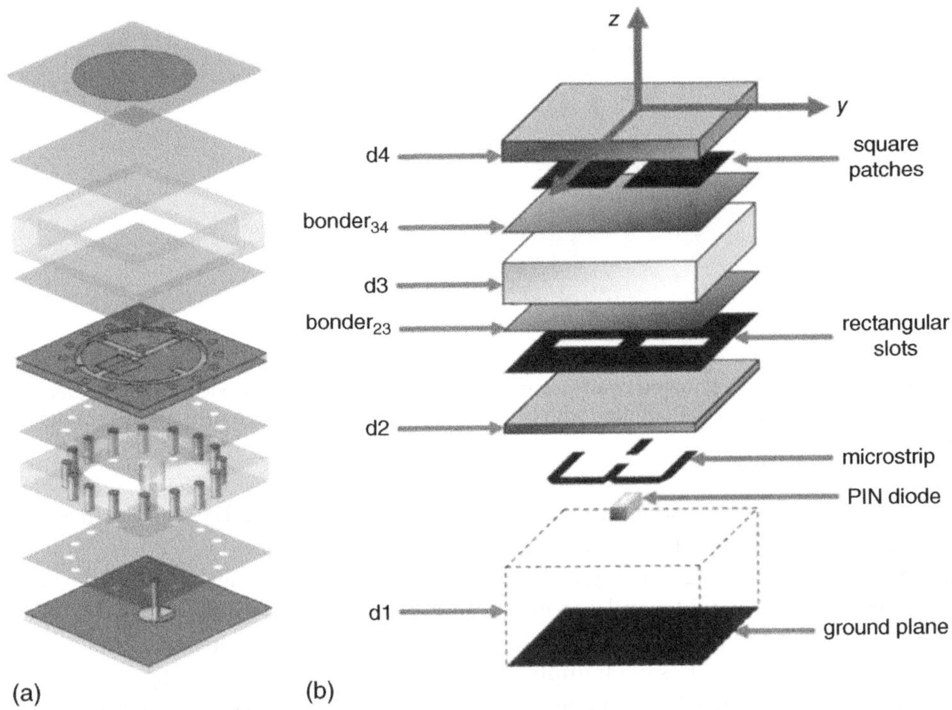

Figure 10.39 Electronic beam-scanning reflectarray elements using PIN diodes. (a) Exploded view of the elementary cell in [75] where phase control is implemented at the sub-array level. Reprinted with permission from [75]. (b) Expanded view of a reflectarray element based on gathered patches aperture-coupled to a common delay line, using a PIN diode as electronic control switch. Reprinted with permission from [76].

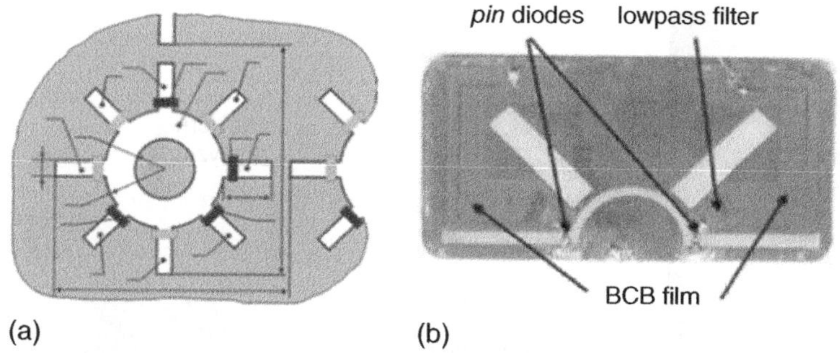

Figure 10.40 Electronic element rotation technique for beam-scanning reflectarray antennas: (a) schematic, (b) element prototype. Reprinted with permission from [78].

Another very interesting application of PIN diodes is in electronic element rotation, which was initially proposed in [77] where selective activation of the lumped elements of the rotationally symmetric geometry mimics a rotation of the reflected electromagnetic wave [78], [79], [80]. The concept was demonstrated experimentally using ring slot resonators for a constitutive reflectarray unit-cell [78]. This electronically rotated element is shown in Figure 10.40. Experimental results for full arrays were also given in [80], however, due to the complexity of implementation, only static arrays were demonstrated.

10.3.2.2.2 FET Switches

FET devices, particularly metal semiconductor FET (MESFET) can be used as microwave switches for reflectarray elements. The FET switch draws almost no dc current in either the On or Off state, which gives it an advantage over the PIN diode switch in terms of a lower dc power consumption. However, a comparative study between PIN diodes and MESFET for Ka-band patch elements with 2-bit delay-line phase shifters [81] showed that while the beam-scanning performances for these two designs were quite similar, the PIN diode design showed a better performance in terms of RF insertion loss. As such, FET switches are seldom used for electronic beam-scanning reflectarrays.

10.3.2.2.3 MEMS Switches

MEMS RF switches show a momentous promise for reconfigurable reflectarray antennas [82]–[91]. In comparison with PIN diodes or other solid-state switches, they have lower insertion loss, high linearity, very low dc power consumption, and higher power handling capability. Moreover, they can be readily integrated with antennas in a monolithic fabrication process. These integrated electronic devices use micro-machined cantilevers or membranes that can be deflected to provide an open or short for an RF circuit. MEMS devices can be designed for either direct (ohmic) or capacitive-coupled contact with the RF circuit. Different techniques can be used for mechanical actuation in MEMS, such as electrostatic, electromagnetic, and piezoelectric, where electrostatic beam actuation is the most common technique. In this section, we will take a closer look at some of the reported beam-scanning reflectarray antennas that use MEMS switches.

Several designs of reflectarray elements have been proposed for dynamic phase control using MEMS devices. A patch element loaded with two slots controlled by a set of 10 MEMS devices was demonstrated in [18]. Another reconfigurable reflectarray

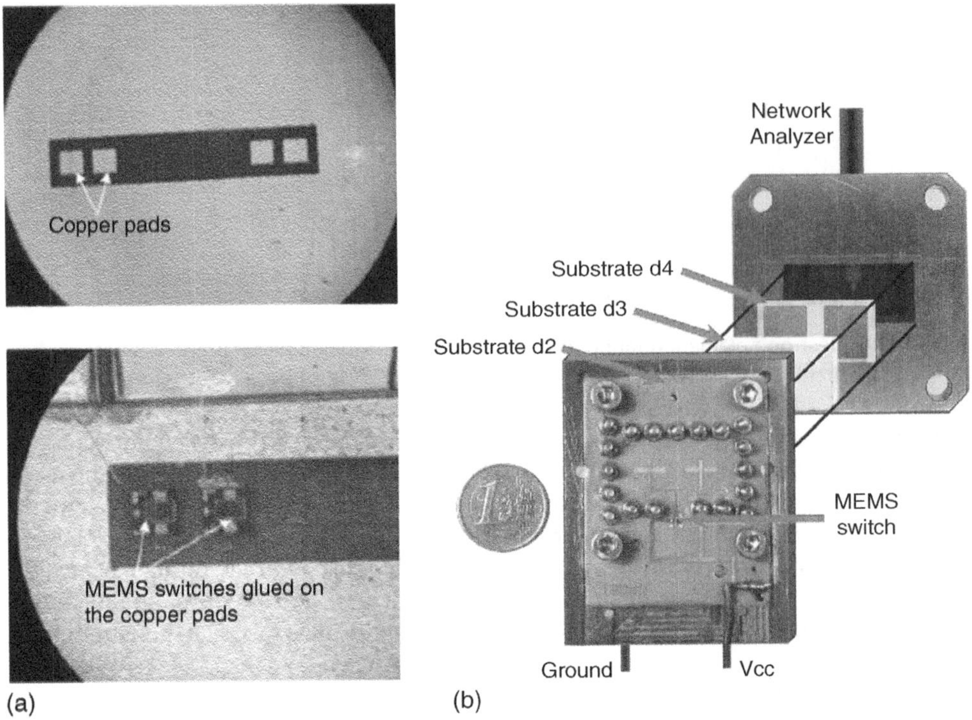

Figure 10.41 Beam-scanning reflectarray elements using MEMS switches: (a) Variable length slot configuration. Reprinted with permission from [86]. (b) Aperture-coupled patch configuration. Reprinted with permission from [87].

element consisting of a patch and variable length slots using MEMS switches was demonstrated in [86]. In this design, 4 RF MEMS switches were mounted on the slot symmetrically about the center. Photos of the fabricated element prototype are given in Figure 10.41(a). A 2-bit X-band reflectarray element with electronic phase control implemented by ohmic MEMS switches was also demonstrated in [87]. The element is based on two patches aperture-coupled to a microstrip network with a common delay line in the sub-array configuration. The experimental model of the element in the waveguide simulator is shown in Figure 10.41(b).

The first functional MEMS reflectarray prototype operating at 26.5 GHz was demonstrated in [88] using aperture-coupled microstrip patch elements. This 100-element beam-switching reflectarray antenna is monolithically integrated with 90 RF MEMS switches. The series RF MEMS switches provide only two states, corresponding to a broadside beam and a 40° scanned beam in the H-plane. The schematic model of the element and photos of the fabricated prototype are shown in Figure 10.42.

A dual-band CP reflectarray antenna using one ring for each band, actuated with six RF MEMS switches per ring has also been demonstrated [89]. This design utilizes the element rotation technique for circular polarization and provides independent beam-switching in both K and Ka bands. Experimental results, however, were only given for static (frozen) MEMS states.

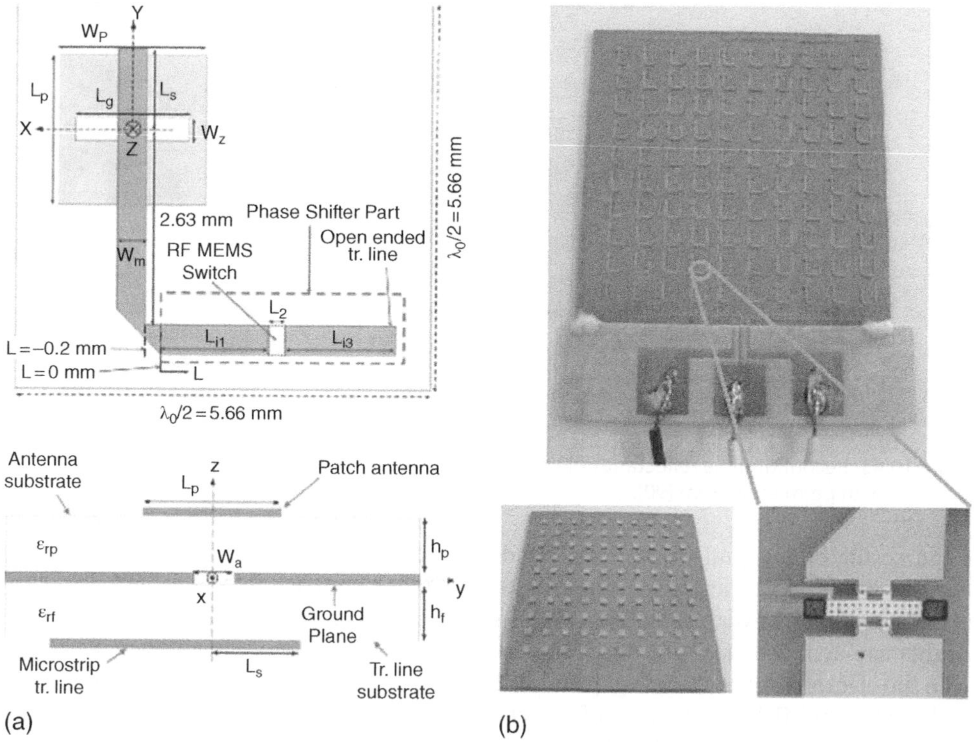

Figure 10.42 A beam-scanning reflectarray antenna using MEMS demonstrating a scan range from 0° to 40° off-broadside: (a) element model, (b) 100 element prototype. Reprinted with permission from [88].

Reflectarray elements integrated with MEMS capacitors are also a very practical application of MEMS technology for beam-scanning designs. Most MEMS variable capacitors have been developed as digitally switched capacitors; however, analog MEMS designs have also been explored, which will be addressed later on. The basic idea of using capacitors in the reflectarray elements is that an element loaded with an electronically controlled capacitor can achieve frequency agility. In these designs, this property is utilized to provide phase agility. A 5-bit digital controlled monolithic MEMS capacitor reflectarray element was demonstrated in [90]. Each digital capacitor provides an Up-state corresponding to zero voltage, and a Down-state corresponding to a pulled-in MEMS membrane. In this design, the capacitors were integrated with ring resonators providing 32 states with 12° steps between successive phases. At 12 GHz the loss of this design was less than 0.3 dB. A photograph of this digital MEMS capacitor reflectarray element is given in Figure 10.43.

It is important to point out here that while MEMS switches have many advantages over PIN diodes and FET switches, they typically have a relatively slower switching speed. In addition, they are usually more expensive than other electronic switches and require higher biasing voltages.

10.3.2.2.4 MEMS Tunable Devices

Tunable MEMS capacitors have significant potential in addressing the loss and linearity issues, and if built correctly, they exhibit very high-quality factors [91]–[94]. For these analog designs, the continuous tuning range is limited in some simpler designs, which

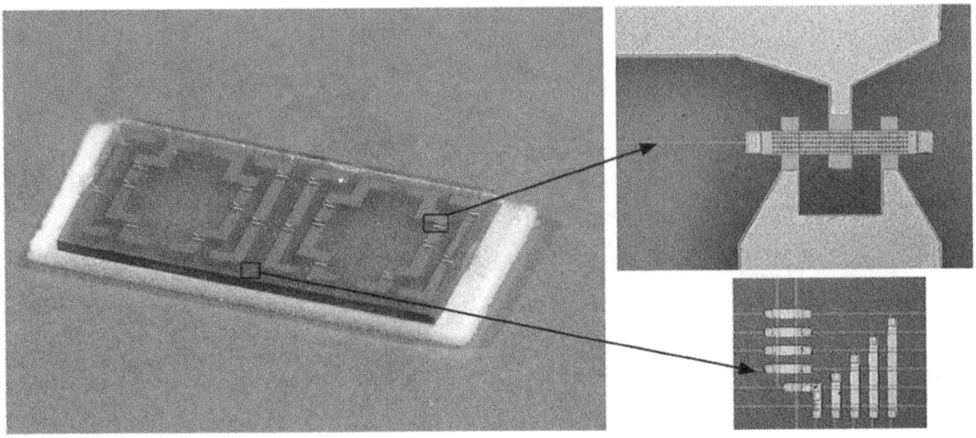

Figure 10.43 Reconfigurable reflectarray elements using digitally switched MEMS capacitors. Reprinted with permission from [90].

is the result of the pull-in phenomena due to the imbalance between the squared increase in electrostatic force compared to the linear increase in restoring force with displacement; however, advanced deigns such as two-layered bridges with dual gap heights can avoid the pull-in tuning limitation by separating the capacitance plate gap from the electrostatic electrode gap and increase the tuning range. A two-layered bridge dual gap height MEMS capacitor reflectarray element was experimentally demonstrated in [91]. This design provided a phase agility of about 230° for the elements at 5.8 GHz with a loss about 3 dB at resonance. The authors pointed out that the technology is capable of achieving a better performance and this can be alleviated in the future with electroplating. A photograph of this tunable MEMS capacitor reflectarray element is given in Figure 10.44.

10.3.2.2.5 Varactor Diodes

The varactor diode is a semiconductor p-n junction device, which can provide variable junction capacitance by controlled applied bias voltage. Integrated with reflectarray phasing element, varactor diodes can provide the means to control the capacitance of the elements. A phase-agile reflectarray element using two varactor diodes that serially connect two halves of a microstrip patch was demonstrated in [94]. The phase range of these elements was 320°. A 10×7 element prototype reflectarray was designed for the operating frequency of 5.8 GHz and experimentally demonstrated a scan range from −50° to +50°. Another C-band reconfigurable reflectarray using varactor diodes was demonstrated in [95]. This design consists of a microstrip patch, printed on a flexible membrane substrate and aperture coupled to a transmission line loaded with two varactor diodes. A full phase range was achieved with this element. A 30-element prototype was fabricated and wide-angle 2D beam-scanning was experimentally demonstrated. The maximum loss of the elements was 2.4 dB for this design. Photographs of these two-varactor loaded beam-scanning reflectarray antenna prototypes are given in Figure 10.45.

While several designs of varactor diode loaded reflectarray elements have been demonstrated over the years [92]–[99], the focus in the recent years has been on performance improvement of these configurations. In [98] the study revealed that a

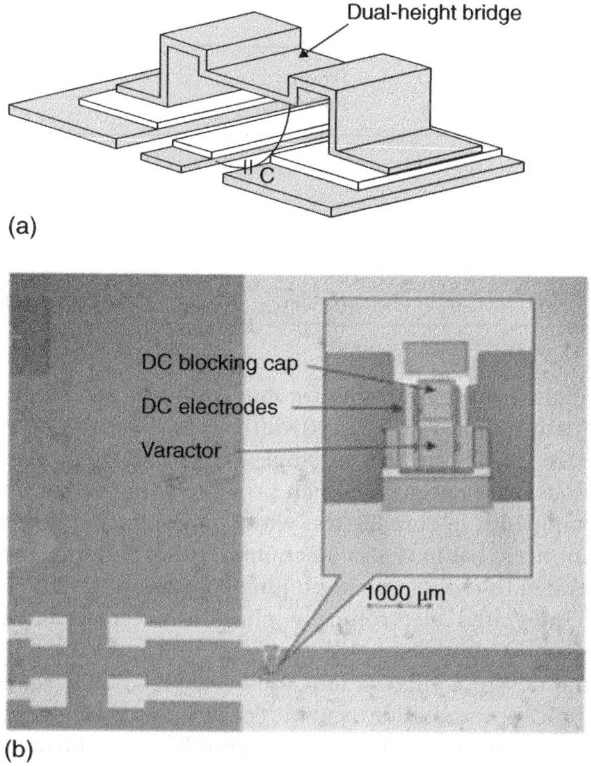

Figure 10.44 Beam-scanning reflectarray element using a tunable MEMS device: (a) schematic model of the MEMS capacitor, (b) element prototype. Reprinted with permission from [91].

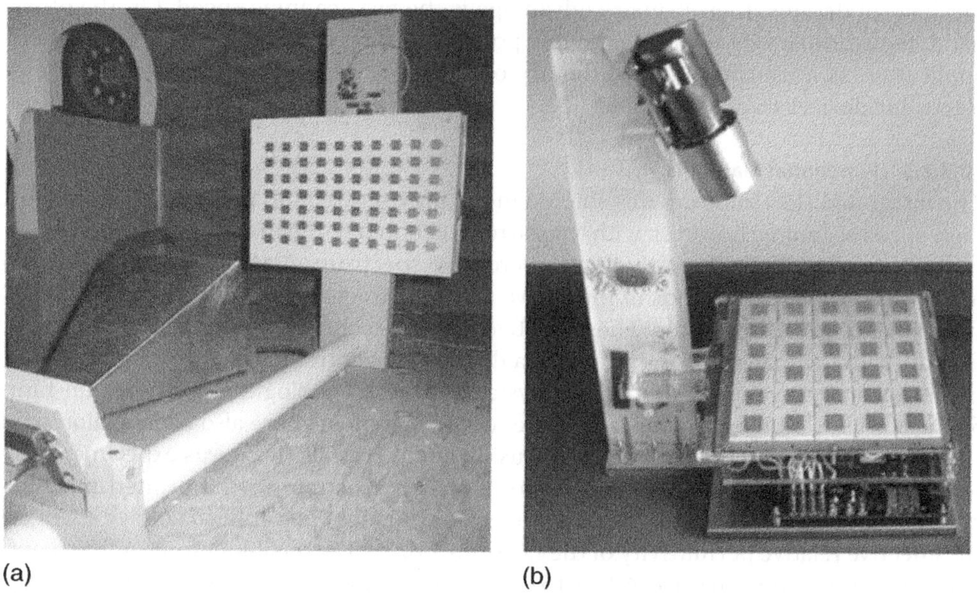

Figure 10.45 Electronic beam-scanning reflectarray antennas using varactor diodes. (a) A 70-element prototype demonstrating a 100° azimuth scan coverage. Reprinted with permission from [94]. (b) A 30-element prototype demonstrating 2D beam-scanning. Reprinted with permission from [95].

Table 10.2 Key characteristics of some electronic devices used in reconfigurable reflectarrays.

Device	Tuning	Power Handling	Bias	DC Power Consumption	RF Loss
PIN	Discrete	Low	Low voltage	Moderate	Moderately low
FET	Discrete	Low	Low voltage	Low	Moderately high
MEMS	Discrete/ continuous	Moderate	High activation voltage	Low	Low
Varactor	Continuous	Moderate	Moderately high-voltage	Low	Moderate

single patch with a shunt-connected varactor diode can be designed to attain the full phase-tuning range. Bandwidth improvements of varactor diode tuned reflectarrays were studied in [99]. A single-layer double-square loop element using four varactor diodes was designed, which provides over 380° of phase agility, and nearly doubles the bandwidth in comparison with single-resonance designs. It is also important to point out here that in the same context, tunable impedance surfaces have also been demonstrated to redirect an impinging plane wave [100].

Integrated electronically phase-tuned reflectarray antennas show a momentous promise for wide-angle beam-scanning applications, and a myriad of electronic devices and design configurations are available. The designs reviewed in this section are just some representative examples of integrated electronic devices that have been proposed for use in electronically reconfigurable reflectarray antennas. In general, the choice of switch or tunable device strongly depends on the application. In comparison with mechanical designs, their primary advantage is the fast scanning speed. On the other hand, their primary drawback is the low power handling capability, which limits their application as a transmitter. A summary of some of the key characteristics of these electronic devices is given in Table 10.2.

10.3.2.3 Functional Materials

The integrated electronic devices studied in the previous section provide the means for high-speed scanning; however, with the current technology, the majority of these devices cannot operate efficiently at the high microwave spectrum. A promising approach to address the growing demands in mm-wave and THz frequencies is based on the use of functional or exotic materials. These relatively newer technologies are based on reconfiguration of the material, for example, via distributed control of the dielectric constant of the substrate in the case of liquid crystals. Other technologies based on thin-film ferroelectrics, ceramics, plasma, and more recently graphene have also been explored.

A beam-scanning reflectarray antenna using liquid crystals (LCs) was experimentally demonstrated in [19]. The cell consists of a carrier substrate with a printed metallic patch on the lower side, and an LC cavity beneath the patch. The applied voltage changes the effective relative permittivity of the LC from 2.62 to 3.04, enabling some adjustment of the phase of a reflected wave. A 16 × 16 reflectarray prototype, shown in Figure 10.46(a), with unidirectional bias control was demonstrated for Ka-band operation. Two other similar designs operating at 102 and 130 GHz were also demonstrated in [101]. Another design combining LCs and multilayer technology of low temperature co-fired ceramics

Figure 10.46 Liquid crystal beam-scanning reflectarray antennas. (a) A Ka-band prototype demonstrating a 40° scan coverage. Reprinted with permission from [19]. (b) A 256 element 77 GHz prototype. Reprinted with permission from [102].

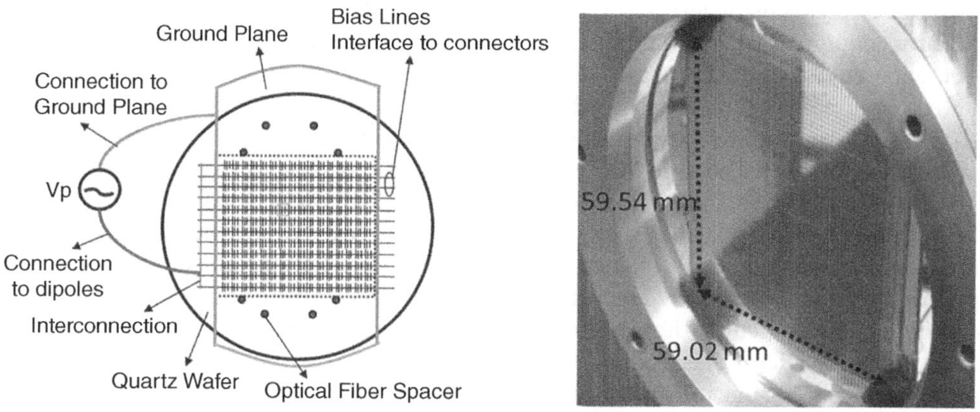

Figure 10.47 Liquid crystal reflectarray prototype. Reprinted with permission from [104].

(LTCC) was also demonstrated in [102], and a 256-element array prototype operating at 77 GHz was demonstrated. The prototype is shown in Figure 10.46(b).

Due to the limited tunability of the material properties in LCs, most of these designs exhibit some limitations in the phase agility of the elements. To mitigate these problems, a multi-resonant unit-cell made up of three parallel dipoles in the same cell and a tunable liquid crystal as substrate was proposed in [103], [104]. The reconfigurable reflectarray [104], shown in Figure 10.47, operates in the frequency range from 96 to 104 GHz, and consists of 52 × 54 identical cells. The multi-resonant element configuration also demonstrated an improved bandwidth. The authors mention that the technology and the manufacturing process used could be applied to produce reconfigurable antennas in frequencies from 60 to 500 GHz.

Thin-film ferroelectric-based devices [105] have matured greatly over the past decade. While ferrite phase shifters are still considered the main workhorse in military phased

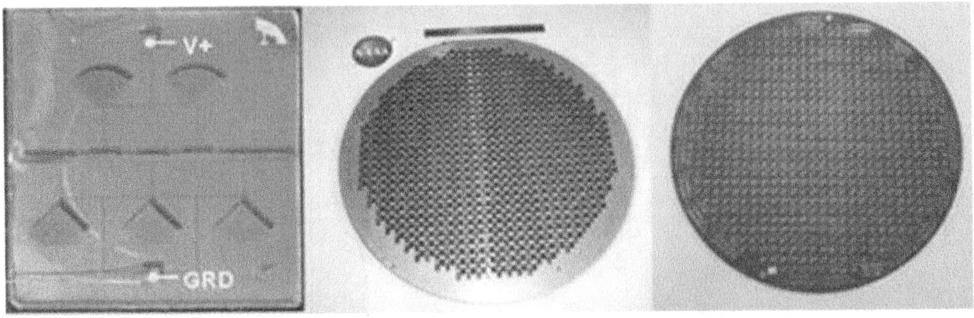

Figure 10.48 A 19-GHz beam-scanning reflectarray antenna using thin-film ferroelectric phase shifters. Reprinted with permission from [106].

arrays, interest in ferroelectric-based devices is mounting, because of their high-power handling capability, negligible DC power consumption, and potential for low loss and cost [7]. In particular the high-power handling capability of these devices cannot be matched by any of the other designs presented in this section. A 19 GHz reflectarray with 615 elements using ferroelectric thin-film device was demonstrated in [106]. Patch radiators equipped with reflection-mode ferroelectric phase shifters were used for the elements. Photos of the fabricated prototype are given in Figure 10.48.

Reconfigurable unit-cells using thick-film ceramic materials were also studied in [107]. A very new and yet promising material for THz beam-scanning applications is graphene [108]. In contrast to LC design, the substrate material is fixed, and resonance is achieved by changing the complex conductivity of the graphene patch [109]. Plasma based phase control is also a very interesting approach for designing beam-scanning reflectarray antennas. In [110], surface PIN (SPIN) diodes are proposed that generate plasma on the diode surface due to the injected carriers. Patterns of antenna elements can then be generated on the aperture by selectively activating the SPIN diodes on the grid. In another design laser is used to induce plasma [111], where the pattern of the photo-induced plasma inside the silicon was changed by using optical illumination, which generated plasma that caused the effective length of the slot to shrink. More recently, fluidic phase tuning has also been explored [112]. Also, effective material implementations using metallic conductors have recently been proposed that seem to provide wide scanning ranges [113].

Addressing the challenges at high frequencies still remains as a fundamental concern for researchers in this field. While these new technologies are very encouraging, they are certainly not mature enough yet for practical applications. Moreover, they exhibit losses that are significantly higher than RF electronic devices. Nonetheless, with the potential capabilities that these new technologies can offer, it seems that they will be the state-of-the-art for THz and optical beam-forming applications in the near future. Some key characteristics of these functional material beam-scanning reflectarrays are summarized here.

Advantages:
- Potential for THz, infrared, and optical applications
- Potentially offer low-loss continuous phase tuning

Limitations:

- Technology is not mature
- High-cost prototyping

10.4 Frontiers in Beam-Scanning Reflectarray Research

In the previous sections, the basic design methodologies and enabling technologies for beam-scanning reflectarrays were outlined, and some recent examples were reviewed. However, in comparison with the very mature technology in beam-forming reflector systems and phased array antennas, the reflectarray technology is still in its development stage, and many advanced design concepts are just recently being explored. In this section, we will review some of these advanced concepts.

10.4.1 Active Reflectarrays

Integration of active devices, mainly amplifiers, with reconfigurable reflectarray elements is an area of considerable interest that has very recently been explored. The concept is similar to active phased array antennas [6]. In the case of reflectarrays, however, the primary motivations are increasing the overall gain of the antenna, or compensating for phase shifter losses. The first active reflectarray antenna was demonstrated in [114] where the X-band amplifying elements were designed using FET transistors. The prototype is shown in Figure 10.49(a). In the same work, the authors demonstrated another design for active spatial power combining. A few other designs for active reflectarrays have also been demonstrated [115], [116]. The first active reconfigurable design was demonstrated in [117], where a 48-element beam-scanning reflectarray prototype operating at 5.7 GHz was demonstrated. Photographs of the active reflectarray prototype are given in Figure 10.49(b).

While these active designs have great potential, very little work has been done on active reconfigurable reflectarrays. In comparison with active phased arrays that use a very mature T/R module technology, active reconfigurable reflectarrays are still in a very primitive state.

10.4.2 Comparison Between Analog and Digital Phase Control

In aperture phase tuned designs, scan resolution and pattern control typically depend on the phase range and available phase states of the elements, that is phase quantization effects in array antennas [1], [2]. As such the choice of analog or digital phase control (and the number of bits) is generally a primary concern. While continuous phase tuning is certainly advantageous for small array configurations, in most cases for large arrays (such as reflectarrays), discrete phase control provides satisfactory results due to the phase front averaging of the collimated beam. Moreover, low-bit discrete phase control is generally less complicated to implement, which is quite advantageous for large arrays. In the same context, increasing the number of bits (or phase states) in digital phase shifters increases the complexity of the design. As a result, in practice for a specific application, one must determine the minimum requirements of the element before selecting the optimal enabling technology.

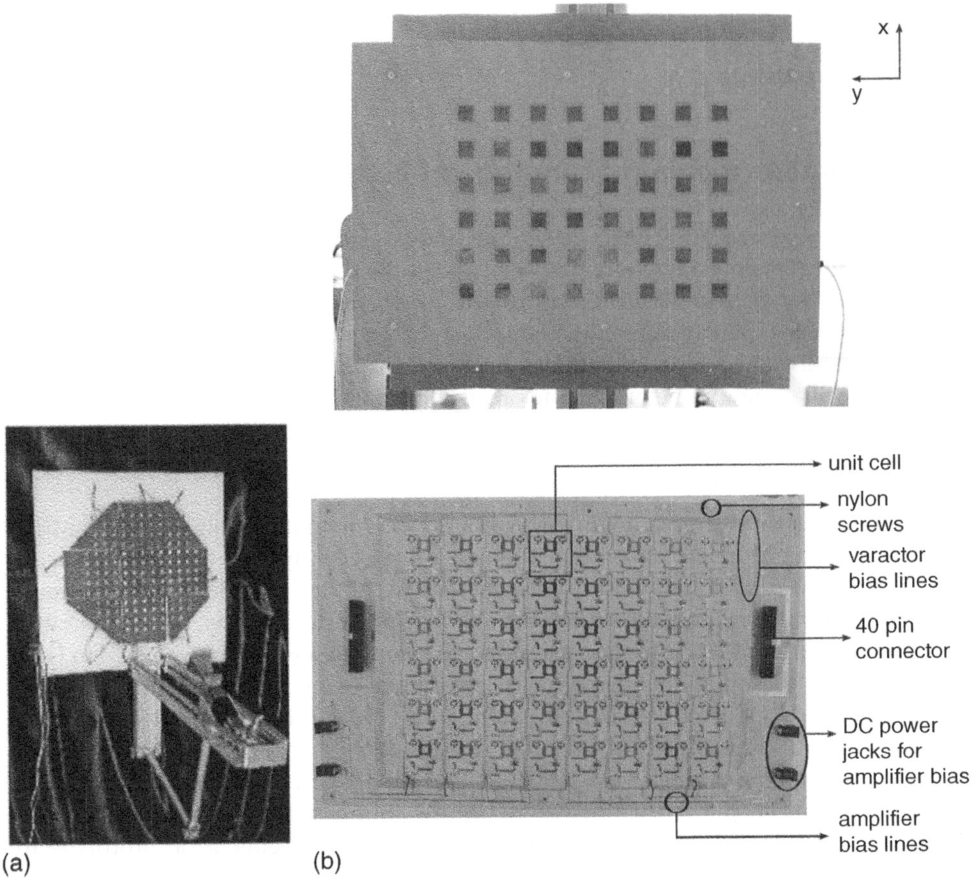

Figure 10.49 Active reflectarray prototypes. (a) A fixed beam design. Reprinted with permission from [114]. (b) A reconfigurable design. Reprinted with permission from [117].

To quantitatively illustrate these design considerations, we study the performance of a reflectarray antenna with a diameter of 20λ with ideal phase tuned elements. The edge taper is set to −10 dB for this reflectarray system. Broadside and scanned patterns are obtained as described in [118] and are given in Figure 10.50 using different phase control capabilities.

The radiation patterns show that while the performance of the 1-bit design is rather poor, 2- and 3-bit configurations can provide acceptable performance [119]. It is interesting to notice that the beamwidths are similar in all these cases, since they are determined by the aperture size. In general, while for wide scan-angle capability or better SLL control, increasing the number of bits is desirable, for large arrays, 3-bit phase control provides comparable performance with the analog case over all scan angles, as shown in Figure 10.51.

Combined with the practical issues of cost, volume, and manufacturability for large arrays, it is expected that similar to phased arrays, digitally controlled designs with 3 or 4 discrete bits will be more desirable for electronically reconfigurable reflectarrays [14], [120].

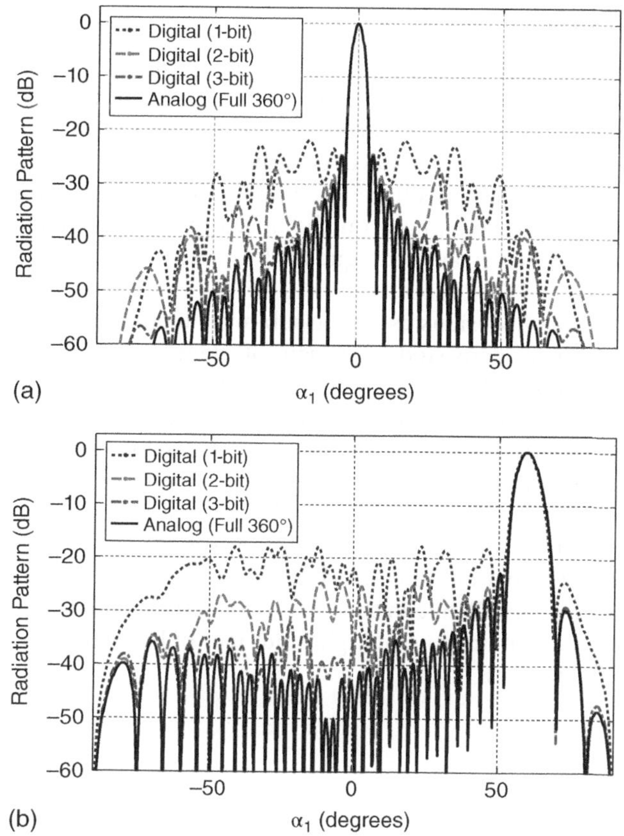

Figure 10.50 Comparison of reflectarray radiation pattern with digital and analog phase control: (a) broadside beam, (b) scanned beam. *Source:* Nayeri 2015 [8]. Reproduced with permission of IEEE.

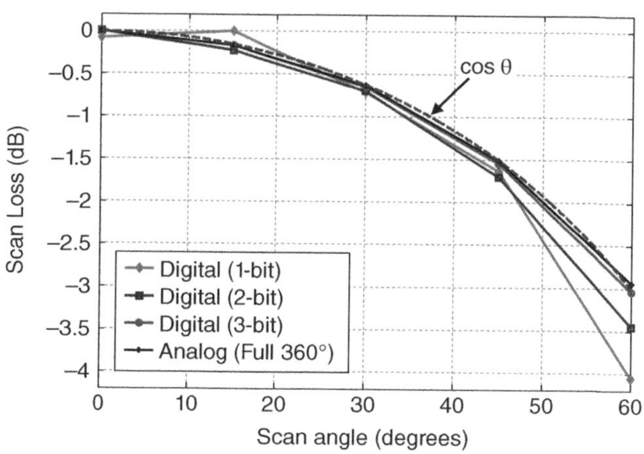

Figure 10.51 Scan-loss performance of reflectarrays with digital and analog phase control.

10.4.3 Sub-Array Techniques

Organizing an array into equally spaced sub-arrays is a well-known concept in phased array antennas. The primary advantage of this technique is reduction of the number of control bits, which are a considerable portion of the array cost. A number of beam-scanning reflectarrays using the sub-array technique have recently been demonstrated. In contrast to phased arrays, reflectarray elements have to compensate for the spatial delay, in addition to providing the progressive phase on the aperture for a scanned beam. Consequently, implementing the sub-array technique for beam-scanning reflectarrays is more complicated.

Recently, two reconfigurable reflectarray designs using the sub-array technique were demonstrated [76], [87]. Each sub-array consisted of a pair of elements. A more advanced sub-array technique was proposed in [121], where delay lines were used to compensate for the spatial delay from the feed. A 400-element prototype with only 40 control lines was demonstrated with this design; however, only a static case was experimentally studied. The prototype is shown in Figure 10.52.

In general, for these configurations, control at row or column level (and in general the sub-array level) significantly reduces the scan capability. A fully controlled sub-reflectarray configuration was studied in [122] that demonstrated the beam-scanning operation in a range ±6° using 3-bit ideal switches. While in this case quantization effects did not result in a significant pattern distortion, in general higher number of bits would be required for reconfigurable sub-reflectarrays.

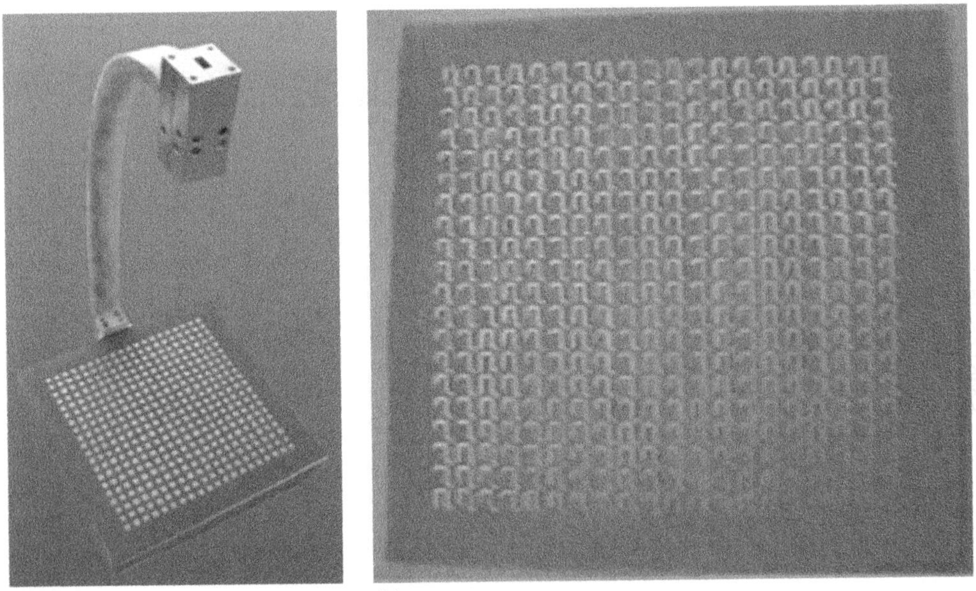

(a) (b)

Figure 10.52 A reflectarray prototype with reduced number of control bits: (a) reflectarray and feed horn, (b) phase shifting layer. Reprinted with permission from [121].

10.4.4 Hybrid Configurations

Given the hybrid nature of reflectarray antennas, one can combine both scanning methodologies, that is feed displacement and aperture phases tuning, to improve the scan performance and reduce the system complexity. While very little work has been done in this context, a dual-reflector configuration using a phase-tunable sub-reflectarray has been proposed. Dual-reflector/array configurations with a phase-tunable sub-reflectarray can be a very suitable choice for beam-scanning, since the use of sub-reflectarray reduces the number of control elements. A beam-scanning 1.5 m parabolic reflector antenna using a sub-reflectarray at 11.95 GHz was studied in [24]. In this work, the scanning was achieved by introducing a progressive phase shift along the reflectarray sub-reflector, which allows for 1D scanning. Simulated results were given using ideal phase shifters, which showed a scan range from −2° to +2° (how many beamwidth) with a gain reduction less than 1 dB in all the cases.

References

1 R. C. Hansen, *Phased Array Antennas.* 2nd Edn, Wiley Series in microwave and Optical Engineering, Hoboken, NJ: John Wiley & Sons, Inc., 2009.

2 R. J. Mailloux, *Phased Array Antenna Handbook*, 2nd Edn, Artech House, 2005.

3 M. I. Skolnik, *Radar Handbook*, 3rd Edn, McGraw-Hill, 2008.

4 M. E. Cooley, and D. Davis, "Reflector antennas," in *Radar Handbook*, 3rd Edn, McGraw-Hill, 2008.

5 A. D. Olver, P. J. B. Clarricoats, A. A. Kishk, and L. Shafai, *Microwave Horns and Feeds*, The Institution of Electrical Engineers, 1994.

6 A. J. Fenn, D. H. Temme, W. P. Delaney, and W. E. Courtney, "The development of phased-array radar technology," *Lincoln Laboratory Journal*, Vol. 12, No. 2, 2000.

7 R. R. Romanofsky, "Array phase shifters: theory and technology," in *Antenna Engineering Handbook*, 4th Edn, J. Volakis, McGraw-Hill, 2007.

8 P. Nayeri, F. Yang, and A. Z. Elsherbeni, "Beam-scanning reflectarray antennas: a technical overview and state-of-the-art," *IEEE Antennas Propag. Magazine*, vol. 57, no. 4, pp. 32–47, Aug. 2015.

9 W. Menzel, D. Pilz, and M. Al-Tikriti, "Millimeter-wave folded reflector antennas with high gain, low loss, and low profile," *IEEE Antennas Propag. Mag.*, Vol. 44, No. 3, pp. 24–29, Jun. 2002.

10 S. J. Blank and W. A. Imbriale, "Array feed synthesis for correction of reflector distortion and vernierbeamsteering," *IEEE Trans. Antennas Propag.*, Vol. 36, No. 10, pp. 1351–1358, Oct. 1998.

11 Y. Rahmat-Samii, "Array feeds for reflector surface distortion compensation: concepts and implementation," *IEEE Antennas and Propagation Magazine*, Vol. 32, No. 4, pp. 20–26, Aug. 1990.

12 B. Saka and E. Yazgan, "Pattern optimization of a reflector antenna with planar-array feeds and cluster feeds," *IEEE Trans. Antennas Propag.*, Vol. 45, No. 1, pp. 93–97, Jan. 1997.

13 L. Shafai, S. K. Sharma, and S. Rao, *Handbook of Reflector Antennas and Feed Systems: Volume 2 – Feed Systems*, Artech House, 2013.

14 J. Huang, "Bandwidth study of microstrip reflectarray and a novel phased reflectarray concept," presented at the *IEEE Antennas and Propagation Society International Symposium, California*, USA, June 1995.
15 A. Yu, F. Yang, A. Z. Elsherbeni, J. Huang, and Y. Rahmat-Samii, "Aperture efficiency analysis of reflectarray antennas," *Microwave and Optical Technology Letters*, Vol. 52, No. 2, pp. 364–372, Feb. 2010.
16 V. F. Fusco, "Mechanical beam scanning reflectarray," *IEEE Trans. Antennas Propagat.*, Vol. 53, No. 11, pp. 3842–3844, Nov. 2005.
17 S. Hum, M. Okoniewski, and R. Davies, "Realizing an electronically tunable reflectarray using varactor diode-tuned elements," *IEEE Microw. Wireless Compon. Lett.*, Vol. 15, No. 6, pp. 422–424, 2005.
18 H. Legay, B. Pinte, M. Charrier, A. Ziaei, E. Girard, and R. Gillard, "A steerable reflectarray antenna with MEMS controls," in *Proc. 2003 IEEE Int. Symp. Phased Array Syst. Tech.*, 2003, pp. 494–499.
19 A. Moessinger, R. Marin, S. Mueller, J. Freese, and R. Jakoby, "Electronically reconfigurable reflectarrays with nematic liquid crystals," *Electron. Lett.*, Vol. 42, No. 16, pp. 899–900, Aug., 2006.
20 T. S. Chu, "Polarization properties of offset dual-reflector antennas," *IEEE Trans. Antennas Propag.*, Vol. 39, No. 12, pp. 1753–1756, Dec. 1991.
21 Y. Rahmat-Samii, "Reflector antennas," in *Antenna Handbook: Theory, Applications, and Design*, Y. T. Lo and S. W. Lee (eds), Van Nostrand Reinhold, 1988.
22 B. L. J. Rao, "Bifocal dual reflector antenna," *IEEE Trans. Antennas Propag.*, Vol. 22, No. 5, pp. 711–714, Sep. 1974.
23 F. S. Holt, and E. L. Bouche, "A Gregorian corrector for spherical reflectors," *IEEE Trans. Antennas Propag.*, Vol. 12, No. 1, pp. 44–47, Jan. 1964.
24 C. Tienda, M. Arrebola, and J. A. Encinar, "Recent developments of reflectarray antennas in dual-reflector configurations," *International Journal of Antennas and Propagation.*, Vol. 2012, pp. 1–10, 2012.
25 D. M. Pozar, S. D. Targonski, and H. D. Syrigos, "Design of millimeter wave microstrip reflectarrays," *IEEE Trans. Antennas Propag.*, Vol. 45, No. 2, pp. 287–296, Feb. 1997.
26 T. Milligan, *Modern Antenna Design*, 2nd Edn, Chichester, UK: John Wiley & Sons, Ltd, 2005.
27 S. R. Rengarajan, "Scanning and defocusing characteristics of microstrip reflectarrays," *IEEE Antennas and Wireless Propag. Letters*, Vol. 9, pp. 163–166, 2010.
28 J. Lanteri, J. Y. Dauvignac, Ch. Pichot, and C. Migliaccio, "Beam-scanning improvement of reflectarrays by reducing the cell size at millimeter waves," *Microwave and Optical Technology Letters*, Vol. 48, No. 5, pp. 966–968, May 2006.
29 S. Rengarajan, "Scanning properties of faceted reflectarrays," presented at the IEEE Antennas and Propagation Society International Symposium, Illinois, USA, July 2012.
30 A. Roederer, "Reflector antenna comprising a plurality of panels," USA Patent no. US 2001/0020914 A1, Sep. 13, 2001.
31 D. M. Pozar, "Bandwidth of reflectarrays," *Electron. Lett.*, Vol. 39, No. 21, Oct. 2003.
32 J. A. Encinar, "Design of two-layer printed reflectarrays using patches of variable size," *IEEE Trans. Antennas Propag.*, Vol. 49, No. 10, pp. 1403–1410, Oct. 2001.
33 M. R. Chaharmir, J. Shaker, M. Cuhaci, and A. Ittipiboon, "A broadband reflectarray antenna with double square rings," *Microwave and Optical Technology Letters*, Vol. 48, No. 7, pp. 1317–1320, July 2006.

34 D. M. Pozar, "Wideband reflectarrays using artificial impedance surfaces," *Electron. Lett.*, Vol. 43, No. 3, Feb. 2007.
35 E. Carrasco, J. A. Encinar, and M. Barba, "Bandwidth improvement in large reflectarrays by using true-time delay," *IEEE Trans. Antennas Propag.*, Vol. 56, No. 8, pp. 2496–2503, Aug. 2008.
36 P. Nayeri, F. Yang, and A. Z. Elsherbeni, "A broadband microstrip reflectarray using sub-wavelength patch elements," presented at the *IEEE Antennas and Propagation Society International Symposium, South Carolina*, USA, June 2009.
37 P. Nayeri, F. Yang, and A. Z. Elsherbeni, "Broadband reflectarray antennas using double-layer subwavelength patch elements," *IEEE Antennas Propag. Lett.*, Vol. 9, pp. 1139–1142, 2010.
38 L. Liang, and S. V. Hum, "Design of an ultra-wideband reflectarray using transformation optics," presented at the *IEEE Antennas and Propagation Society International Symposium, Florida*, USA, July 2013.
39 J. Huang and J. A. Encinar, *Reflectarray Antennas*. Hoboken, NJ: Wiley-IEEE, 2008.
40 P. Nayeri, F. Yang, and A. Z. Elsherbeni, "Bifocal design and aperture phase optimizations of reflectarray antennas for wide-angle beam-scanning performance," *IEEE Trans. Antennas Propag.*, Vol. 61, No. 9, pp. 4588–4597, Sep. 2013.
41 G. R. Fowles, *Introduction to Modern Optics*, 2nd Edn, New York: Dover Publications, 1992.
42 D. McGrath, "Planar three-dimensional constrained lenses," *IEEE Trans. Antennas Propag.*, Vol. 34, No. 1, pp. 46–50, Jan. 1986.
43 W. Menzel, M. Al-Tikriti, and R. Leberer, "A 76 GHz multiple-beam planar reflector antenna," presented at the *32nd European Microwave Conference, Milano*, Italy, Sep. 2002.
44 J. Rodriguez-Alvarez, M. Arrebola, C. Tienda, J. A. Encinar and F. Las-Heras "Bifocal antenna based on dual-reflectarray dual-offset configuration" *6th European Conference on Antennas and Propagation, Prague*, Czech Republic, pp. 2348–2352, March, 2012.
45 J. Shaker, A. Petosa, A. Ittipiboon, and M. Cuhaci, "An overview of reflectarray technology: History, achievements, and future challenges," *Symposium on Antenna Technology and Applied Electromagnetics*, pp. 283–285, Winnipeg, Canada, 2000.
46 F. Arpin, J. Shaker, and D. A. McNamara, "Multi-feed single-beam power-combining reflectarray antenna," *Electron. Lett.*, Vol. 40, No. 17, pp. 1035–1037, Aug. 2004.
47 P. Nayeri, F. Yang, and A. Z. Elsherbeni, "Design and experiment of a single-feed quad-beam reflectarray antenna," *IEEE Trans. Antennas Propag.*, Vol. 60, No. 2, pp. 1166–1171, Feb. 2012.
48 J. Robinson, and Y. Rahmat-Samii, "Particle swarm optimization in electromagnetics," *IEEE Trans. Antennas Propag.*, Vol. 52, No. 2, pp. 397–407, Feb. 2004.
49 D. W. Boeringer and D. H. Werner, "Particle swarm optimization versus genetic algorithms for phased array synthesis," *IEEE Trans. Antennas Propag.*, Vol. 52, No. 3, pp. 771–779, Mar. 2004.
50 Y. Rahmat-Samii, J. Huang, B. Lopez, M. Lou, E. Im, S. L. Durden, and K. Bahadori, "Advanced precipitation radar antenna: array-fed offset membrane cylindrical reflector antenna," *IEEE Trans. Antennas Propag.*, Vol. 53, No. 8, Aug. 2005, pp. 2503 – 2515.
51 S. Hsu, C. Han, J. Huang, and K. Chang, "An offset linear-array-fed Ku/Ka dual-band reflectarray for planet cloud/precipitation radar," *IEEE Trans. Antennas Propag.*, Vol. 55, No. 11, Nov. 2007, pp. 3114 – 3122.
52 F. Guo, et al., "A modified parabolic-cylindrical reflectarray antenna for wide beam scanning capability," in *Proc. IEEE AP-S/URSI, Memphis*, TN, July 2014, pp. 1023–1024.

53 T. S. Chu, and P. P. Iannone, "Radiation properties of parabolic torus reflector," *IEEE Trans. Antennas Propag.*, Vol. 37, No. 7, pp. 865–874, July 1989.

54 C. J. Sletten, *Reflector and Lens Antennas: Analysis and Design Using Personal Computers*, Artech House, Inc., 1988.

55 P. Nayeri, F. Yang, and A. Z. Elsherbeni, "The offset parabolic-torus-phase planar reflectarray antenna," presented at the *IEEE Antennas and Propagation Society International Symposium, Tennessee*, USA, July 2014.

56 T. Li, "A study of spherical reflectors as wide-angle scanning antennas," *IRE Trans. Antennas Propag.*, Vol. 7, No. 3, pp. 223–226, July 1959.

57 P. Nayeri, F. Yang, and A. Z. Elsherbeni, "Studies on planar reflectarrays with spherical-phase distribution for 2D beam-scanning," presented at the *IEEE Antennas and Propagation Society International Symposium, Florida*, USA, July 2013.

58 B. Khayatian and Y. Rahmat-Samii, "Characterizing reflectarray antennas radiation performance," *IEEE AP – S/URSI Symposium, Columbus, Ohio*, June 2003, pp. 298–301.

59 B. Khayatian, Y. Rahmat-Samii, and J. Huang, "Radiation characteristics of reflectarray antennas: methodologies and applications to dual configurations," in *Proc. 2006 Europ. Conf. Antennas Propag. (EuCAP2006)*, Nov. 2006.

60 W. Hu, M. Arrebola, R. Cahill, J. A. Encinar, V. Fusco, H. S. Gamble, Y. Álvarez, and F. Las-Heras, "94 GHz dual-reflector antenna with reflectarray subreflector," *IEEE Trans. Antennas Propagat.*, Vol. 57, No. 10, pp. 3043–3050, Oct. 2009.

61 W. Menzel, M. Al-Tikriti, and R. Leberer, "A 76 GHz multiple-beam planar reflector antenna," presented at the *32nd European Microwave Conference, Milano*, Italy, Sep. 2002.

62 J. Rodriguez-Alvarez, M. Arrebola, C. Tienda, J. A. Encinar and F. Las-Heras "Bifocal antenna based on dual-reflectarray dual-offset configuration" Proceedings of the *6th European Conference on Antennas and Propagation, Prague*, Czech Republic, pp. 2348–2352, March, 2012.

63 S. Xu and Y. Rahmat-Samii, "A compensated spherical reflector antenna using subreflectarrays," *Microwave and Optical Technology Letters*, Vol. 51, No. 2, pp. 577–582, Feb. 2009.

64 D. G. Berry, R. G. Malech, and W. A. Kennedy, "The reflectarray antenna," *IEEE Trans. Antennas Propag.*, Vol. 11, No. 6, pp. 645–651, Nov. 1963.

65 S. V. Hum and J. Perruisseau-Carrier, "Reconfigurable reflectarrays and array lenses for dynamic antenna beam control: a review (invited paper)," *IEEE Trans. Antennas Propag.*, Vol. 62, No. 1, pp. 183–198, Jan. 2014.

66 A. Petosa, "An overview of tuning techniques for frequency-agile antennas," *IEEE Antennas and Propagation Magazine*, Vol. 54, No. 5, pp. 271–296, Oct. 2012.

67 M. E. Cooley, J. F. Walker, D. G. Gonzalez, and G. E. Pollon, "Novel reflectarray element with variable phase characteristics," *IEE Proc. Microwaves Antennas Propag.*, Vol. 144, No. 2, pp. 149–151, May 1997.

68 R. Phillion and M. Okoniewski, "Improving the phase resolution of a micromotor-actuated phased reflectarray," in *2008 Microsystems and Nanoelectronics Research Conference (MNRC 2008)*, 2008, pp. 169–172.

69 X. Yang, "Design of a circularly polarized reconfigurable reflectarray using micromotors," presented at the *IEEE Antennas and Propagation Society International Symposium, Canada*, July 2015.

70 J. P. Gianvittorio, and Y. Rahmat-Samii, "Reconfigurable patch antennas for steerable reflectarray applications," *IEEE Trans. Antennas Propagat.*, Vol. 54, No. 5, pp. 1388–1392, May 2006.
71 J. Shaker, M. Chaharmir, M. Cuhaci, and A. Ittipiboon, "Reflectarray research at the communications research centre Canada," *IEEE Antennas and Propagation Magazine*, Vol. 50, No. 4, pp. 31–52, Aug. 2008.
72 G. M. Rebeiz, *RF MEMS: Theory, Design, and Technology*, Chichester, UK: John Wiley & Sons, Ltd, 2003.
73 R. J. Richards, E. W. Dittrich, O. B. Kesler, and J. M. Grimm, "Microstrip phase shifting reflect array antenna," USA Patent 6,020,853, Feb. 1, 2000.
74 H. Kamoda, T. Iwasaki, J. Tsumochi, T. Kuki, and O. Hashimoto, "60-GHz electronically reconfigurable large reflectarray using single bit phase shifters," *IEEE Trans. Antennas Propag.*, Vol. 59, No. 7, pp. 2524–2531, 2011.
75 S. Montori, F. Cacciamani, R. Gatti, E. Carrasco, M. Barba, J. Encinar, and R. Sorrentino, "Wideband dual-polarization reconfigurable elementary cell for electronic steerable reflectarray at Ku-band," in *Proc. 2010 Europ. Conf. Antennas Propag.* (EuCAP2010), 2010.
76 E. Carrasco, M. Barba, and J. Encinar, "X-band reflectarray antenna with switching-beam using pin diodes and gathered elements," *IEEE Trans. Antennas Propag.*, Vol. 60, No. 12, pp. 5700–5708, 2012.
77 H. R. Phelan, "Spiraphase reflectarray for multitarget radar," *Microwave Journal*, Vol. 20, p. 67, Jul. 1977.
78 A. E. Martynyuk, A. G. Martinez-Lopez, and J. Rodriguez-Cuevas, "Spiraphase-type element with optimal transformation of switch impedances," *Electron. Lett.*, Vol. 46, No. 10, pp. 673–675, May 2010.
79 A. E. Martynyuk, J. I. Martinez Lopez, and N. A. Martynyuk, "Spiraphase-type reflectarrays based on loaded ring slot resonators," *IEEE Trans. Antennas Propag.*, Vol. 52, No. 1, pp. 142–153, 2004.
80 J. Rodriguez-Zamudio, J. I. Martinez-Lopez, J. Rodriguez-Cuevas, and A. E. Martynyuk, "Reconfigurable reflectarrays based on optimized spiraphase-type elements," *IEEE Trans. Antennas Propag.*, Vol. 60, No. 4, pp. 1821–1830, Apr. 2012.
81 J. O. McSpadden, F. Lu, C. Kai, and J. Huang, "Ka-band beam steering reflectarray study," presented at the *IEEE Antennas and Propagation Society International Symposium, Florida*, USA, July 1999.
82 R. J. Richards, "Integrated microelectromechanical phase shifting reflect array antenna," USA Patent 6,195,047, Feb. 27, 2001.
83 R. Gilbert, "Dipole tunable reconfigurable reflector array," USA Patent 6,426,727, Jul. 30, 2002.
84 H. P. Hsu and T. Y. Hsu, "Optically controlled RF MEMS switch array for reconfigurable broadband reflective antennas," USA Patent 6,417,807, Jul. 9, 2002.
85 H. Salti, E. Fourn, R. Gillard, and H. Legay, "Minimization of MEMS breakdowns effects on the radiation of a MEMS based reconfigurable reflectarray," *IEEE Trans. Antennas Propag.*, Vol. 58, No. 7, pp. 2281–2287, 2010.
86 H. Rajagopalan, Y. Rahmat-Samii, and W. Imbriale, "RF MEMS actuated reconfigurable reflectarray patch-slot element," *IEEE Trans. Antennas Propag.*, Vol. 56, No. 12, pp. 3689–3699, 2008.
87 E. Carrasco, M. Barba, B. Reig, C. Dieppedale, and J. Encinar, "Characterization of a reflectarray gathered element with electronic control using ohmic RF MEMS and patches

aperture-coupled to a delay line," *IEEE Trans. Antennas Propag.*, Vol. 60, No. 9, pp. 4190–4201, 2012.
88 O. Bayraktar, O. Civi, and T. Akin, "Beam switching reflectarray monolithically integrated with RF MEMS switches," *IEEE Trans. Antennas Propag.*, Vol. 60, No. 2, pp. 854–862, 2012.
89 C. Guclu, J. Perruisseau-Carrier, and O. Civi, "Proof of concept of a dual-band circularly-polarized RF MEMS beam-switching reflectarray," *IEEE Trans. Antennas Propag.*, Vol. 60, No. 11, pp. 5451–5455, 2012.
90 J. Perruisseau-Carrier and A. K. Skrivervik, "Monolithic MEMS-based reflectarray cell digitally reconfigurable over a 360° phase range," *IEEE Antennas and Wireless Propag. Letters*, Vol. 7, pp. 138–141, 2008.
91 S. V. Hum, G. McFeetors, and M. Okoniewski, "Integrated MEMS reflectarray elements," in *Proc. 2006 Europ. Conf. Antennas Propag. (EuCAP2006)*, Nov. 2006.
92 L. Boccia, F. Venneri, G. Amendola, and G. Di Massa, "Application of varactor diodes for reflectarray phase control," in *Proc. 2002 Antennas Propag. Soc. Int. Symp.*, Vol. 3, Jun. 2002, pp. 132–135.
93 O. Vendik and M. Parnes, "A phase shifter with one tunable component for a reflectarray antenna," *IEEE Antennas Propag. Mag.*, Vol. 50, No. 4, pp. 53–65, 2008.
94 S. V. Hum, M. Okoniewski, and R. J. Davies, "Modeling and design of electronically tunable reflectarrays," *IEEE Trans. Antennas Propagat.*, Vol. 55, No. 8, pp. 2200–2210, Aug. 2007.
95 M. Riel and J. Laurin, "Design of an electronically beam scanning reflectarray using aperture-coupled elements," *IEEE Trans. Antennas Propag.*, Vol. 55, No. 5, pp. 1260–1266, 2007.
96 M. Hajian, B. Kuijpers, K. Buisman, A. Akhnoukh,M. Plek, L. C. N. de Vreede, J. Zijdeveld, and L. P. Ligthart, "Active scan-beam reflectarray antenna loaded with tunable capacitor," in Proc. 3rd EuCAP, pp. 1158–1161, Berlin, Germany, March, 2009.
97 F. Venneri, S. Costanzo, and G. Di Massa, "Design and validation of a reconfigurable single varactor-tuned reflectarray," *IEEE Trans. Antennas Propag.*, Vol. 61, No. 2, pp. 635–645, 2013.
98 L. Boccia, G. Amendola, and G. Di Massa, "Performance improvement for a varactor-loaded reflectarray element," *IEEE Trans. Antennas Propag.*, Vol. 58, No. 2, pp. 585–589, 2010.
99 C. Liu and S. Hum, "An electronically tunable single-layer reflectarray antenna element with improved bandwidth," *IEEE Antennas Wireless Propag. Lett.*, Vol. 9, pp. 1241–1244, 2010.
100 D. Sievenpiper, J. Schaffner, R. Loo, G. Tangonan, S. Ontiveros, and R. Harold, "A tunable impedance surface performing as a reconfigurable beam steering reflector," *IEEE Trans. Antennas Propag.*, Vol. 50, No. 3, pp. 384–390, Mar. 2002.
101 W. Hu, R. Cahill, J. Encinar, R. Dickie, H. Gamble, V. Fusco, and N. Grant, "Design and measurement of reconfigurable millimeter wave reflectarray cells with nematic liquid crystal," *IEEE Trans. Antennas Propag.*, Vol. 56, No. 10, pp. 3112–3117, 2008.
102 A. Gaebler, A. Moessinger, F. Goelden, A. Manabe, M. Goebel, R. Follmann, et al., "Liquid crystal-reconfigurable antenna concepts for space applications at microwave and millimeterwaves," *International Journal of Antennas and Propagation*, Vol. 2009, pp. 1–8, 2009.

103 G. Perez-Palomino, J. Encinar, M. Barba, and E. Carrasco, "Design and evaluation of multi-resonant unit cells based on liquid crystals for reconfigurable reflectarrays," *IEE Proc. Microwaves Antennas Propag.*, Vol. 6, No. 3, pp. 348–354, 2012.

104 G. Perez-Palomino, P. Baine, R. Dickie, M. Bain, J. Encinar, R. Cahill, et al., "Design and experimental validation of liquid crystal-based reconfigurable reflectarray elements with improved bandwidth in F-band," *IEEE Trans. Antennas Propag.*, Vol. 61, No. 4, pp. 1704–1713, 2013.

105 R. Romanofsky, J. Bernhard, F. Van Keuls, F. Miranda, G. Washington, and C. Canedy, "K-band phased array antennas based on Ba0.60Sr0.40TiO3 thin-film phase shifters," *IEEE Trans. Microw. Theory Tech.*, Vol. 48, No. 12, pp. 2504–2510, 2000.

106 R. Romanofsky, "Advances in scanning reflectarray antennas based on ferroelectric thin-film phase shifters for deep-space communications," *Proc. IEEE*, Vol. 95, No. 10, pp. 1968–1975, 2007.

107 M. Sazegar, A. Giere, Y. Zheng, H. Maune, A. Moessinger, and R. Jakoby, "Reconfigurable unit cell for reflectarray antenna based on barium-strontium-titanate thick-film ceramic," in *Europ. Microwave Conf. (EuMC 2009)*, 2009, pp. 598–601.

108 E. Carrasco and J. Perruisseau-Carrier, "Reflectarray antenna at terahertz using graphene," *IEEE Antennas Wireless Propag. Lett.*, Vol. 12, pp. 253–256, 2013.

109 E. Carrasco, M. Tamagnone, and J. Perruisseau-Carrier, "Tunable graphene reflective cells for thz reflectarrays and generalized law of reflection," *Appl. Phys. Lett.*, Vol. 102, No. 10, pp. 103–104, 2013.

110 S. Kanamaluru, A. E. Fathy, and A. Rosen, "Reconfigurable adaptive wideband antenna," USA Patent no. US 2002/0101391, Aug. 2002.

111 M. Chaharmir, J. Shaker, M. Cuhaci, and A. Sebak, "Novel photonically-controlled reflectarray antenna," *IEEE Trans. Antennas Propag.*, Vol. 54, No. 4, pp. 1134–1141, 2006.

112 S. Long and G. Huff, "A fluidic loading mechanism for phase reconfigurable reflectarray elements," *IEEE Antennas Wireless Propag. Lett.*, Vol. 10, pp. 876–879, 2011.

113 L. Liang, and S. V. Hum, "Wide-angle scannable reflector design using conformal transformation optics," *Opt. Express*, Vol. 21, No. 2, pp. 2133–2146, Jan. 2013.

114 M. E. Bialkowski, A. W. Robinson, and H. J. Song, "Design, development, and testing of X-band amplifying reflectarrays," *IEEE Trans. Antennas Propag.*, Vol. 50, No. 8, pp. 1065–1076, Aug. 2002.

115 R. W. Clark, G. H. Huff, and J. T. Bernhard, "An integrated active microstrip reflectarray element with an internal amplifier," *IEEE Trans. Antennas Propag.*, Vol. 51, No. 5, pp. 993–999, May 2003.

116 L. Cabria, J. A. Garcia, J. Gutierrez-Rios, A. Tazon, and J. Vassallo, "Active reflectors: possible solutions based on reflectarrays and fresnel reflectors," *Int. Jour. Antennas Propag.*, Vol. 2009, Article ID 653952, 2009.

117 K. Kishor and S. Hum, "An amplifying reconfigurable reflectarray antenna," *IEEE Trans. Antennas Propag.*, Vol. 60, No. 1, pp. 197–205, 2012.

118 P. Nayeri, A. Z. Elsherbeni, and F. Yang, "Radiation analysis approaches for reflectarray antennas," *IEEE Antennas and Propagation Magazine*, Vol. 55, No. 1, pp. 127–134, Feb. 2013.

119 P. Nayeri "Advanced design methodologies and novel applications of reflectarray antennas," Ph.D. dissertation, Department of Electrical Engineering, University of Mississippi, University, MS, 2012.

120 B. Wu, A. Sutinjo, M. E. Potter, and M. Okoniewski, "On the selection of the number of bits to control a dynamic digital MEMS reflectarray," *IEEE Antennas Wireless Propag. Lett.*, Vol. 7, pp. 183–186, 2008.

121 B. Mencagli, R. V. Gatti, L. Marcaccioli, and R. Sorrentino, "Design of large mm-wave beam-scanning reflectarrays," presented at the *European Conference on Wireless Technology, Paris*, France, Oct. 2005.

122 M. Arrebola, E. Carrasco, and J. A. Encinar, "Beam scanning antenna using a reflectarray as sub-reflector," *ACES Journal*, Vol. 26, No. 6, 2011.

11

Reflectarray Engineering and Emerging Applications

As discussed in the previous chapters, planar reflectarray antennas combine the favorable features of both printed arrays and parabolic reflectors and create a low-profile, low-mass, and low-cost antenna. This unique technology has rapidly been evolving over the past few decades and a multitude of new potential applications have been envisioned for reflectarrays. In this chapter, we will take a closer look at some of the advanced configurations and applications that have been developed over the years.

11.1 Advanced Reflectarray Geometries

11.1.1 Conformal Reflectarrays

11.1.1.1 Analysis of Conformal Reflectarrays

The feasibility of designing reflectarray antennas on conformal surfaces was studied in [1]. In general, a reflectarray antenna imitates the conventional parabolic reflectors, with the added advantage of having a flat surface instead of the curved reflecting surface. The surface of the reflectarray antenna consists of phase changing elements, which create a collimated beam and are usually distributed on a planar aperture. A notable advantage of individual control over the phase of the elements is that, the reflectarray surface does not have to be restricted to a planar geometry.

Several applications require the use of conformal antenna technology. While the need for a conformal antenna system generally depends on the specific application, the advantages can usually be attributed to either mechanical (aerodynamic, hydrodynamic) or electrical benefits [2]. In terms of the mechanical advantages, a conformal antenna is designed according to a prescribed shape, which can be some part of an airplane, a high-speed train, or other vehicle. The main purpose is to build an antenna that becomes integrated with the structure and causes less disturbance such as extra wind drag. The electrical advantages of conformal antennas are typically wide-angle scanning capabilities. Conformal beam-scanning array antennas can achieve full azimuth and wide elevation coverage, which in most cases is not possible with planar designs.

A major step in design of a reflectarray is analyzing the radiation characteristics of the antenna. The formulation developed for planar reflectarray antennas can be generalized to analyze the performance of conformal reflectarrays. With the array theory formulation, as long as the elements on the conformal surface are correctly modeled with mutual coupling included, array summation can provide accurate results. Therefore, an

Reflectarray Antennas: Theory, Designs, and Applications, First Edition.
Payam Nayeri, Fan Yang, and Atef Z. Elsherbeni.

important consideration in radiation analysis of conformal designs is the accurate modeling of element radiation characteristics. In addition to the shape of the element pattern, one should realize that the peak of the element pattern is normal to the local surface, so the peaks will not all point in the same direction for a conformal array [3]. Furthermore, a reflectarray antenna is functioning in the reflection mode, which is different from conventional antenna arrays. Scattering properties of curved surfaces are more complicated than planar surfaces and, in general, high order diffraction effects such as generation of creeping waves and depolarization should be taken into account [4]. Nonetheless, if the conformal surface is not highly curved and edge diffraction is minimized, microwave optic approaches are suitable for the analysis.

The geometry of a conformal reflectarray system is given in Figure 11.1. The radiation pattern of a conformal reflectarray with $M \times N$ elements can be calculated using the vectorized array summation formulation, which was described in Chapter 4, albeit with some modifications.

Similar to planar arrays, a cosine q_e model is usually used for the element pattern function A, which in general is not symmetric here due to the surface curvature. Some highly curved conformal arrays may have a $\cos^{1.5}$ behavior in the strong coupling plane and a $\cos^{1}$ behavior in the weak coupling plane [6]. As discussed earlier, in the studies reported in [1] and [5], the conformal reflectarray surface is not highly curved, so each unit-cell can be approximated with a planar surface that is tangential to the conformal surface. The element pattern can then be approximated with a symmetric $\cos^{1}$ model with its peak normal to the unit-cell surface. Therefore, the important considerations for element pattern modeling are to:

1) Determine the exact angle of incidence for each element that controls the receive mode element pattern,
2) Determine the orientation of the local cell coordinates with respect to the global coordinates, which controls the transmit mode element pattern.

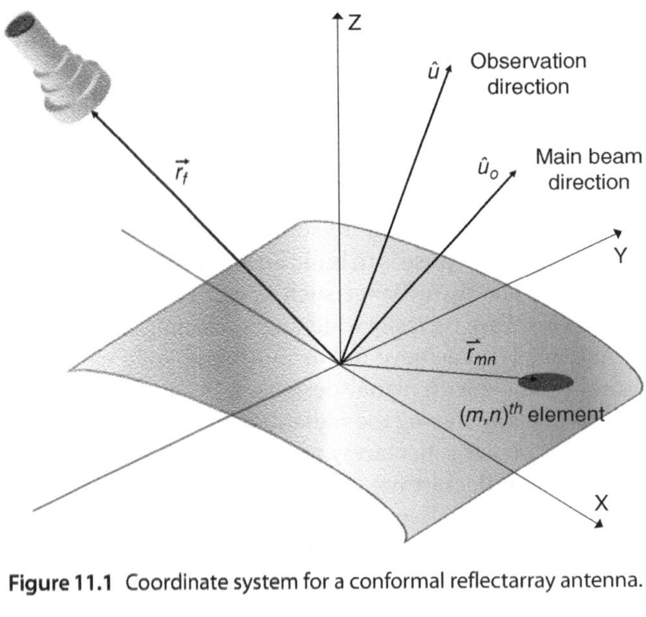

Figure 11.1 Coordinate system for a conformal reflectarray antenna. Reprinted with permission from [5].

The transmit element pattern function can be written as

$$A_{mn}(\theta,\varphi) \approx \cos^{q_e}\theta_{local} \cdot e^{jk(\vec{r}_{mn}\cdot\hat{u})}. \qquad (11.1)$$

In comparison with planar designs, this function takes into account the effect of surface curvature on the transmit element pattern by using local element coordinates for each radiating element.

As discussed in Chapter 4, the element excitation function I, is determined by both the incident field and element property. The element receive pattern is essentially part of the element excitation function, which determines the reflected power from the element based on the angle of excitation and element pattern shape, that is,

$$|\Gamma_{mn}| = Cos^{q_e}\theta_e(m,n). \qquad (11.2)$$

While the function here is similar to the planar case, the difference is that for conformal reflectarrays the element angle θ_e depends on both the element location and the local surface orientation.

It is worthwhile pointing out again that for planar designs, the element patterns can be obtained using two approaches, namely a unit-cell model and a transmitting antenna model, and it was shown in Chapter 2 that both approaches show a close agreement with each other. In general, for conformal designs, a periodic boundary model may not exist, and one may need to utilize the transmitting antenna model for the analysis, which can be implemented on any type of curved platform.

With these approximations, the radiation pattern of a conformal reflectarray antenna can be simplified to the scalar form

$$E(\theta,\varphi) = \sum_{m=1}^{M}\sum_{n=1}^{N} Cos^{q_e}\theta_{local}\, Cos^{q_e}\theta_e(m,n) e^{j\phi_{mn}} \frac{Cos^{q_f}\theta_f(m,n)}{|\vec{r}_{mn}-\vec{r}_f|} e^{-jk(|\vec{r}_{mn}-\vec{r}_f| - \vec{r}_{mn}\cdot\hat{u})}. \qquad (11.3)$$

In this equation θ_f is the spherical angle in the feed coordinates and $\vec{r}_f$ is the position vector of the feed. The required phase delay ϕ_{mn} of the mn^{th} element is adjusted to set the main beam in the $\hat{u}_o$ direction.

It is also important to note that for an accurate gain calculation it is necessary to determine the overall aperture efficiency of the antenna, where the major influential terms are the spillover and taper efficiency. The effect of taper efficiency is taken into account when one calculates the radiation pattern using (11.3), however, the spillover efficiency, on the other hand, requires some attention. While the analysis approach is generally the same as for the planar case, it should be noted that for conformal designs the integration has to be performed over a curved surface.

11.1.1.2 Radiation Characteristics of Conformal Reflectarrays on Cylindrical Surfaces

Low profile printed antennas flush mounted on singly curved surfaces, such as cylinders, represent an important class of conformal arrays that have been used in many experimental radar and communication systems. Cylindrical array antennas are mainly utilized for either full azimuth scanning or sector arrays covering a specific subtended angle. The aim of this study is to investigate the radiation characteristics of reflectarrays on conformal sector cylindrical surfaces. The analysis of this singly curved canonical problem can provide an insight into many radiation characteristics of conformal reflectarrays.

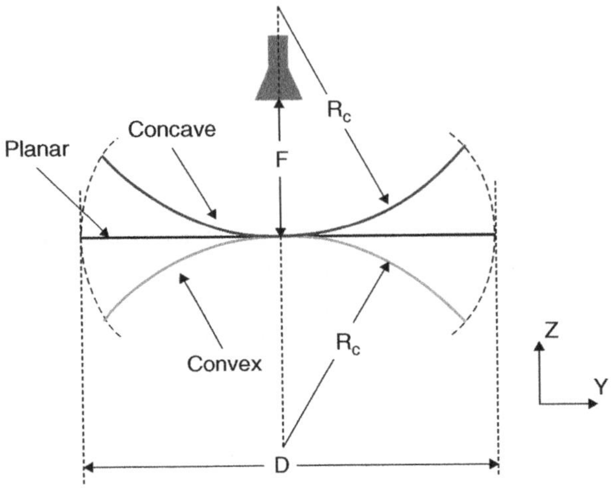

Figure 11.2 Cross-section of the conformal cylindrical reflectarray systems. Reprinted with permission from [5].

11.1.1.2.1 Radiation Pattern Characteristics

The cross-section of the conformal reflectarray system is shown in Figure 11.2, where concave and convex surfaces are compared with their planar counterpart with the same surface area. In this figure, R_c is the radius of the cylinder for the conformal systems, and D represents the physical aperture size.

The numerical approach described before was applied to the radiation analysis of sector reflectarray antennas mounted on cylindrical surfaces [1]. For a meaningful comparison of the surface curvature effect, the feed properties and the physical aperture size were kept constant, and the only parameter that was changed was the cylinder radius R_c. The reflectarray antennas have a $20\lambda \times 20\lambda$ aperture and are designed to generate a beam in the broadside direction. Ideal phasing elements are used in each design and the element pattern is modeled as a $\cos^1$ function with no azimuth dependence. It should be noted that the aperture phase distribution for each reflectarray is designed according to its corresponding platform shape. The feed is prime focus, positioned with an F/D = 0.75 and the power q of radiation pattern model is 6.5.

The radiation patterns of the reflectarray systems in the curvature plane (yz-plane) are given in Figure 11.3 for various cases. It can be seen that due to the effects of the conformal surface, the radiation patterns show a noticeable difference with planar designs ($R_c \to \infty$). In particular for the concave design, the side lobes are higher than the planar case. On the other hand, the convex designs show a lower side-lobe level (SLL), but a much wider beamwidth. For both designs, however, as the cylinder radius increases, the radiation patterns of the conformal designs approach that of a planar case. A table summarizing the SLLs of the designs is presented later on in this section. While the minimum cylinder radius that mimics a planar aperture generally depends on the aperture size, for the cases studied here it is about $R_c = 40\lambda$; that is, the patterns for $R_c = 40\lambda$ and $R_c = \infty$ are almost identical. Note that the patterns in the cylindrical plane (xz-plane) are almost identical for all cylinder radiuses studied here and for both concave and convex configurations.

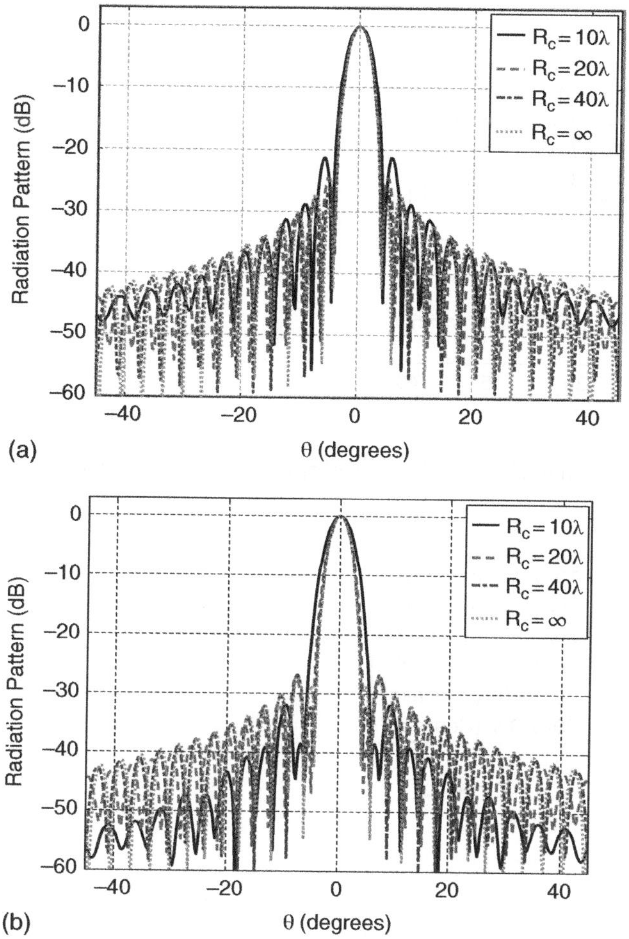

Figure 11.3 Radiation patterns of cylindrical reflectarrays: (a) concave, (b) convex. Reprinted with permission from [5].

The conformal cylindrical systems studied so far were defined based on the size of the cylinder radius (R_c). The aperture size (D), however, also plays an important role. For a fixed cylinder radius, a larger aperture means a wider subtended angle ($\alpha = D/R_c$) and a larger curved surface. To study this effect three different aperture sizes were also considered, and the beamwidth and SLL of these designs were compared together [1].

The normalized half-power beamwidth (HPBW) in the curvature plane as a function of cylinder radius is given in Figure 11.4 for different aperture sizes. It can be seen that in comparison between the two conformal designs, a convex reflectarray shows a much wider beamwidth. In general, however, both conformal designs show defocusing effects, that is, a wider beamwidth in comparison with a planar design. This is mainly attributed to the facts that first, the elements on the conformal surface radiate in the outward (convex) or inward (concave) radial directions of the cylinder, which defocuses the main beam. The second factor is the smaller projected aperture size of the conformal designs as seen in Figure 11.2. It should be noted that for a concave design, an optimum position

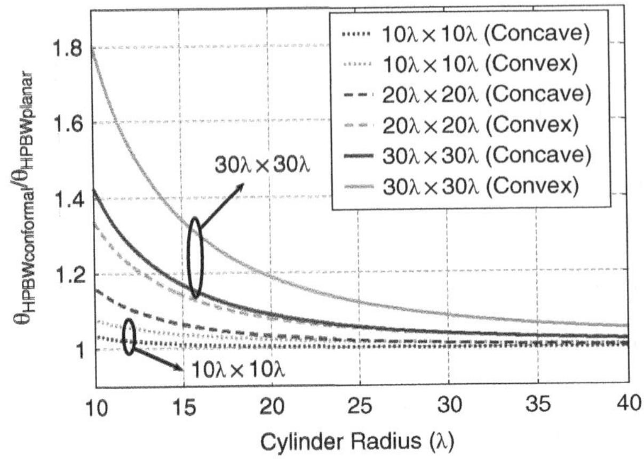

Figure 11.4 Normalized HPBW in the curvature plane as a function of cylinder radius. Reprinted with permission from [5].

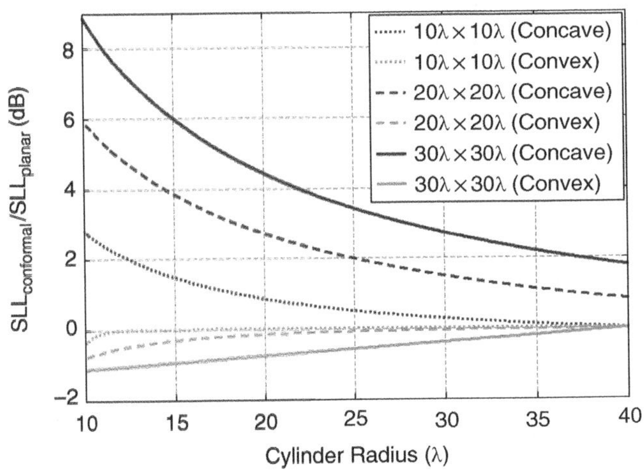

Figure 11.5 Normalized SLL in the curvature plane as a function of cylinder radius. Reprinted with permission from [5].

exists for the feed that can achieve maximum gain, which is due to the focusing properties of the conformal surface.

The change in SLL as a function of cylinder radius is also given in Figure 11.5 for different aperture sizes. Comparisons of these results also show that in all cases, for a fixed cylinder radius, as the aperture size increases, the SLL increases for the concave design and decreases for the convex design. In other words, the effect of the conformal surface on the radiation performance becomes more noticeable as the aperture size increases.

The results in Figures 11.4 and 11.5 indicate that the radiation performance of conformal cylindrical reflectarrays is a function of both cylinder radius and aperture size. However, it will be quite advantageous if one can determine the overall performance of the system based on a single conformal parameter. The subtended angle α, takes into

Table 11.1 Change in radiation performance of conformal cylindrical reflectarrays with $D/R_c = 1$.

Aperture size	Design	$10\lambda \times 10\lambda$	$20\lambda \times 20\lambda$	$30\lambda \times 30\lambda$
HPBW ratio	Concave	0.036	0.033	0.037
	Convex	0.077	0.080	0.083
SLL ratio (dB)	Concave	2.81	2.73	2.73
	Convex	−0.40	−0.25	−0.38

account the aperture size, D, which controls the gain, in addition to the cylinder radius, R_c, which controls the curvature of the conformal surface. As such it can be a suitable measure to evaluate the performance of the system. Table 11.1 summarizes the change in radiation characteristics of some conformal designs with a fixed subtended angle of 1 radian.

From these results, it can be seen that the ratio of the change in HPBW and SLL is almost identical for all three cases. Similar results were observed for other values of the subtended angle. In other words, it is possible to correctly specify the radiation performance of a conformal reflectarray system based on the value of α. This is quite advantageous from a system design view point, since for any available curved platform the radiation characteristics of the conformal reflectarray can be determined with only one conformal parameter. As an example, for a 10% acceptable increase in HPBW in comparison with a planar design, α is equal to 1.62 and 1.13 radians for concave and convex designs, respectively, for any aperture size. It is important to point out that the subtended angle (α) decreases as the cylinder radius increases. For the smallest cylinder radius, the aperture size increase corresponds to an increase in α from almost $\pi/3$ to π. Larger apertures will not be practical here since (1) the elements cannot be properly excited (shadow region of feed), and (2) the pattern of the elements at the sector edges point far away from the broadside direction.

11.1.1.2.2 Gain and Bandwidth Characteristics

Conformal reflectarray gain and bandwidth characteristics were also studied in [1]. The reflectarray antenna gain is directly proportional to its projected aperture. Thus, for a fixed aperture size, the projected aperture and the corresponding antenna gain of conformal designs will be slightly smaller than that of a planar design. For a $20\lambda \times 20\lambda$ aperture with a fixed value of $\alpha = 1$, this will result in a gain reduction of 0.1 and 0.6 dB for concave and convex designs, respectively. This is consistent with the HPBW broadening effects in Table 11.1. It is possible to mitigate this slight difference in antenna gain by increasing the aperture size.

The bandwidth study reported in [1] compared the normalized antenna gain as a function of frequency. The bandwidth of the reflectarray antenna is affected by the phase error associated with the spatial delay as well as the bandwidth of elements [7]. While the element bandwidth generally depends on the design methodology, for this comparative study we consider a frequency independent reflection phase for the elements, that is, only the bandwidth limitation associated with spatial phase delay is taken into account. Figure 11.6 shows the normalized gain bandwidth of a $20\lambda \times 20\lambda$ reflectarray antenna with planar and conformal apertures ($\alpha = 1$ radian). It can be seen

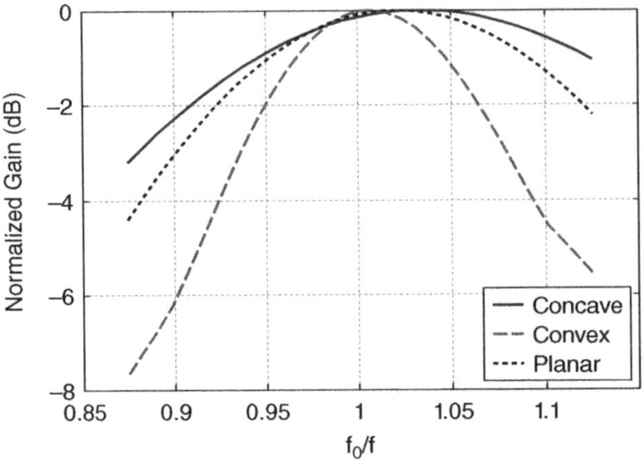

Figure 11.6 Normalized gain versus frequency for the reflectarray antennas. Reprinted with permission from [5].

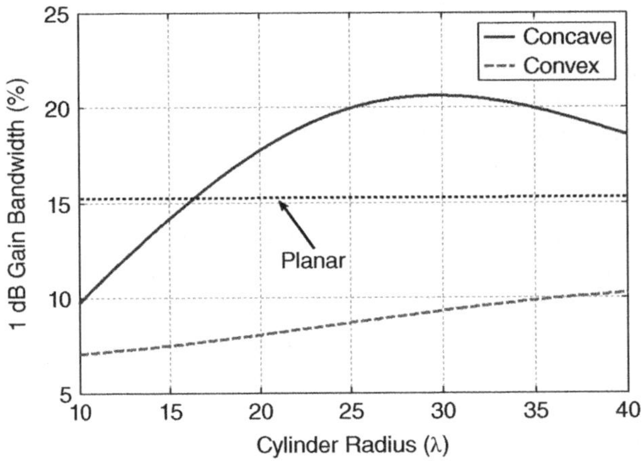

Figure 11.7 Gain bandwidth of the reflectarray antennas as a function of cylinder radius. Reprinted with permission from [5].

that a concave design shows a notable improvement in gain bandwidth in comparison with the planar case. On the other hand, as expected, a convex design shows a reduction in bandwidth.

As discussed earlier, for a concave reflectarray an optimum position exists for the feed so that the antenna achieves its maximum gain. A similar argument is attributed to the bandwidth performance of concave reflectarrays where the maximum bandwidth is attained at a certain radius.

Gain bandwidths of these conformal reflectarrays were also studied as a function of cylinder radius, for the same $20\lambda \times 20\lambda$ reflectarray. These results are given in Figure 11.7. It can be seen that for a convex design, the bandwidth of the reflectarray increases monotonically with R_c. For a concave design however, an optimum subtended angle exists where the antenna achieves the maximum bandwidth. For the design here, this optimum subtended angle is 0.67 radians. Note that the F/D for the studies reported here is 0.75.

11.1.2 Dual-Reflectarrays

The reflectarray antenna has similar optical properties as the parabolic reflector, and as such it can be used in dual-reflector configurations [8]. These dual configurations can offer several advantages such as reduction of overall antenna profile height, or a longer equivalent focal length compared to a single reflectarray configuration, which can improve the antenna bandwidth [9]. Two different dual-reflector configurations exist. The Cassegrain approach which generally uses a convex hyperboloid sub-reflector, or the Gregorian approach which uses a concave ellipsoid sub-reflector [10]–[12]. There are several configurations for these dual-reflectarray/reflector systems that use reflectarrays as the primary reflector or sub-reflector. There are three main categories as shown in Figure 11.8:

1) Use a planar reflectarray as the main reflector, and a hyperboloidal surface as the sub-reflector;
2) Use a parabolic surface as the main reflector, and a reflectarray as the sub-reflector;
3) Use reflectarrays as both the main reflector and the sub-reflector.

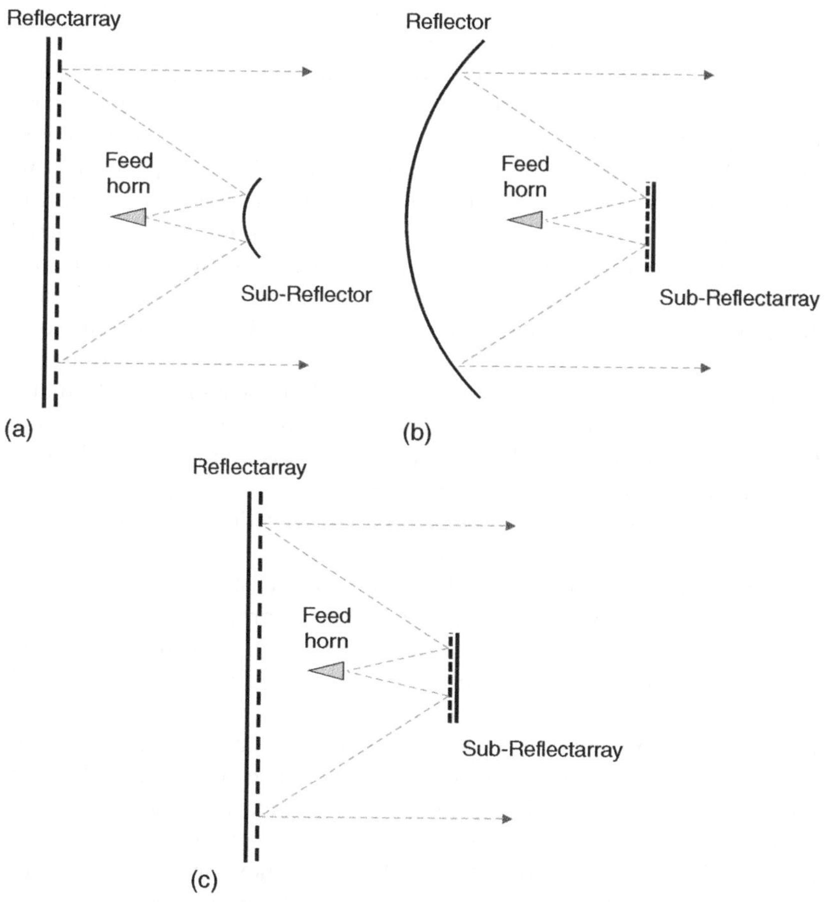

Figure 11.8 Dual-reflector/reflectarray configurations: (a) reflectarray as the primary reflector, (b) reflectarray as the sub-reflector, (c) reflectarray as both the primary and sub-reflector.

Figure 11.9 Photograph of a 77 GHz dual-reflectarray using a hyperboloidal sub-reflector. Reprinted with permission from [13].

The first dual-reflectarray prototype was reported in [13], where a reflectarray was designed for the primary aperture and a hyperboloidal sub-reflector was used in a Cassegrain configuration. The reflectarray aperture was a 6" square, and operated at 77 GHz. The antenna was fabricated on 0.005" Duroid substrate, with 5776 linearly polarized rectangular patch elements. The sub-reflector was 1.61 cm in diameter, and the feed was a conical horn with an aperture diameter of 1.25 cm. A photo of this prototype is given in Figure 11.9. It should be noted that while the aperture size was too small for optimum aperture efficiency in this dual-reflector configuration, the reported measured gain was 36.9 dB. An offset dual-reflector/reflectarray with a Cassegrain configuration was reported in [14]. The antenna was developed for an X/Ka dual-band operation and has a main circular aperture with a diameter of 75 cm and uses annular ring elements with each element having two capacitive gaps for circular polarization. The sub-reflector was a convex hyperboloid with a projected elliptical aperture with major axis dimension of 150 mm and minor axis dimension of 84 mm. The measured radiation patterns at both bands showed a good performance in terms of side-lobe and cross-polarization levels with aperture efficiencies close to 50%.

Several other designs of dual reflector/reflectarray configurations have since been developed. One of the very practical configurations is a sub-reflectarray with a parabolic main reflector as shown in Figure 11.10. This configuration provides better bandwidth than a single reflectarray antenna and is very appropriate for electronic beam-scanning over a limited field of view, because the beam can be scanned or reconfigured by controlling the phase at the elements of a small-size sub-reflectarray. A detailed overview of this configuration was reported in [15]. From an analysis view point, in a dual-reflector antenna, the sub-reflector is normally illuminated in the near-field region of the primary feed. As such the angles of incidence are moderately wide, and therefore they should be considered in the analysis of the sub-reflectarray.

A very cost-effective approach to compensate for the surface distortion of large reflector antennas was reported in [16]. Thermal effects inevitably cause certain reflector surface distortions, thus degrading the overall antenna performance.

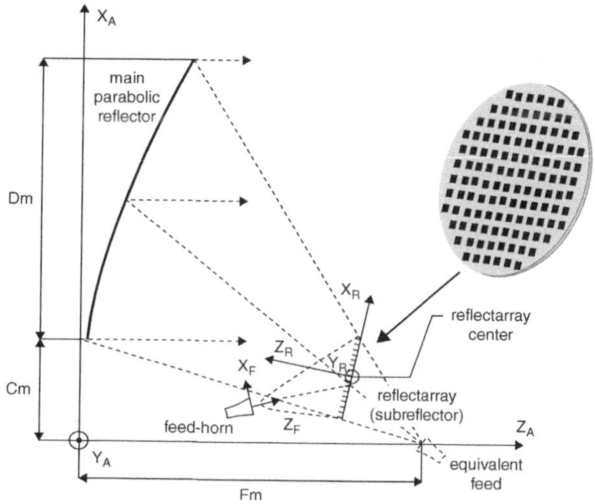

Figure 11.10 Schematic model of a sub-reflectarray/parabolic reflector configuration. Reprinted with permission from [15].

A novel reflector surface distortion compensation technique using a microstrip sub-reflectarray was presented in [16], where by properly adjusting the additional phase shift provided by the sub-reflectarray, the aperture phase errors caused by the distorted 20-m offset parabolic reflector antenna operating at the X band were compensated, resulting in a considerably improved antenna performance. An image of this surface distortion compensation scheme along with the far-field radiation patterns of the reflector are given in Figure 11.11.

While these dual-reflector antennas comprised of a reflectarray sub-reflector and a parabolic classic reflector offer several advantages, dual-reflectarray antennas comprising a main reflectarray and a sub-reflectarray can offer additional advantages [17]. Namely, the additional degrees of freedom of adjusting the phase on both surfaces can be used to improve the antenna performance, and also from a mechanical viewpoint having two flat reflectarrays is more suitable to be easily folded and deployed.

This configuration of dual-reflectarray antenna was utilized in [18] to provide broadband operation and very low levels of cross-polarization by exploiting the degrees of freedom of adjusting the patch dimensions on both reflectarray surfaces. The authors mention that in this configuration, the conventional far field models or ray tracing cannot be used to analyze the antenna because of the proximity of the feed, sub-reflectarray, and main reflectarray being in the near-field zone in all cases. As such, for this design, the near-fields radiated by the feed-horn [19] was used to determine the required phase distribution on the sub-reflectarray, and the contributions from all the elements in the sub-reflectarray are taken into account to compute the phase required at each element on the main reflectarray.

The antenna geometry was derived from a dual offset Cassegrain reflector antenna configuration with the aim of replacing each reflector with an equivalent flat reflectarray. The main parabolic reflector has an aperture diameter of 450 mm and a hyperbolic sub-reflector with an eccentricity of 2 and an aperture diameter of 380 mm. The focal

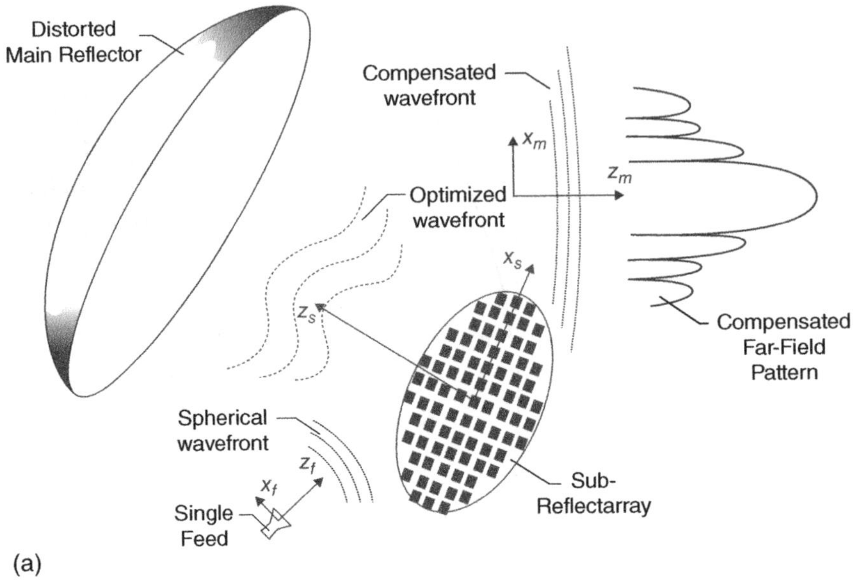

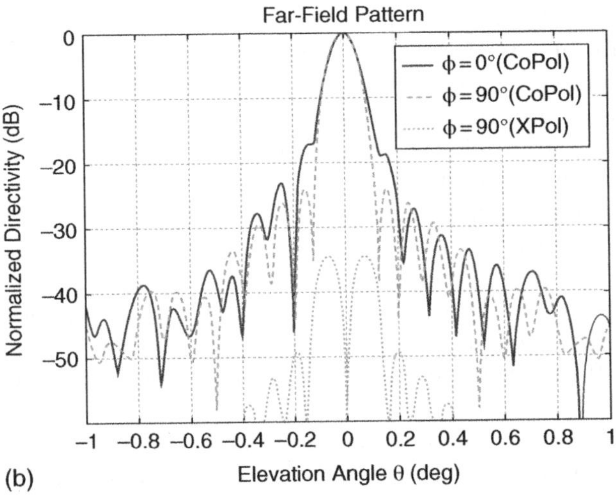

Figure 11.11 (a) Concept of sub-reflectarray compensation. (b) Compensated far-field pattern using sub-reflectarray compensation. Reprinted with permission from [16].

distance of the paraboloid reflector was 1800 mm and the offset distance was 1700 mm. The geometrical model of the antenna is shown in Figure 11.12.

Photographs of the fabricated prototype and measured radiation patterns for the vertical polarization are given in Figures 11.13 and 11.14, respectively. The measured gain of the reflectarray at 14 GHz is 34.56 dB and the cross-polarization discrimination (XPD) is 37.12 dB. The XPD is better than 30 dB at all frequencies from 12.2 to 15 GHz, and the measured gain varies from 32.12 dB to 34.51 dB across this frequency band.

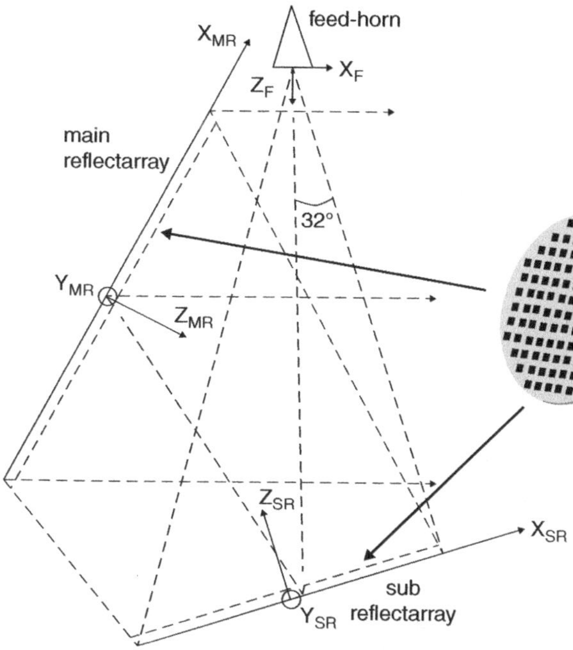

Figure 11.12 Top view of the antenna configuration. Reprinted with permission from [18].

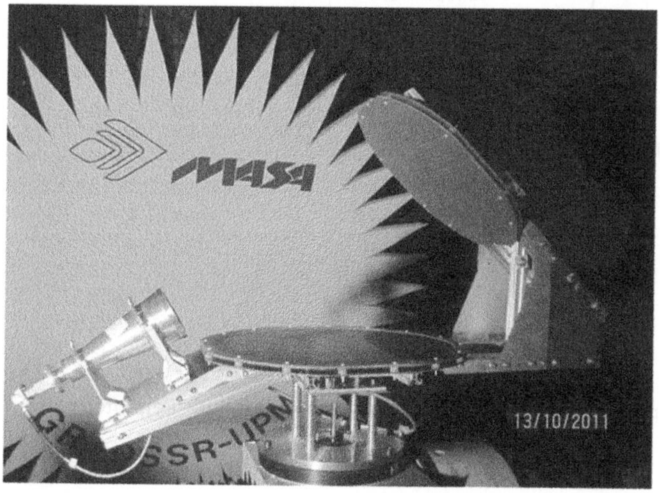

Figure 11.13 A Cassegrain dual-reflectarray antenna. Reprinted with permission from [18].

11.2 Reflectarrays for Satellite Applications

The numerous advantages of reflectarray antennas also make them a very practical choice for satellite communications [9]. Numerous satellite applications require high-gain antennas with shaped beams that can be easily achieved using reflectarrays.

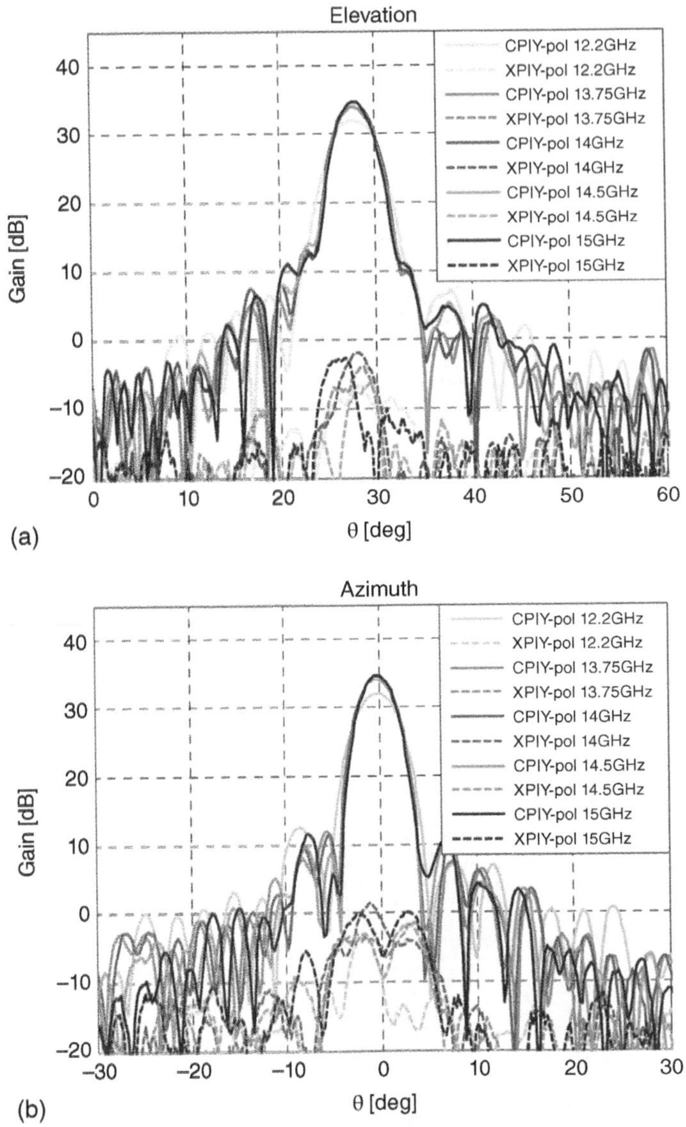

Figure 11.14 Measured radiation patterns of the antenna: (a) elevation, (b) azimuth. Reprinted with permission from [18].

Shaped or contoured beam reflectarrays have been demonstrated for a cosecant squared pattern in [20], and also for Direct Broadcast Satellites (DBS) applications in [21]. Several other examples of shaped beam reflectarrays for satellite communication were also given in [9], and in Chapter 8 of this book.

On the other hand, many satellite applications require high-gain beam-scanning antennas. For instance, the positioning accuracy of a satellite needs to be improved constantly by monitoring its signal. Therefore high-gain antennas with beam-scanning capabilities are usually employed to track the satellites along various orbits.

Traditionally, large-aperture parabolic reflector antennas are extensively used mainly because of their excellent radiation performance. However, have distinct disadvantages, such as large size, heavy weight, and demanding requirements for the servo system for beam-scanning. A detailed review of beam-scanning reflectarrays was given in Chapter 10 for readers' reference.

In this section, two representative reflectarray designs will be discussed. The first one is used as a ground station antenna to monitor the orbit of navigation satellite; and the second one is design for onboard application, which integrates the antenna aperture with solar panels.

11.2.1 An L-Band Reflectarray for the Beidou Satellite System

One interesting topic is to accurately monitor the operation of navigation satellites. Typically, these satellites operate at lower frequencies, thus requiring a very large physical aperture, making them a challenge to realize in practice. Although reflectarray antennas covering different frequencies have been reported in literature, only a few L-band reflectarray antennas have been proposed [22], [23] for low-gain applications. Here, we will summarize the design of an L-band reflectarray antenna for a geostationary navigation satellite at the operating frequency of 1.56 GHz [24].

The reflectarray antenna is designed for the Beidou GEO satellite system, and because of the high orbiting altitude, the required gain of the reflectarray is 30 dBc such that the monitored signal level is adequate for direct demodulation process. A pictorial model of the satellite in earth orbit is given in Figure 11.15. To satisfy the relatively high gain required for this system a square aperture with side length of 2.88 m was selected for the design. Although the design procedure for conventional reflectarrays at high frequencies is quite mature and various examples have been reported in literature, the large reflectarray aperture required for high-gain reflectarray antennas operating at the L-band does impose certain practical challenges, especially when the fabrication cost, assembly accuracy, and transportability are considered.

The reflectarray is placed on a flat rooftop to ease the requirements on assembly and maintenance. Therefore, the main beam is tilted 45.5° off the zenith direction. Note that this unconventional large offset main beam angle also causes a challenge and must be taken into account in the design. A schematic view of the reflectarray system is shown in Figure 11.16. The reflectarray aperture is divided into four 1.44-m square panels so

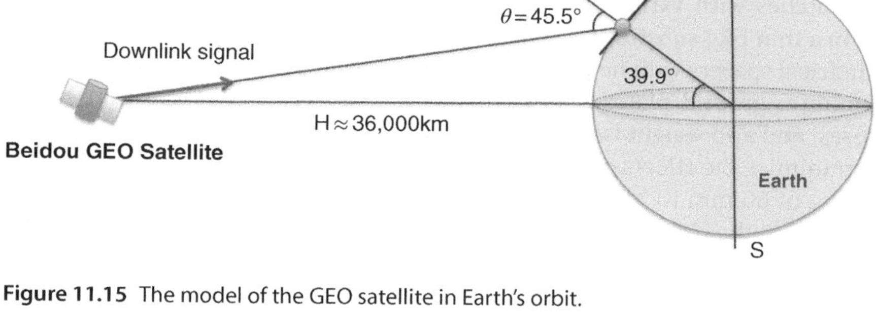

Figure 11.15 The model of the GEO satellite in Earth's orbit.

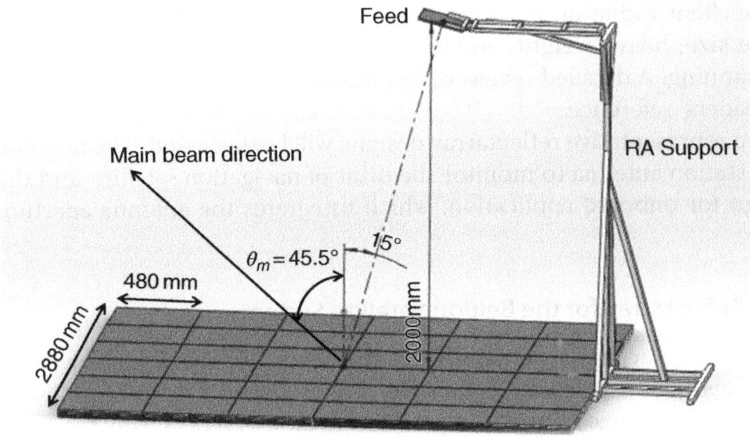

Figure 11.16 Schematic view of the L-band reflectarray. Reprinted with permission from [24].

Figure 11.17 Zoomed view of a small section of the L-band reflectarray.

that it can be easily assembled and transported if necessary. Each panel consists of nine pieces of subarrays with side length of 0.48 m for the ease of fabrication. The feeding antenna is supported by an aluminum frame with height of 2 m, tilted 15° off the normal direction. The main beam direction is tilted 45.5° opposite to the feed.

Microstrip patches with variable sizes were used for the phasing elements, which were etched on a thin FR4 substrate ($\varepsilon_r = 4.4, tan\delta = 0.025$) with thickness of 2 mm, and polymer cylindrical spacers with height of 8 mm were placed between the FR4 substrate and the aluminum ground plate. This element topology reduces fabrication costs, dielectric losses, and also weight issues in comparison with using a thick higher quality laminate. To minimize the effects of large tilted angle of the main beam, sub-wavelength element spacing of 60 mm (0.31 λ_0) was used in the design to smooth the phase distribution and reduce the element sensitivity for oblique incidence. A zoomed view of a section of the reflectarray aperture is given in Figure 11.17.

A horn antenna is usually used as the feed in reflectarrays. At L-band, however, it is bulky and heavy, therefore, a left-handed circularly polarized (LHCP) microstrip array antenna was designed to provide a proper illumination on the reflectarray aperture. The same double-layer structure with FR4 and air was utilized for the feed array. The probe-fed 2 × 2 array with side length of 255 mm was excited with equal amplitudes but with different phases (0°, 90°, 180°, 270°) [25] to realize a LHCP wave. Three Wilkinson power dividers were placed on the back layer to provide the required excitations and good port isolations. The fabricated feed array and measured radiation patterns at 1.56 GHz are shown in Figure 11.18.

Photograph of the fabricated L-band reflectarray antenna is given in Figure 11.19. The reflectarray antenna weights about 120 kg with all the supporting frames. Also note

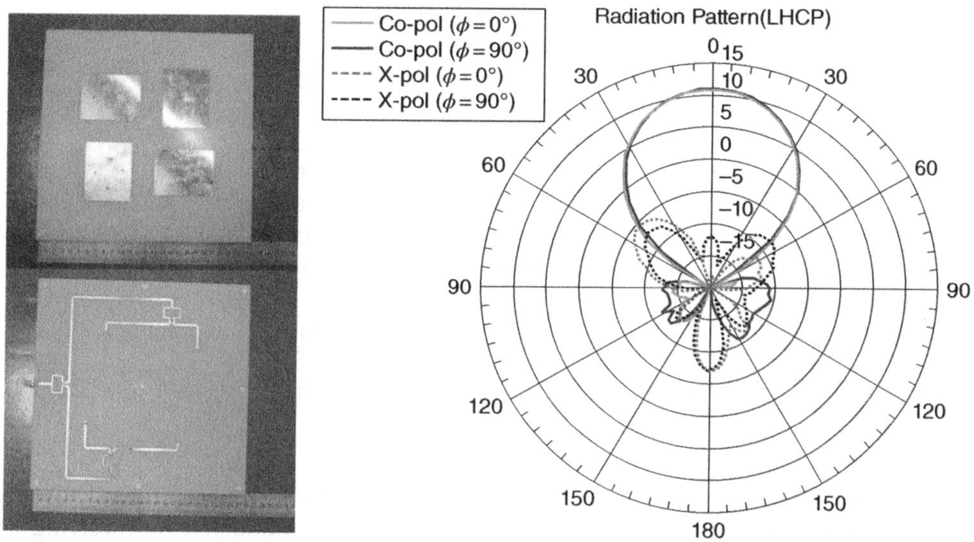

Figure 11.18 Feed antenna for the L-band reflectarray and measured radiation patterns. Reprinted with permission from [24].

Figure 11.19 Photos of the L-band reflectarray antenna system designed for the Beidou navigation satellite system.

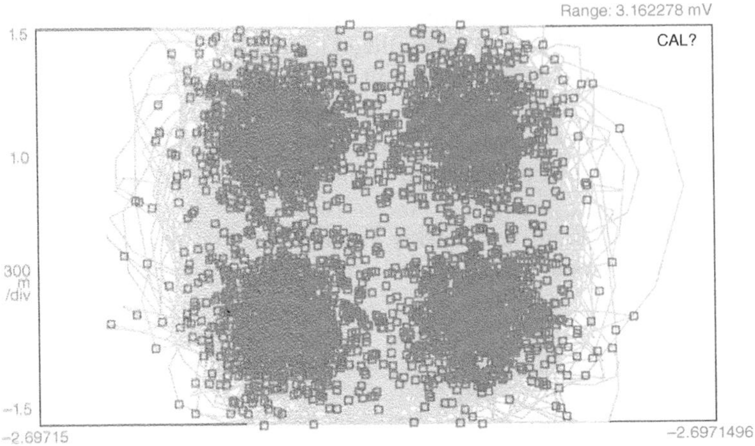

D: Ch1 QPSK Syms/Errs

EVM	=46.833	%rms	139.73	% pk	at	sym	1653
Mag Err	=32.268	%rms	127.98	% pk	at	sym	1653
Phase Err	=18.708	deg	57.432	deg pk	at	sym	1136
Freq Err	=4.2384	Hz	SNR(MER) = 7.438				dB
IQ Offset	=−39.238	dB	Rho	=0.82733			
Quad Err	=2.4633	deg	Gain Imb	=0.272			dB

Figure 11.20 Constellation diagram for the L-band signal received by the reflectarray.

that since the reflectarray is permanently placed outdoors, its surface is protected by an anti-corrosion paint. Testing of this satellite antenna was conducted by receiving a modulated QPSK signal from Beidou G3 at the ground station. The measured constellation diagram is given in Figure 11.20, where it can be seen that the signal received is nearly demodulated.

11.2.2 Reflectarrays Integrated with Solar Cells

Besides the base station antennas, reflectarrays are also promising candidates for onboard satellite communications. For instance, due to the limited payload space and mass on satellites, especially small cube-satellites, it is desirable to integrate antennas with solar panels. The concept of solar antennas has been developed since the 1990s [26]–[34]. They primarily focused on designing low- or medium-gain antennas, such as microstrip patch, dipole, and slot antennas, on or between solar cells. Recent developments in small satellites demand high-gain antennas in order to accommodate fast data transfer rate. It is, however, difficult to have large reflectors onboard. Solar panels provide large flat surfaces, which reflectarray antennas could utilize as a radiating aperture. Moreover, the primary feed is separated and only simple passive reflectarray elements are to be printed on the surface. It is a natural decision to integrate high-gain reflectarray antennas with solar cells. Even for large satellites equipped with high-gain reflector antennas, it is also desirable to reduce the volume, mass and cost by combining the two largest structures onboard, antennas and solar panels, into one.

A reflectarray antenna integrated with solar cells was first proposed in [35] and later developed in [36], [37] but the measured aperture efficiency was only about 10% without referring solar energy efficiency. In [38] the design of a reflectarray element integrated on solar cells was studied. In 2012 NASA/JPL initiated an ISARA project [39], which will test a Ka-band high-gain reflectarray antenna integrated on the back of the solar panel of a cubesat. For some small satellites and space vehicles, solar cells are mounted directly on the main structure, which makes the back-to-back approach unsuitable.

Here we will summarize the design and development of a Ka-band reflectarray antenna integrated on the front of solar cells for satellite applications which was reported in [40]. The solar cells are employed as the reflectarray antenna substrate with a multi-layered structure. The solar cell is manufactured using germanium-based technology with an octagonal shape of 80 mm × 80 mm. An image of the solar cell along with its detailed multi-layered structure is given in Figure 11.21. An epitaxial layer, which mainly consists of InGaAs and InGaP for solar energy harvesting, is grown from a 4-inch silver-backed germanium wafer. Grid electrodes are deployed to collect converted electricity. Coverglass, bonded to the grid electrodes and the epitaxial layer, serves to protect them in space. On top of the coverglass, an array of gilded crossed-dipoles is printed. Therefore, the solar cells are directly used as the substrate for reflectarray, both

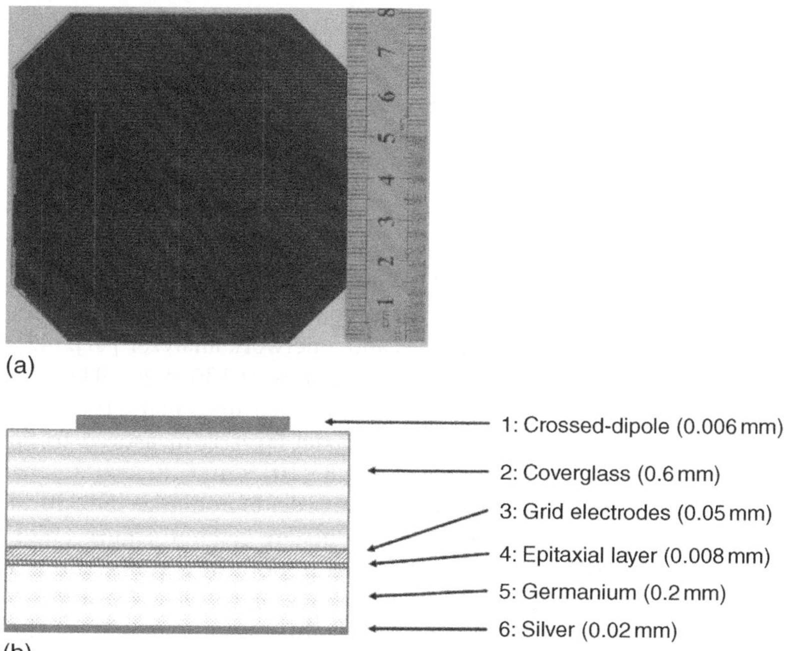

Figure 11.21 (a) Photograph of an octagonal solar cell. (b) Multi-layered structure of the solar cell with integrated reflectarray elements. Reprinted with permission from [40].

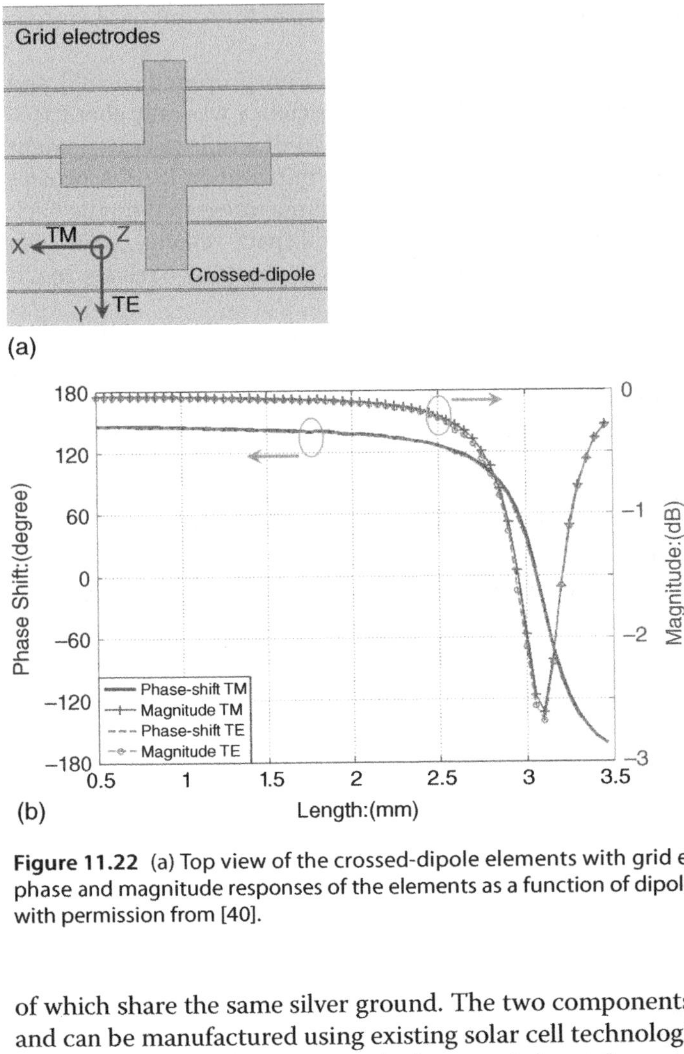

Figure 11.22 (a) Top view of the crossed-dipole elements with grid electrodes. (b) Simulated phase and magnitude responses of the elements as a function of dipole length. Reprinted with permission from [40].

of which share the same silver ground. The two components are completely integrated and can be manufactured using existing solar cell technologies.

The coverglass's electromagnetic characteristics at the operating frequency band were measured using a standard method with waveguides and a network analyzer [41], [42]. The measured dielectric constant is 6.39 and the loss tangent is 0.0135 at 20 GHz. The thin-film resistance of individual layer of the solar cell was measured using the four-terminal sensing method. The conductivity of the epitaxial layer is measured to be 2.6×10^5 S/m, and that of the doped germanium layer is 2.3×10^3 S/m.

Crossed-dipole elements were used for the reflectarray, in order to minimize the optical blockage. The element's top view and reflection coefficients are given in Figure 11.22. A sub-wavelength element spacing of 3.81 mm was used in both directions, about a quarter of free-space wavelength at the operating frequency of 20 GHz, in order to achieve a more stable phase response and better bandwidth performance [43]. The width of the crossed-dipoles is 0.5 mm, and the length is varied from 0.5 mm to 3.5 mm. The grid electrodes are deployed horizontally below the cover glass with line spacing of 0.8 mm.

Figure 11.23 Prototype of a Ka-band reflectarray antenna integrated with solar cells. Reprinted with permission from [40].

A prototype of a reflectarray antenna integrated with solar cells operating at 20 GHz was designed and fabricated, which is shown in Figure 11.23. The dimension of the square aperture is 160 mm × 160 mm. Four 80 mm × 80 mm octagonal solar cells are used as reflectarray antenna substrate, with copper foil filled at the four corners of each solar cell. A linearly polarized corrugated horn was utilized as the feed. The measured and simulated patterns of the antenna are given in Figure 11.24. The measured gain of the antenna is 26.3 dBi, and the antenna achieves an aperture efficiency of 29.8%. The 1dB gain bandwidth is 8.75%.

In order to successfully integrate a reflectarray on the front of solar cells and keep both function properly, its impact on solar energy efficiency must be minimized. In [40], a set of comparative experiments were performed to measure the light transmittance through the cover glass with and without crossed-dipoles in the proposed integrated reflectarray design. The output power from the solar cell (P_{out}) is proportional to its solar energy efficiency (η) when the incident optical power (P_{in}) is fixed. The light transmittance is defined as the ratio of the solar energy efficiencies when the cover glasses are with and without crossed-dipoles under the same illumination, which can be readily calculated using the output powers from the solar cell. An image of the test setup and the computed output power is given in Figure 11.25. It was observed that the transmittance is about 60% at peak output powers and gradually increases to above 90% as the output power decreases. The average transmittance is 87.7% and 74.7%, respectively, at the separation distance of 100 and 200 mm. It is worthwhile pointing out that the crossed-dipoles occupy 17.6% of the aperture area. Note that the blockage ratio is relatively high due to the wide dipole width and the small element spacing used in the design. This can be reduced by adopting narrow dipoles and regular element spacing (that is half-wavelength) at the expense of some reduction in antenna bandwidth.

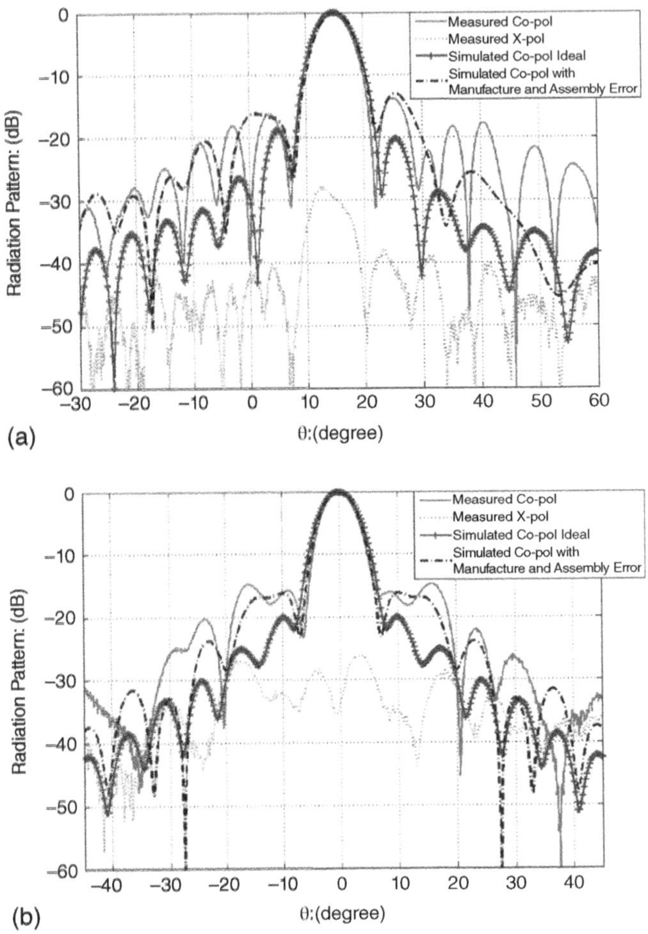

Figure 11.24 Measured and simulated radiation patterns of the reflectarray. Reprinted with permission from [40].

11.3 Power Combining and Amplifying Reflectarrays

As discussed in earlier chapters active elements can be readily integrated into the structure of reflectarray unit-cells. Advances in this area led to novel applications such as power combining and beam splitting [44]–[48]. Spatial power combining at high frequencies has the potential to avoid the complexity of the design and operation of high-power amplifiers.

A 37-element spatial power combining reflectarray was demonstrated in [47]. The feed array is a corporate-fed dual-polarized four element microstrip patch with transmit and receive capabilities. An image of the feed array is shown in Figure 11.26.

In the transmit mode, the feed array sends vertically polarized signal to the reflectarray. These signals get amplified and their polarization is changed to horizontal on the reflectarray surface, as shown in Figure 11.27. Then instead of focusing a far-field beam, these reflectarray elements reradiate the energy back to the feed array in the receive

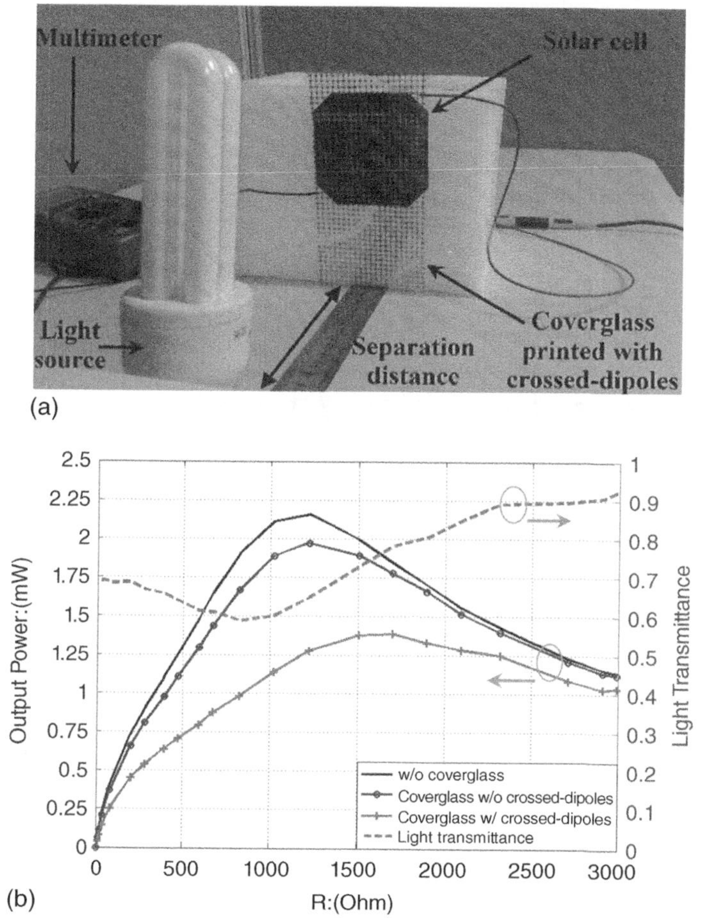

Figure 11.25 (a) Measurement setup. (b) Calculated output power and light transmittance for a separation distance of 200 mm. Reprinted with permission from [40].

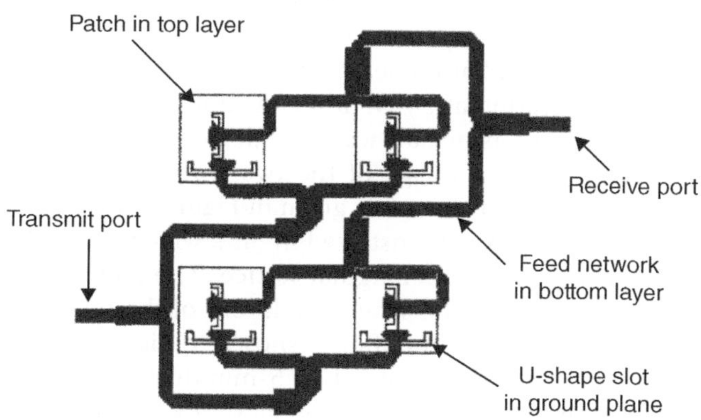

Figure 11.26 The feed array configuration for the power combining reflectarray. Reprinted with permission from [47].

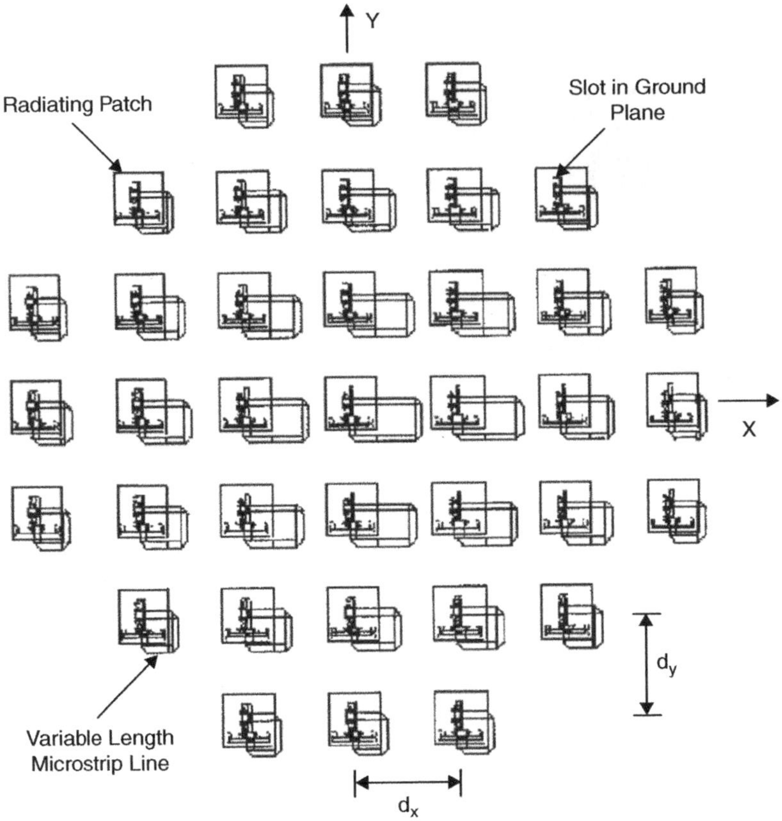

Figure 11.27 The 37-element power combining reflectarray. Reprinted with permission from [47].

mode. This process allows for a spatial power amplification with low loss. The power gain achieved by this reflectarray was about 7 dB.

Power combining can also be achieved at the feed level by using multiple feeds. The phase shift on the aperture of the reflectarray has to be designed to combine the power from the multiple feeds into a single radiation beam. In [48] a two-feed single-beam power combining reflectarray operating at 30 GHz was demonstrated with a 33% radiation efficiency. The advantages of using sub-wavelength elements in these designs were further studied in [49], [50]. An image of the two-feed single-beam reflectarray antenna along with the reflectarray layout is given in Figure 11.28.

The first X-band amplifying reflectarray was demonstrated in [45], which used 137 active microstrip patch antennas arranged in a rectangular lattice. The amplifying elements were designed using FET transistors. The schematic layout of the reflector including the construction of a single unit-cell amplifier is shown in Figure 11.29. The array is about 320-mm in diameter, consisting of 7.4 × 7.5-mm dual-feed aperture-coupled patch antenna elements separated by 24 mm or about 0.92 wavelengths at 11.5 GHz. Passive and active varieties of the reflectarray combiner were fully assembled and tested using a HP8510C vector network analyzer. The distance between the

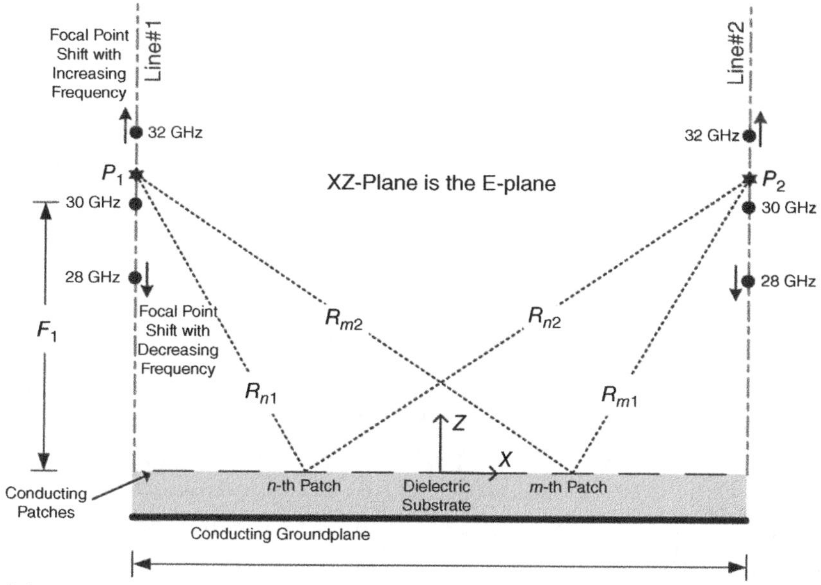

(a)

(b)

Figure 11.28 A two-feed single-beam reflectarray antenna: (a) geometry of the system, (b) photograph of the layout. Reprinted with permission from [50].

illuminating subarray and the reflectarrays was chosen as 108 mm. The results of these tests are shown in Figure 11.30, where note that the maximum small signal gain is approximately 2 dB at 9.93 GHz. The 3dB gain bandwidth is about 10% from 9.60 to 10.55 GHz.

To reduce the complication and cost for large amplifying reflectarrays, a Cassegrain configuration was proposed in [9] where the sub-reflector is a reflectarray that can illuminate a very large aperture. The sub-reflector elements have amplifiers, which allow

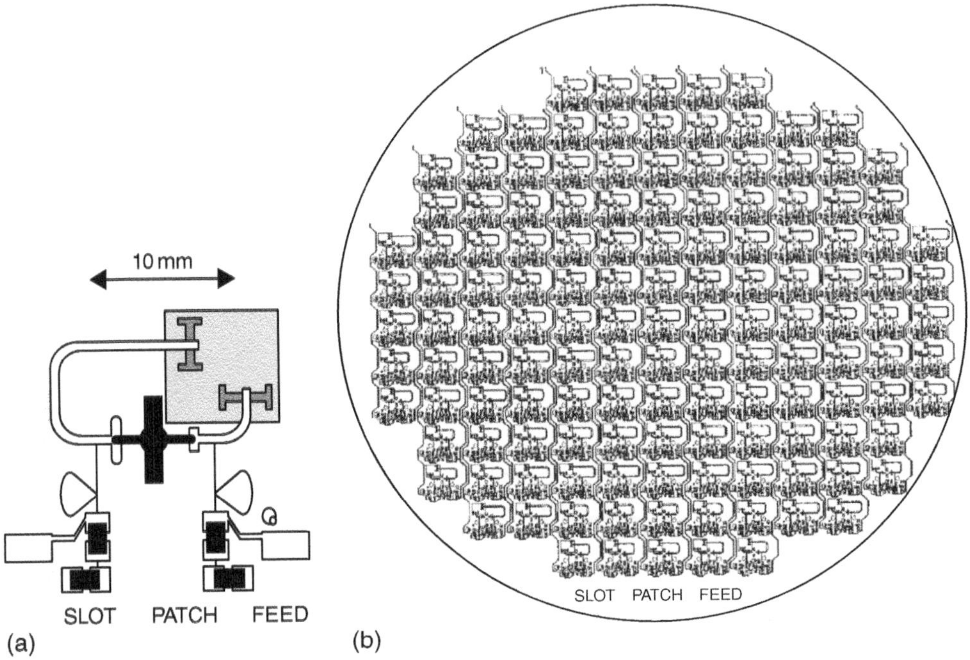

Figure 11.29 (a) Configuration of a single unit-cell amplifier. (b) The layout of the 137-element active reflectarray. Reprinted with permission from [45].

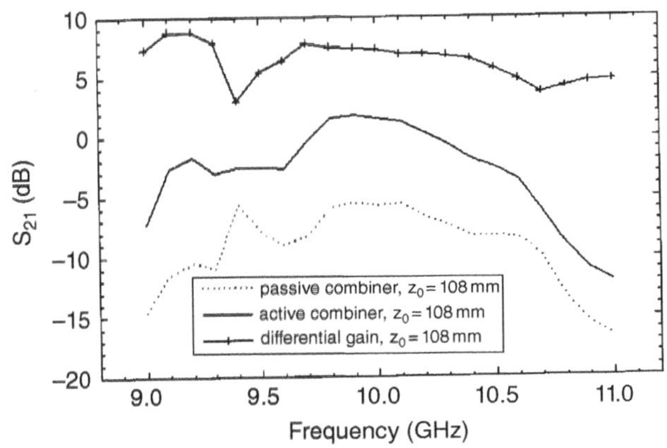

Figure 11.30 Performance comparison between the spatial combiner using passive and active reflectarrays. Reprinted with permission from [45].

relatively low-power, solid-state amplifiers to be combined spatially to provide very high overall antenna radiated power with graceful degradation in performance. In doing so, a relatively small number of amplifiers, such as tens or hundreds, can be used on the sub-reflector, instead of millions on the very large main reflectarray, to provide high power and to minimize complexity and cost.

11.4 A Perspective on Reflectarray Antennas

Although the concept of reflectarray antenna was introduced in the 1960s and it started to gain popularity in the early 1990s, it is still a relatively new kind of antenna comparing to others. In particular, their practical applications in wireless communications, radar, deep space exploration, remote sensing, and wireless power transfer, have to be fully developed in the future. Some representative reflectarrays are discussed in previous sections, and here we briefly envision several interesting topics in future development.

11.4.1 Large-Aperture Planar Reflectarray Antennas

One distinct advantage of reflectarray antennas is the freedom and flexibility in the reflector surface geometry. Using the element phasing approach, the reflector surface is not limited to be a parabolic surface to focus the beam. Various planar and conformal geometries can be adopted, which greatly stimulate potential applications.

An interesting example is a planar reflectarray using inflatable structure and thin membrane developed by Dr. John Huang for NASA space applications. NASA's Earth remote sensing and deep space exploration programs have increasing demands for high gain and large-aperture antennas. Compared to conventional parabolic reflectors, the reflectarray has very low mass and small stowage volume so that the payload weight and size is significantly reduced. Furthermore, for a very large aperture in space, the flat surface is much more reliable and can easily maintain its required surface tolerance compared to a curved surface. Figure 11.31 shows the 3-m Ka-band inflatable reflectarray,

Figure 11.31 A 3-m Ka-band inflatable reflectarray antenna.

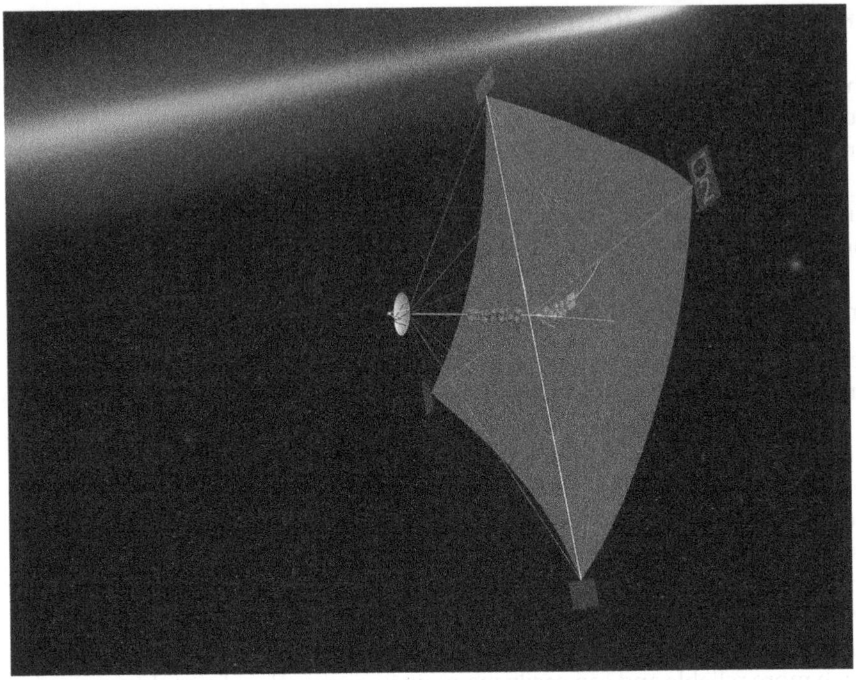

Figure 11.32 Reflectarrays for solar sail applications. Image is from [51]. *Source:* NASA, USA.

which has approximately 200,000 elements and 54.4 dBi gain, and is considered electrically the largest reflectarray ever build.

A more ambitious idea is to use the reflectarray concept in solar sails. Inflatable solar sails have been proposed for interplanetary missions, which collect the solar-flux power to provide thrust for spacecraft. The large surface area of the sails (i.e., 100 m in diameter), as shown in Figure 11.32, which is required to capture enough energy for spacecraft navigation, has inspired a lot of novel antenna designs [51], [52]. Because of the flexibility of the surface geometry, reflectarray antenna is a natural choice for this large solar sail antenna.

Beside the spaceborne antennas, large-aperture reflectarrays can also be used in various ground applications. A potential application is for radio telescope, where a large aperture is required to collect very weak signals from deep space. Figure 11.33 shows a 500-m radio telescope, which is currently being assembled in China. Although the current radio telescope uses a spherical surface design, reflectarrays could be considered for future designs because of the flat surface for easy construction and the low cost for economic deployment and maintenance. Another exciting project is the Ivanpah solar power facility. Operating at the optic frequency to collect solar power, the principle of this solar power facility is similar to reflectarray antennas, where a super large flat area consisting of more than 170,000 heliostats is used instead of a conventional parabolic surface. The Ivanpah facility provides useful guidelines for future reflectarray engineering. Figure 11.34 shows a photo of this facility.

Figure 11.33 The 500-m radio telescope being assembled in China. Image is from [53]. *Source:* http://news.163.com/15/1123/08/B93HU14400014JB6.html.

Figure 11.34 The Ivanpah solar power facility from [54]. *Source:* https://commons.wikimedia.org/wiki/File:Ivanpah_Solar_Power_Facility_(1).jpg. Used under https://creativecommons.org/licenses/by/2.0/deed.en.

11.4.2 Reflectarray Antennas with Broad Bandwidth, Beam-Scanning Capability, and Low Cost

A critical challenge in reflectarray antenna applications is the bandwidth limitation. As discussed in Chapter 3, the bandwidth of a reflectarray is usually a few percent. Over the last decade, a lot of effort has been put into increasing the reflectarray bandwidth, and some designs achieve bandwidths over 10% or even more, as shown in Chapter 5. Although some works continue to further improve the bandwidth, it is more practical to design reflectarrays whose bandwidths satisfy specific application requirements. It should be pointed out that a practical bandwidth requirement is not only for the antenna gain, but also for the beam direction, SLLs, cross-polarization levels, and so on. In some satellite applications where the transmit and receive bands are separated, dual-band reflectarrays may be a better option.

Another grand challenge in reflectarray antenna applications is the beam-scanning capability. Some beam-scanning technologies have been presented in Chapter 10, and it is worthwhile emphasizing the importance of reconfigurable reflectarrays because of the fast scanning speed. In general, reconfigurable reflectarrays can be considered as phased arrays where the phase shifters are directly integrated with the antenna elements. Hence, reconfigurable reflectarrays have lower insertion loss and reduced system cost compared to conventional phased arrays. Therefore, it has a great potential for a wide range of applications in radar and wireless communication systems.

Cost is a practical issue when a commercial reflectarray product is considered. Because of the low loss-tangent property, a high-performance microwave dielectric substrate is used in most of the reflectarray literature. This substrate is an expensive component for large-aperture reflectarrays, and its cost makes massive production infeasible. To reduce the antenna cost, a low-cost substrate such as FR-4 can be chosen as an alternative solution. The sub-wavelength technique was used in order to mitigate the high dielectric loss associated with the FR-4 substrate [55]. To completely eliminate the dielectric loss, metal-only reflectarrays have been proposed. Rectangular groove elements were used in [56] to design a metal-only reflectarray operating at 75 GHz, and slot-type phasing elements were introduced in [57] to design a Ku-band metal-only reflectarray.

11.4.3 From Reflectarray Antennas to Transmitarray Antennas

As the final remark, it is interesting to refer to another emerging antenna concept: the transmitarray antenna [58]. The relation between a reflectarray and a transmitarray is similar to the relation between a mirror and a lens, as shown in Figure 11.35. Although inspired by the reflectarray, the transmitarray, also known as array lens antenna, encounters a great challenge: both magnitude and phase control of the element. In the reflectarray, the reflection magnitude is always close to unity due to the existence of a metal ground plane; thus, one only needs to control the element reflection phase. In the transmitarray, besides the phase control, the magnitude of the transmission coefficient needs to be close to 1 to ensure a high efficiency.

Although the design of a transmitarray antenna is more challenging, it has several distinct advantages, such as no feed blockage, less sensitivity to surface roughness, wide beam scan angle, and so on. Among these advantages, it is worthwhile highlighting the no feed blockage feature. As a consequence, a number of feeds or a feed array can be

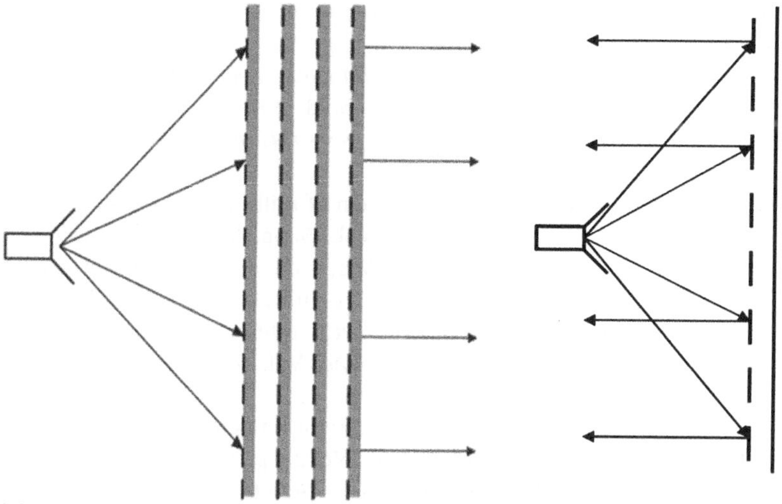

Figure 11.35 Geometries of: (a) a transmitarray antenna, and (b) a reflectarray antenna.

used to illuminate the transmitarray simultaneously, which is quite desirable for space power combination or multi-feed multi-beam applications.

The rapid progress made in reflectarray antennas, together with the transmitarray antennas, will greatly enrich knowledge in the area, enhance the system performance of various wireless communication devices, and, finally, improve our daily lives.

References

1 P. Nayeri, F. Yang, and A. Z. Elsherbeni, "Radiation analysis and characteristics of conformal reflectarray antennas," *International Journal of Antennas and Propagation*, Vol. 2012, pp. 1–6, 2012.

2 L. Josefsson and P. Persson, *Conformal Array Antenna Theory and Design*, IEEE Press Series on Electromagnetic Wave Theory, Chichester, UK: John Wiley & Sons, Ltd, 2006.

3 R. L. Haupt, *Antenna Arrays: A Computational Approach*, John Wiley & Sons, Ltd, 2010.

4 T. M. Smith, and S. L. Borison, "Depolarization of a circularly polarized wave by an infinite cylinder," *IEEE Trans. Antennas Propag.*, pp. 796–798, Nov. 1974.

5 P. Nayeri "Advanced design methodologies and novel applications of reflectarray antennas," Ph.D. dissertation, Department of Electrical Engineering, University of Mississippi, University, MS, 2012.

6 R. C. Hansen, *Phased Array Antennas*, 2nd Edn, Chichester, UK: John Wiley & Sons, Ltd, 2009.

7 D. M. Pozar, "Bandwidth of reflectarrays," *Electron. Lett.*, Vol. 39, No. 21, Oct. 2003.

8 P. W. Hannan, "Microwave antennas derived from the Cassegrain telescope," *IRE Trans. Antennas Propagat.*, Vol. AP - 9, pp. 140–153, March 1961.

9 J. Huang and J. A. Encinar, *Reflectarray Antennas*. Hoboken, NJ: Wiley-IEEE, 2008.

10 A. W. Love, *Reflector Antennas*, New York: Institute of Electrical and Electronics Engineers and John Wiley & Sons, Inc., 1978.
11 Y. Rahmat-Samii, "Reflector antennas," in *Antenna Handbook: Theory, Applications, and Design*, Y. T. Lo and S. W. Lee (eds), Van Nostrand Reinhold, 1988.
12 C. J. Sletten, *Reflector and Lens Antennas: Analysis and Design Using Personal Computers*, Artech House Inc., 1988.
13 D. M. Pozar, S. D. Targonski, and H. D. Syrigos, "Design of millimeter wave microstrip reflectarrays," *Proc. IEEE Trans. Antennas Propagat.*, Vol. 45, pp. 287–295, Feb. 1997.
14 J. Huang, C. Han, and K. Chang, "A Cassegrain offset-fed dual-band reflectarray," *IEEE AP- S/URSI Symposium, Albuquerque*, New Mexico, July 2006, pp. 2439–2442.
15 M. Arrebola, L. Haro, J. Encinar, "Analysis of dual-reflector antennas with a reflectarray as subreflector," *IEEE Antennas and Propag. Magazine*, Vol. 50, No. 6, pp. 39–51, Dec. 2008.
16 S. Xu, Y. Rahmat-Samii, W. A. Imbriale, "Subreflectarrays for reflector surface distortion compensation," *IEEE Trans. Antennas Propag.*, Vol. 57, No. 2, pp. 364–372, Feb. 2009.
17 C. Tienda, J. A. Encinar, S. Montori, R. Vincenti Gatti, M. Arrebola, and R. Sorrentino, "Dual-reflectarray antennas for bidirectional satellite links in Ku-band," presented at the *5th European Conference. on Antennas and Propagation, Rome, Italy*, Apr. 11–15, 2011.
18 C. Tienda, "Design, manufacturing and test of a dual-reflectarray antenna with improved bandwidth and reduced cross-polarization," *IEEE Trans. Antennas Propag.*, Vol. 61, No. 3, pp. 1180–1190, Mar. 2013.
19 M. Arrebola, Y. Alvarez, J. A. Encinar, and F. Las-Heras, "Accurate analysis of printed reflectarrays considering the near field of the primary feed," IET Microw., *Antennas Propag.*, Vol. 3, No. 2, pp. 187–194, Mar. 2009.
20 D.-C. Chang and M -C. Huang. "Feasibility study of erecting cosecant pattern by planar microstrip reflectarray antenna," *AMPC' 93*, Vol. 2 pp. 19.20–24, 1993.
21 J. A. Encinar, "Breadboard of a three-layer printed reflectarray for dual polarisation and dual coverage," *28th ESA Antenna Workshop on Space Antenna Systems and Technologies, Noordwijk* (The Netherlands), 30 May – 3 June, 2005.
22 H. Tseng Chou, P. Hsueh, T. Hung, L. Kuo, and Y. Ze, "Near-zone focused radiations of reflectarray antennas for RFID applications at 0.9 and 2.4 GHz," in *Proc. IEEE Antennas Propagat. Soc. Int. Symp.*, Spokane, WA, pp. 1727–1730, Jul. 2011.
23 T. Smith, U. Gothelf, O. S. Kim, and O. Breinbjerg, "An FSS-backed 20/30 GHz circularly polarized reflectarray for a shared aperture L- and Ka-Band satellite communication antenna," *IEEE Trans. Antennas Propagat.*, Vol. 62, No. 2, pp.661–668, Feb. 2014.
24 X. Zhu, *et al.*, "Design of an L-band reflectarray antenna for BeiDou satellite applications," *IEEE International Symposium on Antennas and Propag.* (ISAP), pp. 297–298, Dec. 2014.
25 J. Huang, "A technique for an array to generate circular polarization with linearly polarized elements," *IEEE Trans. Antennas Propagat.*, Vol. Ap-34, No. 9, pp. 1113–1124, Sep. 1986.
26 M. Tanaka, R Suzuki, Y. Suzuki, and K. Araki, "Microstrip antenna with solar cells for microsatellites," in *Proc. IEEE Int. Symp. Antennas Propag.*, Jun. 1994, pp. 786–789.
27 T. Wu, R.-L. Li, and M. M. Tentzeris, "A scalable solar antenna for autonomous integrated wireless sensor nodes," *IEEE Antennas Wireless Propaga. Lett.*, Vol. 10, pp. 510–513, 2011.

28 S. Vaccaro, C. Pereira, J. R. Mosig, and P. de Maagt, "In-flight experiment for combined planar antennas and solar cells (SOLANT)," *IET Microw. Antennas Propat.*, Vol. 3, No. 8, pp. 1279–1287, 2009.

29 S. V. Shynu, M. J. R. Ons, M. J. Ammann, B. Norton, and S. McCormack, "Dual band a-Si:H solar-slot antenna for 2.4/5.2 GHz WLAN applications," in *Proc. 3rd Eur. Conf. on Antennas and Propagation*, Mar. 23–27, 2009, pp. 408–410.

30 R. Caso, D'Alessandro. A, Michel. A, Nepa. P, "Integration of slot antennas in commercial photovoltaic panels for stand-alone communication systems," *IEEE Trans. Antennas Propag.*, Vol. 61, No. 1, pp. 62–69, Jan. 2013.

31 O. Yurduseven and D. Smith, "A solar cell stacked multi-slot quad-band PIFA for GSM, WLAN and WiMAX networks," *IEEE Microwave and Wireless Components Letters*, Vol. 23, No. 6, pp. 285–287, 2013.

32 M. J. Roo-Ons, S. V. Shynu, M. J. Ammann, S. J. McCormack, and B. Norton, "Transparent patch antenna on a-Si thin-film glass solar module," *Electron. Lett.*, Vol. 47, No. 2, pp. 85–86, Jan. 2011.

33 T. W. Turpin and R. Baktur, "Meshed patch antennas integrated on solar cells," *IEEE Antennas Wireless Propag. Lett.*, Vol. 8, pp. 693–696, 2009.

34 E. H. Lim., K. W. Leung., C. C. Su., and H. Y. Wong, "Green antenna for solar energy collection," *IEEE Antennas Wireless Propag. Lett.*, Vol. 9, pp. 689–692, 2008.

35 J. Huang, "Capabilities of PRINTED REFLECTARRAY ANTENNAS," *IEEE Phased Array Systems and Technology Symposium*, Boston, MA, Oct. 1996, pp. 131–134.

36 M. Zawadzki and J. Huang, "Integrated RF antenna and solar array for spacecraft application," in *Proc. 2000 IEEE Int. Conf. Phased Array Syst. Technol.*, 2000, Dana Point, CA, May 2000, pp. 239–242.

37 J. Huang and M. Zawadzki, "Antennas integrated with solar arrays for space vehicle applications," in *Proc. ISAPE'2000*, pp. 86–89.

38 P. Dreyer, J. S. G. Díaz, and J. Perruisseau-Carrier, "Design of a reflectarray element integrated in a solar cell panel," in *Proc. IEEE Int. Symp. Antennas Propag.*, Jul. 2013, pp. 1558–1559.

39 NASA. Website: www.nasa.gov/directorates/spacetech/small_spacecraft/isara_project.html (accessed July 2017).

40 W. An, S. Xu, F. Yang, and J. Gao, "A Ka-band reflectarray antenna integrated with solar cells," *IEEE Trans. Antennas Propagat.*, Vol. 62, No. 11, pp. 5539–5546, Nov. 2014.

41 K. Sarabandi and F. T. Ulaby, "Technique for measuring the dielectric constant of thin materials," *IEEE Trans. Instrum. Meas.*, Vol. 37, pp. 631–636, Dec. 1988.

42 T. Yasin, R. Baktur, and C. Furse, "A study on the efficiency of transparent patch antennas designed from conductive oxide films," in *Proc. IEEE Int. Symp. Antennas Propag.*, 2011, pp. 3085–3087.

43 P. Nayeri, F. Yang, and A. Z. Elsherbeni, "Broadband reflectarray antennas using double-layer subwavelength patch elements," *IEEE Antennas Wireless Propag. Lett.*, Vol. 9, pp. 1139–1142, 2010.

44 A. W. Robinson, and M. E. Bialkowski, "An X-band active microstrip reflectarray antenna," *Proceedings of Asia-Pacific Microwave Conference*, 1997, pp. 925–928.

45 M. E. Bialkowski, A. W. Robinson, and H. J. Song, "Design, development, and testing of X-band amplifying reflectarrays," *IEEE Trans. Antennas Propag.*, Vol. 50, pp. 1065–1076, August 2002.

46 A. W. Robinson, M. E. Bialkowski, and H. J. Song, "An 137-element active reflect-array with dual - feed microstrip patch elements," *Microwave and Optical Technology Letters*, Vol. 26, No. 3, pp. 147–151, August 2000.
47 M. E. Bialkowski and H. J. Song, "Investigations into a power-combining structure using a reflect array of dual-feed aperture-coupled microstrip patch antennas," *IEEE Trans. Antennas Propag.*, Vol. 50, No. 6, pp. 841–849, Jun. 2002.
48 F. Arpin, J. Shaker, and D. McNamara, "Multifeed single-beam power-combining reflectarray antenna," *Electronics Letters*, Vol. 40, No. 17, August 2004, pp. 1035–1037.
49 E. Almajali, D. A. McNamara, J. Shaker, and M.R. Chaharmir, "Improved dual-feed single-beam power combining reflectarray antenna," *Electronics Letters*, Vol. 49, No. 25, pp. 1595–1597, 2013.
50 E. Almajali, D. A. McNamara, J. Shaker, and M. R. Chaharmir, "Beam squint suppression in offset-fed reflectarrays," *IEEE Antennas Wireless Propag. Lett.*, vol. 12, pp. 587–590, 2013.
51 NASA. Website: Beyond Earth. Available at: www.nasa.gov/centers/marshall/multimedia/exploration_gallery.html (accessed July 2017), 2012.
52 B. Khayatian and Y. Rahmat-Samii, "A novel antenna concept for future solar sails: application of Fresnel antennas," *IEEE Antennas Propag. Magazine*, Vol. 46, No. 2, pp. 50–63, Apr. 2004.
53 News.163.com. Global News, "China is the world's largest radio telescope to protect the world's leading 20 years," http://news.163.com/15/1123/08/B93HU14400014JB6.html (accessed July 2017) [in Chinese].
54 Solar Power Facility Wikipedia entry. Available at: https://en.wikipedia.org/wiki/Ivanpah_Solar_Power_Facility (accessed July, 2017).
55 J. Ethier, M. R. Chaharmir, and J. Shaker, "Loss reduction in reflectarray designs using sub-wavelength coupled-resonant elements," *IEEE Trans. Antennas Propag.*, Vol. 60, No. 11, pp. 5456–5459, Nov. 2012.
56 Y. H. Cho, W. J. Byun, and M. S. Song, "High gain metal-only reflectarray antenna composed of multiple rectangular grooves," *IEEE Trans. Antennas Propag.*, Vol. 59, No. 12, pp. 4559–4568, Dec. 2011.
57 W. An, S. Xu, and F. Yang, "A metal-only reflectarray antenna using slot-type elements," *IEEE Antennas Wireless Propag. Lett.*, Vol. 13, pp. 1553–1556, 2014.
58 A. Abdelrahman, F. Yang, A. Elsherbeni, and P. Nayeri, *Analysis and Design of Transmitarray Antennas*, Morgan & Claypool, 2016.

Index

a

Alternating projection method (APM) 267, 276–279, 282, 286, 289, 291, 298
Amplifying reflectarrays 388–392
Analysis technique, xiv 4–6, 15, 22–26, 79–110, 156
Aperture-coupled 27, 197, 345, 346, 348, 350, 390
Aperture efficiency 49–77, 83, 91, 99, 101, 152, 158, 161, 164, 166, 190–192, 206, 207, 210, 212, 214, 221, 255, 285, 291, 293, 306, 310, 369, 376, 385, 387
Array lens 396

b

Bandwidth
- improvement 4, 188, 191, 192, 194, 196, 352
- limitation 113, 115, 117–122, 180, 192, 197, 312, 373

Beam scanning xiv, 3–7, 303–359, 367, 376, 380, 381, 396
Beam-width (or beamwidth), 29, 52, 55, 56, 82, 103, 105, 109, 149, 168, 190, 221, 261, 274, 277, 278, 284, 285, 287, 293, 310, 312, 316, 318, 325, 326, 328, 329, 331, 337, 356, 359, 370, 371
Bifocal 306, 313–325, 336, 338
Broadband xiii, xiv, 4–7, 37, 121, 179–221, 263, 312, 377
Broadband techniques 37, 179, 192, 194, 197, 312

c

Cassegrain 313, 334, 375–377, 379, 391
Center-fed 63–64
Circular polarization 14, 16, 33, 50–51, 94, 122, 147, 165–168, 209–213, 217, 268, 282, 321, 340, 348, 376, 382
Compact size 202, 210
Contoured-beam 4, 267, 297, 380
Coordinate system 9–11, 13, 17, 50, 51, 55, 71, 72, 81–84, 87, 88, 95, 107, 119, 153, 170, 188, 277, 314, 368
Corrugated 83, 165, 217, 321, 331, 387
Coverage requirements 277
Crossed dipoles 15, 385–387, 389
Cross-polarization 36, 37, 49, 61, 82, 90, 98, 100, 103, 108, 109, 161, 166, 168, 170, 179, 212, 215, 221, 294, 306, 342, 376–378, 396
Curved reflectarrays 89, 325, 367

d

Deep-space 393
Delay lines 3, 4, 14–17, 26–29, 37, 38, 118, 121, 197, 305, 339–341, 344–348, 358
Deployable 197, 268, 377, 385, 386, 394
Design curve 16, 124, 125
Differential spatial phase delay 197
Diffraction 49, 60–70, 96, 108, 109, 168, 368
Diode 306, 341, 344–347, 349–352, 354
Direct broadcast satellite 267, 380
Directivity 1, 49, 52, 53, 55, 66, 69, 79, 80, 90–92, 103–105, 108, 140–144, 150, 158, 258, 274, 282, 318, 325, 335, 336, 339, 378

Reflectarray Antennas: Theory, Designs, and Applications, First Edition.
Payam Nayeri, Fan Yang, and Atef Z. Elsherbeni.

Dual-band 206, 208, 210–213, 334, 335, 348, 376
Dual-gridded reflector 294
Dual-linear 16, 30, 205, 282, 283
Dual-reflector xi, 7, 305, 306, 313, 314, 334–337, 359, 375–377

e
Efficiency xiv, 6, 49–77, 82, 83, 90–94, 99, 101, 152, 158, 161, 164, 166, 171, 176, 190–192, 206, 207, 210, 212, 214, 221, 245, 255, 256, 258, 261–263, 277, 284, 285, 291, 293, 306, 310, 312, 369, 376, 385, 387, 390, 396
Electrical size 49, 262
Element selection 42, 43, 179
Element spacing 12, 18, 77, 170, 196, 200, 204, 216, 245, 287, 382, 386, 387
Elliptical reflectarrays 86, 295
Etching 193, 204, 382

f
Faceted 197
Far-field radiation pattern 28, 81, 83, 86, 92, 161, 275, 280, 287, 317, 318, 377
Fast Fourier transform 95
Feed coordinate system 50, 51, 55, 82, 83
Feed horn 82, 84, 90, 92, 94, 99–101, 103, 107, 119, 148, 149, 160, 165, 188, 192, 194, 200, 205, 219, 268, 271, 274, 283, 284, 287, 315, 321, 331, 343, 358, 375, 377, 379
Feed model 49–52, 107
Ferrite 353
Ferroelectric 305, 352–354
Floquet 19, 20, 22–26, 30, 96
Folded 197, 313, 336, 377
Fourier transform 94, 95, 251, 277
Frequency band 37, 119, 122, 199, 200, 206, 209–212, 214, 229, 240, 286, 294, 378, 386
Frequency scanned 190, 293–294
Frequency selective surface 208
Fresnel 43, 254, 256, 262, 263
Full-wave 4, 22, 26, 79, 80, 84, 96–98, 103, 106–110, 147, 156–158, 161, 165, 206, 242, 243, 256, 280–281

g
Gain 1, 49, 79, 113, 147, 180, 228, 267, 303, 369
Genetic algorithm (GA), 279, 280
Grating lobe 210, 262
Green's functions 96
Gregorian 306, 334, 375
Ground plane 14, 44, 96, 168, 181, 207, 208, 234, 245, 247–249, 263, 342, 343, 345, 346, 349, 353, 389, 390, 396

h
High-frequency 208, 210, 261, 342
High-gain xiii, xiv, 1, 2, 5, 6, 79, 113, 150, 157, 249, 261, 297, 303, 306, 312, 332, 338, 379–381, 384, 385, 393
Horizontal (H-) polarization 298, 299
Horn antenna 83, 92–94, 147–149, 151, 159, 165, 166, 169, 217, 259, 261, 310, 315, 382

i
Illumination efficiency 54–56, 60, 92, 99, 101
Impedance 22, 26, 93, 121, 185, 242, 352
Incident field 2, 17, 32, 33, 82, 83, 103, 108, 238, 250, 369
Infinite array approach 3, 18, 43, 79, 98
Inflatable 181, 393, 394
Infrared xiv, 6, 7, 227–263, 354
Intersection 276, 277
Inverse discrete Fourier transform (IDFT) 94, 95, 278

k
Ka-band 11, 26–28, 30, 31, 34, 98, 107, 115, 119, 147, 165, 166, 188, 190, 192–194, 208, 213, 215–221, 315, 321, 325, 330, 334–337, 347, 348, 352, 353, 385, 387, 393
Ku-band 147–165, 204, 295, 324, 396

l
Linear polarization 209, 297
Losses 15, 22, 25, 44, 45, 53, 92–94, 98, 118, 185, 194, 197, 200, 228, 229, 234–245, 248, 249, 256, 306, 354, 355, 382

Low-frequency 208
Low-loss 4, 185, 228, 229, 236, 242, 243, 248–250, 306, 354, 390, 396
Ludwig 89

m
Mask 4, 99, 101, 102, 155, 156, 275–280, 282, 287, 288, 317, 318
Membrane 325, 347, 349, 350, 393
Metal semiconductor FET (MESFET), 347
Metamaterial 184
Metasurface 262
Micro-electromechanical 344
Micro-electromechanical systems (MEMS), 306, 343, 344, 347–352
Micro-machined motor 347
Microstrip 1, 3, 14, 15, 23, 26–31, 46, 55, 83, 92, 113, 155, 179, 182, 190, 248, 294, 325, 334, 344–346, 348–350, 377, 382, 384, 388, 390
Miniature motors 343
Monolithic 347–349
Multiband xiii, xiv, 4–6, 179–221
Multi-beam xiii, xiv, 3–7, 267–299, 314, 336, 397
Multi-facet 197
Multi-fed 298, 312
Multi-layer 385
Multi-panel 197, 198

o
Offset-fed 2, 66, 100
Ohmic losses 92
Optimization 79, 199, 201, 202, 267, 274–283, 286–289, 297, 317–319, 325, 328

p
Parabolic antenna 1
Parabolic reflector 2, 4, 11, 63, 70–77, 79, 80, 262, 307, 312, 331, 334–336, 359, 367, 375, 377, 381, 393
Particle swarm optimization 202, 267, 279, 317
Patch size 40, 44–46, 115, 119, 122, 180, 181, 185–188, 193, 194, 234, 245–248, 321
Path length 55, 115, 197
Periodic 18, 19, 26, 43, 84, 95, 154, 185, 214, 218, 271, 369
Phase calibration 44
Phase compensation 34, 101, 117, 211, 217, 219, 307, 332
Phase control xiii, 35, 42, 43, 248, 249, 306, 313, 334, 336, 342, 344–348, 354–357, 396
Phase curves 37, 121–125, 127, 128, 130, 131, 133, 136, 138, 141, 143, 170, 194, 197, 202, 204, 216, 236–239, 242, 243, 254
Phased array feed (PAFs), 304
Phase delay xiii, 3, 10–12, 16, 17, 26–29, 37, 82, 114–118, 122, 192, 197, 215, 216, 297, 317, 326, 369, 373
Phase distribution 9–13, 19, 77, 92, 99–101, 103, 109, 117, 154–156, 179, 217, 248, 255, 256, 258, 267, 271, 275, 278–280, 282, 283, 286, 289–291, 295, 297, 298, 304, 306, 308, 313–315, 317, 319, 320, 325–327, 330–332, 370, 377, 382
Phase-only 4, 11, 275, 276, 281, 297
Phase response 15, 98, 179, 180, 182–184, 211, 235, 238, 239, 245, 247, 339, 386
Phase shift xiii, 3, 4, 9–14, 16, 17, 26, 27, 37, 42, 77, 84, 109, 117, 119, 153, 154, 157, 160, 169, 179, 188, 197, 210, 245–248, 253, 255, 256, 267, 275, 280, 289, 308, 317, 332, 334, 345, 359, 377, 386, 390
Phase-shift distribution 9–13, 317
Phase-shifter 4, 303, 306, 341, 344, 347, 349, 353–355, 359, 396
Phase variation 15, 27, 155, 179–181, 187, 215, 216
Photo-etching 204, 382
Piecewise flat 312
PIN diode 344–347, 349, 354
Polarization 17, 19, 30, 33, 36, 37, 39, 41, 49–51, 53, 54, 61, 80, 82, 84, 87, 90, 92, 94, 98–101, 103, 108, 109, 161, 165, 166, 168, 170, 179, 198, 209–212, 215, 216, 221, 228, 230, 254, 281, 294, 297–299, 306, 315, 342, 345, 348, 376–378, 388, 396, 397

Power handling 338, 341, 342, 347, 352, 354
Printed antenna 15, 369
Pyramidal horn antenna 147, 148, 259

r
Radiating element 115, 262, 369
Radiation Pattern 7, 29, 49, 79, 114, 148, 180, 256, 267, 303, 368
Reconfigurable antenna 353
Rectangular patch 15, 26–29, 33, 197, 376
Reflectarray 1–7, 9–46, 49, 79, 113–144, 147–176, 179–221, 227–263, 267–299, 303–359, 367–397
Reflected wave 15, 16, 179, 183, 184, 197, 352
Reflection efficiency 212
Reflection phase 10, 13–15, 23, 30, 32, 33, 36, 39, 41, 44, 98, 99, 101–103, 114–116, 119, 122, 124, 125, 127, 144, 155, 165, 170, 179–182, 184–188, 193, 197, 199–202, 211, 215–217, 241–243, 247, 248, 254–256, 262, 275, 285, 295, 304, 317, 321, 339, 340, 373, 396
Remote sensing xiii, 1, 5, 228, 393
Resonant dimension 124
Rotation technique 3, 33–37, 122, 165–168, 211, 215–217, 342, 347, 348

s
Satellite xiii, 5, 7, 204, 267, 268, 277, 287, 298, 379–385, 396
Scanning xiv, 3–7, 303–359, 367, 369, 376, 380, 381, 396
Scattered field 14, 33
Scattering matrix 84
Scattering parameter 82, 84, 368
S-curve 16, 99, 101, 107, 236, 238, 241, 243
Shaped beam xiii, xiv, 4, 6, 7, 9, 11, 267–299, 379, 380
Simulator 19–24, 26, 348
Solar cell 262, 263, 384–388
Spacecraft 267, 394
Space-deployable 268, 386
Spatial phase delay 10–12, 114–118, 122, 192, 197, 215, 216, 317, 326, 373
Spatial power combining 3, 7, 355, 388
Spillover efficiency 53–54, 60, 92, 306, 312, 369
Spiralphase 3
Square loop 122–124, 130–132, 135–137, 140–144, 169, 171–175, 197, 204, 213, 215, 216, 352
Stacked 113, 122, 181, 182, 184, 194, 209, 297
Subreflector 337, 377
Subwavelength 188, 194
Synthesis 2–4, 11, 197–206, 267, 275–281, 286, 289, 295, 297, 298

t
Tangential 31, 32, 84–86, 96, 148, 271, 368
Terahertz xiv, 5, 7, 227–263
Thin film 234, 305, 352–354, 386
Transmitarray 396–397
Transmit/receive (T/R), 2, 303, 306, 355
Tri-band elements 209, 213–221
True time delay 4, 118, 197, 198, 267

v
Varactor diode 341, 344, 350–352
Variable-size xiii, 3, 14–16, 30–33, 37–39, 41, 43, 45, 97, 98, 100, 101, 107, 115, 121–123, 154, 169, 171, 179–181, 185, 193, 216, 248–251, 282, 289, 291, 295, 297, 305, 321, 326, 339, 340, 382
Vertical (V-) polarization 297–299

w
Waveguide 2, 3, 14, 19–24, 26, 43–45, 83, 159, 227, 326, 327, 340, 348, 386

x
X-band 23–25, 208, 213, 216–221, 298, 312, 334, 335, 345, 348, 355, 377, 390

Printed and bound by CPI Group (UK) Ltd, Croydon, CR0 4YY
16/09/2026
14963479-0007

Printed and bound by CPI Group (UK) Ltd, Croydon, CR0 4YY
16/09/2026
14963392-0001